Proceedings of the Ninth Conference on

QUANTUM FIELD THEORY UNDER THE INFLUENCE OF EXTERNAL CONDITIONS

(QFEXT09)

Proceedings of the Ninth Conference on

QUANTUM FIELD THEORY UNDER THE INFLUENCE OF EXTERNAL CONDITIONS (QFEXT09)

Devoted to the Centenary of H B G Casimir

University of Oklahoma, USA 21–25 September 2009

Editors

Kimball A Milton
University of Oklahoma, USA

Michael Bordag
Universität Leipzig, Germany

World Scientific

NEW JERSEY • LONDON • SINGAPORE • BEIJING • SHANGHAI • HONG KONG • TAIPEI • CHENNAI

Published by

World Scientific Publishing Co. Pte. Ltd.
5 Toh Tuck Link, Singapore 596224
USA office: 27 Warren Street, Suite 401-402, Hackensack, NJ 07601
UK office: 57 Shelton Street, Covent Garden, London WC2H 9HE

British Library Cataloguing-in-Publication Data
A catalogue record for this book is available from the British Library.

Cover images:
Photograph of H. B. G. Casmir used by permission of AIP Emilio Segre Visual Archives.
Background image (courtesy of Liming Dai) is a scanning electron microscope image of vertically aligned multiwalled carbon nanotubes used for dry adhesive applications. [L. Qu, L. Dai, M. Stone, Z. Xia, and Z. L. Wang, *Science* **322**, 238 (2008)].

QUANTUM FIELD THEORY UNDER THE INFLUENCE OF EXTERNAL CONDITIONS (QFEXT09)
Devoted to the Centenary of H B G Casimir
Proceedings of the Ninth Conference

ISBN-13 978-981-4289-85-6
ISBN-10 981-4289-85-X

Printed in Singapore by World Scientific Printers

CONTENTS

Measurement of Casimir Forces

Thermal Casimir Effects

Energy Densities

Solitons and Nanotubes

Schwinger Effect and Quantum Electrodynamics

Field Theory in Backgrounds

PREFACE

The *Ninth Conference on Quantum Field Theory Under the Influence of External Conditions* (QFEXT09) was held at the University of Oklahoma, Norman, USA during the period 21–25 September 2009. This conference celebrated the Centenary of the birth of H. B. G. Casimir (1909–2000), hence the title of this Proceedings. Approximately 108 scientists from more than 25 nations and four continents were participants. 78 talks were given, of which 27 were plenary or invited, and there were 10 posters presented. All presenters were invited to contribute a paper to the Proceedings. All papers were subject to a rigorous reviewing process. The resulting 68 papers appearing here represent well the exciting physics discussed at this meeting.

QFEXT09 was a continuation of a series of workshops of the same title held at the University of Leipzig in 1989, 1992, 1995, 1998, 2001, 2007, at the University of Oklahoma in 2003, and at the University of Barcelona in 2005. The average number of participants has been about 80. The last two meetings have had over 100 participants. Started with the intention of bringing about an East-West scientific dialog, QFEXT developed into one of the most prominent meetings in the field of the Casimir effect, quantum vacuum energy, and related questions in several areas, ranging from quantum field theory and cosmology to atomic, subnuclear, and experimental physics. Typically mathematical questions related to spectral geometry are represented as well. These meetings have created a unique atmosphere where theoreticians, mathematicians and experimentalists are brought together for a week, where talks by colleagues that they might rarely hear otherwise often spark lively debate and result in numerous collaborations.

The past few years have seen spectacular developments in the physics of the quantum vacuum and related fields, both theoretically and experimentally, and therefore this year's meeting was very exciting and timely. Numerous new researches and collaborations are already ensuing as a result. Controversy still boils concerning the temperature correction to the Casimir effect for metals, and whether there is some sort of conflict between fundamental physics of electrodynamics and statistical mechanics and experiment. There have been some remarkable new experiments, such as measuring lateral forces between corrugated materials and demonstrating repulsive Casimir forces. Analytical and numerical work in several different directions have allowed us to transcend the proximity force approximation and numerically calculate Casimir forces for arbitrary geometries at least at medium and large separations.

As noted above, QFEXT09 marked a celebration of the work of Hendrik Casimir, whose idea of quantum vacuum energy turned out to be so very fruitful, laying a foundation for a field of physics which has developed into a major thrust area in both fundamental and applied physics. We recall that Casimir was able to participate in the 1998 QFEXT meeting, where he expressed astonishment at how remarkably this field of research has grown. He would be even more surprised at the vitality of the field today.

It is remarkable that in addition to QFEXT09, two other major conferences in the US took place shortly before and immediately following this meeting: The *Workshop on Casimir Forces and Their Measurement* at Yale (August 10–11, 2009) and the workshop on *New Frontiers in Casimir Force Control* in Santa Fe (September 27–29, 2009). This demonstrates the tremendous interest in the field at present.

The Proceedings of the last six workshops were published as follows:

- M. Bordag and V. M. Mostepanenko, eds., *J. Phys. A* **41,** No. 16 (2008), special issue devoted to papers presented in the 8th workshop on Quantum Field Theory Under the Influence of External Conditions,
- E. Elizalde and S. D. Odintsov, eds., *J. Phys. A* **39,** No.21 (2006), special issue devoted to papers presented in the 7th workshop on Quantum Field Theory Under the Influence of External Conditions,
- K. A. Milton, ed., *Proceedings of the 6th Workshop on Quantum Field Theory Under the Influence of External Conditions* (Rinton Press, Princeton, 2004),
- M. Bordag, ed., *Int. J. Mod. Phys. A* **17,** Nos. 6&7, pp. 711-1064 (2002),
- M. Bordag, ed., *The Casimir Effect 50 Years Later* (World Scientific, Singapore, 1999),
- M. Bordag, ed., *Quantum Field Theory Under the Influence of External Conditions* (Teubner, Stuttgart, 1996).

Here is a link to all these proceedings, as well as to the unpublished proccedings of the 1992 meeting:
http://www.physik.uni-leipzig.de/~bordag/qfext.html

We are very grateful for generous funding that made this conference possible. We are delighted to acknowledge grants from the European Science Foundation, the US National Science Foundation, the US Department of Energy, World Scientific Publishing Company, Journal of Physics A, and very generous support from the University of Oklahoma, through the Homer L. Dodge Department of Physics and Astronomy, Chairs Ryan Doezema and Greg Parker, the College of Arts and Sciences, Dean Paul Bell, and the Vice President for Research, Lee Williams.

We are extremely appreciative of the superb organization and facilities supplied by the Oklahoma Center for Continuing Education, where the meeting was held, and especially of the tremendous assistance of Deb Corley with all arrangements. We thank Mary Kaye Park and Journey House Travel for efficient arrangement of travel to and from Oklahoma, and Sharon Widner for her expert handling of all the complex financial accounts involved. Finally we thank all the participants in the meeting for making QFEXT09 such a rousing success, and for their help afterwards in the sometimes lengthy refereeing process. We hope this resulting volume is a worthy record of this remarkable conference.

M. Bordag University of Leipzig, Germany
Chairman, International Advisory Committee

K. A. Milton University of Oklahoma, USA
Chairman, Local Organizing Committee

23 February 2010

ORGANIZING COMMITTEES

INTERNATIONAL ADVISORY COMMITTEE

for the Series on
Quantum Field Theory Under the Influence of External Conditions

Michael Bordag (Chairman)	– University of Leipzig, Germany
Gerald Dunne	– University of Connecticut, USA
Emilio Elizalde	– IEEC/CSIC, Barcelona, Spain
Kimball Milton	– University of Oklahoma, USA
Umar Mohideen	– University of California, Riverside, USA
Vladimir Mostepanenko	– A. Friedmann Lab., St. Petersburg, Russia
Alexei Starobinsky	– Landau Institute, Moscow, Russia
Herbert Weigel	– University of Stellenbosch, South Africa

LOCAL ORGANIZING COMMITTEE

for QFEXT09

Kimball Milton (Chairman)	– University of Oklahoma, Norman, USA
Steve Fulling	– Texas A&M University, College Station, USA
Brad Abbott	– University of Oklahoma, Norman, USA
Phil Gutierrez	– University of Oklahoma, Norman, USA
Ron Kantowski	– University of Oklahoma, Norman, USA
Chung Kao	– University of Oklahoma, Norman, USA
Mike Strauss	– University of Oklahoma, Norman, USA

SCHEDULE OF TALKS

SUNDAY, SEPTEMBER 20

5:00–8:00pm Registration at Sooner Legends and Suites

MONDAY, SEPTEMBER 21

7:30–8:45 Registration at Forum/Coffee in Conference Room B
8:45–9:00 Welcome: Bell, Parker, Williams, Milton
- Session I. Chair: *Steve Fulling*

9:00–10:00 Michael Bordag, "Casimir Force at Small Separation"
10:00–10:15 *Coffee Break, in Conference Room B*
- Session II. Chair: *Diego Dalvit*

10:15–11:15 Ricardo Decca, "Details on the Experimental Setup for Precision Measurements of the Casimir Force"
11:15–12:15 Umar Mohideen, "Demonstration of the Geometry Dependence through the Lateral Casimir Effect"
12:15–1:30 *Lunch in Corridor A*
- Session III. Chair: *Kim Milton*

1:30–2:30 Jeremy Munday, "Repulsive Casimir-Lifshitz Forces and Torques—A Route Toward Quantum Levitation and Ultra-low Friction Devices"
2:30–3:30 Maarten DeKieviert, "Precision Experiments on Casimir-Polder Forces"
3:30–3:45 *Coffee Break in Conference Room B.*
- Session IV. Chair: *Umar Mohideen*

3:45–4:45 Joel Chevrier, "Measures of Casimir Force and of Near-Field Radiative Heat Transfer"

Parallel Session VA Room A3/5, Chair: *Ricardo Decca*

4:45–5:15 Giuseppe Ruoso, "The PVLAS Experiment and its Results: Probing the Quantum Vacuum with Polarized Light and Magnetic Fields"
5:15–5:45 Caterina Braggio, "Parametric amplification of thermal photons to test the MIR apparatus"
5:45–6:15 Jamir Marino, "Casimir-Polder Force Between Two Uniformly Accelerating Atoms and the Unruh Effect"
6:15–6:45 C. Farina, "Dynamical Casimir Effect with Robin Boundary Condition in a Three Dimensional Open Cavity"

Parallel Session VB Room A4/6, Chair: *Israel Klich*

4:45–5:15 Francesco Intravaia, "Dissecting the Casimir Effect: Getting a Clear Understanding of the Undergoing Physical Mechanisms"
5:15–5:45 Felipe DaRosa, "Electromagnetic Energy, Zero-Point Energy, and Casimir Energy"
5:45–6:15 Paulo Maia Neto, "Non-trivial Geometry Effects in the Atom-Surface Dispersive Interaction"
6:15–6:45 Shang-Yung Wang, "Chiral Symmetry Breaking and Stability of the Magnetized Vacuum"

7:00–9:00 *Reception at Sam Noble Oklahoma Museum of Natural History*

TUESDAY, SEPTEMBER 22

8:00–9:00 Morning coffee in Conference Room B
- Session VI. Chair: *Gerald Dunne*

9:00–10:00 Iver Brevik, "Wedge Casimir Theory: Notes on Thermal Casimir Corrections"
10:00–11:00 Diego Dalvit, "Electrostatic Calibrations and Casimir Force Measurements: The Case of Ge Samples in a Torsion Balance Set-Up"
11:00–11:15 *Coffee Break, in Conference Room B*
- Session VII. Chair: *Michael Bordag*

11:15–12:15 Emil Mottola, "The Trace Anomaly and Dynamical Vacuum Energy in Cosmology"
12:15–1:30 *Lunch in Corridor A*
- Session VIII. Chair: *Bo Sernelius*

1:30–2:30 Lev Pitaevskii, "Casimir-Lifshitz Forces and Entropy"

2:30–3:30 Serge Reynaud, "The Scattering Approach to the Casimir Force"

3:30–3:45 *Coffee Break, in Conference Room B*

Parallel Session IXA Room A3/5, Chair: *Holger Gies*

3:45–4:15 Klaus Kirsten, "Cosmological Pistons"

4:15–4:45 Hongbo Cheng, "The Casimir Effect for Parallel Plates in the Braneworld"

4:45–5:15 Lee-Peng Teo, "Finite Temperature Casimir Effect in the Presence of Extra Dimensions"

5:15–5:45 Francisco Diego Mazzitelli, "Computing the Casimir Force Between Pistons of Arbitrary Shape"

5:45–6:15 Fernando Lombardo, "Computing the Casimir Energy Using the Point-Matching Method"

6:15–6:45 Alexej Weber, "Interplay Between Geometry and Temperature for Inclined Casimir Plates and a Sphere"

Parallel Session IXB Room A4/6, Chair: *Galina Klimchitskaya*

3:45–4:15 Andrea Gambassi, "Controlling and Harnessing Critical Casimir Forces"

4:15–4:45 Francesco Parisen Toldin, "Critical Casimir Forces in the Presence of a Chemically Structured Substrate"

4:45–5:15 Gustavo Moreno, "Bragg Spectroscopy for Measuring Casimir-Polder Interactions with Bose-Einstein Condensates Above Corrugated Surfaces"

5:15–5:45 Luis Reyes Galindo, "Classical Acoustic Casimir Effect"

5:45–6:15 Alexander Studenikin, "Electromagnetic Properties of Neutrinos in Standard Model and Beyond"

6:15–6:45 Yury Sherkunov, "Dispersion Interaction Between Two Atoms in Electromagnetic Fields"

7:00–8:30 *Poster Session in Conference Room B*

WEDNESDAY, SEPTEMBER 23

8:00–9:00 Morning coffee in Conference Room B

- Session X. Chair: *Serge Reynaud*

9:00–10:00 Gerald Dunne, "The Search for the Schwinger Effect: Nonperturbative Vacuum Pair Production"

10:00–11:00 Thorsten Emig, "Casimir Physics: Geometry, Shape and Material"

11:00–11:15 *Coffee Break, in Conference Room B*

- Session XI. Chair: *Emilio Elizalde*

11:15–12:15 Holger Gies, "Geothermal Casimir Phenomena"

12:15–1:30 *Lunch in Corridor A*

- Session XII. Chair: *Iver Brevik*

1:30–2:30 Israel Klich, "Spatial Dispersion and the Interaction Between Quasi One Dimensional Objects"

2:30–3:30 Bo Sernelius, "Possible Saturation Effects in Experiments on the Thermal Casimir Effect"

3:30–3:45 *Coffee Break, in Conference Room B*

Parallel Session XIIIA Room A3/5, Chair: *Thorsten Emig*

3:45–4:15 Martin Schaden, "A Quasi-Local Algorithm for Generating Triangulated d-Dimensional Convex Surfaces with a Particular Measure"

4:15–4:45 Jef Wagner, "Scalar Casimir Energies for Separable Coordinate Systems"

4:45–5:15 James R. Babington, "Casimir Forces in N-Sphere Configurations"

5:15–5:45 Jon Harrison, "Vacuum Energy on Quantum Graphs"

5:45–6:15 Babette Doebrich, "Scalar Casimir-Polder Forces for Arbitrary Uniaxial Corrugations"

6:15–6:45 Saad Zaheer, "Casimir Interactions of an Object Inside a Spherical Metal Shell"

Parallel Session XIIIB Room A4/6, Chair: *Ilya Shapiro*

3:45–4:15 Florian Hebenstreit, "Schwinger Effect in Short Laser Pulses"

4:15–4:45 David Owen, "Fermion-Scalar Bound State with Finite Size Corrections"

4:45–5:15 Vladimir Skalozub, "The Spectrum and Characteristics of Gluon Plasma in Chromomagnetic Field at High Temperature"

5:15–5:45 Juan Mateos Guilarte, "Quantum Fluctuations of S^N-Kinks"

5:45–6:15 Inés Cavero-Peláez, "Green's Function Approach in the Sine-Gordon Kink Background"

6:15–6:45 Nikola Petrov, "Fields in Pulsating Resonators: A Dynamical Systems Approach"

7:00–10:00 *Banquet in Forum Conference Room A*

THURSDAY, SEPTEMBER 24

8:00–9:00 Morning coffee in Conference Room B

- Session XIV. Chair: *Maarten DeKieviert*

9:00–10:00 Davide Iannuzzi, "Casimir Force Experiments in Air: Two Birds with One Stone"

10:00–11:00 Roberto Onofrio, "Macroscopic Quantum Vacuum and Gravitation"

11:00–11:15 *Coffee Break, in Conference Room B*

- Session XV. Chair: *Lev Pitaevskii*

11:15–12:15 Galina Klimchitskaya, "Thermal Casimir Force Between Magnetic Materials"

12:15–1:15 Vladimir Mostepanenko, "The Casimir Effect and the Foundations of Statistical Physics"

1:15–2:30 *Lunch in Corridor A*

- Session XVI. Chair: *Jeremy Munday*

2:30–3:30 Carsten Henkel, "Nonzero Temperature Dispersion Forces: Modes and Matter"

3:30–4:30 Ho Bun Chan, "Experimental Demonstration of the Geometry Dependence of the Casimir Force on Nanostructured Surfaces"

4:30–4:45 *Coffee Break, in Conference Room B*

Parallel Session XVIIA Room A3/5, Chair: *Joel Chevrier*

4:45–5:15 Mauro Antezza, "Casimir-Lifshitz Force Out of Thermal Equilibrium and Asymptotic Nonadditivity"

5:15–5:45 Giuseppe Bimonte, "The Bohr-van Leeuwen Theorem and the Thermal Casimir Effect for Conductors"

5:45–6:15 Simen Ellingsen, "Molecular Guiding with Thermal Casimir-Polder Forces?"

6:15–6:45 Carlos Villarreal, "Bose-Einstein Condensation in Finite-size Regions: Towards a Theory of High-T_c superconductivity"

6:45–7:15 Ryan Behunin, "Non-Equilibrium CP-Force"

Parallel Session XVIIB Room A4/6, Chair: *Steven Johnson*

4:45–5:15 Valery Marachevsky, "Exact Results for the Casimir Lateral Force"

5:15–5:45 Christian Schubert, "Three-Loop Euler-Heisenberg Lagrangian and Asymptotic Analysis in $1+1$ QED"

5:45–6:15 Prachi Parashar, "Non-Contact Gears III. Electromagnetic Case"

6:15–6:45 Hyunsoo Min, "Efficient and Precise Evaluation of Effective Action in Radial Background Fields"

6:45–7:15 Fabrizio Pinto, "Improved Finite-Difference Dispersion Force Computations in Realistic Geometries"

7:30–9:00 *Roundtable on Thermal Casimir Effects, in Conference Room A*

FRIDAY, SEPTEMBER 25

8:00–9:00 Morning coffee in Conference Room B

- Session XVIII. Chair: *Davide Iannuzzi*

9:00–10:00 Steven Johnson, "Geometry-independent Methods to Compute Casimir Forces"

10:00–11:00 Steve Fulling, "Vacuum Energy Density and Pressure Near Boundaries"

11:00–11:15 *Coffee Break, in Conference Room B*

- Session XIX. Chair: *Carsten Henkel*

11:15–12:15 Larry Ford, "Negative Energy Densities in Quantum Field Theory"

12:15–1:15 Ilya Shapiro, "Exact Formfactors in the One-loop Curved-space QED and the Nonlocal Multiplicative Anomaly"

1:15–2:30 *Lunch in Corridor A*

- Session XX. Chair: *Vladimir Mostepanenko*

2:30–3:30 Emilio Elizalde, "Repulsive Casimir Forces from Additional Dimensions"

3:30–3:45 *Coffee Break, in Conference Room B*

Parallel Session XXIA Room A3/5, Chair: *Larry Ford*

3:45–4:15 Oleg Zaslavskii, "Quasi-Black Holes and Lorentz-Abraham Electron in General Relativity"

4:15–4:45 Antonino Flachi, "CFT Quantum Back Reaction and Brane World Black Holes"

4:45–5:15 Alexander Burinskii, "Beam-like Black Hole Radiation and Its Back Reaction on Metric as a Prequantum Kerr-Schild Gravity"

5:15–5:45 Aram Mkhitaryan, "Topological Casimir Effect in Power-law FRW Cosmologies"

Parallel Session XXIB Room A4/6, Chair: *Roberto Onofrio*

3:45–4:15 Marin-Slobodan Tomaš, "Casimir Pressure and Force on a Metal Slab in a Planar Cavity"

4:15–4:45 Ignat Fialkovskiy, "The Casimir Effect for Graphene Described by the Dirac Model"

4:45–5:15 Raul Esquivel-Sirvent, "Controlling Casimir Forces and Torques with External Magnetic Fields. The Role of Magneto-Plasmons"

5:15–5:45 Irina Pirozhenko, "Repulsive Casimir Forces and the Role of Surface Modes"

5:45–6:15 Marco Govoni, "First Principle Calculations of the Casimir Force Between Silicon Films"

Posters

Displayed throughout conference in Conference Room B.

- Antoine Canaguier-Durand, "Casimir Interaction between Plane and Spherical Metallic Surfaces"
- Claudio Ccapa, "Non-superposition Effects in the Dirichlet Casimir Effect"
- Olindo Corradini, "Worldline Approach to QFT on Manifolds with Boundary"
- Ignat Fialkovskiy, "Casimir Type Effects for Scalar Fields Interacting with Material Slabs"
- Guglielmo Fucci, "Non-Perturbative Heat Kernel Asymptotics on Homogeneous Abelian Bundles"
- Toshiyuki Fujii, "Theoretical Studies on Dynamical Casimir Effect for a Quantum Flux in Josephson Artificial Atoms"
- Harald Haakh, "Thermal Effects in the Magnetic Casimir-Polder Interaction"
- Norman Horing, "Graphene van der Waals Interactions"
- Riccardo Messina, "Scattering Approach to Dispersive Atom-Surface Interactions"
- Aram Saharian, "Topological Casimir Effect in Nanotubes and Nanoloops"

CASIMIR FORCES BETWEEN SPHERES AND LOOP INTEGRALS

JAMES BABINGTON*

Quantum Optics and Laser Science, Blackett Laboratory, Imperial College London, Prince Consort Road, London SW7 2AZ, U.K.

**E-mail: j.babington@imperial.ac.uk*

A summary of recent calculations of Casimir forces between a collection of N-dielectric spheres is presented. This is done by evaluating directly the force on a sphere constructed from a stress tensor, rather than an interaction energy. A loop integral formulation is also discussed where we rewrite the expressions for the force in terms of loop integrals for the effective classical propagation of the electric and magnetic fields.

1. Introduction

An important step in better understanding Casimir forces and their application to nano-scale environments is the nature and influence of the geometries used. If we consider a collection of small bodies and their interactions, a quantitative understanding is necessary to address any issue of applications. One can calculate the forces between bodies directly by using a stress tensor evaluated on the body in question. This is an experimentally accessible prediction of the theory e.g. three body forces can be measured between dielectric spheres[1] for critical Casimir forces. The choice of stress tensor in media is not unique and depends on what consistency criteria is used. Typically the differences will show up both in the scale of the forces and the higher order curvature corrections, and one may view this as a way of finding the correct low energy description.

To address the nature of such interactions between bodies[2–5] Casimir interaction energies have been evaluated for collections of compact objects interacting with different force carrying fields (electromagnetic and scalar). The approach taken has been to evaluate a suitable energy functional integral using a T-matrix, whereby an interaction energy can be deduced, normalised with respect to their energy when separated at infinity.

In this talk I summarise recent work we have done on calculating the Casimir force on a single sphere, in an N-sphere setup.[6] By using a multiple scattering approach to evaluate essentially the classical scattering Green's function of the configuration, we are able to evaluate the force directly on the sphere. The total path length plays a fundamental role and what leads one to a loop description.[7] Similar loop descriptions have appeared before in Ref. 8–10 where a semi-classical type expression of the density of states is found, and in Ref. 11 a worldline description for scalar fields is given.

2. The N-Sphere configuration

The question we are addressing is how to calculate the force on a particular sphere as a result of all the interactions with the remaining spheres in a particular static configuration. The Casimir force on a sphere (given by the ball B^2) in the j-direction due to the effects of the N-sphere system is given by

$$F^j(1|N-1) = \int_{B^2} d^3x \nabla_i T^{ij}(x). \tag{1}$$

The stress tensor we choose is the standard vacuum expression (which is consistent with the Lorentz force law[12])

$$T_{ij}(x) = \mathbf{E}_i(x)\mathbf{E}_j(x) + \mathbf{B}_i(x)\mathbf{B}_j(x) - \frac{1}{2}\delta_{ij}(|\mathbf{E}(x)|^2 + |\mathbf{B}(x)|^2), \tag{2}$$

where $x \in B^2$ and it is understood that we are taking the limit for the initial and final points. We then need to evaluate the scattering correlation functions (whilst dropping the direct modes of propagation)

$$\lim_{y\to x} \mathbf{E}_i(x)\mathbf{E}_j(y) = \int_0^\infty \int_0^\infty d\omega d\omega' \langle \mathbf{E}_i^{out}(x;\omega)^\dagger \mathbf{E}_j^{in}(y;\omega')\rangle, \tag{3}$$

and similarly for magnetic fields. To construct the scattering two point function we write the fields in a mode decomposition[13] of spherical vector wave functions that are centred on each sphere centre. Then by applying the standard continuity equations at each of the spheres surfaces, one can calculate the out modes in terms of the in modes and scattering (Mie) coefficients. Assuming that the background in which we are evaluating this is filled with quantum noise such that the noise-current two point function is non-zero we find for the N-body force on a sphere (suppressing the $SO(3)$

indices)

$$\mathbf{F}[1|N-1] = -(-1)^N \frac{\hbar}{4\pi} R[1] \Im \int_0^\infty d\omega k \coth(\hbar\omega/K_B T) \langle \mathbf{1} | [\alpha^1(\omega R[1])$$

$$\times \sum_{i=2}^{N} A^{1,i}(\mathbf{r}[1,i]) \cdot \alpha^i(\omega R[i]) \cdot \cdots \cdot \sum_{j=2}^{N} A^{i,j}(\mathbf{r}[i,j]) \cdot \alpha^j$$

$$\times \sum_{j=2}^{N} \nabla_{\mathbf{r}[j,1]} A^{j,1}(\mathbf{r}[j,1])] j(kR[1]) h^+(kR[1]) W(\omega R[1]) | \mathbf{1} \rangle.$$

$$\equiv -(-1)^N \frac{\hbar}{4\pi} \sum_{i=2}^{N} \nabla^{last}_{\mathbf{r}[i,1]} \int_0^\infty d\Omega \cot\left(\frac{\hbar\Omega}{k_B T}\right) \mathcal{Z}[\alpha, A, W]. \quad (4)$$

Here, $\alpha^i(\omega R[i])$ are the Mie scattering coefficients in the $SO(3)$ basis for sphere i with radius $R[i]$; $A^{i,j}(\mathbf{r}[i,j])$ are the translation matrices mapping the TE and TM vector wave functions between spheres i and j; the vectors $|\mathbf{1}\rangle$ give the truncation in the L angular momentum quantum number (leading to a multipole type expansion); and the W is just the collection of the four different contributions that make up the stress tensor, together with the two necessary Bessel functions evaluated on the spheres surface. In the last line a Wick rotation to imaginary frequencies has been performed and the $\mathcal{Z}$-function has been defined for later reference.

Note the explicit form of the translation matrices involve exponentials of the inter-sphere separations[6] and thus it is the *total path length* that plays the key role in understanding the variables of the system. For simple setups (e.g. two and three sphere systems) we can evaluate Equation (4) in different perturbative regimes e.g. retarded or non-retarded limits using static values for the permittivities (see Ref. 6 for explicit evaluations and force plots of two and three sphere systems).

3. Loop integrals

Following on from the observation that Equation (4) involves the total path length in the form of a loop, in Ref. 7 an attempt is made to develop this further. It is similar to the path integrals used in Refs. 8 and 11 where a fictitious time is introduced as well as mass scale set equal to unity and an appearance of Planck's constant in the particle action. As pointed out in Ref. 10, it is misleading to call this a semi-classical evaluation because of the absence of the dimensionful Planck's constant. Concurrent with this is the absence of a mass or length scale with which to define a dimensionless action.

In Ref. 7 a symbolic expression for the $\mathcal{Z}$-function was found that features the loop structure in an explicit manner

$$\mathcal{Z}^{(s)}[\Omega, x] = \sum_{\mathcal{C}_x} \langle \mathbf{F}_x^s | \exp\left[- \oint_{\mathcal{C}_x} dq^i \hat{D}_i(q, \Omega) \right] | \mathbf{I}_x^s \rangle$$
$$= \sum_{\mathcal{C}_x} \mathbf{Tr} \left(\exp\left[-\hat{S}_R(\mathcal{C}_x, \Gamma, \Omega) \right] \right)_{\{\mathbf{I}_x^s, \mathbf{F}_x^s\}} \tag{5}$$

where x is the initial and final point of the loop, and $|\mathbf{I}\rangle$ and $|\mathbf{F}\rangle$ are the initial and final states (i.e. boundary conditions imposed on the eigenfunctions used to represent in and out modes). The generator $\hat{D}_i$ of translations and the loop $\mathcal{C}_x$ implicitly depend on the background potentials. The connection Γ is formed from the background potentials in which the field propagates. One now needs to find a representation of this object. In fact it can be given a path integral representation, albeit a classical one i.e. no $\hbar$ featuring anywhere. One first needs to invert the Helmholtz operator (here partial derivatives have been promoted to covariant derivatives w.r.t. the permittivity and permeability)

$$\mathbf{\Delta}(z, y) = \langle z | \left[-c^2/\Omega^2 \nabla_\epsilon \wedge \nabla_\mu \wedge -\mathbf{1}\cdot \right]^{-1} |y\rangle$$
$$= \int_0^\infty d\tau \langle z | e^{-\tau[c^2/\Omega^2 \nabla_\epsilon \wedge \nabla_\mu \wedge + \mathbf{1}\cdot]} |y\rangle. \tag{6}$$

Introducing a world line metric e to implement the Helmholtz equation on physical states, and an integration over the $\hat{\nabla}_i$ operators (again acting on physical states), together with an integration over paths one finds for the scalar version of the Helmholtz operator (i.e. two potentials but no spatial indices)

$$\mathcal{Z}[\Omega, x] = \int_0^1 d\tau \sum_{n=1}^{\infty} \langle \mathbf{F}_x | \oint_{q[0]=q[1]=x} [dq] \sqrt{\det[\Omega^2/c^2(\epsilon \cdot \mu)]}$$
$$\cdot \int [d\hat{\nabla}][de] \exp\left(-n \int_0^T dt [\dot{q} \cdot \hat{\nabla} + e(\hat{\nabla}_\epsilon \cdot \hat{\nabla}_\mu + 1)] \right) |\mathbf{I}_x\rangle, \tag{7}$$

and

$$\int_0^\infty d\Omega \mathcal{Z}[\Omega, x] \sim \sum_{loops} \mathcal{Z}[\Omega, x]. \tag{8}$$

If we perform first the integral over the world line metric, the Helmholtz equation is implemented. Performing the integral over the derivatives returns a configuration space path integral which would require gauge fixing the world-line metric. The classical equations of motion then lead to closed geodesics defining the loops. Integration over the frequency provides part of the sum over paths, whilst the tau integration gives the windings of the loops.

4. Conclusions

In this talk I have summarised recent work we have done on calculating Casimir forces between spheres using a multiple scattering approach. The total closed path length plays a key role in understanding the calculated forces and leads to a loop description. By considering the origin of the translation coefficients, together with a path integral representation of the Helmholtz operator, one is able to reformulate the loop integral as a sum over of all possible loops.

Acknowledgments

J. B. wishes to thank Stefan Buhmann, Stefan Scheel, Alex Crosse, Rachele Fermani and John Gracey for numerous helpful and constructive discussions. This work was supported by the SCALA programme of the European commission.

References

1. M. Brunner, J. Dobnikar, H.-H. von Grünberg, and C. Bechinger, Phys. Rev. Lett. **92**, 078301 (2004).
2. T. Emig and R. L. Jaffe, J. Phys. A **41**, 164001 (2008).
3. S. J. Rahi, T. Emig, N. Graham, R. L. Jaffe, and M. Kardar, Phys. Rev. D **80**, 085021 (2009).
4. O. Kenneth and I. Klich, Phys. Rev. B **78**, 014103 (2008).
5. A. Bulgac, P. Magierski and A. Wirzba, Phys. Rev. D **73** 025007 (2006).
6. J. Babington and S. Scheel", arXiv:0909.3285 [quant-ph].
7. J. Babington, arXiv:0909.3315 [quant-ph].
8. M. Schaden and L. Spruch, Phys. Rev. A **58**, 2, (1998).
9. M. Schaden, Phys. Rev A **73**, 4, 042102, (2006).
10. A. Scardicchio and R.L. Jaffe, Nucl. Phys B, **704**, 3, (2005).
11. H. Gies and K. Langfeld, Int. J. Mod. Phys. **A17**, 966-978, (2002).
12. C. Raabe and D.-G. Welsch, Phys. Rev. A **71**, 013814 (2005).
13. D. W. Mackowski, Proc. R. Soc. Lond. A **433**, 599 (1991).

ANALYTIC CORRECTIONS TO THE ELECTROMAGNETIC CASIMIR INTERACTION BETWEEN A SPHERE AND A PLATE AT SHORT DISTANCES

M. BORDAG

Leipzig University, Vor dem Hospitaltore 1, D-04103 Leipzig, Germany
E-mail: bordag@itp.uni-leipzig.de

V. NIKOLAEV

Halmstad University, Box 823, S-30118 Halmstad, Sweden
E-mail: Vladimir.Nikolaev@ide.hh.se

For the vacuum interaction of a sphere in front of a plane, both obeying conductor boundary conditions, we consider the approximation of small separation. We derive the next-to-leading order of the asymptotic expansion in the separation-to-radius ratio ε. This correction is of order ε. In opposite to the scalar cases it contains also contributions proportional to logarithms in first and second order, $\varepsilon \ln \varepsilon$ and $\varepsilon(\ln \varepsilon)^2$. We compare this result with the available findings of numerical and experimental approaches.

Keywords: Casimir effect, proximity force approximation, conductor boundary conditions.

In the presence of material bodies, the vacuum energy of the electromagnetic field results in the Casimir effect. Originally it was derived for plane parallel surfaces. Only quite recently an effective method[1,2] was found to calculate it for the interaction of separated bodies of more complicated geometry. This method is based on a functional determinant representation of the interaction energy and it does not contain ultraviolet divergences in any intermediate steps. Sometimes it is also called TGTG-representation.[3] Several different representations were given in Refs. 4, 5 (see also Ref. 6). The method can be used for a direct numerical evaluation at large and medium separations between the bodies. However, for small separation the numerical effort grows fast preventing calculations below approximately $\varepsilon \lesssim 10^{-2} \cdots 10^{-3}$, where

$$\varepsilon = \frac{d}{R} \tag{1}$$

is the ratio of the separation d between the two surfaces to the radius R of the sphere. Close separation is just the region of most experimental interest since in that case the forces are bigger and higher precision can be reached.

Up to know, for calculation at small ε, the *Proximity Force Approximation* (PFA) was used. Within this approximation, one takes the force density known from parallel planes and integrates it over the surface. This method is universal since it does not depend on the nature of the force. However, this method gives the correct and unique value[a] only in leading order for $\varepsilon \to 0$. Therefore, using this method it is not possible to determine the precision of this approximation or to calculate corrections beyond PFA.

The configuration of a sphere at close separation in front of a plane is most interesting for the experimental setups since it prevents from problems to keep the interacting bodies in parallel position. Therefore it is of primary interest to consider this situation. An approximation at close separation can be obtained in three ways. The first is to make an asymptotic expansion for $\varepsilon \to 0$ in the functional determinant representation. It takes the form

$$E_0(\varepsilon) = E_0^{\rm PFA} \left(1 + \varepsilon \left(\alpha + \beta \ln \varepsilon + \gamma (\ln \varepsilon)^2\right) + \ldots\right) , \tag{2}$$

where the dots denote contributions of order ε^2 and higher. In leading order one reproduces the PFA. This can be viewed also as an independent derivation of PFA,

$$E_0^{\rm PFA} = -\frac{\pi^3 R^{1/2}}{960\sqrt{2} d^{5/2}} \qquad \text{for a cylinder,} \tag{3}$$

$$E_0^{\rm PFA} = -\frac{\pi^3 R}{720 d^2} \qquad \text{for a sphere,} \tag{4}$$

in front of a plane for the electromagnetic interaction. The next order contains the corrections beyond PFA. In (2) these are written as relative corrections. It must be mentioned that, in general, the appearance of logarithmic contributions cannot be excluded. This follows, by hindsight, from the mathematical structure involved in the functional determinant representation.

The first calculation of a correction beyond PFA was done in Ref. 7 for a cylinder in front of a plane. In that case no logarithmic terms appeared, $\beta = \gamma = 0$, and the coefficient α was found to be

$$\begin{aligned} \alpha &= \frac{7}{36}, && \text{(Dirichlet boundary conditions)} \\ \alpha &= \frac{7}{36} - \frac{20}{3\pi^2}. && \text{(Neumann boundary conditions)} \end{aligned} \tag{5}$$

[a]Thereby it makes no difference whether the calculation is made sphere or plane based.

Since in cylindrical geometry the polarizations of the electromagnetic field separate, the half sum of these two gives the correction for the electromagnetic field.

For a sphere in front of a plane the calculation is more involved. First, one needs a specific asymptotic expansion of the Clebsch-Gordan coefficients. It was derived in Ref. 8, where the case of a scalar field is considered. Second, the polarizations of the electromagnetic field do not separate and additional factors following from its vector structure appear. This case was considered in Ref. 9.

For the scalar field in Ref. 8 it was found that there are also no logarithmic contributions. For the different combinations of boundary conditions on the two surfaces the result is shown in the Table 1.

Table 1. The relative correction α beyond PFA for Dirichlet (D) and Neumann (N) boundary conditions an a sphere and on a plane.

plane	sphere	α
D	D	$\frac{1}{3}$
N	N	$\frac{1}{3}-\frac{10}{\pi^2}$
D	N	$\frac{1}{3}-\frac{5}{\pi^2}$
N	D	$\frac{1}{3}$

The second method of calculation is to push the numerical evaluation of the functional determinant as far as possible for small ε, to extrapolate for smaller ε and to fit the result. In this way, for a cylinder in front of a plane, in Ref. 10 agreement was obtained for Dirichlet boundary conditions. For Neumann boundary conditions, a dependence on the kind of the fit was observed and agreement with the analytical result was obtained only for a fit admitting a logarithmic contribution in the order ε^2. For a sphere in front of a plane the numerical results in Ref. 5 are in agreement also for Dirichlet boundary conditions on both, the sphere and the plane, only.

A third method to obtain the Casimir energy at close separation, results from the world line methods. In Ref. 11, again, for Dirichlet boundary conditions, agreement was obtained. Regrettably, so far this method did not give results for other boundary conditions and also not for the electromagnetic field.

There are also experimental efforts[12] to measure the deviation from PFA. There the outcome was that these corrections, for the electromagnetic field, were quite small. Essentially, within the precision archived, no corrections were found.

It is the aim of the present paper to report on the analytical corrections beyond PFA for a sphere in front of a plane in case of the electromagnetic field. The general formula for the interaction energy in the functional determinant representation is

$$E = \frac{1}{2}\int_{-\infty}^{\infty}\frac{d\xi}{2\pi}\,\mathrm{Tr}\ln\left(1-\mathbf{N}\right), \tag{6}$$

where ξ is the imaginary frequency and $\mathbf{N}_{l,l'}^{(s,s')}(\xi)$ is a matrix in the orbital momenta involving the Bessel function. In case of the electromagnetic field it is also a matrix in the polarizations s, s'. For details we refer to the more complete paper[9] and to the general representation in Ref. 6, chapter 10, and in Ref. 5. The trace is over the orbital momentum indices l and l' and over the indices $s, s' = 1, 2$. These are mixed in the given geometry, i.e., terms with $s \neq s'$ are present. This mixing and the specific structure of the matrices $N_{l,l'}^{(s,s')}$ following from the vector structure of the electromagnetic field make the calculation of the asymptotic expansion considerably more involved as compared to the scalar case. These also result in logarithmic contributions. The results found in Ref. 9 are

$$\alpha = -5.2, \quad \beta = -0.0044, \quad \gamma = 8.5\ 10^{-6}. \tag{7}$$

The contributions to α can be split into three parts, $\alpha = a_1 + \alpha_2 + \alpha_3$. The first one,

$$\alpha_1 = \frac{1}{3} - \frac{5}{\pi^2} \simeq -0.173\,, \tag{8}$$

results from the two polarizations taken separately. This is just the half of the sum of the scalar contributions with Dirichlet and Neumann boundary conditions. These are the first and the second lines in Table 1. We remark that, as shown in Ref. 8, in this approximation there is no difference between Neumann boundary conditions and the boundary conditions for the TM mode.

The remaining two contributions are

$$\alpha_2 = \frac{180}{\pi^4}\left[(1+2\ln 2)\zeta(3) - 2(1+\ln 2)\zeta(2)\right] \simeq -4.99 \tag{9}$$

and

$$\alpha_3 = -0.045 - 0.0044\ln\varepsilon + 8.5\ 10^{-6}(\ln\varepsilon)^2. \tag{10}$$

Both result from structures specific for the electromagnetic case. The splitting into α_2 and α_3 was made since α_2 could be calculated analytically, whereas α_3 could be calculated in the final steps only numerically. It must be mentioned that α_2 gives the dominating contribution to the corrections. It is unexpectedly large. The logarithmic contributions appear in α_3, however these go with quite small numerical factors in front.

In must be mentioned that the result (7) is not in agreement with the numerical findings in Refs. 5, 13 who reported $\alpha \sim -1.4$. Also (7) is not in agreement with the experimental result.[12] At the moment this must be considered an opened problem.

Possible explanations for the discrepancy with the numerical method could be in the method of asymptotic expansion. In principle, it is unknown how small ε must be made in order to give an approximation with a given precision. However, since the first coefficient α is of order one, the considered ε of order 10^{-3} can be considered to be sufficiently small. Another problem is that in the numerical method ε cannot be made very small. Therefore the results were extrapolated and fitted. As a consequence the analytical and numerical results may be actually not in a disagreement. There might be also another problem related to the Neumann boundary conditions. As mentioned, in opposite to the case of Dirichlet boundary conditions, here we have weak agreement in the cylindrical case and no agreement in the spherical case between the analytical and the numerical methods. The point is that from the structure of the analytical method it follows that the case of Neumann boundary conditions is somehow more difficult. For the calculation of the asymptotic expansion one starts with an expansion of the logarithm in (6). For Dirichlet boundary conditions, the emerging series, after expanding for small ε, converges like $\sum_s s^{-4}$. For Neumann boundary conditions on the sphere (or on the cylinder) the convergence is weaker, it is like $\sum_s s^{-2}$ only. For the analytical method this is not a problem since the sum does converge. For the numerical approach this seems also not to be a problem since there the logarithm of the determinant is calculated and no expansion of the logarithm is made. However, one could speculate that this slower convergence could show up, for example, somewhere in the asymptotic properties of the extrapolation used in the numerical approach or in some other place.

As for the disagreement with the experimental findings one must take into account that the present calculation is for ideal conductor boundary conditions. Before a meaningful comparizon with the experimental results can be done one would need to calculate or at least to estimate the influence of the real structure of the metallic surfaces on the correction beyond PFA. A first calculation of this kind was done in Ref. 14 for a cylinder in front of a plane. In that paper a dielectric half space described by the plasma model and a dielectric or a semitransparent cylinder were considered. The first correction beyond PFA, which is α in the notation of Eq.(2), is the ratio f_1/f_0 in Ref. 14. In the cases considered in that paper it is larger than one (see figures 5, 6 and 8 there). In this sense the real structure tends to amplify the correction beyond PFA. It would be interesting to look up this in spherical geometry.

Acknowledgments

V.N. was supported by the Swedish Research Council (Vetenskapsrådet), grant 621-2006-3046. The authors benefited from exchange of ideas by the ESF Research Network CASIMIR.

References

1. A. Bulgac, P. Magierski and A. Wirzba, *Phys. Rev.* **D73**, p. 025007 (2006).
2. T. Emig, N. Graham, R. L. Jaffe and M. Kardar, *Phys. Rev. Lett.* **99**, p. 170403 (2007).
3. O. Kenneth and I. Klich (2007), arXiv:0707.4017.
4. S. J. Rahi, T. Emig, N. Graham, R. L. Jaffe and M. Kardar, *Phys. Rev.* **D80**, p. 085021 (2009).
5. T. Emig, *J. Stat. Mech.* **08**, p. P04007 (2008).
6. M. Bordag and G.L. Klimchitskaya and U. Mohideen and V.M. Mostepanenko, *Advances in the Casimir Effect* (Oxford University Press, International Series of Monographs on Physics 145, 2009).
7. M. Bordag, *Phys. Rev.* **D73**, p. 125018 (2006).
8. M. Bordag and V. Nikolaev, *J. Phys. A: Math. Gen.* **41**, p. 164001 (2008).
9. M. Bordag and V. Nikolaev (2009), ArXiv: 0911.0146.
10. F. C. Lombardo, F. D. Mazzitelli and P. I. Villar, *Phys. Rev.* **D78**, p. 085009 (2008).
11. H. Gies and K. Klingmuller, *Phys. Rev.* **D74**, p. 045002 (2006).
12. D. Krause, R. S. Decca, D. Lopez and E. Fischbach, *Phys. Rev. Lett.* **98**, p. 050403 (2000).
13. P. A. Maia Neto, A. Lambrecht and S. Reynaud, *Phys. Rev. A* **78**, p. 012115 (2008).
14. M. Bordag, *Phys.Rev.D* **75**, p. 065003 (2007).

CASIMIR PHYSICS: GEOMETRY, SHAPE AND MATERIAL

T. EMIG

Institut für Theoretische Physik, Universität zu Köln, Zülpicher Strasse 77, 50937 Köln, Germany

Laboratoire de Physique Théorique et Modèles Statistiques, CNRS UMR 8626, Bât. 100, Université Paris-Sud, 91405 Orsay cedex, France

The properties of fluctuation induced interactions like van der Waals and Casimir-Lifshitz forces are of interest in a plethora of fields ranging from biophysics to nanotechnology. Here we describe a general approach to compute these interactions. It is based on a combination of methods from statistical physics and scattering theory. We showcase how it is exquisitely suited to analyze a variety of previously unexplored phenomena. Examples are given to show how the interplay of geometry and material properties helps to understand and control these forces.

1. Introduction

All material objects, even if charge neutral, support instantaneous current fluctuations due to quantum and thermal fluctuations of their charge distribution. The interaction that results from the electromagnetic coupling of these currents on different objects is usually called the Casimir force. Originally, this force has been derived for two parallel perfect metal plates[1] and atoms,[2] and generalized later to two infinite dielectric half-spaces with planar and parallel surfaces.[3–6] The non-additivity of the Casimir force limits these results in their applicability to objects at very short separation via the so-called proximity force approximation which provides only an uncontrolled approximation of surface curvature to lowest order at vanishingly small separations and ignores the global geometrical arrangement of the objects. Generically, one encounters in practice geometries and shapes that are rather distinct from infinite, parallel and planar surfaces. Hence one faces the problem to compute the Casimir force between objects of general shape, arrangement and material decomposition.

This article summarizes recent progress that has been proofed useful in solving this problem for a variety of geometries. (For an overview of the development of related approaches, see Ref. 7.) In order to study Casimir forces in more general geometries, it turns out to be advantageous to describe how fluctuating currents are induced on the objects by the scattering of electromagnetic waves. This representation of the Casimir interaction was developed in Refs. 7–9. Each object is characterized by its on-shell electromagnetic scattering amplitude. The separations and orientations of the objects are encoded in universal translation matrices, which describe how a solution to the source-free Maxwell's equations in the basis appropriate to one object looks when expanded in the basis appropriate to another. These matrices hence describe the electrodynamic interaction of the multipole moments associated with the currents and depend on the displacement and orientation of coordinate systems, but not on the shape and material of the objects themselves. The scattering amplitudes and translation matrices are then combined in a simple formula that allows efficient numerical and, in some cases, analytical calculations of Casimir forces and torques for a wide variety of geometries, materials, and external conditions. The approach applies to any finite number of arbitrarily shaped objects with arbitrary linear electromagnetic response at zero or finite temperature.

To illustrate this general formulation, we provide some sample applications, including results for the interaction between metallic objects for two spheres and for a sphere and a plane, taking into account the combined effect of shape and material properties at large distances. In addition, we provide examples for the non-additivity of the interaction by considering three objects (two spheres and a plane) and for the orientation dependence in the case of spheroids. The results are presented in form of analytical expressions at large distances and as numerical results at smaller separations.

2. Fluctuating Currents and T-operators

We consider the Casimir energy for neutral objects with electric and magnetic susceptibilities. The partition function Z is defined through the path integral, which sums all configurations of the electromagnetic field (outside and inside the objects) with periodic boundary conditions in time between 0 and T. The free energy F of the field at inverse temperature β is

$$F(\beta) = -\frac{1}{\beta} \log Z(\beta). \tag{1}$$

The unrenormalized free energy generally depends on the ultraviolet cutoff, but cutoff-dependent contributions arise from the objects individually and do not depend on their separations or orientations. Since we are only interested in energy differences, we can remove these divergences by subtracting the energy of the system when the objects are in some reference configuration, see below. By replacing the time T by $-i\hbar\beta$, we obtain the partition function $Z(\beta)$ in 4D Euclidean space. In $A^0 = 0$ gauge, the result is simply to replace the Matsubara frequencies $\omega_n = \frac{2\pi n}{T}$ by $i\frac{2\pi n}{\hbar\beta} = ic\kappa_n$, where κ_n is the nth Matsubara frequency divided by c. The action is quadratic, so the modes with different κ_n decouple and the partition function decomposes into a product of partition functions for each mode. In the limit $\beta \to \infty$, the sum $\sum_{n\geq 0}$ turns into an integral $\frac{\hbar c\beta}{2\pi}\int_0^\infty d\kappa$, and we have the ground state energy

$$\mathcal{E}_0 = -\frac{\hbar c}{2\pi}\int_0^\infty d\kappa \, \log Z(\kappa), \tag{2}$$

with

$$Z(\kappa) = \int \mathcal{D}\mathbf{A}\mathcal{D}\mathbf{A}^* \, \exp\left[-\beta \int d\mathbf{x}\, \mathbf{E}^*(\kappa,\mathbf{x})\left(\mathbb{H}_0 + \frac{1}{\kappa^2}\mathbb{V}(\kappa,\mathbf{x})\right)\mathbf{E}(\kappa,\mathbf{x})\right], \tag{3}$$

where we have used $\boldsymbol{\nabla} \times \mathbf{E} = i\frac{\omega}{c}\mathbf{B}$ to eliminate $\mathbf{B}$ in the action, and it is assumed that $\mathbf{E}$ is expressed by $\mathbf{E} = -c^{-1}\partial_t \mathbf{A}$ in terms of the vector potential $\mathbf{A}$. This functional integral sums over configurations of $\mathbf{A}$. This sum must be restricted by a choice of gauge, so that it does not include the infinitely redundant gauge orbits. We will choose to work in the gauge $A^0 = 0$, although of course no physical results depend on this choice. Here we defined the Helmholtz operator

$$\mathbb{H}_0(\kappa) = \mathbb{I} + \frac{1}{\kappa^2}\nabla \times \nabla \times \,, \tag{4}$$

which is inverted by the Green's function that is defined by

$$\kappa^2 \mathbb{H}_0(\kappa)\mathbb{G}_0(\kappa,\mathbf{x},\mathbf{x}') = \mathbb{I}\delta^{(3)}(\mathbf{x}-\mathbf{x}')\,. \tag{5}$$

The potential operator is

$$\mathbb{V}(\kappa,\mathbf{x}) = \mathbb{I}\,\kappa^2\left(\epsilon(ic\kappa,\mathbf{x}) - 1\right) + \boldsymbol{\nabla} \times \left(\frac{1}{\mu(ic\kappa,\mathbf{x})} - 1\right)\boldsymbol{\nabla} \times \,. \tag{6}$$

It is nonzero only at those points in space where the objects are located ($\epsilon \neq 1$ or $\mu \neq 1$). At small frequencies, typical materials have $\epsilon > 1$ and $\mu \approx 1$, and $\mathbb{V}$ can be regarded as an attractive potential.

Next, we transform to a free field (with kernel $\mathbb{H}_0$) by introducing fluctuating currents $\mathbf{J}$ that are confined to the objects. To perform this Hubbard-Stratonovich-like transformation we multiply and divide the partition function of Eq. (3) by

$$W = \int \mathcal{D}\mathbf{J}\mathcal{D}\mathbf{J}^*|_{\rm obj} \exp\left[-\beta \int d\mathbf{x}\, \mathbf{J}^*(\mathbf{x}) \cdot \mathbb{V}^{-1}(\kappa,\mathbf{x})\mathbf{J}(\mathbf{x})\right] = \det \mathbb{V}\,, \tag{7}$$

where $|_{\rm obj}$ indicates that the currents are defined only over the objects, *i.e.* the domain where $\mathbb{V}$ is nonzero (and therefore $\mathbb{V}^{-1}$ exists), and we have represented the local potential as a matrix in position space, $\mathbb{V}(\kappa,\mathbf{x},\mathbf{x}') = \mathbb{V}(\kappa,\mathbf{x})\delta^{(3)}(\mathbf{x}-\mathbf{x}')$. We then change variables in the integration, $\mathbf{J}(\mathbf{x}) = \mathbf{J}'(\mathbf{x}) + \frac{i}{\kappa}\mathbb{V}(\kappa,\mathbf{x})\mathbf{E}(\mathbf{x})$ and $\mathbf{J}^*(\mathbf{x}) = \mathbf{J}'^*(\mathbf{x}) + \frac{i}{\kappa}\mathbb{V}(\kappa,\mathbf{x})\mathbf{E}^*(\mathbf{x})$, to obtain

$$\begin{aligned} Z(\kappa) &= \frac{1}{W}\int \mathcal{D}\mathbf{A}\mathcal{D}\mathbf{A}^*\, \mathcal{D}\mathbf{J}'\mathcal{D}\mathbf{J}'^*|_{\rm obj} \times \\ &\quad \exp\left[-\beta\int d\mathbf{x}\, \mathbf{E}^*(\kappa,\mathbf{x})\left(\mathbb{H}_0(\kappa) + \frac{1}{\kappa^2}\mathbb{V}(\kappa,\mathbf{x})\right)\mathbf{E}(\kappa,\mathbf{x})\right. \\ &\quad \left. + \left(\mathbf{J}'^*(\mathbf{x}) + \frac{i}{\kappa}\mathbb{V}(\kappa,\mathbf{x})\mathbf{E}^*(\kappa,\mathbf{x})\right)\mathbb{V}^{-1}(\kappa,\mathbf{x})\left(\mathbf{J}'(\mathbf{x}) + \frac{i}{\kappa}\mathbb{V}(\kappa,\mathbf{x})\mathbf{E}(\kappa,\mathbf{x})\right)\right], \\ &= \frac{1}{W}\int \mathcal{D}\mathbf{A}\mathcal{D}\mathbf{A}^*\, \mathcal{D}\mathbf{J}'\mathcal{D}\mathbf{J}'^*|_{\rm obj} \times \\ &\quad \exp\left[-\beta\int d\mathbf{x}\, \mathbf{E}^*\mathbb{H}_0\mathbf{E} + \mathbf{J}'^*\mathbb{V}^{-1}\mathbf{J}' + \frac{i}{\kappa}\left(\mathbf{J}'^*\mathbf{E} + \mathbf{J}'\mathbf{E}^*\right)\right]. \end{aligned} \tag{8}$$

Now the free electromagnetic field can be integrated out using $\mathbb{H}_0^{-1} = \kappa^2\mathbb{G}_0$, yielding

$$\begin{aligned} Z(\kappa) &= \frac{Z_0}{W}\int \mathcal{D}\mathbf{J}'\mathcal{D}\mathbf{J}'^*|_{\rm obj} \\ &\quad \exp\left[-\beta\int d\mathbf{x}d\mathbf{x}'\, \mathbf{J}'^*(\mathbf{x})\left(\mathbb{G}_0(\kappa,\mathbf{x},\mathbf{x}') + \mathbb{V}^{-1}(\kappa,\mathbf{x})\delta^3(\mathbf{x}-\mathbf{x}')\right)\mathbf{J}'(\mathbf{x}')\right], \end{aligned} \tag{9}$$

with $Z_0 = \int \mathcal{D}\mathbf{A}\mathcal{D}\mathbf{A}^* \exp[-\beta\int d\mathbf{x}\,\mathbf{E}^*\mathbb{H}_0(\kappa)\mathbf{E}]$. Both factors W and Z_0 contain cutoff-dependent contributions but are independent of the separation of the objects. Hence these factors cancel and can be ignored when we consider a *change* in the energy due to a change of the object's separations with the shape and the material composition of the objects fixed. The kernel of the action in Eq. (9) is the inverse of the T-operator, i.e., $\mathbb{T}^{-1} = \mathbb{G}_0 + \mathbb{V}^{-1}$ which is equivalent to

$$\mathbb{T} = \mathbb{V}(\mathbb{I} + \mathbb{G}_0\mathbb{V})^{-1}\,. \tag{10}$$

The Casimir energy at zero temperature (without the cutoff-dependent parts) is hence

$$\mathcal{E} = -\frac{\hbar c}{2\pi}\int_0^\infty d\kappa \log\det\mathbb{T}\,. \tag{11}$$

The determinant is here taken over the spatial indices $\mathbf{x}$ and $\mathbf{x}'$, which are restricted to the objects since $\mathbb{T}$ vanishes if $\mathbf{x}$ or $\mathbf{x}'$ are not on an object. To compute the determinant we start from the expression for $\mathbb{T}^{-1}$ which yields the reciprocal of the determinant. We decompose $\mathbb{T}^{-1}$ by introducing separate position space basis functions for each object. The projection of the currents onto this basis defines the object's multipole moments. This yields a division of $\mathbb{T}^{-1}$ into blocks where each block is labeled by an object.

The off-diagonal blocks are given by $\mathbb{G}_0$ only and describe the interaction of the multipoles on different objects. To see this we choose for each object individually an eigenfunction basis to expand the free Green's function,

$$\mathbb{G}_0(\kappa,\mathbf{x},\mathbf{x}') = \sum_\alpha \mathbf{E}_\alpha^{\text{out}}(\kappa,\mathbf{x}_>)\otimes\mathbf{E}_\alpha^{\text{reg}*}(\kappa,\mathbf{x}'_<) \tag{12}$$

with regular solutions $\mathbf{E}_\alpha^{\text{reg}}$ and outgoing solutions $\mathbf{E}_\alpha^{\text{out}}$ of the free vector Helmholtz equation, where $\mathbf{x}_<$ and $\mathbf{x}_>$ denote the position with smaller and greater value of the "radial" variable of the separable coordinates. The multipole moments of object j are then $Q_{j,\alpha}(\kappa) = \int d\mathbf{x}\mathbf{J}_j(\kappa,\mathbf{x})\mathbf{E}_\alpha^{\text{reg}*}(\kappa,\mathbf{x})$. Regular solutions form a complete set and hence outgoing solutions can be expanded in terms of regular solutions except in a region (enclosed by a surface of constant radial variable) that contains the origin of the coordinate system of object i. This expansion defines the translation matrices $\mathbb{U}_{\beta,\alpha}^{ji}$ via

$$\mathbf{E}_\alpha^{\text{out}}(\kappa,\mathbf{x}_i) = \sum_\beta \mathbb{U}_{\beta\alpha}^{ji}(\kappa,\mathbf{X}_{ji})\mathbf{E}_\beta^{\text{reg}}(\kappa,\mathbf{x}_j)\,, \tag{13}$$

where the definition of the coordinates is shown in Fig. 1. The free Green's function then becomes

$$\mathbb{G}_0(\kappa,\mathbf{x},\mathbf{x}') = \sum_{\alpha,\beta} \mathbf{E}_\alpha^{\text{reg}}(\kappa,\mathbf{x}_i)\otimes\mathbb{U}_{\alpha\beta}^{ji}(\kappa,\mathbf{X}_{ji})\mathbf{E}_\beta^{\text{reg}*}(\kappa,\mathbf{x}'_j) \tag{14}$$

so that the off-diagonal blocks of $\mathbb{T}^{-1}$ are given by the translation matrices. Equivalent translation matrices can be defined between two sets of regular solutions as is necessary for one object inside another, see Ref. 7.

The diagonal blocks of $\mathbb{T}^{-1}$ are given by the matrix elements of the T-operators $\mathbb{T}_j$ of the *individual* objects. By multiplying $\mathbb{T}^{-1}$ by the T-operator $\mathbb{T}_\infty$ without the off-diagonal blocks which can interpreted as describing a reference configuration with infinite separations between the ob-

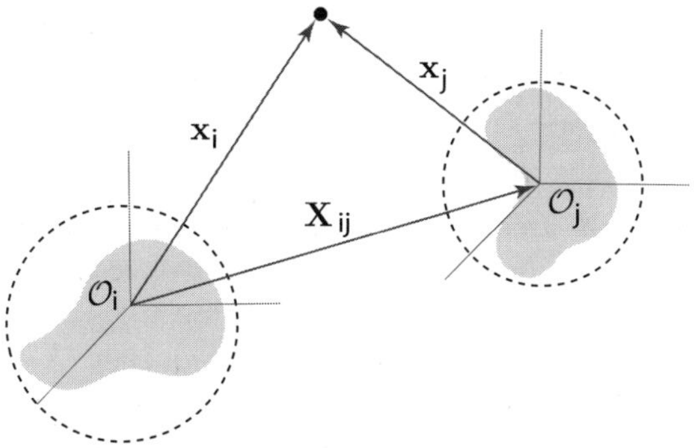

Fig. 1. Geometry of the configuration. The dotted lines show surfaces separating the objects on which the radial variable is constant. The translation vector $\mathbf{X}_{ij} = \mathbf{x}_i - \mathbf{x}_j = -\mathbf{X}_{ji}$ describes the relative positions of the two origins.

jects, one finds that (for objects outside each other) the diagonal blocks are given by the inverse of the matrix representing $\mathbb{T}_j$ in the basis $\mathbf{E}^{\rm reg}_\alpha$.[7] The physical meaning of this matrix follows from the Lippmann-Schwinger equation for the full scattering solution $\mathbf{E}_\alpha(\kappa, \mathbf{x})$,

$$\mathbf{E}_\alpha(\kappa, \mathbf{x}) = \mathbf{E}^{\rm reg}_\alpha(\kappa, \mathbf{x}) - \mathbb{G}_0 \mathbb{V}_j \mathbf{E}_\alpha(\kappa, \mathbf{x}) = \mathbf{E}^{\rm reg}_\alpha(\kappa, \mathbf{x}) - \mathbb{G}_0 \mathbb{T}_j \mathbf{E}^{\rm reg}_\alpha(\kappa, \mathbf{x}) \,. \tag{15}$$

Using the expansion of Eq. (12), the solution sufficiently far away from the object (i.e., for positions that have a radial variable larger than any point on the object) can be expressed as

$$\mathbf{E}_\alpha(\kappa, \mathbf{x}) = \mathbf{E}^{\rm reg}_\alpha(\kappa, \mathbf{x}) - \sum_\beta \mathbf{E}^{\rm out}_\beta(\kappa, \mathbf{x}) \int \mathbf{E}^{\rm reg *}_\beta(\kappa, \mathbf{x}') \mathbb{T}_j(\kappa) \mathbf{E}^{\rm reg}_\alpha(\kappa, \mathbf{x}') d\mathbf{x}' \,, \tag{16}$$

where the integral defines the scattering amplitude $\mathbb{F}_{j,\beta\alpha}(\kappa)$ of object j. It can be obtained, e.g., from matching boundary conditions at the surface of a dielectric object.

The Casimir energy (without cutoff-dependent contributions from W and Z_0) can now be expressed as

$$\mathcal{E} = \frac{\hbar c}{2\pi} \int_0^\infty d\kappa \log \det(\mathbb{M}\mathbb{M}^{-1}_\infty), \tag{17}$$

where

$$\mathbb{M} = \begin{pmatrix} \mathbb{F}^{-1}_1 & \mathbb{U}^{12} & \mathbb{U}^{13} & \cdots \\ \mathbb{U}^{21} & \mathbb{F}^{-1}_2 & \mathbb{U}^{23} & \cdots \\ \cdots & \cdots & \cdots & \cdots \end{pmatrix} \tag{18}$$

and $\mathbb{M}_\infty^{-1}$ is the block diagonal matrix diag$(\mathbb{F}_1, \mathbb{F}_2, \cdots)$. For the case of two objects this expressions simplifies to

$$\mathcal{E} = \frac{\hbar c}{2\pi} \int_0^\infty d\kappa \log\det\left(\mathbb{I} - \mathbb{F}_1 \mathbb{U}^{12} \mathbb{F}_2 \mathbb{U}^{21}\right) . \tag{19}$$

In order to obtain the free energy at nonzero temperature instead of the ground state energy, we do not take the limit $\beta \to \infty$ in Eq. (1).[3] Instead, the integral $\frac{\hbar c}{2\pi}\int_0^\infty d\kappa$ is replaced everywhere by $\frac{1}{\beta}\sum_n'$, where $c\kappa_n = \frac{2\pi n}{\hbar\beta}$ with $n = 0, 1, 2, 3\ldots$ is the nth Matsubara frequency. A careful analysis of the derivation shows that the zero frequency mode is weighted by 1/2 compared to the rest of the terms in the sum; this modification of the sum is denoted by a prime on the summation symbol.

3. Applications

In this section we demonstrate the applicability of the method through some examples. Due to the lack of space, we only present the final analytical and numerical results that all follow from Eq. (17) or Eq. (19) by truncation of the matrices at some order of partial waves, i.e., by considering only a finite set of basis functions. At asymptotically large distances, the interaction only depends on the dipole contribution while with drecreasing distance the number of partial waves has to be increased. Below we will provide results both in form of a asymptotic series in the inverse separation and numerical results for a wide range of distances.

3.1. *Sphere-plane*

First, we consider the sphere-plate geometry that has been employed in the majority of recent experiments. At large distances, the energy can be expanded in an asymptotic series in the inverse separation. For a *dielectric sphere* in front of *perfectly reflecting mirror* with sphere-center to mirror separation L the Casimir energy is

$$\begin{aligned}\mathcal{E} = -\frac{\hbar c}{\pi}\Bigg\{ &\frac{3}{8}(\alpha_1^{\mathrm{E}} - \alpha_1^{\mathrm{M}})\frac{1}{L^4} + \frac{15}{32}(\alpha_2^{\mathrm{E}} - \alpha_2^{\mathrm{M}} + 2\gamma_{13}^{\mathrm{E}} - 2\gamma_{13}^{\mathrm{M}})\frac{1}{L^6} \\ &+ \frac{1}{1024}\left[23(\alpha_1^{\mathrm{M}})^2 - 14\alpha_1^{\mathrm{M}}\alpha_1^{\mathrm{E}} + 23(\alpha_1^{\mathrm{E}})^2 + 2160(\gamma_{14}^{\mathrm{E}} - \gamma_{14}^{\mathrm{M}})\right]\frac{1}{L^7} \\ &+ \frac{7}{7200}\left[572(\alpha_3^{\mathrm{E}} - \alpha_3^{\mathrm{M}}) + 675\left(9(\gamma_{15}^{\mathrm{E}} - \gamma_{15}^{\mathrm{M}}) - 55(\gamma_{23}^{\mathrm{E}} - \gamma_{23}^{\mathrm{M}})\right)\right]\frac{1}{L^8} + \ldots\Bigg\} ,\end{aligned} \tag{20}$$

where α_l^{E}, α_l^{M} are the static electric and magentic multipole polarizabilities of the sphere of order l ($l = 2$ for dipoles), and the coefficients γ_{ln}^{E}, γ_{ln}^{M}

describe finite-frequency corrections to these polarizabilities, i.e., terms $\sim \kappa^{2l+n}$ in the low-κ expansion of the T-matrix element for the lth partial wave. Notice that the first three terms of the contribution at order L^{-7} have precisely the structure of the Casimir-Polder interaction between two atoms with static dipole polarizabilities α_1^{M} and α_1^{E} but it is reduced by a factor of $1/2^8$. This factor and the distance dependence $\sim L^{-7}$ of this term suggests that it arises from the interaction of the dipole fluctuations inside the sphere with those inside its image at a distance $2L$. The additional coefficient of $1/2$ in the reduction factor $(1/2)(1/2^7)$ can be traced back to the fact that the forces involved in bringing the dipole in from infinity act only on the dipole and not on its image. If the sphere is also assumed to be a *perfect reflector*, the energy becomes

$$\mathcal{E} = \frac{\hbar c}{\pi}\frac{1}{L}\sum_{j=4}^{\infty} b_j \left(\frac{R}{L}\right)^{j-1}, \tag{21}$$

where the coefficients up to order $1/L^{11}$ are

$$\begin{aligned} b_4 &= -\frac{9}{16}, \quad b_5 = 0, \quad b_6 = -\frac{25}{32}, \quad b_7 = -\frac{3023}{4096} \\ b_8 &= -\frac{12551}{9600}, \quad b_9 = \frac{1282293}{163840}, \\ b_{10} &= -\frac{32027856257}{722534400}, \quad b_{11} = \frac{39492614653}{412876800}. \end{aligned} \tag{22}$$

Our method can be also employed to study the material dependence of the interaction. When the sphere and the mirror are described by a simple *plasma model*, we can obtain the interaction energy again from Eq. (19) by substituting the dielectric function on the imaginary frequency axis,

$$\epsilon_p(ic\kappa) = 1 + \left(\frac{2\pi}{\lambda_p \kappa}\right)^2, \tag{23}$$

into the T-matrices of sphere and mirror. From this we get at large separations

$$\mathcal{E} = -\frac{\hbar c}{\pi}\left[f_4(\lambda_p/R)\frac{R^3}{L^4} + f_5(\lambda_p/R)\frac{R^4}{L^5} + \mathcal{O}(L^{-6})\right] \tag{24}$$

with the functions

$$\begin{aligned} f_4(z) &= \frac{9}{16} + \frac{9}{64\pi^2}z^2 - \frac{9}{32\pi}z\coth\frac{2\pi}{z} \\ f_5(z) &= -\frac{13}{20\pi}z - \frac{21}{80\pi^3}z^3 + \frac{21}{40\pi^2}z^2\coth\frac{2\pi}{z}. \end{aligned} \tag{25}$$

It is interesting that the amplitude f_4 of the leading term is not universal but depends on the plasma wavelength λ_p. Only in the two limits $\lambda_p/R \to 0$ and $\lambda_p/R \to \infty$ the amplitude assumes material independent values, 9/16 and 3/8, respectively. The first limit describes perfect reflection of electric and magnetic fields at arbitrarily low frequencies and hence agrees with the result of Eq. (21). The change to the second amplitude for large λ_p can be understood when one considers a London superconductor that is described at zero temperature by the plasma dielectric function.[10] If one associates λ_p with the penetration depth, the perfect reflector limit results from the absence of any field penetration while the second limit corresponds to a large penetration depth and hence the suppression of the magnetic mode contribution to the Casimir energy, explaining the reduced amplitude of 3/8. The latter result follows also when the objects are considered to be normal metals, described by the *Drude model* dielectric function

$$\epsilon_p(ic\kappa) = 1 + \frac{(2\pi)^2}{(\lambda_p\kappa)^2 + \pi c\kappa/\sigma}\,. \tag{26}$$

From this function we get for a sphere and a mirror made of a Drude metal the asymptotic energy

$$\mathcal{E} = -\frac{\hbar c}{\pi}\left[\frac{3}{8}\frac{R^3}{L^4} - \frac{77}{384}\frac{R^3}{\sqrt{2\sigma/c}\,L^{9/2}} - \left(\frac{c}{8\pi\sigma} - \frac{\pi}{20}\frac{\sigma R^2}{c}\right)\frac{R^3}{L^5} + \mathcal{O}(L^{-\frac{11}{2}})\right]. \tag{27}$$

In fact, one observes that the leading term is universal and agrees with the $\lambda_p \to \infty$ limit of the plasma model. Note that the result of Eq. (27) does not apply to arbitrarily large dc conductivity σ. The conditions for the validity of Eq. (27) can be written as $L \gg R$, $L \gg c/\sigma$ and $L \gg \sigma R^2/c$. The above results demonstrate strong correlations between shape and material since for two parallel, infinite plates, both the plasma and the Drude model yield at large separations the same (universal) result as a perfect mirror description.

In order to study short separations, Eq. (19) has to be evaluated numerically by including sufficiently many partial waves. The result of an extrapolation from $l = 29$ partial waves is shown in Fig. 2 in the perfect reflection limit.[11] At small separations the result can be fitted to a power law of the form

$$\mathcal{E} = \mathcal{E}_{\rm PFA}\left[1 + \theta_1\frac{d}{R} + \theta_2\left(\frac{d}{R}\right)^2 + \dots\right]. \tag{28}$$

with $\mathcal{E}_{\rm PFA}$ and d defined in Fig. 2. The coefficients θ_j measure corrections to the proximity force approximation and are obtained from a fit of the function of Eq. (28) to the data points for the four smallest studied separations. We find $\theta_1 = -1.42 \pm 0.02$ and $\theta_2 = 2.39 \pm 0.14$. This result is in agreement with numerical findings in Ref. 12 but is in disagreement with an asymptotic expansion for small distances.[13] The latter yields $\theta_1 = -5.2$ and very small logarithmic corrections that however can be ignored at the distances considered here. The origin of this discrepancy is currently unclear but might be related to the applicability of the asymptotic expansion to only much smaller distances than accessible by current numerics.

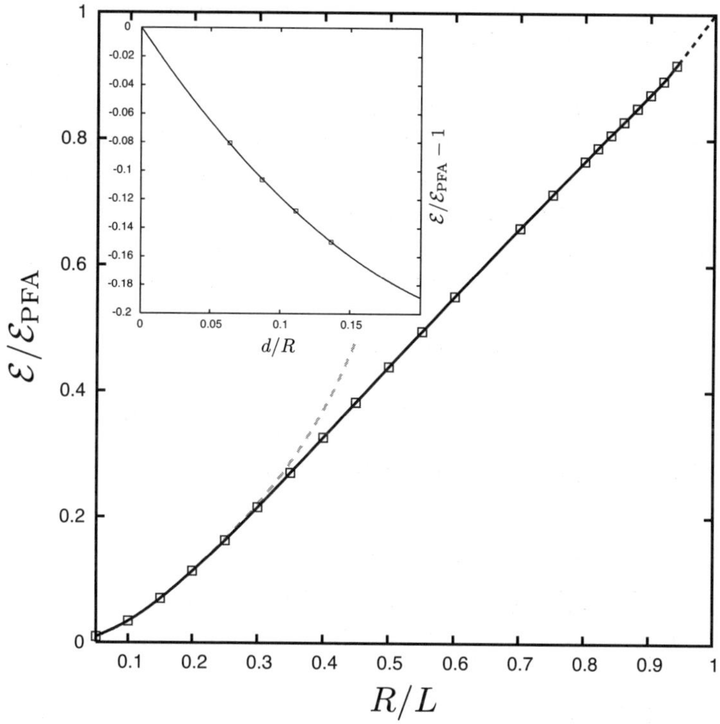

Fig. 2. Electromagnetic Casimir energy for the sphere-plate geometry. The energy is scaled by the proximity force approximation (PFA) energy $\mathcal{E}_{\mathrm{PFA}} = -\frac{\pi^3}{720}\frac{\hbar c R}{d^2}$. The asymptotic expansion of Eq. (21) is shown as dashed line. Inset: Corrections to the PFA at small distances as function of $d = L - R$.

3.2. *Three-body effects*

Casimir interactions are not pair-wise additive. To study the consequences of this property, we consider the case of two identical, general polarizable objects near a perfectly reflecting wall in the dipole approximation, see

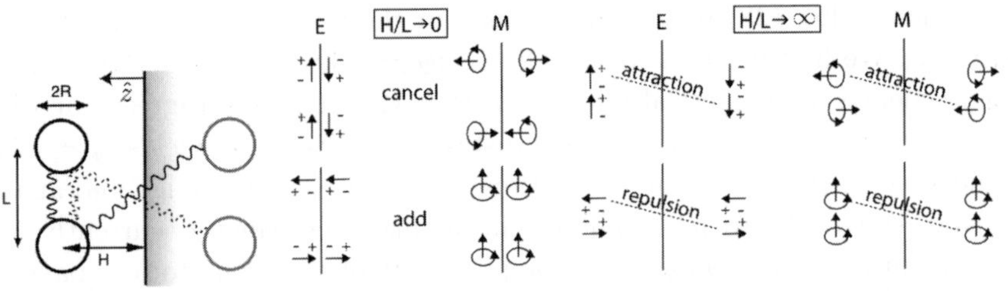

Fig. 3. Left: Geometry of the two-sphere/atom and sidewall system. Shown are also the mirror images (grey) and two- and three-body contributions (solid and dashed curly lines, respectively). Right: Typical orientations of electric (E) and magnetic (M) dipoles and image dipoles for $H/L \to 0$ and $H/L \to \infty$.

Fig. 3. This situation applies to ground state atoms and also to general objects at *large* separations. The separation between the objects is L and the separation of each of them from the wall is H. In dipole approximation, the retarded limit of the interaction is described by the static electric (α_z, $\alpha_\|$) and magnetic (β_z, $\beta_\|$) dipole polarizabilities of the objects which can be different in the directions perpendicular (z) and parallel ($\|$) to the wall. In the absence of the wall the potential for the two polarizable objects is given by the well-known Casimir-Polder (CP) potential

$$\mathcal{E}_{2,|}(L) = -\frac{\hbar c}{8\pi L^7}\Big[33\alpha_\|^2 + 13\alpha_z^2 - 14\alpha_\|\beta_z + (\alpha\leftrightarrow\beta)\Big] \,, \tag{29}$$

The L-dependent part of the interaction energy in the presence of the wall is

$$\mathcal{E}_{\underline{\circ\circ}}(L,H) = \mathcal{E}_{2,|}(L) + \mathcal{E}_{2,\backslash}(D,L) + \mathcal{E}_3(D,L) \tag{30}$$

with $D = \sqrt{L^2+4H^2}$. The change in the relative orientation of the objects with $\ell = L/D$ leads to the modified 2-body CP potential

$$\begin{aligned}\mathcal{E}_{2,\backslash}(D,L) = -\frac{\hbar c}{8\pi D^7}\Big[&26\alpha_\|^2 + 20\alpha_z^2 - 14\ell^2(4\alpha_\|^2 - 9\alpha_\|\alpha_z + 5\alpha_z^2) \\ &+ 63\ell^4(\alpha_\| - \alpha_z)^2 - 14\big(\alpha_\|\beta_\|(1-\ell^2) + \ell^2\alpha_\|\beta_z\big) + (\alpha\leftrightarrow\beta)\Big] \,.\end{aligned} \tag{31}$$

The 3-body energy $\mathcal{E}_3(D,L)$ describes the collective interaction between the two objects and one image object. It is given by

$$\begin{aligned}\mathcal{E}_3(D,L) = \frac{4\hbar c}{\pi}\frac{1}{L^3D^4(\ell+1)^5}\Big[&\Big(3\ell^6 + 15\ell^5 + 28\ell^4 + 20\ell^3 + 6\ell^2 - 5\ell - 1\Big) \\ &\times\left(\alpha_\|^2 - \beta_\|^2\right) - \left(3\ell^6 + 15\ell^5 + 24\ell^4 - 10\ell^2 - 5\ell - 1\right)\left(\alpha_z^2 - \beta_z^2\right) \\ &+ 4\left(\ell^4 + 5\ell^3 + \ell^2\right)\left(\alpha_z\beta_\| - \alpha_\|\beta_z\right)\Big] \,.\end{aligned} \tag{32}$$

It is instructive to consider the two limits $H \ll L$ and $H \gg L$. For $H \ll L$ $\mathcal{E}_{\underline{\circ\circ}}$ turns out to be the CP potential of Eq. (29) with the replacements $\alpha_z \to 2\alpha_z$, $\alpha_\| \to 0$, $\beta_z \to 0$, $\beta_\| \to 2\beta_\|$. The 2-body and 3-body contributions add constructively or destructively, depending on the relative orientation of a dipole and its image which together form a dipole of zero or twice the original strength (see Fig. 3).

For $H \gg L$ the leading correction to the CP potential of Eq. (29) comes from the 3-body energy. The energy then becomes (up to order H^{-6})

$$\mathcal{E}_{\underline{\circ\circ}}(L,H) = \mathcal{E}_{2,|}(L) + \frac{\hbar c}{\pi}\left[\frac{\alpha_z^2-\alpha_\|^2}{4L^3H^4} + \frac{9\alpha_\|^2-\alpha_z^2-2\alpha_\|\beta_z}{8LH^6} - (\alpha \leftrightarrow \beta)\right]. \quad (33)$$

The signs of the polarizabilities in the leading term $\sim H^{-4}$ can be understood from the relative orientation of the dipole of one atom and the image dipole of the other atom, see Fig. 3. If these two electric (magnetic) dipoles are almost perpendicular to their distance vector they contribute attractively (repulsively) to the potential between the two original objects. If these electric (magnetic) dipoles are almost parallel to their distance vector they yield a repulsive (attractive) contribution. For isotropic polarizabilities the leading term of Eq. (33) vanishes and the electric (magnetic) part $\sim H^{-6}$ of the 3-body energy is always repulsive (attractive).

Next, we study the same geometry as before but with the objects assumed to be two perfectly reflecting spheres of radius R. The lengths L and H are measured now from the centers of the spheres, see Fig. 3. Here we do not limit the analysis to large separations but consider arbitrary distances and include higher order multipole moments than just dipole polarizability. For $R \ll L$, H and arbitrary H/L the result for the force can be written as

$$F = \frac{\hbar c}{\pi R^2}\sum_{j=6}^{\infty} f_j(H/L)\left(\frac{R}{L}\right)^{j+2}. \quad (34)$$

The functions f_j can be computed exactly. We have obtained them up to $j = 11$ and the first three are (with $s \equiv \sqrt{1+4h^2}$)

$$f_6(h) = -\frac{1}{16h^8}\Big[s^{-9}(18+312h^2+2052h^4+6048h^6 + 5719h^8) + 18 - 12h^2 + 1001h^8\Big], \quad f_7(h) = 0, \quad (35)$$

$$f_8(h) = -\frac{1}{160h^{12}}\Big[s^{-11}(6210+140554h^2+1315364h^4 + 6500242h^6 + 17830560h^8 + 25611168h^{10} + 15000675h^{12}) - 6210 - 3934h^2 + 764h^4 - 78h^6 + 71523h^{12}\Big]. \quad (36)$$

For $H \gg L$ one has $f_6(h) = -1001/16 + 3/(4h^6) + \mathcal{O}(h^{-8})$, $f_8(h) = -71523/160 + 39/(80h^6) + \mathcal{O}(h^{-8})$ so that the wall induces weak repulsive corrections. For $H \ll L$, $f_6(h) = -791/8 + 6741h^2/8 + \mathcal{O}(h^4)$, $f_8(h) = -60939/80 + 582879h^2/80 + \mathcal{O}(h^4)$ so that the force amplitude decreases when the spheres are moved a small distance away from the wall. This proves the existence of a minimum in the force amplitude as a function of H/R for fixed, sufficiently small R/L. We note that all $f_j(h)$ are finite for $h \to \infty$ but some diverge for $h \to 0$, e.g., $f_9 \sim f_{11} \sim h^{-3}$, making them important for small H.

To obtain the interaction at smaller separations or larger radius, we have computed the energy $\mathcal{E}_{\underline{\text{oo}}}$ and force $F = -\partial\mathcal{E}_{\underline{\text{oo}}}/\partial L$ between the spheres numerically.[14] In order to show the effect of the wall, we plot the energy and force normalized to the results for two spheres without a wall. Fig. 4 shows the force between the two spheres as a function of the wall distance for fixed L. When the spheres approach the wall, the force first decreases slightly if $R/L \lesssim 0.3$ and then increases strongly under a further reduction of H. For $R/L \gtrsim 0.3$ the force increases monotonically as the spheres approach the wall. This agrees with the prediction of the large distance expansion. The expansion of Eq. (34) with $j = 10$ terms is also shown in Fig. 4 for $R/L \leq 0.2$. Its validity is limited to large L/R and not too small H/R; it fails completely for $R/L > 0.2$ and hence is not shown in this range.

3.3. *Orientation dependence*

In this section we investigate the shape and orientation dependence of the Casimir force using Eq. (19). As examples we focus on ellipsoids, computing the orientation dependent force between two spheroids, and between a spheroid and a plane.[15] For two anisotropic objects, the CP potential of Eq. (29) must be generalized. In terms of the Cartesian components of the standard electric (magnetic) polarizability matrix α (β), the asymptotic large distance potential of two objects (with the $\hat{z}$ axis pointing from one object to the other), can be written as

$$\begin{aligned}\mathcal{E} = -\frac{\hbar c}{d^7}\frac{1}{8\pi}\Big\{&13\left(\alpha^1_{xx}\alpha^2_{xx} + \alpha^1_{yy}\alpha^2_{yy} + 2\alpha^1_{xy}\alpha^2_{xy}\right)\\ &+ 20\,\alpha^1_{zz}\alpha^2_{zz} - 30\left(\alpha^1_{xz}\alpha^2_{xz} + \alpha^1_{yz}\alpha^2_{yz}\right) + (\alpha \to \beta)\\ &- 7\left(\alpha^1_{xx}\beta^2_{yy} + \alpha^1_{yy}\beta^2_{xx} - 2\alpha^1_{xy}\beta^2_{xy}\right) + (1 \leftrightarrow 2)\Big\}.\end{aligned} \tag{37}$$

For the case of an ellipsoidal object with static electric permittivity ϵ and magnetic permeability μ, the polarizability tensors are diagonal in a basis

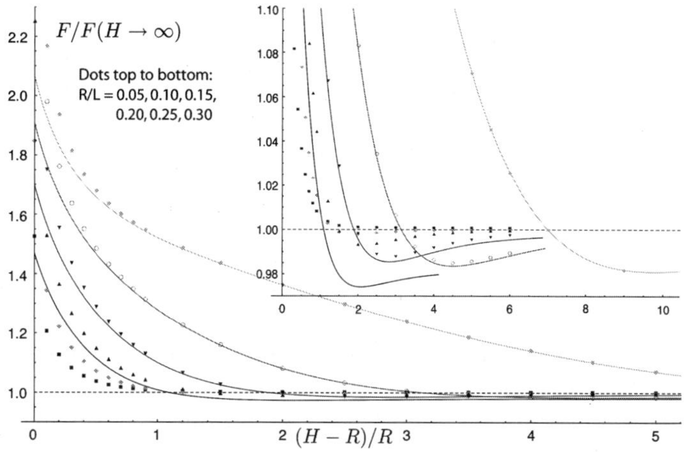

Fig. 4. Numerical results for the force (dots) between two spheres as function of the sidewall separation H/R for different sphere separations R/L. Shown are also the analytical results of Eq. (34), including terms up to $j = 10$ for $R/L \leq 0.2$ (solid curves). Inset: Magnification of the nonmonotonicity.

oriented to its principal axes, with elements (for $i \in \{1, 2, 3\}$)

$$\alpha_{ii}^0 = \frac{V}{4\pi} \frac{\epsilon - 1}{1 + (\epsilon - 1)n_i}, \quad \beta_{ii}^0 = \frac{V}{4\pi} \frac{\mu - 1}{1 + (\mu - 1)n_i}, \tag{38}$$

where $V = 4\pi r_1 r_2 r_3/3$ is the ellipsoid's volume. In the case of spheroids, for which $r_1 = r_2 = R$ and $r_3 = L/2$, the so-called depolarizing factors can be expressed in terms of elementary functions,

$$n_1 = n_2 = \frac{1 - n_3}{2}, \quad n_3 = \frac{1 - e^2}{2e^3} \left(\log \frac{1 + e}{1 - e} - 2e \right), \tag{39}$$

where the eccentricity $e = \sqrt{1 - \frac{4R^2}{L^2}}$ is real for a prolate spheroid ($L > 2R$) and imaginary for an oblate spheroid ($L < 2R$). The polarizability tensors for an arbitrary orientation are then obtained as $\alpha = \mathcal{R}^{-1}\alpha^0\mathcal{R}$, where $\mathcal{R}$ is the matrix that rotates the principal axis of the spheroid to the Cartesian basis, i.e. $\mathcal{R}(1, 2, 3) \to (x, y, z)$. Note that for rarefied media with $\epsilon \simeq 1$, $\mu \simeq 1$ the polarizabilities are isotropic and proportional to the volume. Hence, to leading order in $\epsilon - 1$ the interaction is orientation independent at asymptotically large separations, as we would expect, since pairwise summation is valid for $\epsilon - 1 \ll 1$. In the following we focus on the interesting opposite limit of two identical perfectly reflecting spheroids. We first consider prolate spheroids with $L \gg R$. The orientation of each "needle" relative to the line joining them (the initial z-axis) is parameterized by the two angles (θ, ψ), as depicted in Fig. 5. Then the energy is

$$\mathcal{E}(\theta_1, \theta_2, \psi) = -\frac{\hbar c}{d^7}\left\{\frac{5L^6}{1152\pi\left(\ln\frac{L}{R}-1\right)^2}\left[\cos^2\theta_1\cos^2\theta_2 + \frac{13}{20}\cos^2\psi\sin^2\theta_1\sin^2\theta_2 - \frac{3}{8}\cos\psi\sin 2\theta_1\sin 2\theta_2\right] + \mathcal{O}\left(\frac{L^4R^2}{\ln\frac{L}{R}}\right)\right\}, \tag{40}$$

where $\psi \equiv \psi_1 - \psi_2$. It is minimized for two needles aligned parallel to their separation vector. At almost all orientations the energy scales as L^6, and vanishes logarithmically slowly as $R \to 0$. The latter scaling changes when one needle is orthogonal to $\hat{z}$ (i.e. $\theta_1 = \pi/2$), while the other is either parallel to $\hat{z}$ ($\theta_2 = 0$) or has an arbitrary θ_2 but differs by an angle $\pi/2$ in its rotation about the z-axis (i.e. $\psi_1 - \psi_2 = \pi/2$). In these cases the energy comes from the next order term in Eq. (40), and takes the form

$$\mathcal{E}\left(\frac{\pi}{2}, \theta_2, \frac{\pi}{2}\right) = -\frac{\hbar c}{1152\pi\, d^7}\frac{L^4R^2}{\ln\frac{L}{R}-1}\left(73 + 7\cos 2\theta_2\right), \tag{41}$$

which shows that the least favorable configuration corresponds to two needles orthogonal to each other and to the line joining them.

For perfectly reflecting oblate spheroids with $R \gg L/2$, the orientation of each "pancake" is again described by a pair of angles (θ, ψ), as depicted in Fig. 6. To leading order at large separations, the energy is given by

$$\mathcal{E} = -\frac{\hbar c}{d^7}\left\{\frac{R^6}{144\pi^3}\left[765 - 5(\cos 2\theta_1 + \cos 2\theta_2) + 237\cos 2\theta_1\cos 2\theta_2 + 372\cos 2\psi\sin^2\theta_1\sin^2\theta_2 - 300\cos\psi\sin 2\theta_1\sin 2\theta_2\right] + \mathcal{O}\left(R^5L\right)\right\}. \tag{42}$$

The leading dependence is proportional to R^6, and does not disappear for any choice of orientations. Furthermore, this dependence remains even as the thickness of the pancake is taken to zero ($L \to 0$). This is very different from the case of the needles, where the interaction energy vanishes with thickness as $\ln^{-1}(L/R)$. The lack of L dependence is due to the assumed perfectly reflectivity. The energy is minimal for two pancakes lying on the same plane ($\theta_1 = \theta_2 = \pi/2$, $\psi = 0$) and has energy $-\hbar c\,(173/18\pi^3)R^6/d^7$. When the two pancakes are stacked on top of each other, the energy is increased to $-\hbar c\,(62/9\pi^3)R^6/d^7$. The least favorable configuration is when the pancakes lie in perpendicular planes, i.e., $\theta_1 = \pi/2$, $\theta_2 = 0$, with an energy $-\hbar c\,(11/3\pi^3)R^6/d^7$.

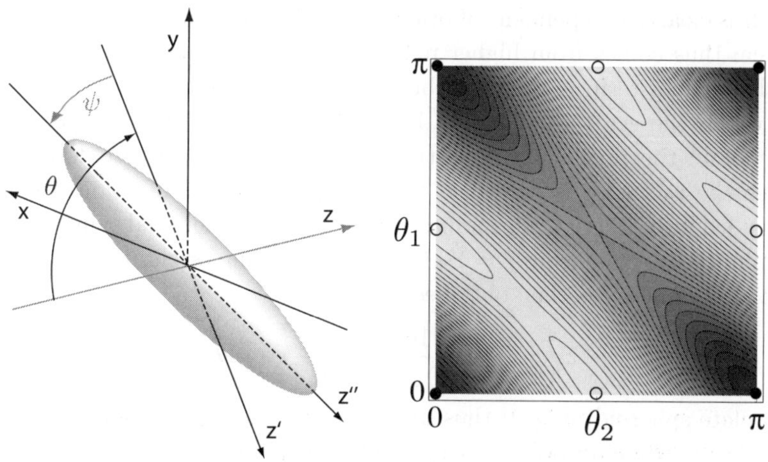

Fig. 5. (Color online) Orientation of a prolate (cigar-shaped) spheroid: The symmetry axis (initially the z-axis) is rotated by θ about the x-axis and then by ψ about the z-axis. For two such spheroids, the energy at large distances is give by Eq. (40). The latter is depicted at fixed distance d, and for $\psi_1 = \psi_2$, by a contour plot as function of the angles θ_1, θ_2 for the x-axis rotations . Minima (maxima) are marked by filled (open) dots.

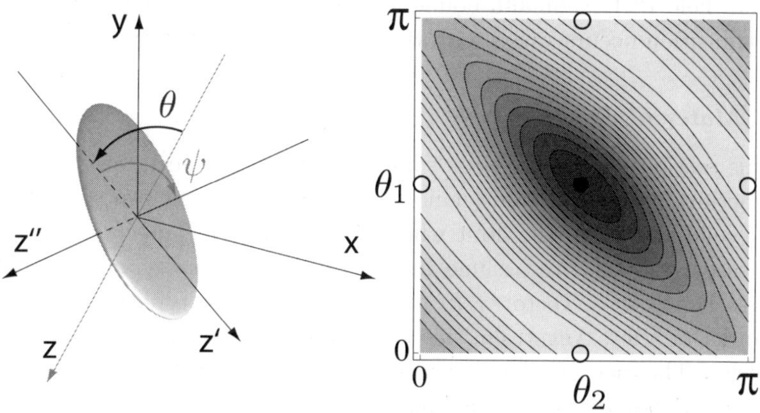

Fig. 6. (Color online) As in Fig. 5 for oblate (pancake-shaped) spheroids, with a contour plot of energy at large separations.

For an anisotropic object interacting with a perfectly reflecting mirror, at leading order the CP potential generalizes to

$$\mathcal{E} = -\frac{\hbar c}{d^4}\frac{1}{8\pi}\mathrm{tr}\,(\alpha - \beta) + \mathcal{O}(d^{-5})\,, \tag{43}$$

which is clearly independent of orientation. Orientation dependence in this system thus comes from higher multipoles. The next order also vanishes, so the leading term is the contribution from the partial waves with $l = 3$ for which the scattering matrix is not known analytically. However, we can obtain the preferred orientation by considering a distorted sphere in which the radius R is deformed to $R + \delta f(\vartheta, \varphi)$. The function f can be expanded into spherical harmonics $Y_{lm}(\vartheta, \varphi)$, and spheroidal symmetry can be mimicked by choosing $f = Y_{20}(\vartheta, \varphi)$. The leading orientation dependent part of the energy is then obtained as

$$\mathcal{E}_f = -\hbar c \frac{1607}{640\sqrt{5}\pi^{3/2}} \frac{\delta R^4}{d^6} \cos(2\theta) . \tag{44}$$

A prolate spheroid ($\delta > 0$) thus minimizes its energy by pointing towards the mirror, while an oblate spheroid ($\delta < 0$) prefers to lie in a plane perpendicular to the mirror. (We assume that the perturbative results are not changed for large distortions.) These configurations are also preferred at small distances d, since (at fixed distance to the center) the object reorients to minimize the closest separation. Interestingly, the latter conclusion is not generally true. In Ref. 15 it has been shown that there can be a transition in preferred orientation as a function of d in the simpler case of a scalar field with Neumann boundary conditions. The separation at which this transition occurs varies with the spheroid's eccentricity.

3.4. *Material dependence*

In this section we shall discuss some characteristic effects of the Casimir interaction between metallic nano-particles by studying two spheres with *finite* conductivity in the limit where their radius R is much smaller than their separation d. We assume further that R is large compared to the inverse Fermi wave vector π/k_F of the metal. Since typically π/k_F is of the order of a few Angstrom, this assumption is reasonable even for nano-particles. Theories for the optical properties of small metallic particles[16] suggest a Drude dielectric function

$$\epsilon(ic\kappa) = 1 + 4\pi \frac{\sigma(ic\kappa)}{c\kappa} , \tag{45}$$

where $\sigma(ic\kappa)$ is the conductivity which approaches for $\kappa \to 0$ the dc conductivity σ_{dc}. For bulk metals $\sigma_{dc} = \omega_p^2 \tau / 4\pi$ where $\omega_p = \sqrt{4e^2 k_F^3 / 3\pi m_e}$ is the plasma frequency with electron charge e and electron mass m_e, and τ is the relaxation time. With decreasing dimension of the particle, $\sigma_{dc}(R)$ is reduced compared to its bulk value due to finite size effects and hence becomes a function of R.[16] In analogy to the result for a sphere and a plate that are described by the Drude model, we obtain for the large distance expansion of the energy the result

$$\mathcal{E} = -\hbar c \frac{23}{4\pi}\frac{R^6}{L^7} - \left(\frac{R\sigma_{dc}(R)}{c} - \frac{45}{4\pi^2}\frac{c}{R\sigma_{dc}(R)}\right)\frac{R^7}{L^8} + \ldots . \tag{46}$$

As in the sphere-plate case, the leading term is material independent but different from that of the perfect metal limit (where the amplitude is $143/16\pi$) since only the electric polarization contributes. At next order, the first and second terms in the parentheses come from magnetic and electric dipole fluctuations, respectively. The term $\sim 1/L^8$ is absent in the interaction between perfectly conducting spheres. The limit of perfect conductivity, $\sigma_{dc} \to \infty$ cannot be taken in Eq. (46) since this limit does not commute with the large L expansion.

In order to estimate the effect of finite conductivity and its dependence on the size of the nano-particle, we have to employ a theory that can describe the evolution of $\sigma_{dc}(R)$ with the particle size. A theory for the dielectric function of a cubical metallic particle of dimensions $R \gg \pi/k_F$ has been developed within the random phase approximation in the limit of low frequencies $\ll c/R$.[16] In this theory it is further assumed that the discreteness of the electronic energy levels, and not the inhomogeneity of the charge distribution, is important. This implies that the particle responds only at the wave vector of the incident field which is a rather common approximation for small particles. From an electron number-conserving relaxation time approximation the complex dielectric function is obtained which yields the size-dependent dc conductivity for a cubic particle of volume a^3.[16] It has been shown that the detailed shape of the particle does not matter much, and we can set $a = (4\pi/3)^{1/3}R$ which defines the volume equivalent sphere radius R. For $\pi/k_F \simeq a$ the nano particle ceases to be conducting, corresponding to a metal-insulator transition due to the localisation of electrons for particles with a size of the order of the mean free path. It is instructive to consider the size dependence of $\sigma_{dc}(R)$ and of the Casimir interaction for a particular choice of material. Following Ref. 16, we focus on small Aluminum spheres with Fermi energy $\epsilon_F = 11.63$eV and $\tau = 0.8 \cdot 10^{-14}$sec. These parameters correspond to $\pi/k_F = 1.8\,\text{Å}$ and a plasma wavelength $\lambda_p = 79$nm. It is useful to introduce the dimensionless conductivity $\hat{\sigma}_{dc}(R)$, which is measured in units of $e^2/2\hbar a_0$ with Bohr radius a_0, so that the important quantity of Eq. (46) can be written as $R\sigma_{dc}(R)/c = (\alpha/2)(R/a_0)\hat{\sigma}_{dc}(R)$ where α is the fine-structure constant. The result is shown in Fig. 7. For example, for a sphere of radius

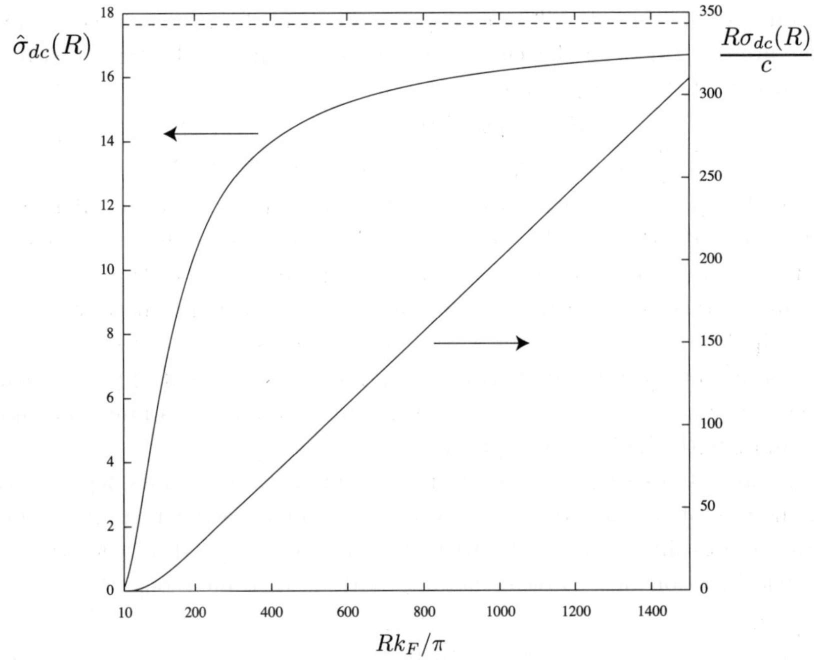

Fig. 7. Dimensionless dc conductivity $\hat{\sigma}_{dc}(R)$ in units of $e^2/2\hbar a_0$ (with Bohr radius a_0) for a Aluminum sphere with $\epsilon_F = 11.63$eV, $\pi/k_F = 1.8\text{Å}$ and $\tau = 0.8 \cdot 10^{-14}$sec as function of the radius R, measured in units of π/k_F. Also shown is the corresponding ratio $R\sigma_{dc}(R)/c$ that determines the Casimir interaction of Eq. (46). The bulk dc conductivity $\hat{\sigma}_{dc}(\infty) = 17.66$ is indicated by the dashed line.

$R = 10$nm, the dc conductivity is reduced by a factor ≈ 0.15 compared to the bulk Drude value. If the radius of the sphere is equal to the plasma wavelength λ_p, the reduction factor ≈ 0.8. These results show that shape and material properties are important for the Casimir interaction between nano-particles. Potential applications include the interaction between dilute suspensions of metallic nano-particles.

3.5. *Further extensions*

The general result of Eq. (17) and its extensions described in Ref. 7 have been recently applied to a number of new geometries and further applications are under way. Examples include so-called interior configurations with an object contained within an otherwise empty, perfectly conducting spherical shell.[17] For this geometry the forces and torques on a dielectric or conducting object, well separated from the cavity walls, have been determined. Corrections to the proximity force approximation for this interior problem have been obtained by computing the interaction energy of a finite-size metal sphere with the cavity walls when the separation between their surfaces tends to zero. Eq. (17), evaluated in parabolic cylinder coordinates, has been used to obtain the interaction energy of a parabolic cylinder and an infinite plate (both perfect mirrors), as a function of their separation and inclination, and the cylinder's parabolic radius.[18] By taking the limit of vanishing radius, corresponding to a semi-infinite plate, the effect of edge and inclination could be studied.

Acknowledgments

The reported results have been obtained in collaboration with N. Graham, R. L. Jaffe, M. Kardar, S. J. Rahi, P. Rodriguez-Lopez, A. Shpunt, S. Zaheer, R. Zandi. This work was supported by the Deutsche Forschungsgemeinschaft (DFG) through grant EM70/3 and Defense Advanced Research Projects Agency (DARPA) contract No. S-000354.

References

1. H. B. G. Casimir, *Proc. K. Ned. Akad. Wet.* **51**, p. 793 (1948).
2. H. B. G. Casimir and D. Polder, *Phys. Rev.* **73**, p. 360 (1948).
3. E. M. Lifshitz, *Dokl. Akad. Nauk SSSR* **100**, p. 879 (1955).
4. E. M. Lifshitz, *Sov. Phys. JETP* **2**, p. 73 (1956).
5. E. M. Lifshitz, *Sov. Phys. JETP* **3**, p. 977 (1957).
6. I. E. Dzyaloshinskii, E. M. Lifshitz and L. P. Pitaevskii, *Advances in Physics* **10**, p. 165 (1961).
7. S. J. Rahi, T. Emig, N. Graham, R. L. Jaffe and M. Kardar, *Phys. Rev. D* **80**, p. 085021 (2009).
8. T. Emig, N. Graham, R. L. Jaffe and M. Kardar, *Phys. Rev. Lett.* **99**, p. 170403 (2007).
9. T. Emig, N. Graham, R. L. Jaffe and M. Kardar, *Phys. Rev. D* **77**, p. 025005 (2008).
10. H. Haakh, F. Intravaia, C. Henkel, S. Spagnolo, R. Passante, B. Power and F. Sols, Temperature dependence of the magnetic casimir-polder interaction, Preprint arXiv:0910.3133, (2009).
11. T. Emig, *J. Stat. Mech.* , p. P04007 (2008).
12. P. A. Maia Neto, A. Lambrecht and S. Reynaud, *Phys. Rev. A* **78**, p. 012115 (2008).
13. M. Bordag and V. Nikolaev, First analytic correction beyond pfa for the electromagnetic field in sphere-plane geometry, Preprint arXiv:0911.0146, (2009).

14. P. Rodriguez-Lopez, S. J. Rahi and T. Emig, *Phys. Rev. A* **80**, p. 022519 (2009).
15. T. Emig, N. Graham, R. L. Jaffe and M. Kardar, *Phys. Rev. A* **79**, p. 054901 (2009).
16. D. M. Wood and N. W. Ashcroft, *Phys. Rev. B* **25**, p. 6255 (1982).
17. S. Zaheer, S. J. Rahi, T. Emig and R. L. Jaffe, Casimir interactions of an object inside a spherical metal shell, Preprint arXiv:0908.3270, (2009).
18. N. Graham, A. Shpunt, T. Emig, S. J. Rahi, R. L. Jaffe and M. Kardar, Casimir force at a knife's edge, Preprint arXiv:0910.4649, (2009).

SEMITRANSPARENT PISTONS

P. MORALES* and K. KIRSTEN†

Department of Mathematics, Baylor University, Waco, Texas 76798-7328, USA
**E-mail: pedro_morales@baylor.edu*
†E-mail: klaus_kirsten@baylor.edu

We consider semitransparent pistons in the presence of extra dimensions. It is shown that the piston is always attracted to the closest wall irrespective of details of the geometry and topology of the extra dimensions and of the cross section of the piston. Furthermore, we evaluate the zeta regularized determinant for this configuration.

Keywords: Pistons, Casimir effect, extra dimensions, semitransparent boundary conditions.

1. Introduction

In this contribution we consider three-dimensional pistons of arbitrary cross section in the context of Kaluza-Klein models. As is well known, pistons have the important advantage that they allow for an unambiguous prediction of Casimir forces.[2] This is the main reason for the recent surge of interest in these configurations; see, e.g., Refs. 3–7. Most of the research done so far has concentrated on a rectangular cross section with boundary conditions that allow for an explicit determination of the energy eigenvalues for the configuration, at least for part of the spectrum. Here we want to investigate further geometries with arbitrary cross section, along the lines of Refs. 5 and 7, with boundary conditions leading to a transcendental equation for the spectrum.[3] Specifically, we will consider semitransparent pistons and we will show that the piston is attracted to the closest wall. This statement holds independently of the cross section of the piston and of the geometry and topology of the additional Kaluza-Klein dimensions.

2. Zeta Function for Semitransparent Pistons

Let $M = [0, L] \times N$, where N represents the cross section of the piston and the additional Kaluza-Klein dimensions assumed to be a smooth

Riemannian manifold possibly with a boundary. We place Dirichlet plates at $x = 0$ and $x = L$ and the semitransparent piston at $x = a$ is modeled by a delta function potential. The energy eigenvalues for a scalar field are then determined by the second order differential operator

$$P = -\frac{\partial^2}{\partial x^2} - \Delta_N + \sigma\delta(x-a), \tag{1}$$

together with Dirichlet boundary conditions at $x = 0$ and $x = L$. The operator Δ_N can be thought of as being the Laplacian on N, but it might also contain a term showing a coupling to the curvature on N as well as a mass term.

Using separation of variables, eigenfunctions, namely solutions of the equation

$$P\phi(x,y) = \lambda^2\phi(x,y),$$

are written in the form

$$\phi(x,y) = X(x)\varphi(y), \qquad x \in [0,L], \quad y \in N.$$

Assuming $\varphi(y)$ to be an eigenfunction of $-\Delta_N$, that is

$$-\Delta_N\varphi_\ell(y) = \eta_\ell^2\varphi_\ell(y),$$

with boundary conditions imposed if $\partial N \neq \emptyset$, the Dirichlet condition along the x-axis implies $X(0) = X(L) = 0$. In addition, we impose continuity at $x = a$, namely $X(a+) = X(a-)$, and the presence of the delta function $\delta(x-a)$ creates a jump in the derivative, $X'(a+) - X'(a-) = \sigma X(a)$. Incorporating all the above information, eigenvalues λ^2 are seen to be of the form $\lambda_{k\ell} = \nu_k^2 + \eta_\ell^2$, where the ν_k satisfy the transcendental equation

$$\sigma\sin(\nu a)\sin(\nu[L-a]) + \nu\sin(\nu L) = 0. \tag{2}$$

The advantage of this particular representation of the secular equation is that in the limit $\sigma \to 0$ we immediately obtain the answer for the configuration of two parallel plates at distance L.

Applying the contour integral formulation of zeta functions put forward in ref. [1], the zeta function reads

$$\zeta(s) = \sum_{k,\ell}(\nu_k^2 + \eta_\ell^2)^{-s} = \frac{1}{2\pi i}\sum_\ell \int_\gamma d\nu(\nu^2 + \eta_\ell^2)^{-s}\frac{d}{d\nu}\ln F(\nu),$$

where γ is a contour enclosing all positive solutions of eq. (2) and

$$F(\nu) = \frac{1}{\nu^2}\left(\sigma\sin(\nu a)\sin(\nu[L-a]) + \nu\sin(\nu L)\right). \tag{3}$$

In writing $F(\nu)$, the transcendental eq. (2) has been divided by ν^2 in order to avoid contributions from the origin in the contour manipulations to come. After deforming the contour to the imaginary axis the representation reads

$$\zeta(s) = \frac{\sin \pi s}{\pi} \sum_\ell \int_{\eta_\ell}^{\infty} dk (k^2 - \eta_\ell^2)^{-s} \frac{d}{dk} \ln F(ik) \tag{4}$$

and our next task is to construct the analytical continuation to a half-plane containing the points $s = -1/2$ (for the Casimir force) and $s = 0$ (for the functional determinant). As usual, the asymptotic $k \to \infty$ behavior of $F(ik)$ plays the dominant role. Furthermore, the zeta function $\zeta_N(s)$ related to the transversal dimensions,

$$\zeta_N(s) = \sum_\ell \eta_\ell^{-2s},$$

will make its appearance. For notational simplicity we assume $\eta_\ell^2 > 0$.

We first note that

$$F(ik) = \frac{1}{k^2} \left(\frac{\sigma}{4} + \frac{k}{2} \right) e^{kL} \left[1 + E_1(k) \right],$$

where $E_1(k)$ is exponentially damped as $k \to \infty$. This shows, as $k \to \infty$,

$$\ln F(ik) = kL - \ln(2k) + \sum_{j=1}^{\infty} (-1)^{j+1} \left(\frac{\sigma}{2k} \right)^j \frac{1}{j} + E_2(k),$$

$E_2(k)$ denoting exponentially damped terms. Subtracting and adding the leading $M + 2$ terms in this expansion, the zeta function $\zeta(s)$ is naturally split into two pieces,

$$\zeta(s) = \zeta_f(s) + \zeta_{as}(s),$$

where

$$\zeta_f(s) = \frac{\sin \pi s}{\pi} \sum_\ell \int_{\eta_\ell}^{\infty} dk (k^2 - \eta_\ell^2)^{-s} \times \\ \frac{d}{dk} \left[\ln F(ik) - kL + \ln(2k) - \sum_{j=1}^{M} (-1)^{j+1} \left(\frac{\sigma}{2k} \right)^j \frac{1}{j} \right], \tag{5}$$

and after performing the k-integration $\zeta_{as}(s)$ reads,

$$\zeta_{as}(s) = \frac{L\Gamma\left(s-\frac{1}{2}\right)}{2\sqrt{\pi}\Gamma(s)}\zeta_N\left(s-\frac{1}{2}\right) - \frac{1}{2}\zeta_N(s) + \sum_{j=1}^{M}(-1)^j\left(\frac{\sigma}{2}\right)^j \frac{\Gamma\left(\frac{j}{2}+s\right)}{\Gamma\left(1+\frac{j}{2}\right)\Gamma(s)}\zeta_N\left(s+\frac{j}{2}\right). \qquad (6)$$

As follows from the $k \to \eta_\ell$ and $k \to \infty$ behavior, the representation for $\zeta_f(s)$ is valid for $1 > \Re s > (n - M - 1)/2$ where $n = \dim(N)$. In the following, choosing $M = n$ respectively $M = n + 1$ we will obtain the results for the functional determinant respectively the Casimir force for the situation under consideration.

Let us first evaluate $\zeta'(0)$ and thus we put $M = n$. The result for $\zeta_f'(0)$ is trivially obtained as the integral occurring in (5) is analytic about $s = 0$. We find

$$\zeta_f'(0) = \sum_\ell \left(\ln F(i\eta_\ell) - \eta_\ell L + \ln(2\eta_\ell) + \sum_{j=1}^{n}(-1)^j\left(\frac{\sigma}{2\eta_\ell}\right)^j \frac{1}{j}\right). \qquad (7)$$

In order to explicitly evaluate this expression, once the manifold N is specified, the eigenvalues η_ℓ, if known explicitly, would be substituted and (7) evaluated numerically. If the eigenvalues η_ℓ can only be determined numerically, then a suitably large number of η_ℓ's needs to be determined numerically and again (7) needs to be evaluated numerically.

For the evaluation of $\zeta_{as}'(0)$ we note that the zeta function $\zeta_N(s)$ has poles at $s = -(2j+1)/2$, $j \in \mathbb{N}$, and at $s = 1/2, 1, ..., n/2$. As is well known, about singular points $s = k$ we have the expansion

$$\zeta_N(s+k) = \frac{1}{s}\text{Res }\zeta_N(k) + \text{FP }\zeta_N(k) + \mathcal{O}(s).$$

As a result we find

$$\zeta_{as}'(0) = -L\left(\text{FP }\zeta_N\left(-\frac{1}{2}\right) - \text{Res }\zeta_N\left(-\frac{1}{2}\right)[-2+\ln 4]\right) - \frac{1}{2}\zeta_N'(0) + 2\sum_{j=1}^{n}(-1)^j\left(\frac{\sigma}{2}\right)^j\frac{1}{j}\left\{\text{FP }\zeta_N\left(\frac{j}{2}\right) + \text{Res }\zeta_N\left(\frac{j}{2}\right)\left[\gamma+\psi\left(\frac{j}{2}\right)\right]\right\}, \qquad (8)$$

with the Euler-Mascheroni constant γ and the psi function $\psi(x) = \Gamma'(x)/\Gamma(x)$. This is as far as we can go without specifying the manifold N. Once N is specified, for example as a torus or sphere, the quantities appearing in (8) can be evaluated explicitly.

The evaluation of the Casimir force

$$F_{Cas} = -\frac{1}{2}\frac{\partial}{\partial a}\zeta\left(-\frac{1}{2}\right)$$

is simplified by the observation that the asymptotic terms do not depend on the distance a. As a result only $\zeta_f(s)$ contributes to F_{Cas} and a representation for the force is

$$F_{Cas} = \frac{1}{2\pi}\sum_\ell \int_{\eta_\ell}^\infty dk (k^2 - \eta_\ell^2)^{1/2} \frac{\partial}{\partial a}\frac{\partial}{\partial k} \ln F(ik).$$

More explicitly we have that

$$\begin{aligned} h(k) = \frac{\partial}{\partial a} \ln F(ik) &= \frac{\sigma k \sinh(k[L-2a])}{\sigma \sinh(ka)\sinh(k[L-a]) + k\sinh(kL)} \\ &= \left(\frac{\sinh(kL)}{\sigma\sinh(k[L-2a])} + \frac{\sinh(ka)\sinh(k[L-a])}{k\sinh(k[L-2a])}\right)^{-1}. \end{aligned} \tag{9}$$

Noting that $g(k) = \sinh(ka)/k$ as well as $f(k) = \sinh(mk)/\sinh(nk)$ for $m > n > 0$ is increasing for $k > 0$, we conclude that for $0 < a < L/2$ the function $h(k)$ is a decreasing function of k, whereas for $L/2 < a < L$ it is an increasing function of k. This shows the piston is always attracted to the closest wall as was found for Dirichlet boundary conditions on the piston and the plates at $x = 0$ and $x = L$.[5]

Acknowledgments

The authors would like to thank Kimball Milton and Steve Fulling for very helpful suggestions. KK is supported by National Science Foundation grant PHY–0554849.

References

1. M. Bordag, E. Elizalde, and K. Kirsten. Heat kernel coefficients of the Laplace operator on the D-dimensional ball. *J. Math. Phys.*, 37:895–916, 1996.
2. R.M. Cavalcanti. Casimir force on a piston. *Phys. Rev.*, D69:065015, 2004.
3. E. Elizalde, S.D. Odintsov, and A.A. Saharian. Repulsive Casimir effect from extra dimensions and Robin boundary conditions: from branes to pistons. *Phys. Rev.*, D79:065023, 2009.
4. M.P. Hertzberg, R.L. Jaffe, M. Kardar, and A. Scardicchio. Attractive Casimir Forces in a Closed Geometry. *Phys. Rev. Lett.*, 95:250402, 2005.
5. K. Kirsten and S.A. Fulling. Kaluza-Klein models as pistons. *Phys. Rev.*, D79:065019, 2009.
6. V. Marachevsky. Casimir interaction of two plates inside a cylinder. *Phys. Rev.*, D75:085019, 2007.
7. L.P. Teo. Finite Temperature Casimir Effect in Kaluza-Klein Spacetime. *Nucl. Phys.*, B819:431–452, 2009.

USING BOUNDARY METHODS TO COMPUTE THE CASIMIR ENERGY

F. C. LOMBARDO[1], F. D. MAZZITELLI[1], and P. I. VILLAR[1,2]

[1] *Departamento de Física Juan José Giambiagi, FCEyN UBA, Facultad de Ciencias Exactas y Naturales, Ciudad Universitaria, Pabellón I, 1428 Buenos Aires, Argentina*

[2] *Computer Applications on Science and Engineering Department, Barcelona Supercomputing Center (BSC), 29, Jordi Girona 08034 Barcelona, Spain*

We discuss new approaches to compute numerically the Casimir interaction energy for waveguides of arbitrary section, based on the boundary methods traditionally used to compute eigenvalues of the 2D Helmholtz equation. These methods are combined with the Cauchy's theorem in order to perform the sum over modes. As an illustration, we describe a point-matching technique to compute the vacuum energy for waveguides containing media with different permittivities. We present explicit numerical evaluations for perfect conducting surfaces in the case of concentric corrugated cylinders and a circular cylinder inside an elliptic one.

1. Introduction

In this paper we will be concerned with the numerical calculation of the Casimir interaction energy in geometries with translational invariance along one direction, i.e. very long cylinders of arbitrary section. For the sake of simplicity, we will first discuss the case of a massless quantum scalar field that satisfies Dirichlet or Neumann boundary conditions on the surfaces of the cylinders. As we will see, in some particular situations the generalization to the electromagnetic field and/or more general boundary conditions will be straightforward.

The Casimir energy is formally given by

$$E_{12}(\sigma) = \lim_{\sigma \to 0} \frac{1}{2} \sum_p (e^{-\sigma w_p} w_p - e^{-\sigma \tilde{w}_p} \tilde{w}_p) \,, \qquad (1)$$

where w_p are the eigenfrequencies of the scalar field satisfying the appropriate boundary conditions on the surfaces of the shells, and $\tilde{w}_p$ are those corresponding to a situation in which the distances between the shells is very large. The subindex p denotes the set of quantum numbers associated to each eigenfrequency. We have introduced an exponential cutoff for high frequency modes.

For the particular geometry considered here, $p = (n, k_z)$ and the eigenfrequencies are of the form $\omega_{n,k_z} = \sqrt{k_z^2 + \lambda_n^2}$, where k_z is a continuous variable associated to the translational invariance along the z-direction and λ^2 are the eigenvalues of the Laplacian on the two-dimensional transversal section Σ contained in the plane (x,y):

$$\Delta^2 u = -\lambda^2 u \, . \tag{2}$$

The eigenfunctions $u(\mathbf{x})$ satisfy Dirichlet or Neumann boundary conditions on Γ, the boundary of Σ.

The Helmholtz equation (2) arises in many branches of physics, from the vibration of membranes to quantum billiards, and there are a plethora of methods to compute numerically its eigenfunctions and eigenvalues.[1] Among them, the "boundary methods" are based on the following strategy: the solution u is written as a (finite) linear combination of basis functions that satisfy Helmholtz equation inside Σ. The coefficients of the linear combination are chosen in such a way that the boundary conditions are satisfied at a finite number of points on Γ. The linear system of equations that determine the coefficients has a non trivial solution only for some particular values of λ, the eigenvalues of the system.

For example, in the Point Matching Method (PMM),[2] one expands the eigenfunction u in terms of a basis of solutions of the Helmholtz equation in free space $\varphi_j^{(\lambda)}(\mathbf{x})$

$$u(\mathbf{x}) = \sum_{j=1}^{\infty} a_j \varphi_j^{(\lambda)}(\mathbf{x}) \, . \tag{3}$$

In the numerical calculation this expansion is truncated at given $j = N$, and the boundary conditions are imposed on N points on Γ. These boundary conditions become a set of homogeneous, linear equations for the unknown coefficients a_j ($Ma = 0$, with M a λ-dependent $N \times N$ matrix) which has nontrivial solutions only when $detM = 0$. The last equation can be used to determine numerically the eigenvalues λ_n.

In a similar approach, known as the Method of Fundamental Solutions (MFS),[3] the eigenfunction u is expanded in terms of solutions of the Helmholtz equation with a point source at an arbitrary location $\mathbf{s}_j$, that we denote by $u_\lambda(\mathbf{x}, \mathbf{s}_j)$

$$u(\mathbf{x}) = \sum_{j=1}^{\infty} b_j u_\lambda(\mathbf{x}, \mathbf{s}_j)\,. \tag{4}$$

If the sources are located *outside* Σ, this is a solution of the homogeneous Helmholtz equation *inside* Σ. Once again, the sum is truncated at $j = N$, and the coefficients b_j are determined by solving the linear system that results when imposing the boundary conditions on a finite number of points on Γ. The roots of the determinant of the associated matrix are the eigenfrequencies of the problem. This is the simplest version of the MFS, in which the locations of the sources are fixed.

One can find in the literature discussions about spurious solutions, improvements and alternative methods to find the eigenvalues. We refer the reader to Refs.[1,4] for more details.

A crucial point is that, at a practical level, the knowledge of the spectrum of the Helmholtz equation is not enough to compute the Casimir energy. The reason is that the numerical evaluation of the sum over modes in Eq.(1) is extremely unstable,[5] and one has to subtract very large numbers to compute the finite interaction energy. The calculation is complicated even for the simplest case of Casimir effect in $1+1$ dimensions.

Instead of performing explicitly the summation, it is far more efficient to combine the methods mentioned previously with the Cauchy's theorem

$$\frac{1}{2\pi i}\int_C dz\, z\, e^{-\sigma z}\frac{d}{dz}\ln f(z) = \sum_i z_i\, e^{-\sigma z_i}\,, \tag{5}$$

where $f(z)$ is an analytic function in the complex z plane within the closed contour C, with simple zeros at $z_1, z_2, \ldots$ within C. We use this result to replace the sum over the eigenvalues of the Helmholtz equation in the Casimir energy Eq.(1) by a contour integral. In this way, it is not necessary to solve numerically the equation $detM = 0$ for the eigenvalues, but to take $f = detM$ in the Cauchy's theorem. In other words, if in the numerical method to solve the Helmholtz equation the eigenvalues are the roots of a given function, one can integrate this function in the complex plane in order to get the Casimir energy. The combination of the use of numerical methods to compute the eigenvalues with the Cauchy's theorem is the main idea we want to put forward in this paper.

In the next Section we will describe the simplest version of the PMM to a situation in which the surfaces separate regions of different permittivities, generalizing our previous results[6] for perfect conductors. In Sections 3 and 4 we will review some numerical evaluations of the Casimir interaction energy for perfect conductors. In Section 5 we include our final remarks.

2. Point-Matching Numerical Approach

A media-separated waveguide presents an interesting setup for the application of the PMM. This technique has been widely used to solve eigenvalue problems in many areas of engineering science.[2] The boundary conditions are imposed at a finite number of points around the periphery of both media.

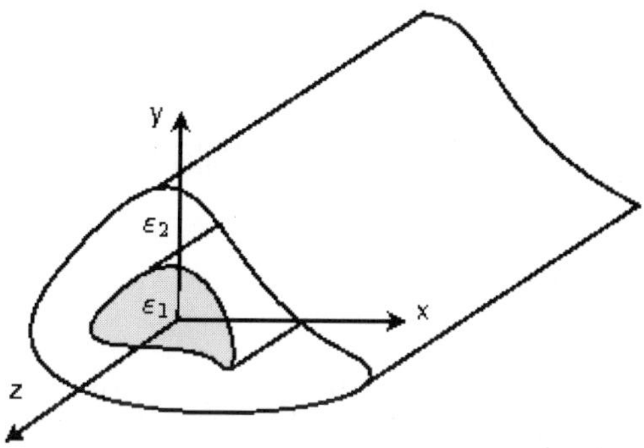

Fig. 1. A two-separated media waveguide in which one conductor encloses two different dielectric media. Each has arbitrary cross section.

For the sake of concreteness, we will bear in mind the situation in which one perfect conductor encloses two dielectric media, as shown in Fig.1, although the method could be applied to more general cases.

The general solution of the Helmholtz equation in region I (inside the inner cylinder) is

$$u = \sum_m A_m J_m(\lambda^{(I)} r) e^{im\theta}, \tag{6}$$

while in region II (annular region)

$$u = \sum_m [B_m J_m(\lambda^{(II)} r) + C_m H_m^{(1)}(\lambda^{(II)} r)] e^{im\theta}, \tag{7}$$

where (r, θ) are polar coordinates, and J_m and $H_m^{(1)}$ are the m-th order Bessel functions. The constants A_m, B_m and C_m are determined by the boundary conditions. In both equations we have defined $\lambda^{(a)} = \sqrt{\epsilon_a \omega^2 - k_z^2}$.

We assume the outer surface to be a "perfect conductor", and impose Dirichlet boundary conditions on a finite number of points (r_q, θ_q) of C_2:

$$0 = \sum_{m=-S}^{S} [B_m J_m(\lambda^{(II)} r_q) + C_m H_m^{(1)}(\lambda^{(II)} r_q)] e^{im\theta_q}, \tag{8}$$

(alternatively, for the TE modes of the electromagnetic field, one should impose Neumann boundary conditions). The surface C_1 as a dielectric interphase separating media ϵ_1 and ϵ_2, and therefore we impose continuity of the field and its derivative:

$$\begin{aligned}
&\sum_{m=-S}^{S} A_m J_m(\lambda^{(I)} r_p) e^{im\theta_p} \\
&\qquad = \sum_{m=-S}^{S} [B_m J_m(\lambda^{(II)} r_p) + C_m H_m^{(1)}(\lambda^{(II)} r_p)] e^{im\theta_p} \\
&\sum_{m=-S}^{S} A_m J'_m(\lambda^{(I)} r_p) e^{im\theta_p} \\
&\qquad = \frac{\lambda^{(II)}}{\lambda^{(I)}} \sum_{m=-S}^{S} [B_m J'_m(\lambda^{(II)} r_p) + C_m H'^{(1)}_m(\lambda^{(II)} r_p)] e^{im\theta_p},
\end{aligned} \tag{9}$$

where (r_p, θ_p) are points on the curve C_1.

The boundary conditions can be written, in matrix form, as

$$\begin{aligned}
0 &= N_1 B + N_2 C, \\
R_1 A &= M_1 B + M_2 C, \\
R_2 A &= M'_1 B + M'_2 C.
\end{aligned} \tag{10}$$

Eliminating the coefficients A_m we end with

$$N_1 B + N_2 C = 0, \quad P_1 B + P_2 C = 0, \tag{11}$$

where P_1 and P_2 can be written as

$$P_1 = M_1 - R_1 R_2^{-1} M'_1, \quad P_2 = M_2 - R_1 R_2^{-1} M'_2 . \tag{12}$$

It is worthy to note that as $R_1 R_2^{-1}$ is proportional to λ_2/λ_1, then $R_1 R_2^{-1} \to 0$ when $\epsilon_1 \to \infty$. Thus, the matrices $P_1 \to M_1$ and $P_2 \to M_2$, re-obtaining in this way, the usual perfect conductor wave-guide case studied in.[6]

For the system of Eq.(11) to have non trivial solutions, the determinant must be zero, i.e.

$$det\begin{bmatrix} N_1 & N_2 \\ P_1 & P_2 \end{bmatrix} = detP_2 \cdot detN_1 \cdot det(1 - N_2 P_2^{-1} P_1 N_1^{-1}) = 0 \; . \qquad (13)$$

This equation determines the eigenfrequencies associated to the geometry. However, as already mentioned, in order to compute the Casimir energy it is not necessary to find each eigenvalue but to integrate the determinant $Q = det(1 - N_2 P_2^{-1} P_1 N_1^{-1})$ in the complex plane.

We have developed a numerical Fortran routine in order to evaluate the Casimir interaction energy in the case in which the field satisfies Dirichlet or Neumann boundary conditions on both curves C_1 and C_2. In this case one should consider the fields only in region II with $\epsilon_2 = 1$. After some straightforward steps one can re-write the Casimir energy as a single integral in the imaginary axis $iy = \lambda^{(II)}$. For Dirichlet boundary conditions the result is

$$E_{12} = \frac{L}{4\pi} \int_0^\infty dy \; y \ln Q(iy) \; , \qquad (14)$$

while for Neumann boundary conditions one can derive a similar expression with a different function Q. It is worth to stress that these Casimir energies correspond to those of TM and TE modes of the electromagnetic field in the presence of perfect conductors.

3. Cylindrical Rack and Pinion

When two concentric cylinders have corrugations, the vacuum energy produces a torque that could, in principle, make one cylinder rotate with respect to the other. This "cylindrical rack and pinion" has been proposed in Ref.,[7] where the torque has been computed using the proximity force approximation. It was further analyzed in,[8] where the authors obtained perturbative results for Dirichlet boundary conditions in the limit of small amplitude corrugations. In this Section, we numerically evaluate the Casimir interaction energy for two concentric corrugated, perfect conductor cylinders. The cylinders have radii a and b, and we will denote by $r_- = b - a$ the mean distance between them and by $r_+ = a + b$ the sum of the radii. We will use the notation $\alpha = b/a$. The points in the mesh, that give us the corrugated cylinder boundaries, are described by the following functions:

$$h_a(\theta) = h \sin(\nu\theta) \quad ; \quad h_b(\theta) = h \sin(\nu\theta + \phi_0), \qquad (15)$$

where h is the corrugation amplitude and ν is the frequency associated with these corrugations. The Casimir torque can be calculated by taking the derivative of the interaction energy with respect to the shifted angle $\mathcal{T} = -\partial E_{12}/\partial\phi_0$.

In Fig.2 we show the numerical evaluation of the TM Casimir interaction energy for this geometry. The plot shows the results obtained using the PMM with $\alpha = 2$ and corrugation frequency $\nu = 3$, for different values of the amplitude of the corrugation h. As expected the amplitude of the oscillations grows with h. For each value of h we have performed a numerical fit of the data in order to compare with the analytical prediction. With dotted lines we have plotted the fit $y(x) = A * cos(x)$ for each curve in Fig.2. The agreement between dots and dotted lines is extremely good. Similar results can be obtained for the Neumann (TE) modes (see[6] for details).

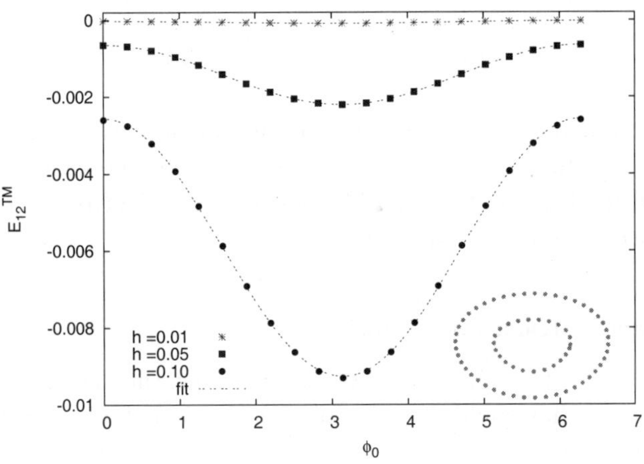

Fig. 2. Casimir interaction energy (TM modes) as a function of ϕ_0 for $\alpha = 2$ and different values of the perturbation h. The different shaped dots are the numerical data obtained with our program while the lines represent the numerical fit of each curve. Energies are measured in units of L/a^2, and distances in units of a.

It is worth to remark that, when the amplitude of the corrugation is not very small, the exact results cannot be reproduced with a simple fit of the form $y(x) = A * \cos(x)$. This is illustrated in Fig.3, where we see that, for the biggest corrugated amplitude $\tilde{h} = h/a = 0.3$, the exact result differs from the cosine function.[6]

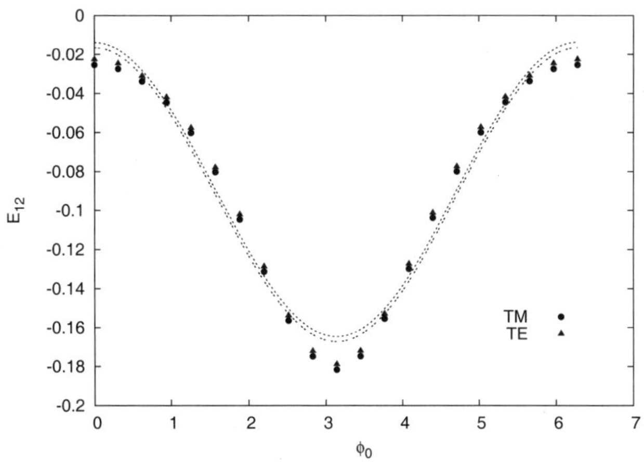

Fig. 3. Casimir interaction energy (TE and TM modes) as a function of ϕ_0 for $\alpha = 2$, $\nu = 3$ and $\tilde{h} = 0.3$. The different shaped dots are the numerical data obtained by our program while the line represents the numerical fit of each curve. In this case, the plot shows that the exact result cannot be fitted by a function $y(x) = A * \cos(x)$. Energies are measured in units of L/a^2.

4. Outer Conductors with Focal Lines: Cylinder Inside an Ellipse

Some time ago, there was a conjeture[9] based on a geometric optics approximation, about the possibility of focusing vacuum fluctuations in parabolic mirrors. It was argued that a parabolic mirror is capable of focusing the vacuum modes of the quantized electromagnetic field, therefore creating large physical effects near the mirror's focus. With this motivation, in this Section we shall evaluate the Casimir interaction energy for configurations in which the outer conducting shell has a cross section that contains focal points.

We will consider one small inner cylinder and an outer ellipse. We will denote by a the radius of the inner cylinder, by b_1 and b_2 the minor and major semiaxes of the ellipse, respectively, and by f the distance between the foci and the center of the ellipse. The coordinates of the center of the cylinder with respect to the center of the ellipse will be (ϵ_x, ϵ_y). We will use an additional tilde to denote adimensional quantities, i.e distances in units of a: $\tilde{b}_i = b_i/a\,, \tilde{f} = f/a$, etc.

For this configuration, we use a mesh where with an inner cylinder, and an outer ellipse with semiaxes $\tilde{b}_1 = 4$ and $\tilde{b}_2 = 4.33$. The ellipse has two focal points at $\tilde{f} = 1.66$. We present the results for the Casimir energy in Fig.4.

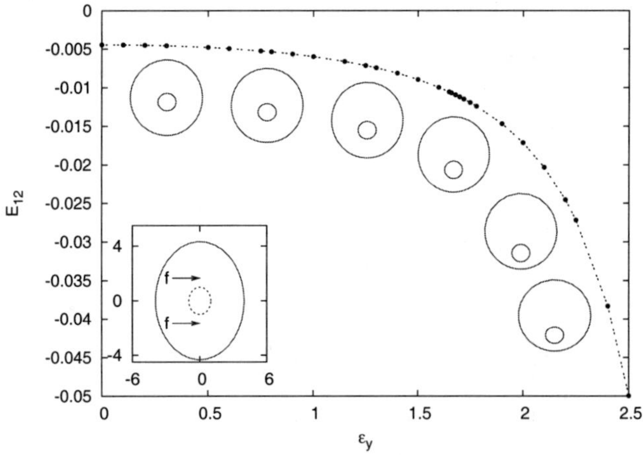

Fig. 4. Numerical evaluation of the Casimir interaction energy for an inner cylinder an eccentric outer ellipse, as a function of the position of the cylinder along the vertical axis. Energies are measured in units of L/a^2.

From Fig.4 it is possible to see that there is an unstable equilibrium position at the origin under displacements of the inner cylinder along the (vertical) ϵ_y direction. As expected, it is also possible to check that the energy grows as well as the cylinder gets closer to the surface of the outer ellipse. Fig.4 also shows a monotonic behaviour of the energy as a function of the position, even when passing through the focus. So we do not see a focusing of vacuum fluctuations near the focus of the ellipse. However, in order to confirm this result one should consider much smaller inner cylinders, in order to explore shorter wavelengths. This will require much more computational effort.

Finally, we have also checked[6] that there is an unstable equilibrium position at the origin when moving the inner cylinder in the (horizontal) ϵ_x direction.

5. Final Remarks

We have presented new numerical methods to compute the vacuum energy for arbitrary geometries with translational invariance. The approach is based on the use of traditional boundary methods to compute eigenvalues of the two dimensional Helmholtz equation, combined with Cauchy's theorem.

As a particular example, we have described a straightforward version of the point-matching method to compute the Casimir interaction energy for a waveguide with different permittivities, and reviewed some numerical calculations for perfect conductors. In all examples, for the numerical calculations we have chosen pair of points with the same angular coordinate with respect to the inner cylinder. For less symmetric configurations, and when the surfaces of both conductors are closer to each other, it will be necessary to consider grids with a larger number of points, and to optimize their positions. As in the applications to acoustic or classical electromagnetism, special care must be taken for surfaces with pronounced edges, clefts or "handles", where the point-matching technique may not be accurate to determine the eigenfrequencies. In these cases, more sophisticated approaches[4] could be necessary to optimize the numerical evaluation and to avoid spurious solutions.

Acknowledgments

We would like to thank Kim Milton for the organization and his kind hospitality during QFEXT09. This work has been supported by CONICET, UBA and ANPCyT, Argentina.

References

1. J.R. Kuttler and V.G. Sigillito, SIAM Review 26, 163 (1984).
2. R Bates, IEEE Trans. on Microwave Theory and Techniques, Vol. MTT-17, 297 (1969); H. Y. Yee and N.F.Audeh, IEEE Trans. on Microwave Theory and Techniques, Vol. MTT-13, 847 (1965); ibidem Vol. MTT-14, 487 (1966); J.R. Kuttler and V.G. Sigillito, SIAM Review bf 26, 163 (1984) and references therein. For a generalization and applications to scattering problems see F.M. Kahnert, J. Quant. Spectrosc. Radiat. Transfer **79-80**, 775 (2003) and references therein.
3. A. Karageorghis, Appl. Math. Lett. 14, 837 (2001); C.C. Tsai et al, Proc. R. Soc. A462, 1442 (2006).
4. J.V. Villadsen and E. Stewart, Chem. Engng. Sci. 22, 1483 (1967); T. Betcke and L.N. Trefethen, SIAM Review 47, 469 (2005)C.J.S. Alves and P.R.S. Antunes, CMC 2, 251 (2005); D. Cohen, N. Lepore and E.J. Heller, J. Phys. A: Math. Gen. 37, 2139 (2004); P. Amore, arXiv 0910.4798v1 [quant-ph].
5. A. Rodriguez, M. Ibanescu, D. Iannuzzi, F. Capasso, J. D. Joannopoulos, and S.G. Johnson, Phys. Rev. Lett. **99**, 080401 (2007).
6. F.C. Lombardo, F.D. Mazzitelli, P.I. Villar, and M. Vázquez, Phys. Rev. D**80**, 0605018 (2009).
7. F. D. Mazzitelli, F. C. Lombardo and P. I. Villar, J. Phys.: Conf. Ser. 161, 012015 (2009).
8. I. Cavero-Peláez, K.A. Milton, P. Parashar and K.V. Shajesh, Phys. Rev. D **78**, 065019 (2008).
9. L.H. Ford and N.F. Svaiter, Phys. Rev. A **62**, 062105 (2000).

ELECTROMAGNETIC NON-CONTACT GEARS: PRELUDE

PRACHI PARASHAR* and KIMBALL A. MILTON†

Oklahoma Center for High Energy Physics and
Homer L. Dodge Department of Physics and Astronomy,
University of Oklahoma, Norman, OK 73019, USA
*E-mail: *prachi@nhn.ou.edu, †milton@nhn.ou.edu*

INÉS CAVERO-PELÁEZ

Theoretical Physics Department, Zaragoza University, Zaragoza 50009, Spain
E-mail: cavero@unizar.es

K. V. SHAJESH

Saint Edward's School, Vero Beach, FL 32963, USA
E-mail: shajesh@nhn.ou.edu

We calculate the lateral Lifshitz force between corrugated dielectric slabs of finite thickness. Taking the thickness of the plates to infinity leads us to the lateral Lifshitz force between corrugated dielectric surfaces of infinite extent. Taking the dielectric constant to infinity leads us to the conductor limit which has been evaluated earlier in the literature.

1. Introduction

In past decade significant attention has been given to evaluation of the lateral force between corrugated surfaces (for example see Ref. 1–5 and references there-in). In an earlier work we calculated the contribution of the next-to-leading order to the lateral Casimir force between two corrugated semi-transparent δ-function plates interacting with a scalar field,[4] and the leading order contribution for the case of two concentric semi-transparent corrugated cylinders[5] using the multiple scattering formalism (see Ref. 6,7 and references there-in). We observed that including the next-to-leading order contribution significantly reduced the deviation from the exact result in the case of weak coupling. Comparison with experiments requires the analogous calculation for the electromagnetic case. Here we present preliminary results of our ongoing work on the evaluation of the lateral Lifshitz force between two corrugated dielectric (non-magnetic) slabs of finite thickness interacting through the electromagnetic field (see Fig. 1).

From the general result it is easy to take various limiting cases. Taking the thickness of the dielectric slabs to infinity leads us to the lateral Lifshitz force between dielectric slabs of infinite extent. The lateral Casimir force between corrugated perfect conductors was evaluated by Emig *et al.*[1] In our situation this is achieved by taking the dieletric constants $\varepsilon_i \to \infty$. Our results agree with the results in Emig *et al.*[1] up to an overall factor of 2. Taking the thin-plate approximation based on the plasma model we have calculated the lateral force between corrugated plasma sheets. Our goal is to extend these results to next-to-leading order. Most of these will appear in a forthcoming paper.

2. Interaction energy

We consider two dielectric slabs of infinite extent in x-y plane, which have corrugations in y-direction, as described in Fig. 1. We describe the dielectric slabs by the potentials

$$V_i(z,y) = (\varepsilon_i - 1)\left[\theta(z - a_i - h_i(y)) - \theta(z - b_i - h_i(y))\right], \tag{1}$$

where $i = 1, 2$, designates the individual dielectric slabs. $\theta(z)$ is the Heaviside theta function defined to equal 1 for $z > 0$, and 0 when $z < 0$. $h_i(y)$ describes the corrugations on the surface of the slabs. We define the thickness of the individual slabs as $d_i = b_i - a_i$, such that $a = a_2 - b_1 > 0$ represents the distance between the slabs. The permittivities of the slabs are represented by ε_i.

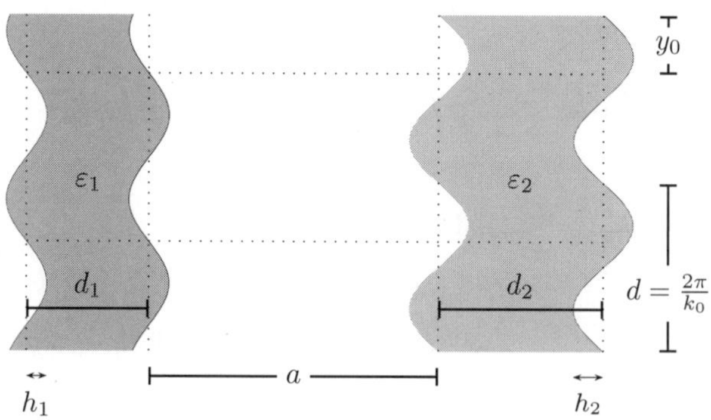

Fig. 1. Parallel dielectric slabs with sinusoidal corrugations.

Using the multiple scattering formalism for the case of the electromagnetic field[8,9] based on Schwinger's Green's dyadic formalism[10] and following the formalism described in Gears-I[4] we can obtain the contribution to the interaction energy between the two slabs in leading order in the corrugation amplitudes to be

$$E_{12}^{(2)} = \frac{i}{2} \int \frac{d\omega}{2\pi} \,\mathrm{Tr}\left[\boldsymbol{\Gamma}^{(0)} \Delta V_1^{(1)} \cdot \boldsymbol{\Gamma}^{(0)} \,\Delta V_2^{(1)}\right], \tag{2}$$

where $\Delta V_i^{(1)}$ are the leading order contributions in the potentials due to the presence of corrugations. In particular, we have

$$\Delta V_i^{(1)}(z,y) = -h_i(y)\,(\varepsilon_i - 1)\,[\delta(z-a_i) - \delta(z-b_i)]\,. \tag{3}$$

Note that $V_i^{(0)}$ describes the potential for the case when the corrugations are absent and represent the background in the formalism. $\boldsymbol{\Gamma}^{(0)} = \boldsymbol{\Gamma}^{(0)}(\boldsymbol{x},\boldsymbol{x}';\omega)$ is the Green's dyadic in the presence of background potential $V_i^{(0)}$ and satisfies

$$\left[-\frac{1}{\omega^2}\boldsymbol{\nabla}\times\boldsymbol{\nabla}\times\; + \mathbf{1} + V_1^{(0)} + V_2^{(0)}\right] \cdot \boldsymbol{\Gamma}^{(0)} = -\mathbf{1}. \tag{4}$$

The corresponding reduced Green's dyadic $\boldsymbol{\gamma}^{(0)}(z,z';k_x,k_y,\omega)$ is defined by Fourier transforming in the transverse variables as

$$\boldsymbol{\Gamma}^{(0)}(\boldsymbol{x},\boldsymbol{x}';\omega) = \int \frac{dk_x}{2\pi}\frac{dk_y}{2\pi}\, e^{ik_x(x-x')} e^{ik_y(y-y')}\,\boldsymbol{\gamma}^{(0)}(z,z';k_x,k_y,\omega). \tag{5}$$

Since our system is translationally invariant in the x-direction, we can write

$$\frac{E_{12}^{(2)}}{L_x} = \int_{-\infty}^{\infty} \frac{dk_y}{2\pi} \int_{-\infty}^{\infty} \frac{dk'_y}{2\pi}\, \tilde{h}_1(k_y - k'_y)\, \tilde{h}_2(k'_y - k_y)\, L^{(2)}(k_y,k'_y), \tag{6}$$

where L_x is the length in the x-direction and $\tilde{h}_i(k_y)$ are the Fourier transforms of the functions $h_i(y)$ describing the corrugations. The kernel $L^{(2)}(k_y,k'_y)$ is given by

$$L^{(2)}(k_y,k'_y) = -\frac{1}{2}\int \frac{d\zeta}{2\pi} \int \frac{dk_x}{2\pi}\, I^{(2)}(k_x,\zeta,k_y,k'_y), \tag{7}$$

where

$$\begin{aligned} I^{(2)}(k_x,\zeta,k_y,k'_y) = (\varepsilon_a - 1)\,(\varepsilon_b - 1) \qquad&\\ \times\Big[\boldsymbol{\gamma}^{(0)}(a_2,a_1;k_x,k_y,\omega)\cdot\boldsymbol{\gamma}^{(0)}(a_1,a_2;k_x,k'_y,\omega)&\\ -\boldsymbol{\gamma}^{(0)}(b_2,a_1;k_x,k_y,\omega)\cdot\boldsymbol{\gamma}^{(0)}(a_1,b_2;k_x,k'_y,\omega)&\\ -\boldsymbol{\gamma}^{(0)}(a_2,b_1;k_x,k_y,\omega)\cdot\boldsymbol{\gamma}^{(0)}(b_1,a_2;k_x,k'_y,\omega)&\\ +\boldsymbol{\gamma}^{(0)}(b_2,b_1;k_x,k_y,\omega)\cdot\boldsymbol{\gamma}^{(0)}(b_1,b_2;k_x,k'_y,\omega)\Big]&. \end{aligned} \tag{8}$$

The reduced Green's dyadics are evaluated after solving Eq. (4). We note that $\gamma^{(0)}(z,z';k_x,k_y,\omega) = \gamma^{(0)\dagger}(z',z;k_x,k_y,\omega)$. Our task reduces to evaluating the reduced Green's dyadic in the presence of the background. The details of this evaluation will be described in the forthcoming paper.

2.1. *Evaluation of the reduced Green's dyadic*

The Green's dyadic satisfies Eq. (4) whose solution can be determined by following the procedure decribed in Schwinger et al.[10] The expression for the reduced Green's dyadic

$$\gamma^{(0)}(z,z';k,0,\zeta) = \begin{bmatrix} \frac{1}{\varepsilon(z)}\frac{\partial}{\partial z}\frac{1}{\varepsilon(z')}\frac{\partial}{\partial z'}\,g^H & 0 & \frac{ik}{\varepsilon(z')}\frac{1}{\varepsilon(z)}\frac{\partial}{\partial z}\,g^H \\ 0 & -\zeta^2\,g^E & 0 \\ -\frac{ik}{\varepsilon(z)}\frac{1}{\varepsilon(z')}\frac{\partial}{\partial z'}\,g^H & 0 & \frac{k^2}{\varepsilon(z)\varepsilon(z')}\,g^H \end{bmatrix} \tag{9}$$

is given in terms of the electric and magnetic Green's functions* $g^E(z,z')$ and $g^H(z,z')$, which satisfy the following differential equations:

$$-\left[\frac{\partial^2}{\partial z^2} - k^2 - \zeta^2\varepsilon(z)\right] g^E(z,z') = \delta(z-z'), \tag{10}$$

$$-\left[\frac{\partial}{\partial z}\frac{1}{\varepsilon(z)}\frac{\partial}{\partial z} - \frac{k^2}{\varepsilon(z)} - \zeta^2\right] g^H(z,z') = \delta(z-z'). \tag{11}$$

We have used the definitions $k^2 = k_x^2 + k_y^2$ and $\varepsilon(z) = 1 + V_1^{(0)}(z) + V_2^{(0)}(z)$.

The reduced Green's dyadic for arbitrary k_y is generated by the rotation

$$\gamma^{(0)}(z,z';k_x,k_y,\zeta) = \mathbf{R}\cdot\gamma^{(0)}(z,z';k,0,\zeta)\cdot\mathbf{R}^T, \tag{12}$$

where

$$\mathbf{R} = \frac{1}{k}\begin{pmatrix} k_x & -k_y & 0 \\ k_y & k_x & 0 \\ 0 & 0 & k \end{pmatrix}. \tag{13}$$

We have dropped delta functions in Eq. (9) because they are evaluated at different points and thus do not contribute. We shall not present explicit solutions to the electric and magnetic Green's functions here which will be presented in our forthcoming paper.

---

*Here we use the notation in Schwinger *et al.*[10] which was reversed in many of Milton's publications, for example in Milton's book.[11]

2.2. *Interaction energy for corrugated dielectric slabs*

Using the solutions to the electric and magnetic Green's function in Eq. (9) we can evaluate $I^{(2)}(k_x, \zeta, k_y, k'_y)$ in Eq. (8) as

$$\begin{aligned}-\frac{1}{k^2}\frac{1}{k'^2}\frac{1}{2\kappa}\frac{1}{2\kappa'}\Bigg[&\frac{1}{\Delta}\frac{1}{\Delta'}M(-\alpha_1,-\alpha'_1)M(-\alpha_2,-\alpha'_2)(k_x^2+k_yk'_y)^2\zeta^4\\ &+\frac{1}{\Delta}\frac{1}{\bar{\Delta}'}M(-\alpha_1,\bar{\alpha}'_1)M(-\alpha_2,\bar{\alpha}'_2)k_x^2(k_y-k'_y)^2\zeta^2\kappa'^2\\ &+\frac{1}{\bar{\Delta}}\frac{1}{\Delta'}M(\bar{\alpha}_1,-\alpha'_1)M(\bar{\alpha}_2,-\alpha'_2)k_x^2(k_y-k'_y)^2\zeta^2\kappa^2\\ &+\frac{1}{\bar{\Delta}}\frac{1}{\bar{\Delta}'}\left\{M(\bar{\alpha}_1,\bar{\alpha}'_1)(k_x^2+k_yk'_y)\kappa\kappa'+M(-\bar{\alpha}_1,-\bar{\alpha}'_1)k^2k'^2\frac{1}{\varepsilon_1}\right\}\\ &\quad\times\left\{M(\bar{\alpha}_2,\bar{\alpha}'_2)(k_x^2+k_yk'_y)\kappa\kappa'+M(-\bar{\alpha}_2,-\bar{\alpha}'_2)k^2k'^2\frac{1}{\varepsilon_2}\right\}\Bigg], \quad (14)\end{aligned}$$

where

$$\begin{aligned}\Delta &= \big[(1-\alpha_1^2\,e^{-2\kappa_1 d_1})(1-\alpha_2^2\,e^{-2\kappa_2 d_2})\,e^{\kappa a}\\ &\qquad -\alpha_1\alpha_2(1-e^{-2\kappa_1 d_1})(1-e^{-2\kappa_2 d_2})\,e^{-\kappa a}\big], \quad (15)\\ M(\alpha_i,\alpha'_i) &= (\varepsilon_i-1)\big[(1-\alpha_i^2)\,e^{-\kappa_i d_i}(1-\alpha_i'^2)\,e^{-\kappa'_i d_i}\\ &\qquad -(1+\alpha_i)(1-\alpha_i\,e^{-2\kappa_i d_i})(1+\alpha'_i)(1-\alpha'_i\,e^{-2\kappa'_i d_i})\big], \quad (16)\end{aligned}$$

where $\kappa_i^2 = k^2 + \zeta^2\varepsilon_i$, $\bar{\kappa}_i = \kappa_i/\varepsilon_i$, and $\alpha_i = (\kappa_i - \kappa)/(\kappa_i + \kappa)$. Quantities with primes are obtained by replacing $k_y \to k'_y$ everywhere, and quantities with bars are obtained by replacing κ_i with $\bar{\kappa}_i$ except in the exponentials.

2.3. *Conductor limit*

In the conductor limit ($\varepsilon_i \to \infty$) the above expression takes the form

$$I^{(2)}_{\varepsilon\to\infty}(\kappa,\kappa',k_y-k'_y) = -\frac{\kappa}{\sinh\kappa a}\frac{\kappa'}{\sinh\kappa' a}\left[1+\frac{\{\kappa^2+\kappa'^2-(k_y-k'_y)^2\}^2}{4\,\kappa^2\kappa'^2}\right]. \quad (17)$$

For the case of sinusoidal corrugations described by $h_1(y) = h_1\sin[k_0(y+y_0)]$ and $h_2(y) = h_2\sin[k_0 y]$ the lateral force can be evaluated to be

$$F^{(2)}_{\varepsilon\to\infty} = 2k_0 a\,\sin(k_0 y_0)\left|F^{(0)}_{\text{Cas}}\right|\frac{h_1}{a}\frac{h_2}{a}A^{(1,1)}_{\varepsilon\to\infty}(k_0 a), \quad (18)$$

where

$$A^{(1,1)}_{\varepsilon\to\infty}(t_0) = \frac{15}{\pi^4}\int_{-\infty}^{\infty}dt\int_0^{\infty}\bar{s}d\bar{s}\frac{s}{\sinh s}\frac{s_+}{\sinh s_+}\left[\frac{1}{2}+\frac{(s^2+s_+^2-t_0^2)^2}{8\,s^2s_+^2}\right], \quad (19)$$

where $s^2 = \bar{s}^2 + t^2$ and $s_+^2 = \bar{s}^2 + (t+t_0)^2$. The first term in Eq. (19) corresponds to the Dirichlet scalar case,[4] which here corresponds to the E mode (referred to in Ref. 1 as TM mode). We note that $A^{(1,1)}_{\varepsilon\to\infty}(0) = 1$. See Fig. 2 for the plot of $A^{(1,1)}_{\varepsilon\to\infty}(k_0 a)$ versus $k_0 a$. We observe that only in the PFA limit is the electromagnetic contribution twice that of the Dirichlet case, and in general the electromagnetic case is less than twice that of the Dirichlet case.

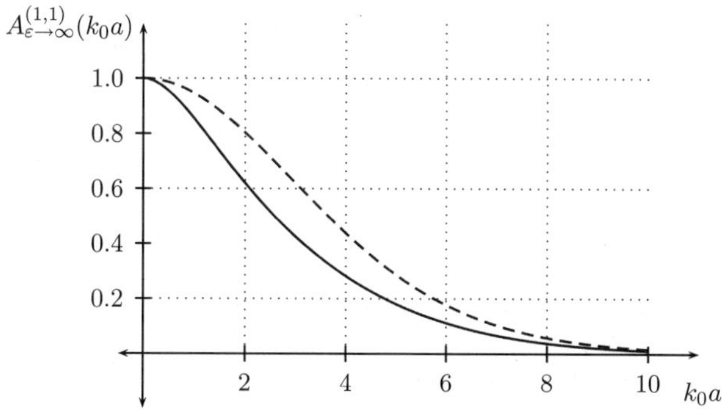

Fig. 2. Plot of $A^{(1,1)}_{\varepsilon\to\infty}(k_0 a)$ versus $k_0 a$. The dotted curve represents 2 times the Dirichlet case.

Since the above expression involves a convolution of two functions we can evaluate one of the integrals to get

$$A^{(1,1)}_{\varepsilon\to\infty}(t_0) = \frac{15}{4}\int_0^\infty du\, \frac{\sin(2t_0u/\pi)}{(2t_0u/\pi)}\left[\frac{\sinh^2 u}{\cosh^6 u}\left(\frac{7}{2} - \sinh^2 u\right)\right.$$
$$\left. -\frac{1}{2}\left(\frac{2t_0}{\pi}\right)^2 \frac{\sinh^2 u}{\cosh^4 u} + \frac{1}{16}\left(\frac{2t_0}{\pi}\right)^4 \frac{\sinh^2 u}{\cosh^2 u}\right], \tag{20}$$

which reproduces the result in Emig *et al*[1] apart from an overall factor of 2, which presumably is a transcription error. Even though Eq. (20) involves only a single integral it turns out that the double integral representation in Eq. (19) is more useful for numerical evaluation because of the oscillatory nature of the function $\sin x/x$ in the former.

3. Conclusion

We have evaluated leading order contribution to the lateral Lifshitz force between two corrugated dielectric slabs. Taking the dielectric constants of the two bodies to infinity gives the lateral Casimir force between corrugated conductors. We shall extend these results to next-to-leading order contribution for a better comparison with experiments in future publication as well as include various other limiting cases, which can be readily obtained from Eq. (14).

Acknowledgments

We thank the US Department of Energy for partial support of this work. We extend our appreciation to Jef Wagner, Elom Abalo and Nima Pourtolami for useful comments throughout the work.

References

1. T. Emig, A. Hanke, R. Golestanian and M. Kardar, Phys. Rev. A **67**, 022114 (2003).
2. F. Chen, U. Mohideen, G. L. Klimchitskaya and V. M. Mostepanenko, Phys. Rev. A **66**, 032113 (2002).
3. A. Lambrecht and V. N. Marachevsky, Phys. Rev. Lett. **101**, 160403 (2008).
4. I. Cavero-Peláez, K. A. Milton, P. Parashar and K. V. Shajesh, Phys. Rev. D **78**, 065018 (2008) [arXiv:0805.2776 [hep-th]].
5. I. Cavero-Peláez, K. A. Milton, P. Parashar and K. V. Shajesh, Phys. Rev. D **78**, 065019 (2008) [arXiv:0805.2777 [hep-th]].
6. K. A. Milton and J. Wagner, J. Phys. A **41**, 155402 (2008) [arXiv:0712.3811 [hep-th]].
7. K. A. Milton, J. Phys. Conf. Ser. **161**, 012001 (2009) [arXiv:0809.2564 [hep-th]].
8. K. A. Milton, P. Parashar and J. Wagner, Phys. Rev. Lett. **101**, 160402 (2008) [arXiv:0806.2880 [hep-th]].
9. K. A. Milton, P. Parashar and J. Wagner, in *The Casimir Effect and Cosmology*, ed. S. D. Odintsov, E. Elizalde, and O. B. Gorbunova, in honor of Iver Brevik (Tomsk State Pedagogical University), pp. 107-116 (2009) [arXiv:0811.0128 [math-ph]].
10. J. S. Schwinger, L. L. . DeRaad and K. A. Milton, Annals Phys. **115**, 1 (1979).
11. K. A. Milton, *The Casimir Effect: Physical Manifestations of Zero-Point Energy* (World Scientific, Singapore, 2001).

THE SCATTERING APPROACH TO THE CASIMIR FORCE

S. REYNAUD*, A. CANAGUIER-DURAND, R. MESSINA and A. LAMBRECHT
Laboratoire Kastler Brossel, ENS, UPMC, CNRS, Jussieu, 75252 Paris, France
**E-mail: serge.reynaud@upmc.fr*

P. A. MAIA NETO
Instituto de Física, UFRJ, CP 68528, Rio de Janeiro, RJ, 21941-972, Brazil

We present the scattering approach which is nowadays the best tool for describing the Casimir force in realistic experimental configurations. After reminders on the simple geometries of 1d space and specular scatterers in 3d space, we discuss the case of stationary arbitrarily shaped mirrors in electromagnetic vacuum. We then review specific calculations based on the scattering approach, dealing for example with the forces or torques between nanostructured surfaces and with the force between a plane and a sphere. In these various cases, we account for the material dependence of the forces, and show that the geometry dependence goes beyond the trivial *Proximity Force Approximation* often used for discussing experiments.

1. The Many Facets of the Casimir Effect

The Casimir effect [1] is a jewel with many facets. First, it is an observable effect of vacuum fluctuations in the mesoscopic world, which deserves careful attention as a crucial prediction of quantum field theory [2–7].

Then, it is also a fascinating interface between quantum field theory and other important aspects of fundamental physics. It has connections with the puzzles of gravitational physics through the problem of vacuum energy [8,9] as well as with the principle of relativity of motion through the dynamical Casimir-like effects [10–12]. Effects beyond the Proximity Force Approximation also make apparent the extremely rich interplay of vacuum energy with geometry (references and more discussions below).

Casimir physics also plays an important role in the tests of gravity at sub-millimeter ranges [13,14]. Strong constraints have been obtained in short range Cavendish-like experiments [15]: Should an hypothetical new force have a Yukawa-like form, its strength could not be larger than that of gravity if the range is larger than 56μm. For scales of the order of the micrometer, similar tests are performed by comparing with theory the results of Casimir force measurements [16,17]. At even shorter scales, the same can be done with atomic [18] or nuclear [19] force measurements.

Finally, the Casimir force and closely related Van der Waals force are dominant at micron or sub-micron distances, which entails that they have strong connections with various important domains, such as atomic and molecular physics, condensed matter and surface physics, chemical and biological physics, micro- and nano-technology [20].

2. Comparison of the Casimir Force Measurements with Theory

In short-range gravity tests, the new force would appear as a difference between the experimental result F_{exp} and theoretical prediction F_{th}. This implies that F_{th} and F_{exp} have to be assessed independently from each other and should forbid anyone to use theory-experiment comparison for proving (or disproving) some specific experimental result or theoretical model.

Casimir calculated the force between a pair of perfectly smooth, flat and parallel plates in the limit of zero temperature and perfect reflection. He found universal expressions for the force F_{Cas} and energy E_{Cas}

$$F_{\mathrm{Cas}} = \frac{\hbar c \pi^2 A}{240 L^4} \quad , \quad E_{\mathrm{Cas}} = -\frac{\hbar c \pi^2 A}{720 L^3} \tag{1}$$

with L the distance, A the area, c the speed of light and $\hbar$ the Planck constant. This universality is explained by the saturation of the optical response of perfect mirrors which reflect 100% (no less, no more) of the incoming fields. Clearly, this idealization does not correspond to any real mirror. In fact, the effect of imperfect reflection is large in most experiments, and a precise knowledge of its frequency dependence is essential for obtaining a reliable theoretical prediction for the Casimir force [21].

The most precise experiments are performed with metallic mirrors which are good reflectors only at frequencies smaller than their plasma frequency ω_{P}. Their optical response is described by a reduced dielectric function usually written at imaginary frequencies $\omega = i\xi$ as

$$\varepsilon[i\xi] = \hat{\varepsilon}[i\xi] + \frac{\sigma[i\xi]}{\xi} \quad , \quad \sigma[i\xi] = \frac{\omega_{\mathrm{P}}^2}{\xi + \gamma} \tag{2}$$

The function $\hat{\varepsilon}[i\xi]$ represents the contribution of interband transitions and it is regular at the limit $\xi \to 0$. Meanwhile $\sigma[i\xi]$ is the reduced conductivity (σ is measured as a frequency and the SI conductivity is $\epsilon_0\sigma$) which describes the contribution of the conduction electrons.

A simplified description corresponds to the lossless limit $\gamma \to 0$ often called the plasma model. As γ is much smaller than ω_P for a metal such as Gold, this simple model captures the main effect of imperfect reflection. However it cannot be considered as an accurate description since a much better fit of tabulated optical data is obtained with a non null value of γ [21]. Furthermore, the Drude model meets the important property of ordinary metals which have a finite static conductivity $\sigma_0 = \frac{\omega_P^2}{\gamma}$, in contrast to the lossless limit which corresponds to an infinite value for σ_0.

Another correction to the Casimir expressions is associated with the effect of thermal fluctuations [22,23] which is correlated to the effect of imperfect reflection [24]. Bostrom and Sernelius have remarked that the small non zero value of γ had a significant effect on the force evaluation at $T \neq 0$ [25]. This remark has led to a blossoming of contradictory papers (see references in [26–28]). The current status of Casimir experiments appears to favor predictions obtained with $\gamma = 0$ rather than those corresponding to the expected $\gamma \neq 0$ (see Fig.1 in [29]). Note that the ratio between the prediction at $\gamma = 0$ with that at $\gamma \neq 0$ reaches a factor 2 at the limit of large temperatures or large distances, although it is not possible to test this striking prediction with current experiments which do not explore this domain.

At this point, it is worth emphasizing that microscopic descriptions of the Casimir interaction between two metallic bulks lead to predictions agreeing with the lossy Drude model rather than the lossless plasma model at the limit of large temperatures or large distances [30–32]. At the end of this discussion, we thus have to face a worrying situation with a lasting discrepancy between theory and experiment. This discrepancy may have various origins, in particular artefacts in the experiments or inaccuracies in the calculations. A more subtle but maybe more probable possibility is that there exist yet unmastered differences between the situations studied in theory and the experimental realizations.

3. The Role of Geometry

The geometry of Casimir experiments might play an important role in this context. Precise experiments are indeed performed between a plane and a sphere whereas calculations are often devoted to the geometry of two parallel planes. The estimation of the force in the plane-sphere geometry involves the so-called *Proximity Force Approximation* (PFA) [33] which amounts to averaging over the distribution of local inter-plate distances the force calculated in the two-planes geometry, the latter being deduced from the Lifshitz formula [34,35].

This trivial treatment of geometry cannot reproduce the rich interconnection expected to take place between the Casimir effect and geometry [36]. In the plane-sphere geometry in particular, the PFA can only be valid when the radius R is much larger than the separation L [37]. But even if this limit is met in experiments, the PFA does not tell one what is its accuracy for a given value of L/R or whether this accuracy depends on the material properties of the mirror. Answers to these questions can only be obtained by pushing the theory beyond the PFA, which has been done in the past few years (see references in [38–42]). In fact, it is only very recently that these calculations have been done with plane and spherical metallic plates coupled to electromagnetic vacuum [43], thus opening the way to a comparison with experimental studies of PFA in the plane-sphere geometry [44].

Another specific geometry of great interest is that of surfaces with periodic corrugations. As lateral translation symmetry is broken, the Casimir force contains a lateral component which is smaller than the normal one, but has nevertheless been measured in dedicated experiments [45]. Calculations beyond the PFA have first been performed with the simplifying assumptions of perfect reflection [46] or shallow corrugations [47–49]. As expected, the PFA was found to be accurate only at the limit of large corrugation wavelengths. Very recently, experiments have been able to probe the beyond-PFA regime [50,51] and it also became possible to calculate the forces between real mirrors with deep corrugations [52]. More discussions on these topics will be presented below.

4. Introduction to the Scattering Approach

The best tool available for addressing these questions is the scattering approach. We begin the review of this approach by an introduction considering the two simple cases of the Casimir force between 2 scatterers on a 1-dimensional line and between two plane and parallel mirrors coupled through specular scattering to 3-dimensional electromagnetic fields [53].

The first case corresponds to the quantum theory of a scalar field with two counterpropagating components. A mirror is thus described by a 2x2 $S-$matrix containing the reflection and transmission amplitudes r and t. Two mirrors form a Fabry-Perot cavity described by a global $S-$matrix which can be evaluated from the elementary matrices S_1 and S_2 associated with the two mirrors. All $S-$matrices are unitary and their determinants are shown to obey the simple relation

$$ln \det S = \ln \det S_1 + \ln \det S_2 + i\Delta \tag{3}$$
$$\Delta = i \ln \frac{d}{d^*} \quad , \quad d(\omega) = 1 - r_1 r_2 \exp\left(\frac{2i\omega L}{c}\right)$$

The phaseshift Δ associated with the cavity is expressed in terms of the denominator d describing the resonance effect. The sum of all these phaseshifts over the field modes leads to the following expression of the Casimir free energy $\mathcal{F}$

$$\mathcal{F} = -\hbar \int \frac{\mathrm{d}\omega}{2\pi} N(\omega)\Delta(\omega) \quad , \quad N(\omega) = \frac{1}{\exp\left(\frac{\hbar\omega}{k_\mathrm{B}T}\right) - 1} + \frac{1}{2} \tag{4}$$

Here N is the mean number of thermal photons per mode, given by the Planck law, augmented by the term $\frac{1}{2}$ which represents the contribution of vacuum [53].

This phaseshift formula can be given alternative interpretations. In particular, the Casimir force

$$F = \frac{\partial \mathcal{F}(L,T)}{\partial L} \tag{5}$$

can be seen as resulting from the difference of radiation pressures exerted onto the inner and outer sides of the mirrors by the field fluctuations [53]. Using the analytic properties of the scattering amplitudes, the free energy may be written as the following expression after a Wick rotation ($\omega = i\xi$ are imaginary frequencies)

$$\mathcal{F} = \hbar \int \frac{\mathrm{d}\xi}{2\pi} \cot\left(\frac{\hbar\xi}{2k_\mathrm{B}T}\right) \ln d(i\xi) \tag{6}$$

Using the pole decomposition of the cotangent function and the analytic properties of $\ln d$, this can finally be expressed as the Matsubara sum ($\sum'_m$ is the sum over positive integers m with $m = 0$ counted with a weight $\frac{1}{2}$)

$$\mathcal{F} = k_\mathrm{B}T \sum_m{}' \ln d(i\xi_m) \quad , \quad \xi_m \equiv \frac{2\pi m k_\mathrm{B} T}{\hbar} \tag{7}$$

The same lines of reasoning can be followed when studying the geometry of two plane and parallel mirrors aligned along the axis x and y. Due to the symmetry of this configuration, the frequency ω, transverse vector $\mathbf{k} \equiv (k_x, k_y)$ and polarization $p = \mathrm{TE}, \mathrm{TM}$ are preserved by all scattering processes. The two mirrors are described by reflection and transmission amplitudes which depend on frequency, incidence angle and polarization p. We assume thermal equilibrium for the whole "cavity + fields" system, and calculate as in the simpler case of a 1-dimensional space. Care has however to be taken to account for the contribution of evanescent waves besides that of ordinary modes freely propagating outside and inside the cavity [7,54]. The properties of the evanescent waves are described through an analytical continuation of those of ordinary ones, using the well defined analytic behavior of the scattering amplitudes. At the end of this derivation, the free energy has the following form as a Matsubara sum [55]

$$\mathcal{F} = \sum_{\mathbf{k}} \sum_{p} k_{\mathrm{B}} T {\sum_{m}}' \ln d(i\xi_m, \mathbf{k}, p) \quad , \quad \xi_m \equiv \frac{2\pi m k_{\mathrm{B}} T}{\hbar} \tag{8}$$

$$d(i\xi, \mathbf{k}, p) = 1 - r_1(i\xi, \mathbf{k}, p) r_2(i\xi, \mathbf{k}, p) \exp^{-2\varkappa L} \quad , \quad \varkappa \equiv \sqrt{\mathbf{k}^2 + \frac{\xi^2}{c^2}}$$

$\sum_{\mathbf{k}} \equiv A \int \frac{\mathrm{d}^2\mathbf{k}}{4\pi^2}$ is the sum over transverse wavevectors with A the area of the plates, $\sum_p$ the sum over polarizations and ${\sum_m}'$ the Matsubara sum.

This expression reproduces the Casimir ideal formula in the limits of perfect reflection $r_1 r_2 \to 1$ and null temperature $T \to 0$. But it is valid and regular at thermal equilibrium at any temperature and for any optical model of mirrors obeying causality and high frequency transparency properties. It has been demonstrated with an increasing range of validity in [53], [54] and [7]. The expression is valid not only for lossless mirrors but also for lossy ones. In the latter case, it accounts for the additional fluctuations accompanying losses inside the mirrors.

It can thus be used for calculating the Casimir force between arbitrary mirrors, as soon as the reflection amplitudes are specified. These amplitudes are commonly deduced from models of mirrors, the simplest of which is the well known Lifshitz model [34,35] which corresponds to semi-infinite bulk mirrors characterized by a local dielectric response function $\varepsilon(\omega)$ and reflection amplitudes deduced from the Fresnel law.

In the most general case, the optical response of the mirrors cannot be described by a local dielectric response function. The expression (8) of the free energy is still valid in this case with some reflection amplitudes to be determined from microscopic models of mirrors. Recent attempts in this direction can be found for example in [56–58].

5. The Non-specular Scattering Formula

We now present a more general scattering formula allowing one to calculate the Casimir force between stationary objects with arbitrary non planar shapes. The main generalization with respect to the already discussed cases is that the scattering matrix $\mathcal{S}$ is now a larger matrix accounting for non-specular reflection and mixing different wavevectors and polarizations while preserving frequency [7,47]. Of course, the non-specular scattering formula is the generic one while specular reflection can only be an idealization.

As previously, the Casimir free energy can be written as the sum of all the phaseshifts contained in the scattering matrix $\mathcal{S}$

$$\mathcal{F} = i\hbar \int_0^\infty \frac{\mathrm{d}\omega}{2\pi} N(\omega) \ln \det \mathcal{S} = i\hbar \int_0^\infty \frac{\mathrm{d}\omega}{2\pi} \, N(\omega) \mathrm{Tr} \ln \mathcal{S} \tag{9}$$

The symbols det and Tr refer to determinant and trace over the modes of the matrix $\mathcal{S}$. As previously, the formula can also be written after a Wick rotation as a Matsubara sum

$$\mathcal{F} = k_\mathrm{B} T \sum_m {}' \mathrm{Tr} \ln \mathcal{D}(i\xi_m) \quad , \quad \mathcal{D} = 1 - \mathcal{R}_1 \exp^{-\mathcal{K}L} \mathcal{R}_2 \exp^{-\mathcal{K}L} \tag{10}$$

The matrix $\mathcal{D}$ is the denominator containing all the resonance properties of the cavity formed by the two objects 1 and 2 here written for imaginary frequencies. It is expressed in terms of the matrices $\mathcal{R}_1$ and $\mathcal{R}_2$ which represent reflection on the two objects 1 and 2 and of propagation factors $\exp^{-\mathcal{K}L}$. Note that the matrices $\mathcal{D}$, $\mathcal{R}_1$ and $\mathcal{R}_2$, which were diagonal on the basis of plane waves when they described specular scattering, are no longer diagonal in the general case of non specular scattering. The propagation factors remain diagonal in this basis with their diagonal values written as in (8). Clearly the expression (10) does not depend on the choice of a specific basis. Remark also that (10) takes a simpler form at the limit of null temperature (note the change of notation from the free energy $\mathcal{F}$ to the ordinary energy $\mathcal{E}$)

$$F = \frac{\mathrm{d}\mathcal{E}}{\mathrm{d}L} \quad , \quad \mathcal{E} = \hbar \int_0^\infty \frac{\mathrm{d}\xi}{2\pi} \ln \det \mathcal{D}(i\xi) \tag{11}$$

Formula (11) has been used to evaluate the effect of roughness or corrugation of the mirrors [47–49] in a perturbative manner with respect to the roughness or corrugation amplitudes (see the next section). It has clearly a larger domain of applicability, not limited to the perturbative regime, as soon as techniques are available for computing the large matrices involved in its evaluation. It has also been used in the past years by different groups using different notations [39–41,59,60]. The relation between these approaches is reviewed for example in [61].

6. The Lateral Casimir Force between Corrugated Plates

As already stated, the lateral Casimir force between corrugated plates is a topic of particular interest. This configuration is more favorable to theory/experiment comparison than that met when studying the normal Casimir force. It could thus allow for a new test of Quantum ElectroDynamics, through the dependence of the lateral force to the corrugation wavevector [48,49]. Here, we consider two plane mirrors, M1 and M2, with corrugated surfaces described by uniaxial sinusoidal profiles (see Fig. 1 in [49]). We denote h_1 and h_2 the local heights with respect to mean planes $z_1 = 0$ and $z_2 = L$

$$h_1 = a_1 \cos(k_C x) \quad , \quad h_2 = a_2 \cos\left(k_C(x-b)\right) \tag{12}$$

h_1 and h_2 have null spatial averages and L is the mean distance between the two surfaces; h_1 and h_2 are both counted as positive when they correspond to separation decreases; λ_C is the corrugation wavelength, $k_C = 2\pi/\lambda_C$ the corresponding wave vector, and b the spatial mismatch between the corrugation crests.

At lowest order in the corrugation amplitudes, when $a_1, a_2 \ll \lambda_C, \lambda_P, L$, the Casimir energy may be obtained by expanding up to second order the general formula (11). The part of the Casimir energy able to produce a lateral force is thus found to be

$$F^{\mathrm{lat}} = -\frac{\partial \delta\mathcal{E}}{\partial b} \quad , \quad \delta\mathcal{E} = -\hbar \int_0^\infty \frac{\mathrm{d}\xi}{2\pi} \mathrm{Tr}\left(\delta\mathcal{R}_1 \frac{\exp^{-\mathcal{K}L}}{\mathcal{D}_0} \delta\mathcal{R}_2 \frac{\exp^{-\mathcal{K}L}}{\mathcal{D}_0}\right) \tag{13}$$

$\delta\mathcal{R}_1$ and $\delta\mathcal{R}_2$ are the first-order variation of the reflection matrices $\mathcal{R}_1$ and $\mathcal{R}_2$ induced by the corrugations; $\mathcal{D}_0$ is the matrix $\mathcal{D}$ evaluated at zeroth order in the corrugation; it is diagonal on the basis of plane waves and commutes with $\mathcal{K}$.

Explicit calculations of (13) have been done for the simplest case of experimental interest, with two corrugated metallic plates described by the plasma dielectric function. These calculations have led to the following expression of the lateral energy

$$\delta E = \frac{A}{2} G_C(k_C) a_1 a_2 \cos(k_C b) \tag{14}$$

with the function $G_C(k_C)$ given in [49]. It has also been shown that the PFA was recovered for long corrugation wavelengths, when $G_C(k_C)$ is replaced by $G_C(0)$ in (14). This important argument can be considered as a properly formulated "Proximity Force Theorem" [49]. It has to be distinguished from the approximation (PFA) which consists in an identification of $G_C(k_C)$ with its limit $G_C(0)$. For arbitrary corrugation wavevectors, the deviation from the PFA is described by the ratio

$$\rho_C(k_C) = \frac{G_C(k_C)}{G_C(0)} \tag{15}$$

The variation of this ratio ρ_C with the parameters of interest has been described in a detailed manner in [48,49]. Curves are drawn as examples in the Fig. 1 of [48] with $\lambda_P = 137$nm chosen to fit the case of gold covered plates. An important feature is that ρ_C is smaller than unity as soon as k_C significantly deviates from 0. For large values of k_C, it even decays exponentially to zero, leading to an extreme deviation from the PFA.

Other situations of interest have also been studied. When the corrugation plates are rotated with respect to each other, a torque appears to be induced by vacuum fluctuations, tending to align the corrugation directions [62]. In contrast with the similar torque appearing between misaligned birefringent plates [63], the torque is here coupled to the lateral force. The advantage of the configuration with corrugated plates is that the torque has a larger magnitude. Another case of interest may be designed by using the possibilities offered by cold atoms techniques. Non trivial effects of geometry should be visible in particular when using a Bose-Einstein condensate as a local probe of vacuum above a nano-grooved plate [64,65].

These results suggested that non trivial effects of geometry, *i.e.* effects beyond the PFA, could be observed with dedicated lateral force experiments. It was however difficult to achieve this goal with corrugation amplitudes a_1, a_2 meeting the conditions of validity of the perturbative expansion. As already stated, recent experiments have been able to probe the beyond-PFA regime with deep corrugations [50,51] and it also became possible to calculate the forces between real mirrors without the perturbative assumption. In particular, an exact expression has been obtained for the force between two nanostructured surfaces made of real materials with arbitrary corrugation depth, corrugation width and distance [52].

7. The Plane-Sphere Geometry Beyond PFA

In the plane-sphere geometry, it is also possible to use the general scattering formula (11) to obtain explicit evaluations of the Casimir force. The reflection matrices may here be written in terms of Fresnel amplitudes on the plane mirror and of Mie amplitudes on the spherical one. The scattering formula is then obtained by writing also transformation formulas from the plane waves basis to the spherical waves basis and conversely. The result takes the form of a multipolar expansion with spherical waves labeled by quantum numbers ℓ and m ($|m| \leq \ell$). For doing the numerics, the expansion is truncated at some maximum value $\ell_{\max}$, which restricts accurate evaluations to a domain $x \equiv L/R > x_{\min}$ with $x_{\min}$ proportional to $1/\ell_{\max}$.

Such calculations have first been performed for perfectly reflecting mirrors [59,66]. It was thus found that the Casimir energy was smaller than expected from the PFA and, furthermore, than the result for electromagnetic fields was departing from PFA more rapidly than was expected from previously existing scalar calculations [40,41]. It is only very recently that the same calculations have been done for the more realistic case of metallic mirrors described by a plasma model dielectric function [43]. Results of these evaluations are expressed in terms of reduction factors defined for the force F or force gradient G with respect to the PFA expectations F^{PFA} and G^{PFA} respectively

$$\rho_F = \frac{F}{F^{\text{PFA}}} \quad , \quad \rho_G = \frac{G}{G^{\text{PFA}}} \tag{16}$$

Examples of curves for ρ_F and ρ_G are shown on Fig.2 of [43] for perfect and plasma mirrors.

Using these results, it is possible to compare the theoretical evaluations to the experimental study of PFA in the plane-sphere geometry [44]. In this experiment, the force gradient is measured for various radii of the sphere and the results are used to obtain a constraint $|\beta_G| < 0.4$ on the slope at origin β_G of the function

$$\rho_G(x) = 1 + \beta_G x + O(x^2) \tag{17}$$

Now the comparison of this experimental information to the slope obtained by interpolating at low values of x the theoretical evaluations of ρ_G reveals a striking difference between the cases of perfect and plasma mirrors. The slope β_G^{perf} obtained for perfect mirrors is larger than that β_G^{Gold} obtained for gold mirrors by a factor larger than 2

$$\beta_G^{\text{perf}} \sim -0.48 \quad , \quad \beta_G^{\text{Gold}} \sim -0.21 \tag{18}$$

Meanwhile, β_G^{Gold} is compatible with the experimental bound obtained in [44] (see [43]) whereas β_G^{perf} lies outside this bound (see also [59]).

The lesson to be learned from these results is that more work is needed to reach a reliable comparison of experiment and theory on the Casimir effect. Experiments are performed with large spheres for which the parameter L/R is smaller than 0.01, and efforts are devoted to calculations pushed towards this regime [67].

Meanwhile, the effect of temperature should also be correlated with the plane-sphere geometry. The first calculations accounting simultaneously for plane-sphere geometry, temperature and dissipation have been published very recently [68] and they show several striking features. The factor of 2 between the long distance forces in Drude and plasma models is reduced to a factor below 3/2 in the plane-sphere geometry. Then, PFA underestimates the Casimir force within the Drude model at short distances, while it overestimates it at all distances for the perfect reflector and plasma model. If the latter feature were conserved for the experimental parameter region R/L ($> 10^2$), the actual values of the Casimir force calculated within plasma and Drude model could turn out to be closer than what PFA suggests, which would diminish the discrepancy between experimental results and predictions of the thermal Casimir force using the Drude model.

Acknowledgments

The authors thank I. Cavero-Pelaez, D. Dalvit, G.L. Ingold, M.-T. Jaekel and I. Pirozenkho for fruitful discussions. A.C. and R.M. acknowledge support from the ESF Research Networking Programme CASIMIR (www.casimir-network.com). P.A.M.N. thanks CNPq, CAPES and Faperj for financial support. A.L. acknowledges support from the French Contract ANR-06-Nano-062.

References

1. H.B.G. Casimir, Proc. K. Ned. Akad. Wet. **51** 793 (1948).
2. P.W. Milonni, *The quantum vacuum* (Academic, 1994).
3. S.K. Lamoreaux, Resource Letter in Am. J. Phys. **67** 850 (1999).
4. M. Bordag, U. Mohideen and V.M. Mostepanenko, Phys. Rep. **353** 1 (2001) and references therein.
5. K.A. Milton, J. Phys. **A20** 4628 (2005).
6. R.S. Decca, D. López, E. Fischbach *et al*, Annals Phys. **318** 37 (2005).
7. A. Lambrecht, P.A. Maia Neto and S. Reynaud, New J. Phys. **8** 243 (2006).
8. C. Genet, A. Lambrecht and S. Reynaud, in *On the Nature of Dark Energy* eds. U. Brax, J. Martin, J.P. Uzan, 121 (Frontier Group, 2002) [arXiv:quant-ph/0210173] and references therein.
9. M.-T. Jaekel and S. Reynaud, Proc. of the Orleans School on Mass, to appear (2009) [arXiv:0812.3936] and references therein.
10. M.-T. Jaekel and S. Reynaud, Reports on Progress in Physics, **60** 863 (1997) and references therein.
11. A. Lambrecht, Journal of Optics, **B7** 3 (2005) and references therein.
12. C. Braggio *et al*, in the present volume (2009).

13. E. Fischbach and C. Talmadge, *The Search for Non Newtonian Gravity* (AIP Press/Springer Verlag, 1998).
14. E.G. Adelberger, B.R. Heckel and A.E. Nelson, Ann. Rev. Nucl. Part. Sci. **53**, 77, (2003).
15. D.J. Kapner, T.S. Cook, E.G. Adelberger *et al*, Phys. Rev. Lett. **98**, 021101, (2007).
16. A. Lambrecht and S. Reynaud, Poincaré Seminar on Vacuum Energy and Renormalization **1** 107 (2002) [arXiv:quant-ph/0302073] and references therein.
17. R. Onofrio, New J. Phys. **8** 237 (2006).
18. S. Lepoutre *et al*, EPL **88**, 20002, (2009).
19. V. Nesvizhevsky *et al*, Phys. Rev. **D77**, 034020, (2008).
20. V.A. Parsegian, *Van der Waals Forces : a Handbook for Biologists, Chemists, Engineers and Physicists* (Cambridge UP, 2006).
21. A. Lambrecht and S. Reynaud, Euro. Phys. J. **D8** 309 (2000).
22. J. Mehra, Physica **37** 145 (1967).
23. J. Schwinger, L.L. de Raad and K.A. Milton, Ann. Phys. **115** 1 (1978).
24. C. Genet, A. Lambrecht and S. Reynaud, Phys. Rev. **A62** 012110 (2000) and references therein.
25. M. Bostrm and Bo E. Sernelius, Phys. Rev. Lett. **84** 4757 (2000).
26. S. Reynaud, A. Lambrecht and C. Genet, in *Quantum Field Theory Under the Influence of External Conditions*, ed. K.A.Milton (Rinton Press, 2004) p.36, also in arXiv:quant-ph/0312224 .
27. I. Brevik, S.A. Ellingsen and K. Milton, New J. Phys. **8** 236 (2006).
28. G.-L. Ingold, A. Lambrecht and S. Reynaud, Phys. Rev. **E80** 041113 (2009).
29. R. Decca *et al*, Phys. Rev. **D75** 077101 (2007).
30. B. Jancovici and L. Šamaj, Europhys. Lett. **72** 35 (2005).
31. P. R. Buenzli and Ph. A. Martin, Europhys. Lett. **72** 42 (2005).
32. G. Bimonte, Phys. Rev. **A79** 042107 (2009).
33. B.V. Deriagin, I.I. Abrikosova and E.M. Lifshitz, Quart. Rev. **10** 295 (1968).
34. E.M Lifshitz, Sov. Phys. JETP **2** 73 (1956).
35. I.E. Dzyaloshinskii, E.M. Lifshitz and L.P. Pitaevskii, Sov. Phys. Uspekhi **4** 153 (1961).
36. R. Balian and B. Duplantier, Ann. Phys. NY **104** 300 (1977); **112** 165 (1978); in *15th SIGRAV Conference on General Relativity and Gravitation*, [arXiv:quant-ph/0408124].
37. R.L. Jaffe and A. Scardicchio, Phys. Rev. Lett. **92**, 070402 (2004).
38. S. Reynaud, P.A. Maia Neto and A. Lambrecht, J. Phys. **A41** 164004 (2008).
39. T. Emig and R.L. Jaffe, J. Phys. **A41** 164001 (2008).
40. M. Bordag and V. Nikolaev, J. Phys. **A41** 164002 (2008).
41. A. Wirzba, J. Phys. **A41** 164003 (2008).
42. K. Klingmller and H. Gies J. Phys. **A41** 164042 (2008).
43. A. Canaguier-Durand, P.A. Maia Neto, I. Cavero-Pelaez, A. Lambrecht and S. Reynaud, Phys. Rev. Lett. **102** 230404 (2009).
44. D.E. Krause, R.S. Decca, D. Lopez and E. Fischbach, Phys. Rev. Lett. **98** 050403 (2007).

45. F. Chen, U. Mohideen, G.L. Klimchitskaya and V. Mostepanenko, Phys. Rev. Lett. **88** 101801 (2002); Phys. Rev. **A66** 032113 (2002).
46. R. Bscher and T. Emig, Phys. Rev. Lett. **94** 133901 (2005).
47. P.A. Maia Neto, A. Lambrecht and S. Reynaud, Europhys. Lett. **69** 924 (2005); Phys. Rev. **A72** 012115 (2005).
48. R.B. Rodrigues, P.A. Maia Neto, A. Lambrecht and S. Reynaud, Phys. Rev. Lett. **96** 100402 (2006); Phys. Rev. Lett. **98** 068902 (2007).
49. R.B. Rodrigues, P.A. Maia Neto, A. Lambrecht and S. Reynaud, Phys. Rev. **A75** 062108 (2007).
50. H.B. Chan *et al*, Phys. Rev. Lett. **101** 030401 (2008).
51. H.C. Chiu *et al*, Phys. Rev. **B80** 121402 (2009).
52. A. Lambrecht and V.N. Marachevsky, Phys. Rev. Lett. **101** 160403 (2008); A. Lambrecht Nature **454** 836 (2008).
53. M.T. Jaekel and S. Reynaud, J. Physique **I-1** 1395 (1991) [arXiv:quant-ph/0101067].
54. C. Genet, A. Lambrecht and S. Reynaud, Phys. Rev. **A67** 043811 (2003).
55. The question had been asked in [26] whether the regularity conditions needed to write the Matsubara sum were met for the Drude model; this question has been answered positively in [28].
56. L.P. Pitaevskii, Phys. Rev. Lett. **101** 163202 (2008); Phys. Rev. Lett. **102** 189302 (2009); B. Geyer *et al*, Phys. Rev. Lett. **102** 189301 (2009).
57. D.A.R. Dalvit and S.K. Lamoreaux, Phys. Rev. Lett. **101** 163203 (2008); Phys. Rev. Lett. **102** 189304 (2009); R.S. Decca *et al*, Phys. Rev. Lett. **102** 189303 (2009).
58. V.B. Svetovoy, Phys. Rev. Lett. **101** 163603 (2008); Phys. Rev. Lett. **102** 219903 (E) (2009).
59. T. Emig, J. Stat. Mech.: Theory Exp. P04007 (2008).
60. O. Kenneth and I. Klich, Phys. Rev. **B78** 014103 (2008).
61. K.A. Milton and J. Wagner, J. Phys. **A41** 155402 (2008).
62. R.B. Rodrigues, P.A. Maia Neto, A. Lambrecht and S. Reynaud, Europhys. Lett. **76** 822 (2006).
63. J. Munday, D. Ianuzzi, Y. Barash and F. Capasso, Phys. Rev. **A71** 042102 (2005).
64. D.A.R. Dalvit, P.A. Maia Neto, A. Lambrecht and S. Reynaud, Phys. Rev. Lett. **100** 040405 (2008).
65. R. Messina, D.A.R. Dalvit, P.A. Maia Neto, A. Lambrecht and S. Reynaud, Phys. Rev. **A80** 022119 (2009).
66. P.A. Maia Neto, A. Lambrecht and S. Reynaud, Phys. Rev. **A78** 012115 (2008).
67. M. Bordag and V. Nikolaev, arXiv:0911.0146.
68. A. Canaguier-Durand, P.A. Maia Neto, A. Lambrecht and S. Reynaud, accepted for publication in Phys. Rev. Lett. (2010), arXiv:0911.0913.

SCALAR CASIMIR ENERGIES FOR SEPARABLE COORDINATE SYSTEMS: APPLICATION TO SEMI-TRANSPARENT PLANES IN AN ANNULUS

J. WAGNER* and K. A. MILTON†

University of Oklahoma,
Homer L. Dodge Department of Physics and Astronomy,
Norman, OK, 73019
**E-mail: wagner@nhn.ou.edu*
†E-mail: milton@nhn.ou.edu

K. KIRSTEN‡

Baylor University, Department of Mathematics,
One Bear Place # 97328, Waco, TX 76798-7328
‡E-mail: Klaus_Kirsten@baylor.edu

We derive a simplified general expression for the two-body scalar Casimir energy in generalized separable coordinate systems. We apply this technique to the case of radial semi-transparent planes in the annular region between two concentric Dirichlet cylinders. This situation is explored both analytically and numerically.

1. Introduction

In 1948 Casimir[1] predicted that two parallel perfectly reflecting mirrors would attract each other with a pressure of $P = \pi^2/240a^4$. Since then much work has been done studying a variety of geometries and materials. Much of this work has been summarized and referenced in review articles by M. Bordag *et al*[2] and K. A. Milton,[3] and more completely in two books by the same authors.[4,5]

This work only concerns itself with the Casimir effect for a massless scalar field. In order to proceed we will start with the multiple scattering expression for the Casimir energy

$$E = \frac{1}{4\pi} \int_{-\infty}^{\infty} \mathrm{d}\zeta \, \mathrm{Tr} \ln(1 - G_1 V_1 G_2 V_2). \tag{1}$$

Here ζ is the imaginary frequency, and G_i is the Green's function referring to a single potential V_i. An equivalent expression was first used by Renne[6] in 1971, and more recently by many others.[7–9] A very good derivation is given by Kenneth and Klich.[10]

2. Separation of Variables

Equation (1) is a fairly complicated formula to work with. We have to perform a 3-dimensional trace of the logarithm of the $1 - G_1V_1G_2V_2$ operator. We also have to solve a partial differential equation to find G_1 and G_2. However, by working in a coordinate system in which the Helmholtz equation is separable we can greatly simplify this approach. The result will allow us to move the trace inside the logarithm, where it will become a simple integral, and we will only have to solve an ordinary differential equation to find a reduced Green's function for a single coordinate.

In this section we will find a simplified expression based on a general separation of variables using the Stäckel determinant. We will follow the notation of Morse and Feshbach.[11]

We write the Green's function as a sum of eigenfunctions times a reduced Green's function,

$$G(\vec{x}, \vec{x}') = \sum_{\alpha_2} \sum_{\alpha_3} \frac{\rho}{M_1 f_2 f_3} \chi_2(\xi_2)\chi_3(\xi_3)\chi_2(\xi_2')\chi_3(\xi_3')g(\xi_1, \xi_1'). \qquad (2)$$

The $M_1(\xi_2, \xi_3)$ is the minor of the Stäckel determinant, and the $f_i(\xi_i)$ functions are functions of a single variable related to the scale factors of the generalized coordinate system as defined in Morse and Feshbach.[11] The $\chi_2(\xi_2)$ and $\chi_3(\xi_3)$ and α_2 and α_3 are the eigenfunctions and eigenvalues determined by the simultaneous set of equations,

$$\left(-\frac{1}{f_2}\frac{\partial}{\partial \xi_2} f_2 \frac{\partial}{\partial \xi_2} + \Phi_{21}\zeta^2 + \Phi_{22}\alpha_2^2 + \Phi_{23}\alpha_3^2\right)\chi_2(\xi_2; \zeta, \alpha_2, \alpha_3) = 0, \quad (3a)$$

$$\left(-\frac{1}{f_3}\frac{\partial}{\partial \xi_3} f_3 \frac{\partial}{\partial \xi_3} + \Phi_{31}\zeta^2 + \Phi_{32}\alpha_2^2 + \Phi_{33}\alpha_3^2\right)\chi_3(\xi_3; \zeta, \alpha_2, \alpha_3) = 0. \quad (3b)$$

The χ eigenfunctions are orthogonal with respect to some weighting function $\rho(\xi_2, \xi_3)$,

$$\int d\xi_2 d\xi_3 \rho \chi_2(\alpha_2, \alpha_3)\chi_2(\alpha_2', \alpha_3')\chi_3(\alpha_2, \alpha_3)\chi_3(\alpha_2', \alpha_3') = \delta_{\alpha_2, \alpha_2'}\delta_{\alpha_3, \alpha_3'}. \quad (4)$$

Using (3), we find that the reduced Green's function in (2) satisfies the differential equation in the single remaining coordinate,

$$\Big(-\frac{1}{f_1}\frac{\partial}{\partial \xi_1} f_1 \frac{\partial}{\partial \xi_1} + \Phi_{11}\zeta^2 \\ + \Phi_{12}\alpha_2^2 + \Phi_{13}\alpha_3^2 + v(\xi_1) \Big) g(\xi_1, \xi_1'; \zeta, \alpha_2, \alpha_3) = \frac{\delta(\xi_1 - \xi_1')}{f_1}. \quad (5)$$

Working with the Casimir energy written as (1), by expanding the log we can write

$$E = -\frac{1}{4\pi}\int_{-\infty}^{\infty} \mathrm{d}\zeta \sum_{s=1}^{\infty} \frac{1}{s} \mathrm{Tr}(G_1 V_1 G_2 V_2)^s. \quad (6)$$

The simplification comes if the potentials are functions of only the single coordinate ξ_1, with the form $V_i(\vec{x}) = v_i(\xi_1)/h_1^2$. The scale factor h_1 is exactly what is needed to apply the orthogonally condition (4) in performing the trace. Finally if the potential consists of two separate non-overlapping potentials, we can show

$$\mathrm{Tr}(G_1 V_1 G_2 V_2)^s = \sum_{\alpha_2,\alpha_3} \mathrm{tr}(g_1 v_1 g_2 v_2)^s = \sum_{\alpha_2,\alpha_3} (\mathrm{tr}\, g_1 v_1 g_2 v_2)^s. \quad (7)$$

The interaction Casimir energy can now be written in general separable coordinates as

$$E = \frac{1}{4\pi}\int_{-\infty}^{\infty} \mathrm{d}\zeta \sum_{\alpha_2,\alpha_3} \ln(1 - \mathrm{tr}\, g_1 v_1 g_2 v_2). \quad (8)$$

3. Casimir Energy for Planes in an Annular Cavity

As an application we will proceed for the case of two semitransparent radial planes in the region between two concentric cylinders, as shown in figure 1.

This geometry is similar to the wedge geometry first studied in 1978,[12,13] with a good review by Razmi and Modarresi.[14] However here we include circular boundaries in addition to the wedge boundaries. We will enforce Dirichlet boundary condition on the inner and outer cylinder. This is similar to situations studied by Nesterenko *et al*[15,16] for global Casimir energies for the case of one circular boundary and by Saharian *et al*[17,18] for the local properties of the stress energy tensor for the case of both one and two circular boundaries. The radial potentials will be semi-transparent delta-function potentials in the angular coordinates, $v_1(\theta) = \lambda_1\delta(\theta)$ and $v_2(\theta) =$

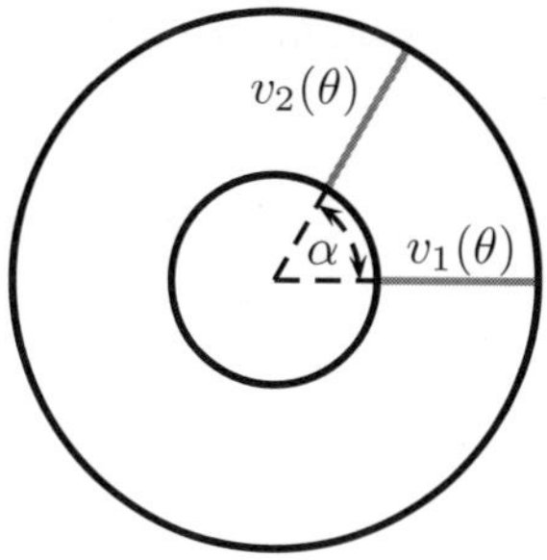

Fig. 1. An annulus with inner radius a, outer radius b, and two semitransparent potentials at $\theta = 0$ and $\theta = \alpha$.

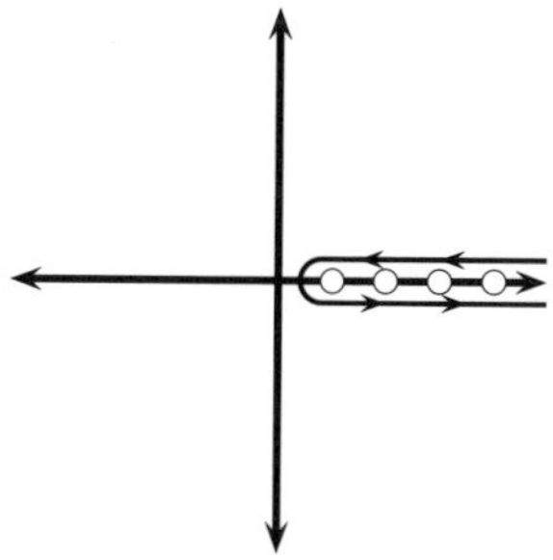

Fig. 2. The contour γ is defined around the positive real line, while not enclosing zero.

$\lambda_2\delta(\theta - \alpha)$. This is most similar to the recent work by Brevik *et al*,[19,20] and Milton *et al.*[21]

This problem can be solved using separation of variables, leaving ξ_1 as the azimuthal coordinate θ. This means we will write our reduced Green's function in the azimuthal coordinate, which is different from the traditional way of writing the reduced Green's function in terms of the radial coordinate. From equation (8) we can immediately write

$$\frac{E}{L_z} = \frac{1}{4\pi}\int_0^\infty d\zeta \sum_\eta \ln(1 - \operatorname{tr} g_\eta^{(1)} v_1 g_\eta^{(2)} v_2). \tag{9}$$

The Green's function is written in terms of exponential functions that, due to the periodicity requirement, give the expression

$$\operatorname{tr} g_\eta^{(1)} v_1 g_\eta^{(2)} v_2 = \frac{\lambda_1\lambda_2 \cosh^2\left(\eta(\pi - \alpha)\right)}{(2\eta \sinh \eta\pi + \lambda_1 \cosh \eta\pi)(2\eta \sinh \eta\pi + \lambda_2 \cosh \eta\pi)}. \tag{10}$$

The ηs are the eigenvalues of the modified Bessel equation of purely imaginary order,

$$\left[-r\frac{\partial}{\partial r}r\frac{\partial}{\partial r} + \kappa^2 r^2\right] R_\eta(\kappa r) = \eta^2 R_\eta(\kappa r). \tag{11}$$

Using the argument principle we can take a complicated sum over eigenvalues and turn it into a contour integral around the real line as shown in figure 2. For this we need a secular function $D(\eta)$, which is analytic along the real line and has the value zero at the eigenvalues. In this case we define $R_\eta(\kappa a) = 0$ then the eigenvalue condition is given by $D(\eta) = R_\eta(\kappa b)$. The eigenfunction R_η can be written in terms of modified Bessel functions

$$R_\eta(\kappa r) = K_{i\eta}(\kappa a)\widetilde{I}_{i\eta}(\kappa r) - \widetilde{I}_{i\eta}(\kappa a)K_{i\eta}(\kappa r), \tag{12}$$

where we define $\widetilde{I}_\eta(x)$ as the part of the modified Bessel function $I_\eta(x)$ even in η.

The energy per unit length L_z can be written as

$$\frac{E}{L_z} = \frac{1}{8\pi^2 i}\int_0^\infty \kappa d\kappa \int_\gamma d\eta \left[\frac{\partial}{\partial\eta}\ln\left(K_{i\eta}(\kappa a)\widetilde{I}_{i\eta}(\kappa b) - \widetilde{I}_{i\eta}(\kappa a)K_{i\eta}(\kappa b)\right)\right]$$
$$\times \ln\left(1 - \frac{\lambda_1\lambda_2\cosh^2\left(\eta(\pi-\alpha)\right)}{(2\eta\sinh\eta\pi + \lambda_1\cosh\eta\pi)(2\eta\sinh\eta\pi + \lambda_2\cosh\eta\pi)}\right). \quad (13)$$

A quick check of this answer is to look at the limit of large inner and outer radius, as shown in figure 3. This should then give the answer for a rectangular piston. For this limit we need the uniform asymptotic expansions of

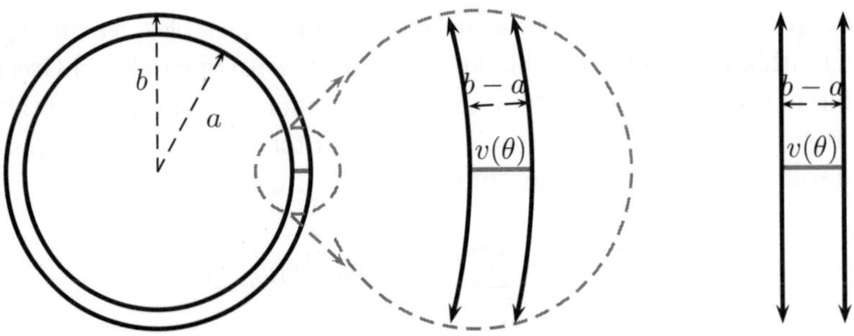

Fig. 3. If the inner and outer radii are both large in comparison to their separation, we should recover the case of a rectangular piston.

$K_{i\eta}$ and $\widetilde{I}_{i\eta}$, which are worked out by Dunster.[22,23] We should also redefine our dimensionless variables in terms of the dimensionful quantities that will appear in the rectangular piston case, $\widetilde{\eta} = \eta/a$, $\widetilde{\lambda} = \lambda/a$, and $d = \alpha a$. In this asymptotic region we recover the formula for a rectangular piston,

$$\frac{E}{L_z} = \frac{1}{8\pi^2 i}\int_0^\infty \kappa d\kappa \int_\gamma d\widetilde{\eta}\left[\frac{\partial}{\partial\widetilde{\eta}}\ln\frac{\sin\left(\sqrt{\widetilde{\eta}^2-\kappa^2}(b-a)\right)}{\sqrt{\widetilde{\eta}^2-\kappa^2}}\right]$$
$$\times \ln\left(1 - \frac{\widetilde{\lambda}_1\widetilde{\lambda}_2 e^{-2\widetilde{\eta}d}}{(2\widetilde{\eta}+\widetilde{\lambda}_1)(2\widetilde{\eta}+\widetilde{\lambda}_2)}\right). \quad (14)$$

The contour integral over $\widetilde{\eta}$ simply ensures that $\eta^2 = \kappa^2 + (m\pi/(b-a))^2$.

4. Numerical Results for Dirichlet Planes

The Casimir energy in equation (9) is a quickly converging function so it should be easy to evaluate. However it can be difficult to evaluate the η eigenvalues, which become functions of the wavenumber κ and a natural number m. We can get around this problem by using (13). We cannot integrate along the real line because of the poles introduced when we used the argument principle, and we cannot distort the contour to one running along the imaginary axis because the integral then becomes divergent. So a simple choice is then to let the η integration run along the angles of $\pi/4$ and $-\pi/4$. Writing $\operatorname{tr} g_\eta^{(1)} v_1 g_\eta^{(2)} v_2 = A(\eta)$ we have

$$\frac{E}{L_z} = -\frac{1}{4\pi^2}\int_0^\infty \kappa d\kappa \int_0^\infty \mathrm{d}\nu \times \left\{ \frac{\operatorname{Re} R_{\sqrt{i\nu}} \partial_\nu \operatorname{Re} R_{\sqrt{i\nu}} + \operatorname{Im} R_{\sqrt{i\nu}} \partial_\nu \operatorname{Im} R_{\sqrt{i\nu}}}{\left|R_{\sqrt{i\nu}}\right|^2} \arctan\left(\frac{\operatorname{Im} A(\sqrt{i\nu})}{1-\operatorname{Re} A(\sqrt{i\nu})}\right) - \frac{\operatorname{Re} R_{\sqrt{i\nu}} \partial_\nu \operatorname{Im} R_{\sqrt{i\nu}} - \operatorname{Im} R_{\sqrt{i\nu}} \partial_\nu \operatorname{Re} R_{\sqrt{i\nu}}}{2\left|R_{\sqrt{i\nu}}\right|^2} \times \ln\left(1 - 2\operatorname{Re} A(\sqrt{i\nu}) + \left|A(\sqrt{i\nu})\right|^2\right) \right\}. \tag{15}$$

Here we have used the property that $R_{\eta^*} = R_\eta^*$, and $A(\eta^*) = A^*(\eta)$. The value of $R_{\sqrt{i\nu}}(b, \kappa)$ is obtained as the numerical solution of the differential equation. Using this technique we can obtain a numerical energy in about 1 cpu-second. The results of this calculation are found in figure 4.

Again we would like to compare to known results, so figure 5 is a graph of the ratio of the energies of an annular piston, and a rectangular piston of similar dimension. The rectangular piston is constructed so it has the same finite width $b-a$ as the annular piston, and the separation distance is the mean distance between the annular plates,

$$d = \frac{b+a}{2} 2\sin\left(\frac{\alpha}{2}\right). \tag{16}$$

The results make a certain amount of physical sense. The energy of the annular piston is greater than that of the rectangular piston for small separation because the inner edge of the annular piston is closer, and will contribute more to the energy. However as the annular piston gets further away, the other side of the piston will start to contribute and lower the overall energy. In addition we see that the energy for a small piston is much closer to that of the rectangular piston for small separations than for a larger piston, $E_{\text{ann}}/E_{\text{rect}} \approx 1.004$ for $b/a = 1.1$ vs. $E_{\text{ann}}/E_{\text{rect}} \approx 1.23$ for

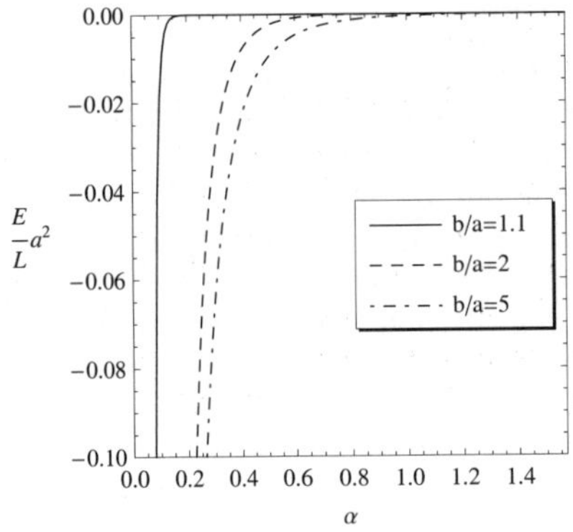

Fig. 4. This figure shows the energy per length vs the angle between the plates. The energy is scaled by the inner radius a.

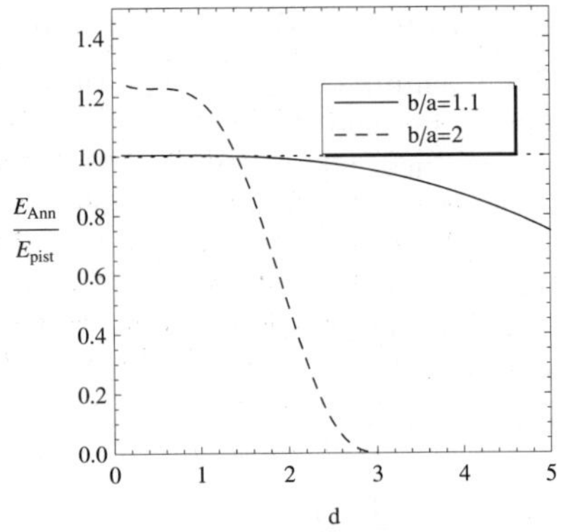

Fig. 5. This figure shows the ratio of the energies of an annular piston to a rectangular piston of similar dimension vs average separation distance between the plates. The separation distance is scaled by the finite size of the piston $b - a$. For $b/a = 2$ only the result for $\alpha \in [0, \pi]$ is shown.

$b/a = 2$. In both cases the value approached in the plateau in figure 5 is very close to the ratio of the energies of a flat plate to that of a tilted plate predicted by using the proximity force approximation.

Acknowledgments

This material is based upon work supported by the National Science Foundation under Grants Nos. PHY-0554926 (OU) and PHY-0757791 (BU) and by the US Department of Energy under Grants Nos. DE-FG02-04ER41305 and DE-FG02-04ER-46140 (both OU). We thank Simen Ellingsen, Iver Brevik, Prachi Parashar, Nima Pourtolami, and Elom Abalo for collaboration. Part of the work was done while KK enjoyed the hospitality and partial support of the Department of Physics and Astronomy of the University of Oklahoma. Thanks go in particular to Kimball Milton and his group who made this very pleasant and exciting visit possible.

References

1. H. B. G. Casimir, *Proc. K. Ned. Akad. Wet.* **60**, 793 (1948).
2. M. Bordag, U. Mohideen and V. M. Mostepanenko, *Phys. Rept.* **353**, 1 (2001).
3. K. A. Milton, *J. Phys.* **A37**, R209 (2004).

4. K. A. Milton, *The Casimir effect: Physical manifestations of zero-point energy* (World Scientific, River Edge, USA, 2001).
5. M. Bordag, G. L. Klimchitskaya, U. Mohideen and V. M. Mostepanenko, *Advances in the Casimir Effect* (Oxford University Press, New York, 2009).
6. M. J. Renne, *Physica* **56**, 125 (1971).
7. T. Emig, *Europhys. Lett.* **62**, 466 (2003).
8. A. Bulgac, P. Magierski and A. Wirzba, *Phys. Rev.* **D73**, 025007 (2006).
9. T. Emig, R. L. Jaffe, M. Kardar and A. Scardicchio, *Phys. Rev. Lett.* **96**, 080403 (2006).
10. O. Kenneth and I. Klich, *Phys. Rev. B* **78**, 014103 (2008).
11. P. M. Morse and H. Feshbach, *Methods of Theoretical Physics: Part I* (McGraw-Hill, 1953).
12. J. S. Dowker and G. Kennedy, *J. Phys.* **A11**, 895 (1978).
13. D. Deutsch and P. Candelas, *Phys. Rev.* **D20**, 3063 (1979).
14. H. Razmi and S. M. Modarresi, *Int. J. Theor. Phys.* **44**, 229 (2005).
15. V. V. Nesterenko, G. Lambiase and G. Scarpetta, *Annals Phys.* **298**, 403 (2002).
16. V. V. Nesterenko, I. G. Pirozhenko and J. Dittrich, *Class. Quant. Grav.* **20**, 431 (2003).
17. A. A. Saharian and A. S. Tarloyan, *J. Phys.* **A38**, 8763 (2005).
18. A. A. Saharian and A. S. Tarloyan, *Annals Phys.* **323**, 1588 (2008).
19. I. Brevik, S. A. Ellingsen and K. A. Milton, *Phys. Rev. E* **79**, 041120 (2009).
20. S. A. Ellingsen, I. Brevik and K. A. Milton, *Phys. Rev. E* **80**, 021125 (2009).
21. K. A. Milton, J. Wagner and K. Kirsten, *Phys. Rev. D* in press, `arXiv:0911.2688`.
22. T. M. Dunster, *SIAM J. Math. Anal.* **21**, 995 (1990).
23. F. W. J. Olver, *Asymptotics and Special Functions* (Academic Press, New York, 1974).

DEMONSTRATING THE STRONG GEOMETRY DEPENDENCE OF THE CASIMIR FORCE ON A SURFACE WITH DEEP, NANOSCALE CORRUGATIONS

H. B. CHAN[1,2*], Y. BAO[1], J. ZOU[1], R. A. CIRELLI[3], F. KLEMENS[3], W. M. MANSFIELD[3] and C. S. PAI[3]

[1]*Department of Physics, University of Florida, Gainesville, FL 32611, USA*

[2]*Department of Physics, the Hong Kong University of Science and Technology, Clear Water Bay, Kowloon, Hong Kong, China*

[3]*Bell Laboratories, Lucent Technologies, Murray Hill, New Jersey 07974, USA*

[*]*E-mail: hochan@phys.ufl.edu*

We measure the Casimir force gradient between silicon surfaces with nanoscale, rectangular corrugations and a gold sphere attached to a micromechanical torsional oscillator. By comparing the force gradients on the corrugated surfaces to that on a smooth, flat surface of the same material, we demonstrate that the Casimir force deviates from the value expected from the pairwise additive approximation and the proximity force approximation. The observed deviation qualitatively agrees with calculations that take into account the interplay between finite conductivity and geometry effects. However, the agreement is not exact, possibly due to uncertainties in the optical properties of the silicon substrate.

Keywords: Casimir force; quantum fluctuations.

1. Introduction

Improvements in force detection and advances in material characterization have made it possible to measure the Casimir force in both conventional and novel arrangements with unprecedented sensitivity.[1–10] Nevertheless, the Casimir force continues to offer challenges in the theoretical and experimental investigation of the fluctuations of quantum fields. For instance, ongoing and future experiments[11–13] are designed to address the controversy on how to account for the temperature corrections of the Casimir force. There has also been much progress in controlling the Casimir force through the optical properties of the material.[10,14,15] In addition, significant efforts are directed towards the goal of generating repulsive Casimir forces in vacuum.[16–18]

One of the most remarkable properties of the Casimir force is its strong dependence on the shape of the interacting bodies. For the common configuration of two perfect metallic planar surfaces, quantum zero point fluctuations give rise to an attractive force that increases rapidly with decreasing distance. However, for closed geometries such as a thin, perfectly conducting spherical shell[19] or a rectangular box with certain aspect ratios,[20] the Casimir energy has been calculated to have opposite sign to parallel plates. Whether such geometries exhibit repulsive Casimir forces remains a topic of current interest. Recently, a number of theoretical approaches have been developed to calculate the Casimir force in various geometries.[21–25] To estimate the Casimir force for relatively smooth objects that only deviates weakly from the planar configuration, the proximity force approximation (PFA) and the pairwise additive approximation (PAA) are commonly used. Typically, these schemes are applicable when the radii of curvature of the surfaces are large compared to their separation. For example, the $1/z^3$ dependence of the Casimir force for the sphere-plate arrangement used in most experiments can be obtained with the PFA or the PAA when the radius of the sphere R is much larger than the separation z.

So far, most experiments that measure the Casimir force involve smooth surfaces where the PAA and the PFA are valid. Demonstration of the strong geometry dependence of the Casimir force requires introduction of artificial deformations to the surfaces. For example, Mohideen et al.[26,27] demonstrated that the lateral Casimir force between sinusoidal corrugated surfaces deviates from the PAA and the PFA. Similar surfaces with small deformations were also used in earlier measurements of the Casimir force normal to the surfaces.[28] Here, we describe experimental details of our recent measurement of the Casimir force between a gold sphere and a nanostructured silicon surface.[29] Nanoscale, high aspect ratio rectangular corrugations were fabricated on a silicon wafer, aiming to generate Casimir forces that deviate strongly from predictions of the PAA and the PFA. A micromechanical torsional oscillator is used to measure the Casimir force gradient. Deviations of up to 20% from the PAA and the PFA are observed, demonstrating the strong geometry dependence of the Casimir force. The measured deviation is, however, about 30% smaller than deviations expected for perfectly conducting surfaces.[30] When compared to recent calculations that include both geometry and finite conductivity effects,[25] the experimental results are found to be $\sim 20\%$ smaller. The discrepancy likely results from uncertainties in the optical properties of the surfaces, in particular the silicon wafer, used in the experiments.

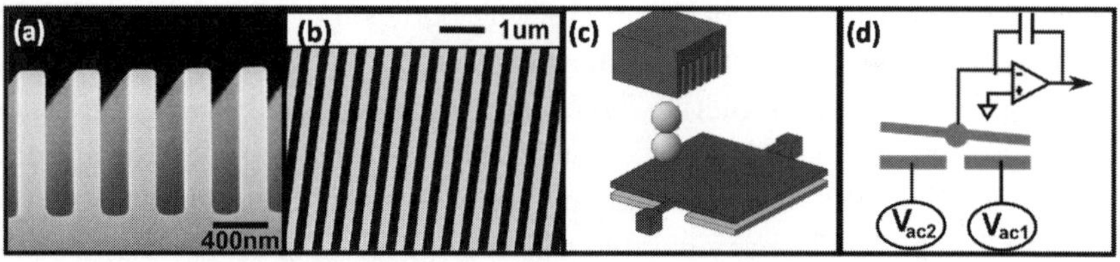

Fig. 1. Scanning electron micrograph of (a) the cross sectional view and (b) the top view of the rectangular trench arrays (sample B). (c) Schematic of the experimental setup (not to scale). (d) Measurement scheme with electrical connections.

2. Rectangular Trench Arrays: Comparison of the Casimir Force to the PAA and the PFA

The most important element of this experiment is the deep rectangular trench arrays. Figure 1(a) shows a scanning electron micrograph cross section of a trench array with period of 400 nm (sample B). The trenches were fabricated by dry etching into a strongly p-doped silicon substrate. Two other samples, one with a period of 1 μm (sample A) and the other with a flat surface, are also fabricated using the same wafer. Fabrication of the corrugated structures started with depositing a 0.2 μm layer of silicon oxide on to a blank silicon wafer by chemical vapor deposition. Photoresist was then spun on the wafer, followed by deep ultra-violet lithography that exposed the resist with a resolution better than 200 nm. The resist pattern was then transferred into the silicon oxide by dry etching with a combination of C_2F_6 and O_2 gas. Before using the patterned silicon oxide as an etch mask in deep reactive ion etching, the photoresist is removed. A continuous etch and deposition recipe involving C_2F_6 and SF_6 was used in a deep reactive ion etcher to create the smooth and near vertical sidewalls with no scalloping. The depth of the trenches $t = 2a$ are about 1 μm in both samples A and B, limited mainly by consumption of the oxide mask in the etching process. Residual hydrocarbons, including the passivation layers on the sidewalls, were removed with an oxygen plasma etch. Finally, the remaining oxide etch mask is removed using hydrofluoric acid. In order to ensure that the optical properties of the silicon are identical, all samples were fabricated on the same silicon wafer and later diced into 0.7 by 0.7 mm pieces for the force measurement.

Our goal is to demonstrate the strong geometry dependence of the Casimir force. The nanoscale, deep rectangular trenches were chosen because the Casimir force on such structures is expected to exhibit large deviations from the PFA and the PAA.[30] We consider the interaction between the trench array with solid fraction $p = 1/2$ and a parallel flat surface at distance z from the top surface of the trenches. Assuming that the PFA is valid, half of the flat surface interacts with the top of the trench array at distance z and the other half interacts with the bottom of the trench array at distance $z + 2a$. Since these two interactions are independent of each other under the PFA, the total force is given by $F_{PFA} = \frac{1}{2}F_{flat}(z) + \frac{1}{2}F_{flat}(z+2a) \sim \frac{1}{2}F_{flat}(z)$. The term $\frac{1}{2}F_{flat}(z+2a)$ is negligible in our structure because $z + 2a > 1\ \mu$m, a distance at which the Casimir force can no longer be resolved with our set up. In other words, under PFA the force on the corrugated structure is exactly half the force on a flat surface made of the same material.

Another approximation commonly used to estimate the Casimir force for non-planar geometries is the PAA. Under the PAA, the force on a rectangular trench array can be obtained with the following procedure. Suppose that the Casimir force is measured in two separate experiments. In both experiments, a flat surface is placed at the top and a rectangular trench array is placed at the bottom. All parameters and material properties in the two experiments are identical except that array II in the second experiment is laterally shifted from array I in the first experiment by half the period. The forces F_I on array I and F_{II} on array II are the same as the lateral shift does not change the magnitude of the force. If array I and array II are superimposed on each other, we recover a solid flat surface. Assuming that the PAA is valid, $F_{flat}(z) = F_I(z) + F_{II}(z) + F_{bottom} = 2F_{PAA}$. We again make the assumption that the trench depth $2a$ is much larger than z such that the force F_{bottom} on the trench bottom can be neglected. Therefore, under the PAA, the force on the corrugated surface is half the force on the flat surface made of the same material. The PFA and the PAA involve different assumptions and in general yield different results.[27] However, for the geometry of deep trenches in this experiment, the PAA and the PFA predicts the same force of $\frac{1}{2}F_{flat}(z)$ regardless of the periodicity lambda and material of the trench arrays.

While the PFA and the PAA are adequate for corrugation periodicity $\lambda >> z$, they no longer give the correct Casimir force when λ/z goes to zero. The Casimir force for rectangular corrugations made of perfect metal was calculated by Büscher and Emig using a path integrals approach.[30]

They predicted strong deviations from the PAA and the PFA when λ/z is small. In the limit when λ/z goes to zero, the Casimir force on the trench array approaches the value between flat surfaces, leading to deviations from the PAA and the PFA by a factor of 2. Such large deviations occur because the Casimir force is associated with quantum fluctuations of electromagnetic modes with characteristic wavelength comparable to the separation z. When $\lambda << z$, these modes fail to penetrate into the trenches. The force on the corrugations is therefore identical to the force on a solid flat surface, as if the corrugations are entirely absent. Recently, Lambrecht and Marachevsky[25] used scattering theory to calculate the Casimir force on corrugated surfaces without assuming perfect conductivity. It is found that interplay of geometry effects and imperfect conductivity reduces the deviations of the Casimir force from the PAA and the PFA values compared to perfect metals.

3. Experimental Procedures

We measure the gradient of the Casimir force on the silicon trench arrays using gold-coated spheres attached to a micromechanical torsional oscillator.[4,31] As shown in Fig.1(c), the oscillator consists of a 3.5μm thick, 500μm square polycrystalline silicon plate suspended by two torsional rods, each with dimensions 20 μm by 3 μm by 2 μm. The plate rotates about the axis defined by the torsional rods in response to external torques. A layer of gold with thickness 50 nm is deposited onto the top surface of the silicon plate. Two glass spheres, each with radius R of 50 μm, are stacked and attached by conductive epoxy onto the oscillator at a distance of b = 210 μm from the rotation axis. Before attachment, a layer of gold with thickness 4000 A is sputtered onto the spheres.

For excitation and detection of rotations of the top plate, two fixed electrodes are fabricated on the substrate, one on each side of the torsional rods. Torsional oscillations of the top plate is excited by a small ac voltage $V_d = V_o cos(\omega_d t)$ on top of a much larger dc voltage V_{dc1} applied to the left electrode, where the driving frequency $\omega_d/2\pi$ is close to the resonant frequency of the oscillator (f_o = 1783 Hz, Q = 32000). Rotation of the top plate is detected capacitively through induced modulations on a carrier voltage signal. The top plate is connected to the virtual ground of a charge sensitive amplifier. Two ac voltages, V_{ac1} and V_{ac2}, with the same frequency (f_c = 100 kHz $>> f_o$) and amplitude (100 mV) but opposite phase are applied to the two electrodes respectively. Rotation of the top plate leads to changes in the capacitances between the top plate and the two fixed electrodes. As a result, the amplitude of the ac current from the top plate is modulated by the plate motion. The current is measured with a lockin amplifier referenced to f_c, yielding an output that is proportional to the rotation angle $\theta(t)$.

A phase-locked loop is used to maintain oscillations of the top plate at the resonance frequency as the sphere approaches the other silicon plate through extension of a closed-loop piezoelectric actuator. As shown in Fig. 1(c), the orientation of the trench array is chosen to be perpendicular to the torsional axis of the oscillator. Such an arrangement eliminates motion of the movable plate in response to lateral Casimir forces[32] because the spring constant for translation along the torsional axis is orders of magnitude larger than the orthogonal direction in the plane of the substrate.

Since the native oxide on the silicon surfaces can trap charges, careful preparation of the surfaces must be performed prior to measurement.[33] The silicon samples are placed in 100:1 hydrofluoric acid to remove the native oxide layer. This step also passivates the silicon surface to avoid re-oxidation for a few hours. To eliminate the possibility of water accumulation in the deep trenches, the silicon samples are baked at 120 °C for 15 minutes. The silicon sample is then positioned to within a few micrometers from the gold sphere and placed in a vacuum chamber that is immediately evacuated by a dry roughing pump and turbo pump to a pressure of 10^{-6} torr.

As long as the oscillation amplitudes are small so that nonlinear effects are negligible, the gradient of the force $F'(z)$ between the surfaces depends linearly on the shifts in the resonant frequency of the oscillator[31]

$$\Delta f = CF'(z), \tag{1}$$

where $C = -b^2/8\pi^2 f_0 I$ and I is the moment of inertia of the oscillator. The distance z is given by $z_0 - z_{piezo} - b\theta$, where z_0 is the initial separation between the two surfaces, z_{piezo} is the extension of the piezoelectric actuator and $b\theta$ is the modification of the separation due to rotation of the top plate to angle θ. Calibration of z_0 and C is performed by applying a dc voltage V to the silicon plate while the gold sphere is electrically grounded. The electrostatic force between a sphere and a flat plate is given by:

$$F_e = 2\pi\epsilon_0 (V - V_0)^2 \sum_{n=1}^{\infty} \frac{[\coth(\alpha) - n\coth(n\alpha)]}{\sinh(n\alpha)}, \tag{2}$$

where ϵ_0 is the permittivity of vacuum and $\cosh\alpha = 1 + d/R$. The residual voltage V_0 arises due to the difference in work function of the silicon and gold surfaces. V_0 is measured to be $\sim -0.43V$ by identifying V at which Δf

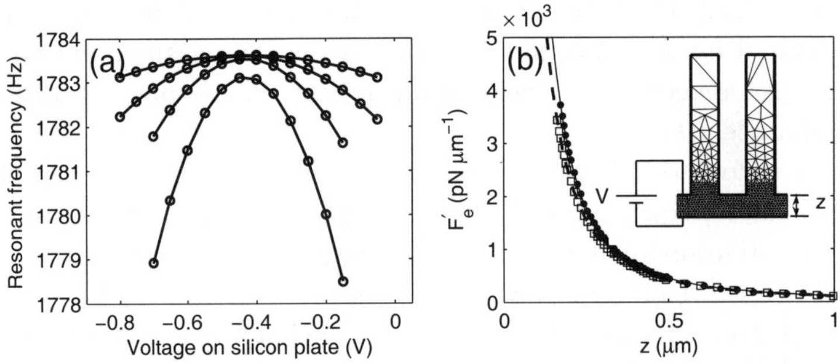

Fig. 2. (a) Measured resonant frequency as a function of voltage V on the corrugated surface, at $z = 525$ nm, 325 nm, 225 nm and 165 nm from top to bottom. The lines are parabolic fits. (b) Measured gradient of the electrostatic force at $V = V_0 + 0.283\,V$ on the flat silicon surface (solid circles) and corrugated sample B (hollow squares). The solid line is a fit using Eq. 2 for a flat surface. The dashed line is a fit using the force gradient for the corrugated surface, obtained from dividing the space between the corrugated structure and a flat surface into triangular mesh (inset) to solve the Poisson equation in 2D ($z = 150$ nm).

attains minimum. As shown in Fig. 2(a), the measured value of V_0 varies by less than 3 mV, when z is increased from 165 nm to 525 nm.

In Fig. 2(b), the solid circles represent the measured electrostatic force gradient, with the proportionality constant C and z_o determined from fitting to using Eq. 2 (solid line). Before fitting the data, we subtract the contribution of the measured Casimir force gradient to the frequency shift. The modification is less than 4% at the smallest z. Six sets of measurement with V ranging from $V_0 + 245$ mV to $V_0 + 300$ mV are taken, yielding an average fitted value of C of $628 \pm 5\,\mathrm{m}\,N^{-1}\,s^{-1}$. Only voltages larger than V_0 are used to avoid depleting the surface of the p-doped silicon with charge carriers. Uncertainties in the distance z is found to be ~ 0.2 nm from fitting to the electrostatic force at different voltages.

A similar calibration procedure is performed on the corrugated silicon surface (dashed line in Fig. 2(b)). Unlike the flat surface, there is no analytical expression for the electrostatic force on a trench array. Instead, the electrostatic force gradient is obtained by solving the 2D Poisson's equation using finite element analysis. As shown in the inset to Fig. 2(b), the boundary conditions are established by maintaining a fixed voltage between the trench array and a flat surface. The volume between them is divided into $N > 10,000$ triangles. Since $R >> z$, the proximity force approximation $F_{sc} = 2\pi R E_{fc}$ is used to relate the force F_{sc} between a sphere and a corrugated surface to the energy E_{fc} between a flat surface and a corrugated surface. To ensure that N is sufficient large, we checked that the calculated force changes by $< 0.1\%$ even when N is doubled.

Obtaining accurate sample dimensions is crucial to the calculation of the electrostatic and Casimir forces on the trench arrays. The fraction of solid volume p is obtained from the top views of the sample in a scanning electron microscope. Ten pictures at different locations with area 30 μm by 50 μm for sample A and 20 by 25 μm for sample B were taken. Figure 1(b) shows a typical picture of sample B. By comparing the area under the two peaks in the histogram of the pixel intensity, p_A and p_B are determined to be 0.478 ± 0.002 and 0.510 ± 0.001 respectively. The depth of the trench is obtained from a cross-sectional view similar to Fig. 1(a) to be $t_A = 0.98\ \mu$m and $t_B = 1.07\ \mu$m. While the top of the trenches have a sharp rectangular shape and the sidewalls are near vertical, the bottom sections show certain degree of rounding. In all calculations reported here, the trenches are assumed to have perfect rectangular shape. The validity of such approximation is justified by the insensitivity of the calculated electrostatic force to the depth of the trenches t. Varying t by 10% produces less than 0.01% change in the calculated force.

4. Measurement of the Casimir Force on Flat and Corrugated Silicon Surfaces

The Casimir force gradient $F'_{C,flat}$ between the gold sphere and the flat silicon surface is measured by setting V equal to V_0 and is plotted as solid circles in Fig. 3. Uncertainties in the measurement ($\sim$0.64 pN μm^{-1} at $z = 300$ nm) arise mainly from thermomechanical fluctuations of the oscillator. As the distance decrease, it is necessary to reduce the oscillation amplitude to avoid nonlinear effects[31] induced by the Casimir force. Measurements were not performed for distances smaller than 150 nm because the oscillation amplitude becomes too small for the phase lock loop to operate reliably. In Fig. 3, the solid line represents the theoretical force gradient between the gold sphere and the flat silicon surface, including both the finite conductivity and roughness corrections. Lifshitz's expressions[33–35] yield the force based on the frequency dependent dielectric properties of the materials. The optical properties of gold and silicon are obtained from tabulated values.[36] For the silicon surface, the dielectric permittivity $\epsilon_{Si}(i\xi)$ derived from tabulated values is further modified by the imaginary part of the Drude dielectric function:[33]

$$\tilde{\epsilon}_{Si}(i\xi) = \epsilon_{Si}(i\xi) + \omega_p^2/[\xi(\xi+\gamma)] \tag{3}$$

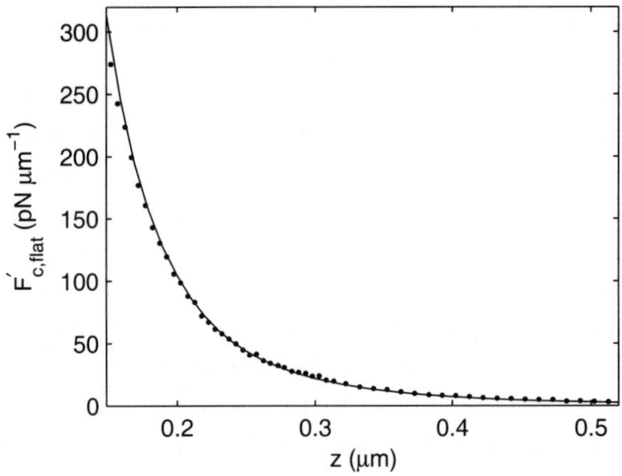

Fig. 3. Measured Casimir force gradient between a gold sphere and a flat silicon surface (circles) and the theoretical values including finite conductivity and surface roughness corrections (line).

where $\omega_p = e\sqrt{n/\epsilon_0 m^*} = 1.8\times10^{14}$rad s^{-1} and $\gamma = \epsilon_0\omega_p^2/\sigma = 7.6\times10^{13}$rad s^{-1}. The concentration of carriers $n = 2\times10^{18}\,\text{cm}^{-3}$ is obtained from measuring the dc conductivity of the wafer ($\sigma = 357.1\Omega^{-1}\text{cm}^{-1}$). m^* represents the effective mass of $0.206\,m_e$ in silicon. An atomic force microscope was used to measure the roughness of the surfaces. The contribution to the roughness correction is dominated by the gold surface (~ 4 nm rms) rather than the silicon surface (~ 0.6 nm rms). As outlined in Ref. [33], the force calculated using Lifshitz expressions are further modified by the roughness correction[35,37] using the geometrical averaging method.[33]

In the next step, the Casimir force gradients $F'_{C,a}$ and $F'_{C,b}$ between the same gold sphere and the corrugated samples A and B were measured using procedures similar to the flat sample to yield Figs. 4(a) and (b). As we described earlier, under the PAA and the PFA the forces on the trench arrays (with z measured from the top of the corrugated surface) are equal to the force on a flat surface multiplied by the fractional solid volumes p_A and p_B, regardless of the period λ of the structure. The solid lines in Figs. 4(a) and 4(b) represent the corresponding force gradients $p_A F'_{C,flat}$ and $p_B F'_{C,flat}$. For each sample, measurement of the Casimir force gradient was repeated in three separate runs, the results of which are consistent within the measurement uncertainty. The deviations of the data points from the solid line in Figs. 4(a) and 4(b) provides experimental evidence for the strong geometry dependence of the Casimir force.

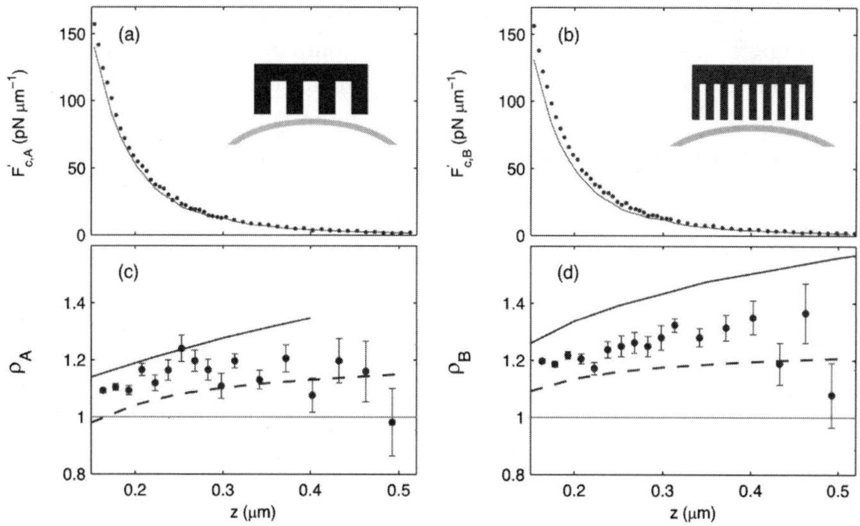

Fig. 4. Measured Casimir force gradient between the same gold sphere and (a) sample A, $F'_{C,a}$ ($\lambda = 1\ \mu$m) and (b) sample B, $F'_{C,b}$ ($\lambda = 400$ nm). The purple lines represent the force gradient expected from the PAA/PFA ($pF'_{C,flat}$). (c) Ratio ρ of the measured Casimir force gradient to the force gradient expected from the PAA/PFA for sample A. The thin purple line is the value for the PFA/PFA. The thick solid line is the theoretical values of ρ for trenches made of perfect conductors.[30] The dashed line is the theoretical values including both geometry effects and finite conductivity.[25] (d) Similar plot for sample B.

To compare the deviations of the measured Casimir force gradient from the PFA and the PAA, we plot $\rho_A = F'_{C,A}/(p_A F'_{C,flat})$ and $\rho_B = F'_{C,B}/(p_B F'_{C,flat})$ in Fig. 4(c) and 4(d) respectively. If the PFA or the PAA is valid, ρ is equal to one, as indicated by the purple thin solid lines. For sample A with $\lambda/a = 1.87$ and sample B with $\lambda/a = 0.82$, where a is half the depth of the trenches, the measured force gradient deviates from the PAA and the PFA by $\sim 10\%$ and $\sim 20\%$ respectively. In the distance range of $150\text{ nm} < z < 250\text{ nm}$, both ρ_A and ρ_B show clear deviations from the PAA and the PFA. At larger distances, the uncertainty increases considerably as the force gradient rapidly decreases.

We compare the measured Casimir force gradient to theoretical results from two different calculations.[25,30] In both calculations, the force between a corrugated surface and a flat surface were obtained. Since $R >> z$, the proximity force approximation allows a direct comparison of the theoretical force on a flat surface and our measured force gradient on a sphere.

The solid lines in Figs. 4(c) and 4(d) correspond to calculations by Büscher and Emig[30] on perfect conductors with λ/a comparable to the experimental geometry. The dashed lines represent calculations by Lambrecht and Marachevsky[25] between a flat gold surface and a corrugated undoped silicon surface with geometry identical to our experiment. Before a quantitative comparison, we first focus on two qualitative trends in the experimental data that can be identified to be consistent with the theoretical results. First, the deviations from the PAA and the PFA in sample A is larger than in sample B. Second, the deviations from the PAA and the PFA show a gradual increase with distance in both samples. Both observations are consistent with the notion that the deviations from the PAA or the PFA increases when λ_f/λ increases, where λ_f is the characteristic wavelength of the fluctuating electromagnetic field in the gap. In other words, it is more difficult for the fluctuating electromagnetic field to penetrate into the trenches for large λ_f/λ.

While the qualitative trends in our data agrees with the theoretical predictions, the exact value of ρ falls between the solid and dashed lines. The discrepancy of measurement from prediction based on perfect metal is about 5% in sample B and 10% in sample A. Such deviation is expected due to the interplay between geometry effects and finite conductivity. Intuitively, the relatively large value of the skin depth in silicon ($\sim$ 11 nm at wavelength of 300 nm) could reduce the deviations from the PAA and the PFA. The measured ρ, however, exceeds the predicted value between gold and undoped silicon by $\sim$ 10% in both samples. Such imperfect agreement requires more investigation. While the shape of the corrugations is directly measured, the optical properties are only taken from tabulated values.[36] As shown in Fig. 3, there exists a $\sim$ 5% discrepancy between the measured Casimir force on a flat surface without corrugations and the values calculated from the Lifshitz expression. Including the effect of dopants produces only a slight change in the predicted force that is negligible compared to the discrepancy. Future efforts will aim at determining if the imperfect agreement between measurement and theory can be accounted for by the optical properties of our samples.

Acknowledgments

This work was supported by DOE No. DE-FG02-05ER46247 and NSF No. DMR-0645448. We thank T. Emig, A. Hanke, A. Lambrecht and V. N. Marachevsky for useful discussions.

References

1. S. Lamoreaux, *Phys. Rev. Lett.* **78**, 5 (1997).
2. U. Mohideen and A. Roy, *Phys. Rev. Lett.* **81**, 4549 (1998).
3. T. Ederth, *Phys. Rev. A* **62**, 062104 (2000).
4. H. B. Chan *et al.*, *Science* **291**, 1941 (2001).
5. G. Bressi *et al.*, *Phys. Rev. Lett.* **88**, 041804 (2002).
6. R. S. Decca *et al.*, *Phys. Rev. Lett.* **91**, 050402 (2003).
7. R. S. Decca *et al.*, *Phys. Rev. Lett.* **94**, 240401 (2005).
8. W. J. Kim *et al.*, *Phys. Rev. A* **78**, 020101 (2008).
9. G. Jourdan *et al.*, *Europhys. Lett.* **85**, 31001 (2009).
10. J. N. Munday, F. Capasso and V. A. Parsegian, *Nature* **457**, 170 (2009).
11. R. S. Decca *et al.*, *Eur. phys. J. C* **51**, 963 (2007).
12. A. Lambrecht *et al.*, *Class. Quantum Grav.* **22**, 5397 (2005).
13. S. K. Lamoreaux and W. T. Buttler, *Phys. Rev. E* **71**, 036109 (2005).
14. F. Chen *et al.*, *Phys. Rev. B* **76**, 035338 (2007).
15. S. de Man *et al.*, *Phys. Rev. Lett.* **103**, 040402 (2009).
16. U. Leonhardt and T. G. Philbin, *New J. Phys.* **9**, 254 (2007).
17. F. S. S. Rosa, D. A. R. Dalvit and P. W. Milonni, *Phys. Rev. Lett.* **100**, 183602 (2008).
18. R. Zhao *et al.*, *Phys. Rev. Lett.* **103**, 103602 (2009).
19. T. H. Boyer, *Phys. Rev.* **174**, 1764 (1968).
20. G. J. Maclay, *Phys. Rev. A* **61**, 052110 (2000).
21. A. Rodriguez *et al.*, *Phys. Rev. Lett.* **99**, 080401 (2007).
22. K. A. Milton and J. Wagner, *J. Phys. A: Math. Theor.* **41**, 155402 (2008).
23. M. T. H. Reid *et al.*, *Phys. Rev. Lett.* **103**, 040401 (2009).
24. T. Emig and R. L. Jaffe, *J. Phys. A: Math. Theor.* **41**, 164001 (2008).
25. A. Lambrecht and V. N. Marachevsky, *Phys. Rev. Lett.* **101**, 160403 (2008).
26. F. Chen *et al.*, *Phys. Rev. Lett.* **88**, 101801 (2002).
27. H. C. Chiu *et al.*, *Phys. Rev. B* **80**, 121402 (2009).
28. A. Roy and U. Mohideen, *Phys. Rev. Lett.* **82**, 4380 (1999).
29. H. B. Chan *et al.*, *Phys. Rev. Lett.* **101**, 030401 (2008).
30. R. Büscher and T. Emig, *Phys. Rev. A* **69**, 062101 (2004).
31. H. B. Chan *et al.*, *Phys. Rev. Lett.* **87**, 211801 (2001).
32. G. L. Klimchitskaya, S. I. Zanette and A. O. Caride, *Phys. Rev. A* **63**, 014101 (2000).
33. F. Chen *et al.*, *Phys. Rev. A* **74**, 022103 (2006).
34. E. M. Lifshitz, *Sov. Phys. JETP* **2**, 73 (1956).
35. G. L. Klimchitskaya *et al.*, *Phys. Rev. A* **60**, 3487 (1999).
36. E. D. Palik, *Handbook of Optical Constants of Solids* (Academic Press, New York, 1985).
37. P. A. M. Neto, A. Lambrecht and S. Reynaud, *Phys. Rev. A* **72**, 012115 (2005).

NEW RESULTS FOR THE CASIMIR INTERACTION: SAMPLE CHARACTERIZATION AND LOW TEMPERATURE MEASUREMENTS

R. S. DECCA

Department of Physics, Indiana University-Purdue University Indianapolis, Indianapolis, IN 46202, USA
E-mail: rdecca@iupui.edu

D. LÓPEZ

Center for Nanoscale Materials, Argonne National Laboratory, Argonne, IL 60439, USA
E-mail: dlopez@anl.gov

E. OSQUIGUIL

Laboratorio de Bajas Temperaturas, Centro Atómico Bariloche, S. C. de Bariloche, RN 8400, Argentina
E-mail: osquigui@cab.cnea.gov.ar

We describe our latest results in the separation dependence of the Casimir interaction in the sphere-plane geometry for two Au-coated surfaces. All results are obtained by measuring the change in the resonant frequency of a sensitive microelectromechanical torsional oscillator as the separation between the sphere and the plane is changed. By means of the proximity force approximation, the change in resonant frequency yields the Casimir pressure between two parallel plates at the same separation. We present results for a new sample at room temperature, where the dielectric function has been measured in the 190-825 nm range. We show that the results of the Casimir force in this sample and in previous samples are virtually indistinguishable. Furthermore, the observed differences between measured and tabulated optical properties data do not show any effect on the calculation of the Casimir interaction. We also present results of the measurement of the Casimir force between a sphere and a plane at 300, 77, 4.2 and 2.1 K. While low temperature results are noisier than room temperature ones, precluding a direct exclusion of either the Drude or the plasma model, the average of the measurements coincide at all temperatures.

Keywords: Casimir force; MEMS; Low-temperature measurements.

1. Introduction

The experimental observation of the Casimir interaction between dielectrics has observed an incredible growth in the last thirteen years.[1,2] Among the different observations, two main strategies have been followed: either the groups have measured the interaction between parallel plates[3] (where the signal is larger, but alignment problems are serious), or the interaction between a sphere and a plate[4–10] (which are self aligned, but the signal's strength is reduced due to the decrease in the interacting area). Other approaches, like cylinder-plane[11] and cylinder-cylinder[12] have also been attempted. While most of the experiments have been performed in vacuum with surfaces covered with good conductors (typically Au-covered surfaces), also important progress has been made when the interacting surfaces are dielectrics,[8,13] they are in air,[9] or the dielectrics are immersed in a fluid.[14]

While there is still quite some ground to cover to achieve a complete understanding of the Casimir interaction, and both geometry and composition effects are very important, the most stringent comparisons between experimental data and theoretical models have been achieved in previous experiments from our own group.[7,15,16] In these experiments, performed at room temperature between a Au-coated sapphire sphere and a Au-coated plane, it was observed that the best agreement between theory and experiment is obtained when the material is theoretically described using a generalized plasma model,[16] which takes into account the deep valence electrons, but disregards the dissipation of the conduction electrons. Models that take into account this dissipation, for example a Drude model, have been excluded by the experiment. The reasons behind why a dissipative medium description does not yield the correct answer in a Casimir interaction geometry remain elusive. Several arguments have been brought forward to explain this difference. Among these, the two most recurrent ones are that the intrinsic characteristics of the Au used in the experiment need to be taken into account[17] (instead of the tabular values reported in Ref. 18), and that there are systematics not yet discovered in the experiment that are masking the true result.

In our attempt to elucidate this conundrum, we have performed two new experiments as reported in this paper. In one of them, we performed the experiment at different temperatures, to see if the natural quenching of the phonon-induced dissipation in the Au conductivity was observed. In the other experiment we performed a room temperature measurement of a sample deposited using a different technique. Furthermore, in this last case, the optical properties of the as deposited sample were determined by ellipsometry.

2. Experimental Details

Casimir interaction measurements were performed in a similar fashion as our previous measurements, where the force-sensitive part of our setup is comprised by a microelectromechanical torsional oscillator (MTO). In this paper's room temperature measurements, the position of the sphere and the plate has been swaped. A sapphire sphere of radius $R \sim 150\mu$m was coated inside a thermal evaporator with ~ 10 nm of Cr and ~ 200 nm of Au and then glued to the Au-covered plate of the MTO. Using a combination of mechanical and piezo driven stages, this was brought in close proximity to a [111] Si wafer which was electroplated with 10 nm Cr-200 nm Au. Except for this difference the rest of the setup, as well as the calibrations performed, is identical to the one described in Ref. 15. As a consequence of mounting the sphere on the MTO, a reduction of its resonance frequency from ~ 700 Hz to ~ 300 Hz was observed. A reduction on the quality factor Q of the oscillator was also observed. Q at room temperature and in vacuum ($P < 10^{-6}$ torr) changes from ~ 8000 to ~ 7500 after gluing the sphere. While probably due to the extra stress induced in the torsional serpentine, more work needs to be performed to understand the changes in Q.

Low temperature measurements were performed in a home built cryostat, where the Au-coated sphere is positioned on top of the Au-coated MTO. The coarse mechanical actuation was performed from the outside, and a piezo-tube and home built slip-stick piezo driven stages were used to provide the fine positioning inside the cryostat. Once the coarse positioning was achieved, the external driving rods were decoupled, and the inner part of the cryostat remained suspended from springs and magnetically damped to reduce vibrations. The inner part of the system was kept at a low He pressure, on the order of 10^{-3} torr, to homogenize the temperature. While the system consists of a sensor and a heater, allowing in principle for temperature dependent measurements, it was found that the thermal drift was so severe that it precluded any measurements where the separation was kept constant and the temperature was changed. Instead, all the calibrations and measurements were performed at constant temperature, provided by the fixed points of liquid nitrogen, liquid helium, and the lambda point of helium. While room temperature measurements for this sample were performed in a different system, it is worth mentioning that room temperature measurements performed in this system with no cryogenic liquids present, showed an increase on the experimental noise on the order of a factor of 2.5. This points out that while the cryostat is not as well isolated from vibrations as the table-top system, the increase in the experimental noise observed at low temperatures is associated with vibrations arising from within the cryostat.

In all our measurements of the Casimir interaction, the MTO's resonance frequency $\omega_{\text{res}}(z)$ was monitored as the separation z between the sphere and the plate is changed. In the linear regime (i.e. small amplitude of oscillation), $\omega_{\text{res}}(z)$ is given by

$$\omega_{\text{res}}^2 = \omega_o^2 \left[1 - \frac{b^2}{I\omega_o^2} \frac{\partial F_C}{\partial z} \right] \tag{1}$$

where ω_o is the MTO's resonant frequency when no interaction is present, b is the lever arm between the point of interaction and the torsional axis, I is the MTO's moment of inertia, and F_C is the Casimir force between the sphere and the plane. By means of the proximity force approximation[a]

$$\frac{\partial F_C}{\partial z} = 2\pi R P_C(z), \tag{2}$$

where $P_C(z)$ is the Casimir force per unit area between two infinite plates separated by a distance z.

3. Results

3.1. *Optical properties of the sample used*

The ellipsometric results for the dielectric function obtained on the Au electroplated Si wafer are shown in Fig. 1. These results are compared with tabulated data. While Fig. 1 shows the frequency dependence for both the real and imaginary parts of the dielectric function, when calculating the Casimir interaction using Lifshitz approach,[19] only the imaginary part is needed. This comes about because in Lifshitz approach the material information enters through an evaluation of its dielectric function along imaginary frequencies, $\varepsilon(i\omega)$, which is given by[20]

$$\varepsilon(i\omega) = 1 + \frac{2}{\pi} \int_0^\infty \frac{x\varepsilon''(x)}{x^2 - \omega^2} dx. \tag{3}$$

Hence, to compare the dielectric function evaluated at imaginary frequencies from tabulated data and from our sample, we first calculated the results

[a] We consider throughout the paper that the attractive Casimir interaction has a positive sign.

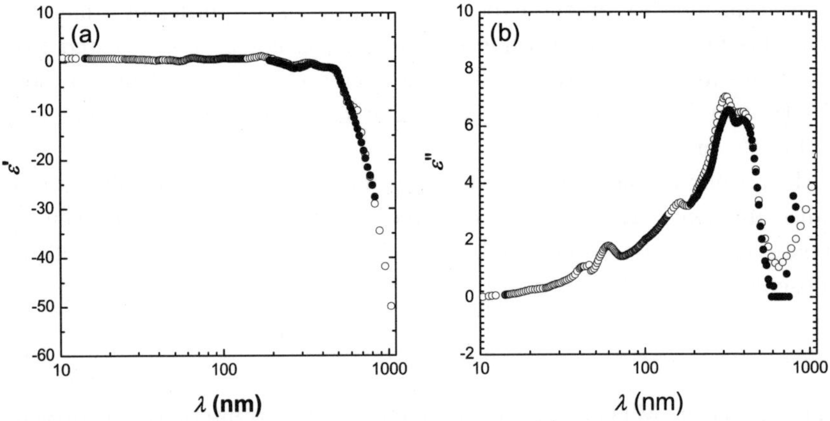

Fig. 1. The filled circles (•) show (a)Obtained real (ε') and (b) imaginary (ε'') parts of the dielectric function from the ellipsometry measurements, between 190 and 825 nm. Tabulated data are displayed as open circles (○).

using Eq. (3) where tabulated data was used between 10^{14} and 10^{19} rad/s. At lower frequencies a Drude model with a value of the plasma frequency $\omega_P = 9$eV and a relaxation parameter $\gamma = 35$ meV was used. For frequencies above 10^{19} rad/s, $\varepsilon''(\omega) \equiv 0$. For the electrodeposited sample, we repeated the same calculation, but tabulated data for ε'' was replaced by our measurements in the relevant frequency range, i.e. between 2×10^{15} and 10^{16} rad/s. The results obtained are shown in Fig. 2.

Unfortunately, the range of frequencies covered by our measurements of the dielectric properties does not seem to be large enough to provide a significant effect on the calculation of the Casimir force. When the Lifshitz expression is used to calculate the Casimir interaction, the difference in P_C obtained when using tabulated data for both the plate and the sphere, or our optical data for the plate and tabulated data for the sphere is about an order of magnitude smaller than the experimental error on P_C. Measurements of the Casimir interaction, however, can be compared between two sets of data. In Fig. 3 we plot the Casimir pressure as a function of separation, $P_C(z)$ for the sample reported in Ref. 16, and the electrodeposited sample. As observed, differences between both data sets are in general smaller than the experimental error. The difference between data sets was obtained at each separation where P_C was measured for the new sample. The values of P_C at these separations for the previously measured sample were obtained by linear interpolation.

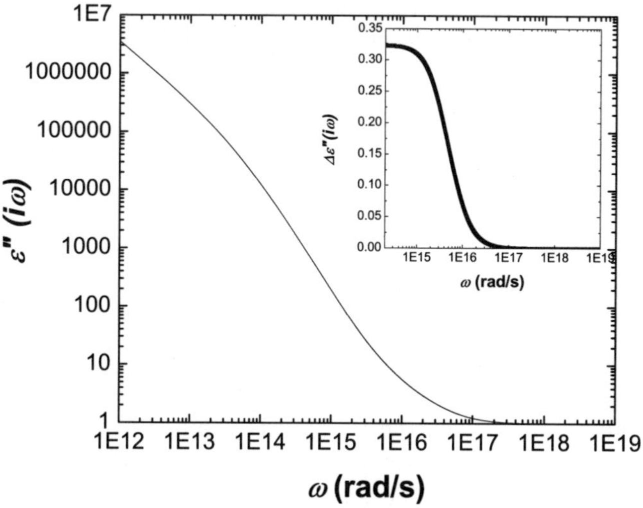

Fig. 2. Dielectric function along the imaginary frequency axis, as obtained from Eq. (3). Results from tabulated data and our sample are shown. They are not distinguishable in the plot. Inset: difference between both calculations. Note the linear vertical axis.

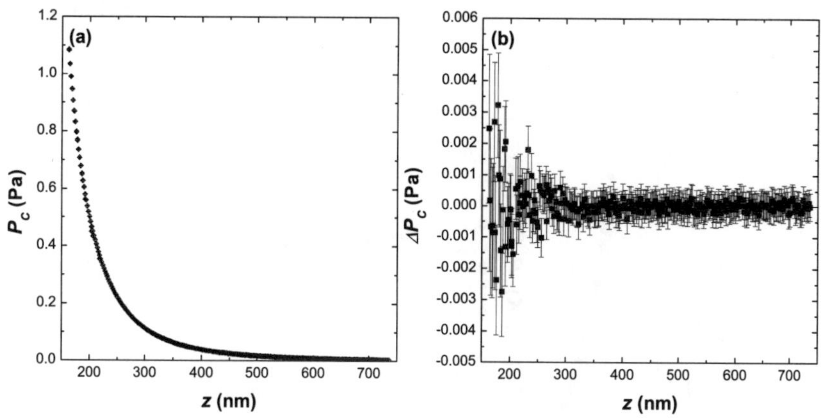

Fig. 3. (a)Measured Casimir pressure as a function of separation for the current run (•) and previously published data (∘). (b) Difference between both sets of data. The error bars represent the 95% confidence level in the experimental measurements, both in position and pressure.

3.2. *Low temperature measurements*

When cooling down the MTO, an increase on both ω_o and Q are observed, as shown in Table 1. More relevant to this work, however, are the low

Table 1. Values of $f_o = \omega_o/(2\pi)$, Q, and the noise δP_C at 300 nm at the different temperatures.

Temperature (K)	f_o (Hz)	Q	δP_C (mPa)
2.1	785.01	11345	22
4.2	766.28	11355	63
77.0	733.33	10222	56
296.5	702.103	7325	2

temperature measurements $P_C(z)$, which are shown in Fig. 4. It is evident from the data that the noise is greater at low temperatures, being the largest at the 77 and 4.2 K. When the He bath is pumped below its λ-point the noise decreases, as shown in Table 1, where we report the noise at different temperatures and 300 nm. The relative increase of the noise with respect to the room temperature measurements are similar for other separations as well. While it is not shown in Fig. 4, when the sample is measured at room temperature inside the cryostat, an increase in the noise of about a factor of 2.5 is observed when compared to $\delta P_C(300nm, 300K)$. From these observations we conclude that although the vibration isolation of the

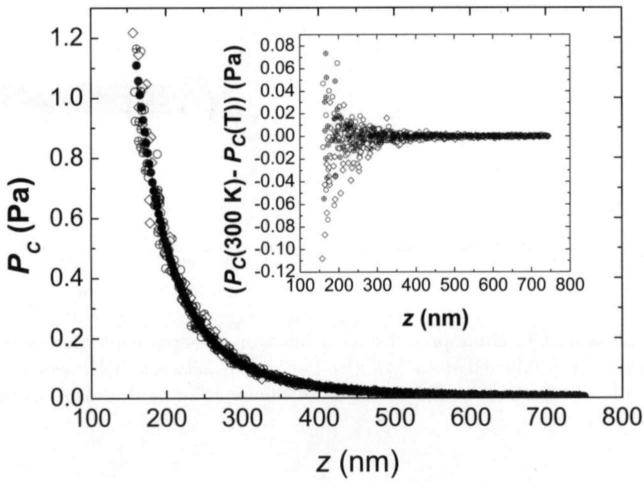

Fig. 4. $P_C(z)$ at different temperatures. (•) T = 300 K, (○) T = 77 K, (◇) T = 4.2 K, and (⊕) T = 2.1 K. Inset: $P_C(z, 300K) - P_C(z, T)$, where T = 77 (○), 4.2 (◇), and 2.1 K (⊕).

cryostat is not as good as in the room temperature system, the main source of noise is due to internal vibrations when the cryogenic liquids are present, most likely due to their boiling. The reduction in the noise when below the λ-point seems to emphasize the importance of cryogenic liquids bubbling. Long term measurements to try to alleviate the random noise induced by vibrations are not possible, since the change in the cryogenic liquid levels induce a temperature gradient in the two-color fiber interferometer, producing a time dependent change in its calibration. This is a result of the different temperature dependance of the fiber's index of refraction at the two wavelengths employed. For this reason, the separation error is also larger at low temperatures.

In spite of the many difficulties presented in the first low temperature measurements of the Casimir interaction, we would like to emphasize that the average P_C at any separation seems to coincide for all investigated temperatures. This is better seen in the inset, where we have subtracted from the experimental values at 300 K the experimental values at the other measured temperatures. The subtraction was performed at the separations measured at low temperatures, and the room temperature value was found by linear interpolation.

4. Conclusions

In conclusion, we have performed two sets of measurements which, while promising, are not conclusive. In one experiment, we observed a clear difference in the optical constants between our sample and tabulated data. Over the range of frequencies of the optical data measurements, this does not translate in a significant effect on the calculated $P_C(z)$. Along the same lines, the observed difference in the measured $P_C(z)$ between samples made by different methods is smaller than the experimental error. On the other experiment, for our measurements of P_C at different temperatures, we were not able to see any difference between the room temperature and low temperature measurements, due in large part to the significant increment in vibrational noise in our low temperature setup. We are hopeful that an improvement on the experimental setup will yield a definite experimental answer to the role played by dissipation in the Casimir interaction.

Acknowledgments

RSD acknowledges NSF support through grants No. CCF-0508239 and PHY-0701236, LANL support through contract No. 49423-001-07. He is also indebted to the Nanoscale Imaging Center at IUPUI for the liberal use of the installations. DL and RSD acknowledge support from DARPA grant No. 09-Y557. EO is a member of CONICET, Argentina, and is indebted to the Fulbright Visiting Scholar program which allowed him to spend three months at IUPUI.

References

1. G. L. Klimchitskaya, U. Mohideen and V. M. Mostepanenko, *Rev. of Mod. Phys.* **81**, 1827 (2009).
2. K. A. Milton, *The Casimir Effect* (World Scientific Publishing Company (2001)).
3. G. Bressi, G. Carugno, R. Onofrio and G. Ruoso, *Phys. Rev. Lett.* **88**, 041804 (2002).
4. S. K. Lamoreaux, *Phys. Rev. Lett.* **78**, 5 (1997); *Phys. Rev. Lett.* **81**, 5475(E)(1998).
5. U. Mohideen and A.Roy, *Phys. Rev. Lett.* **81**, 4549(1998); G. L. Klimchitskaya, A. Roy, U. Mohideen and V. M. Mostepanenko, *Phys.Rev.A* **60**, 3487(1999); A. Roy, C.-Y. Lin and U. Mohideen, *Phys. Rev. D* **60**, 111101(R)(1999); B. W. Harris, F. Chen and U. Mohideen, *Phys.Rev.A* **62**, 052109(2000); F. Chen, G. L. Klimchitskaya, U. Mohideen and V. M. Mostepanenko, *Phys. Rev. A* **69**, 022117(2004).
6. H. B. Chan, V. A. Aksyuk, R. N. Kleiman, D. J. Bishop and F. Capasso, *Science* **291**, 1941 (2001); H. B. Chan, V. A. Aksyuk, R. N. Kleiman, D. J. Bishop and F. Capasso, *Phys. Rev. Lett.* **87**, 211801 (2001).
7. R. S. Decca, D. López, E. Fischbach and D. E. Krause, *Phys. Rev. Lett.* **91**, 050402 (2003).
8. D. Iannuzzi, M. Lisanti and F. Capasso, *Proc. Nat. Acad. Sci.* **101**, 4019 (2004).
9. S. de Man, K. Heeck, R. J. Wijngaarden and D. Iannuzzi, *Phys. Rev. Lett.* **103**, 040402 (2009).
10. W. J. Kim, M. Brown-Hayes, D. A. R. Dalvit, J. H. Brownell and R. Onofrio, *J. Phys.: Conference Series* **161** 012004 (2009).
11. M. Brown-Hayes, D. A. R. Dalvit, F. D. Mazzitelli, W. J. Kim and R. Onofrio, *Phys. Rev. A* **72**, 052102 (2005).
12. T. Ederth, *Phys. Rev. A* **62**, 062104(2000).
13. G. L. Klimchitskaya, U. Mohideen and V. M. Mostepanenko, *J. Phys. A: Math. Theor.* **40**, F841 (2007).
14. J. N. Munday and F. Capasso, *Phys. Rev. A* **75**, 060102(R) (2007).
15. R. S. Decca, E. Fischbach, G. L. Klimchitskaya, D. E. Krause, D. López and V. M. Mostepanenko, *Phys. Rev. D* **68**, 116003 (2003); R. S. Decca, D. López, E. Fischbach, G. L. Klimchitskaya, D. E. Krause and V. M. Mostepanenko, *Ann. Phys. (New York)* **318**, 37 (2005); *Phys. Rev. D* **75**, 077101 (2007).
16. R. S. Decca, D. López, E. Fischbach, G. L. Klimchitskaya, D. E. Krause and V. M. Mostepanenko, *Eur. Phys. J. C***51**, 963 (2007).

17. V. B. Svetovoy, P. J. van Zwol, G. Palasantzas and J. Th. M. De Hosson, *Phys. Rev. B* **77**, 035439 (2008).
18. E. D. Palik (ed.), *Handbook of Optical Constants of Solids* (Academic, New York, 1985).
19. E. M. Lifshitz, *Sov Phys. JETP* **2**, 73 (1956).
20. L. Landau, E. M. Lifshitz, *Landau and Lifshitz Course of Theoretical Physics: Electrodynamics in Continuous Media* (Butterworth-Heinemann, 1980), Chap IX.

CASIMIR FORCE EXPERIMENTS IN AIR: TWO BIRDS WITH ONE STONE

S. DE MAN, K. HEECK, K. SMITH, R. J. WIJNGAARDEN and D. IANNUZZI*

Department of Physics, VU University Amsterdam,
Amsterdam, 1081HV, The Netherlands
**E-mail: iannuzzi@few.vu.nl*
http://www.nat.vu.nl/CondMat/iannuzzi/

We present a short overview of the recent efforts of our group in the design of high precision Casimir force setups. We first describe our Atomic Force Microscope based technique that allows one to simultaneously and continuously calibrate the instrument, compensate for a residual electrostatic potential, measure the Casimir force, and, in the presence of a fluid in the gap between the interacting surfaces, measure the hydrodynamic force. Then we briefly discuss a new force sensor that adapts well to Casimir force measurements in critical environments.

Keywords: Casimir force; Hydrodynamic force; Modulation; Fiber-top technology.

1. Introduction

The mechanical parts of Micro- and NanoElectroMechanical Systems (MEMS and NEMS) are often designed to work at separations where the interaction mechanisms caused by quantum fluctuations of the electromagnetic field cannot be neglected. It is well known, for example, that non-retarded van der Waals forces play a central role in interfacial adhesion and friction between surfaces in contact.[1] At larger separations, long-range retarded van der Waals forces can still affect the performance of MEMS and NEMS.[2–4] It is thus now commonly believed that, if one could suitably engineer the strength of those interactions with techniques that can be easily reproduced in micro- and nanofabrication processes, unprecedented opportunities would come available for the development of conceptually new MEMS and NEMS.[5–10] For this reason, previous studies on dispersion forces in colloidal suspensions[11,12] are currently receiving renewed attention,[7,13,15] because they prove that the retarded van der Waals attraction can be largely reduced or even switched to repulsive if the gap between the surfaces is filled with a liquid.[16–18] Since MEMS and NEMS are often designed to work in air, it is now interesting to investigate whether similarly dramatic effects can be observed in the absence of liquids, where, due to historical reasons, the long-range retarded van der Waals force is called Casimir force.[19]

Driven by these considerations, over the last four years we have developed an experimental setup designed to carry out precise measurements of the Casimir force between a sphere and a plate in the presence of a gaseous intervening medium.[20,21] In this paper we describe the main features of the instrument and show its versatility as a general tool for the investigation of forces between surfaces at sub-micron separation. In the last section, we also introduce a new approach for the development of a Casimir force setup based on fiber-top technology.[22]

2. Casimir Force Measurements: Common Challenges

In a Casimir force setup, there are at least three technical issues that must always be carefully addressed:

- *Calibration:* Casimir force experiments necessarily rely on a force sensor. The force sensor must be calibrated with a known force. Furthermore, one must be sure that, during the measurements, the calibration parameters do not vary significantly.
- d_0: in most of Casimir force setups, the two interacting surfaces are initially put at a given separation d_0, and then moved closer and closer with calibrated steps or ramps. The initial value of d_0 is not known *a priori*. d_0 must thus be independently measured, and must remain constant within one run. It is thus highly desirable to measure how much thermal drifts influence the measurement of d_0.[20,21,23]
- V_0: the electrostatic potential difference between two juxtaposed surfaces connected to ground is, in general, not zero. The residual potential V_0 can give rise to electrostatic forces that easily overcome the Casimir force. It is thus important to measure V_0 and apply a counter-bias potential to compensate for this effect. Recently, it has been emphasized that V_0, in general, varies with surface separation and time.[20,24,25] A single measurement for one single separation is thus not sufficient for accurate Casimir force experiments.

Recently, we presented an experimental method that allows one to simultaneously (i.e., in the same run): (i) calibrate the force sensor, (ii) measure

d_0, (iii) compensate for V_0, (iv) measure the Casimir force.[20,21] The instrument works in air, giving also the opportunity to measure, while measuring the Casimir force, the hydrodynamic force due to the gas present in the gap (hence the title of this paper). In the next section, we report the main idea behind the experimental method.

3. Our New Experimental Method

The experimental setup we designed to perform precise measurements of the Casimir force is a custom made Atomic Force Microscope (AFM). The measurement head is from a Veeco Multimode AFM (with low-noise laser upgrade). The force sensor is a 1 N/m Si cantilever with a 200 μm diameter sphere attached to its free end. This sensor is mounted into the AFM head, and its deflection is measured with an optical lever, where a light beam reflected from the back of the cantilever is detected by a position sensitive photodetector. The surface of the sphere has to be brought in close proximity to another surface, in order for the Casimir effect to become measurable. We mount a planar sample on a piezoelectric stage that is feedback controlled with an internal capacitive sensor (resolution on the order of 50 pm). The calibrated translator is used to accurately vary the separation between the surfaces. We also added a stick-slip piezoelectric stepper motor to perform the coarse adjustment of the sample position. The measurement head and piezoelectric positioning stage (feedback controlled piezo and stepper motor combined) are mounted on an aluminum block that functions as a temperature stabilized heat reservoir roughly 10 K above room temperature; mechanical drifts due to expansion and contraction of the setup is hereby strongly reduced. The AFM is then placed on an active anti-vibration platform inside an anechoic chamber to isolate the setup from mechanical and acoustic vibrations (see Fig. 1a). The anechoic chamber is placed on an optical table inside a temperature controlled laboratory. In Fig. 1b, we show a close-up of our custom designed AFM. Apart from the AFM head and positioners, an optical fiber mounted on a mechanical translation stage is visible on the forefront of the photo. This fiber is pointed towards the bottom of the sample, and is used as an interferometer to measure the phase and amplitude of a small modulation of the planar sample position, as will become clear later in the text.

The technique employed to tackle the technical issues common to Casimir force experiments is based on Kelvin probe force microscopy.[26] The trick is to apply an AC potential difference across the sphere and planar sample surfaces, which will give rise to two signals at different oscillation

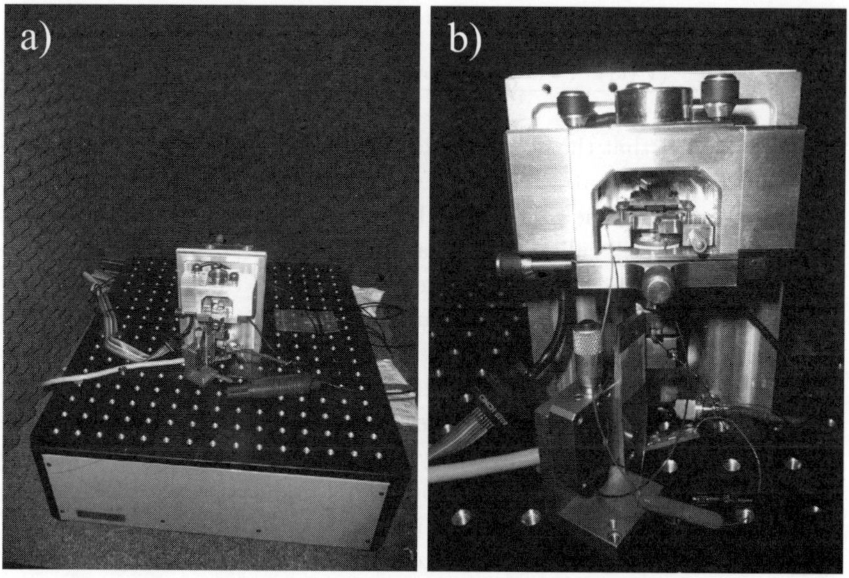

Fig. 1. Photographs of the Casimir force setup. a) Our custom built Atomic Force Microscope (AFM) mounted on an active anti-vibration stage inside an acoustic isolation box. b) Close-up of the AFM with the Veeco Multimode measurement head mounted at the upper part of an aluminum block and the piezoelectric translators mounted underneath the head. In front of the AFM, an optical fiber is mounted on a manual translation stage pointed at the piezoelectric stage to interferometrically measure the phase and amplitude of the separation modulation (see text for details).

frequencies that will allow us to calibrate the instrument and extract d_0, and compensate the residual potential V_0 independently.

The electrostatic force between a conductive sphere and a conductive plate at separation d is

$$F = \frac{\varepsilon_0 \pi R V^2}{d} \tag{1}$$

where ε_0 is the permittivity of vacuum, R is the radius of the sphere, V is the net potential difference across the surfaces, and $d \ll R$ (Proximity Force Approximation[27]). Since the photodetector signal, S, depends linearly on the deflection of the cantilever, it is clear that S is proportional to the applied force.

Let's now suppose that there exists no residual contact potential difference between the two surfaces (i.e. $V_0 = 0$). Then the signal arising from the oscillating potential difference $V(t) = V_{AC}\cos(\omega_1 t)$ satisfies

$$S_{V_0=0}(t) \propto \frac{(V(t))^2}{d} = \frac{(V_{AC}\cos(\omega_1 t))^2}{d} = \frac{V_{AC}^2}{2d} + \frac{V_{AC}^2}{2d}\cos(2\omega_1 t) \qquad (2)$$

in which the fact that the electrostatic force depends quadratically on voltage results in a signal at $2\omega_1$, twice the driving frequency (with $\omega_1 \ll \omega_{res}$, the free resonance of the cantilever). The amplitude of the oscillating signal at $2\omega_1$, which we will call $S_{2\omega_1}$, is measured by a lock-in amplifier and used to extract d_0 and to calibrate the force sensitivity of the instrument as explained in a previous paper.[20] By examining the value of d_0 for consecutive measurement runs, we established that the mechanical drift of our setup is ≤ 0.2 nm per measurement run.[20,21] If one would keep V_{AC} constant during a measurement run in which the separation d is varied, the amount of cantilever oscillation would strongly vary as the force diverges for small d (see Eq. 1 and 2). To avoid that, the lab computer is reducing V_{AC} as the separation decreases (the computer symbol in Fig. 2a).[20] In this way, we typically keep the electrostatic calibration force constant at roughly 100 pN RMS.[20]

However, even when both surfaces are coated with the same material, there generally is a contact potential difference V_0 between the surfaces. We have to compensate this residual potential in order to have no background electric force in our Casimir force measurement. The counter bias voltage is a DC voltage that we call V_{DC} (see Fig. 2a). Then the presence of both potentials V_0 and V_{DC} implies

$$\begin{aligned} S(t) &\propto \frac{(V(t) + V_0 + V_{DC})^2}{d} \qquad (3) \\ &= \frac{(V_0 + V_{DC})^2}{d} + \frac{2(V_0 + V_{DC})V_{AC}}{d}\cos(\omega_1 t) + \frac{V_{AC}^2}{2d} + \frac{V_{AC}^2}{2d}\cos(2\omega_1 t) \end{aligned}$$

The cross-product appears at the frequency ω_1 and is proportional to $(V_0 + V_{DC})$. By measuring the amplitude S_{ω_1} of this signal with a lock-in amplifier we can create a negative feedback loop that generates V_{DC} in such a way that S_{ω_1} becomes negligible, i.e. $V_0 + V_{DC} = 0$;[20,25] this feature is indicated in Fig. 2 by the dashed arrow that point from Fig. 2a to Fig. 2b and eliminates the presence of the signal peak at ω_1. We have thus successfully compensated the contact potential difference between the interacting surfaces. Note that the addition of both V_0 and V_{DC} in the treatment does not alter the amplitude of the oscillation at $2\omega_1$ ($S_{2\omega_1}$).

With the residual potential compensated, the force sensitivity calibrated, and the absolute surface separation determined, we can now turn to the actual measurement of the Casimir force. As it is hard to measure

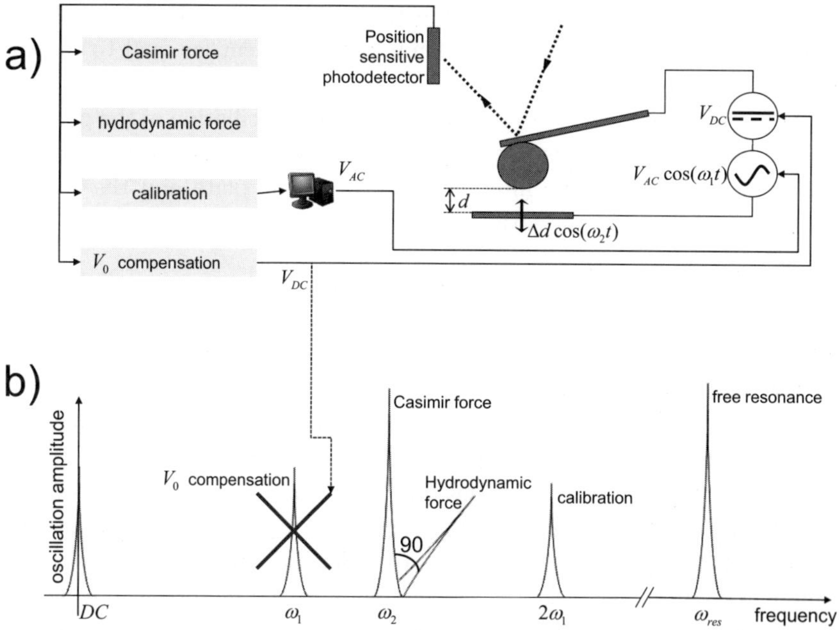

Fig. 2. a) Schematic of the experimental setup and working principle. b) Spectrum of the motion of the cantilever in response to the electric, Casimir and hydrodynamic forces, as a result of an oscillating potential difference across the interacting surfaces at ω_1 and the small modulation of the separation d at ω_2. The dashed arrow between figures a) and b) indicates that the V_0 compensation mechanism effectively zeros the signal at ω_1. See text for details.

the static deflection of the cantilever resulting from the Casimir force (the force is small, the spring constant is relatively high, and the read-out is not very stable in DC), we resort again to a modulation technique. We let the planar sample execute a very small oscillatory motion $\Delta d\cos(\omega_2 t)$, with $\Delta d \simeq 2$ nm and $\omega_2 \ll \omega_{res}$. As a result, there will be an in-phase oscillatory response of the cantilever because the Casimir force is separation dependent:

$$S^I_{\omega_2}(t) \propto \frac{\partial F}{\partial d}\Delta d\cos(\omega_2 t) \tag{4}$$

where $\partial F/\partial d$ represents the gradient of the force at the current separation. By measuring the amplitude of $S^I_{\omega_2}(t)$ with another lock-in amplifier, we measure directly the gradient of the Casimir force[a]. In the sphere-plane

[a]There is also a small contribution of the gradient of the electrostatic calibration force to $S^I_{\omega_2}(t)$, but we can correct for that with the simultaneous measurement of $S_{2\omega_1}$. The derivation of this correction goes beyond the scope of the paper.

geometry that we employ, and within the validity of the Proximity Force Approximation,[27] the measured Casimir force gradient is equal to the expected force between parallel plates.

As the measurements are performed in air, the oscillation of the position of the planar sample gives rise to a moving column of gas above it. This results in a hydrodynamic interaction between the plate and the sphere, resulting in a signal of the form:

$$S^Q_{\omega_2}(t) \propto \frac{v(t)}{f(d)} = \frac{\omega_2 \Delta d \sin(\omega_2 t)}{f(d)} \tag{5}$$

where $v(t)$ is the velocity of the plate, and $f(d)$ is constant for separations $d \gg R$, is proportional to d for separations for which both $d \ll R$ and the non-slip boundary conditions for the Navier-Stokes equations are valid, and has a more complicated dependence on distance for values of d close to the mean free path of the air molecules due to the gas slip at the surfaces.[28,29] From Eqs. 4 and 5, it is clear that both signals at ω_2 are orthogonal (one is a cosine, the other a sine), and can thus be independently measured with the same lock-in amplifier locked at ω_2 [b]. It is of crucial importance, however, that the phase of the lock-in amplifier is aligned perfectly with the actual phase of the oscillatory motion of the plate, because otherwise both signals $S^I_{\omega_2}(t)$ and $S^Q_{\omega_2}(t)$ get mixed by the lock-in. The fiber optic interferometer is used to secure the proper phase alignment of the ω_2 lock-in amplifier.

Using the setup described above, we have recently explored the role of conductive oxides in Casimir force experiments. We refer the reader to our original paper[21] for further details.

4. Fiber-Top Casimir Force Setup

The investigation of the dependence of the Casimir force on the boundary conditions is currently pushing experimentalists to rely on instruments that can measure surface forces under extreme conditions, such as low temperatures, conductive liquids, et cetera. Adapting existing setups to those situations is often not an option, and new technologies have to be developed to extend the experimental studies beyond standard environments.

In 2005, our group has introduced a new all-optical device that represents an interesting platform for the implementation of Casimir force setups in critical environments: the fiber-top cantilever.[22] Fiber-top cantilevers are

[b]Note that the implicit time dependence of d in $f(d)$ (due to the separation modulation at ω_2) does not invalidate eq. 5, as it will only cause signals at higher harmonics of ω_2.

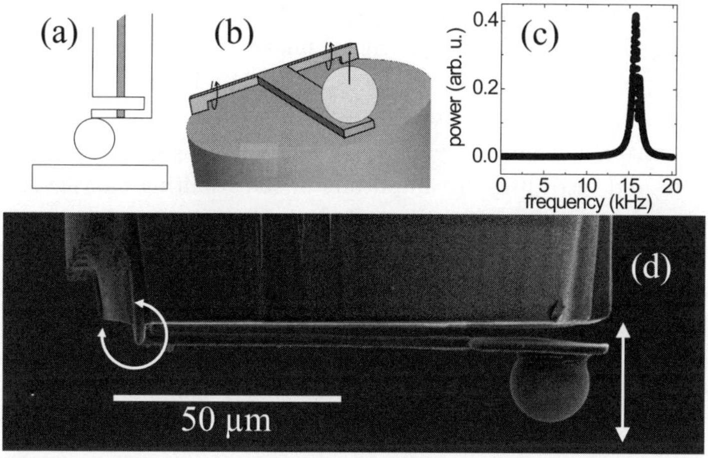

Fig. 3. a) Drawing of a fiber-top based Casimir force setup. A sphere is attached to the free end of a fiber-top cantilever. The deflection of the cantilever is measured by coupling light in the fiber. b) Schematic drawing of a torsional fiber-top device. Instead of a normal cantilever, the torsional device is anchored to the fiber with two thin rods that can rotate around their central axes in response to external forces. c) Power spectrum of a fiber-top torsional device, as measured by driving the cantilever with an electrostatic force. d) Scanning electron microscope image of a fiber-top torsional device equipped with a sphere for performing force measurements. The pivot point of the cantilever is at the left of the figure.

obtained by carving a thin mechanical beam out of the cleaved end of a single mode optical fiber. The light coupled from the opposite end can then be used to detect tiny deflections of the cantilever, and, thus, tiny forces applied to the cantilever's free hanging end.[22,30] This force sensor has no electronic contacts on the sensing head and is entirely fabricated on top of a 125 μm diameter optical fiber. It thus easily fits in small volumes (a good option, for example, for cryogenic temperatures) and still works properly in conductive liquids (where force sensors with electronic readouts would fail). It is thus reasonable to envision that a fiber-top cantilever equipped with a sphere on its free hanging end could perform well as the force sensor in a Casimir force setup (see Fig. 3a).

It is however important to note that, even with the carving resolution of the most sophisticated tools currently available (i.e., Focused Ion Beam milling), it is not possible to fabricate fiber-top cantilevers with dimensions small enough to achieve spring constants smaller than 1 N/m, as required in Casimir force experiments. To solve this problem, we decided to fabricate a slightly different fiber-top device, where the cantilever is suspended over the fiber via two torsional beams, as reported in Fig. 3b. In Fig. 3d, we report an Scanning Electron Microscope (SEM) image of a fiber-top torsional force sensor equipped with a 18 μm diameter sphere on its free hanging end.

In Fig. 3c, we plot the power spectrum of the torsional device. This power spectrum is acquired by applying an oscillating electrostatic force to the sphere and recording the amplitude of the resulting cantilever motion with a lock-in amplifier. It is clear that the force sensor has a resonance frequency of roughly 16 kHz. Because we measured the dimensions of the cantilever and the diameter of the sphere with the SEM, we can calculate the mass of the cantilever (30 nanograms) using the density of the fiber material (2200 kg/m^3). With the parallel axis theorem, we then obtain the total moment of inertia of the cantilever sphere combination around the axis that goes through the torsional beams ($1.6 \cdot 10^{-19}$ kg m^2). Combining the latter with the measurement of the resonance frequency, we arrive at a torsional spring constant of $1.6 \cdot 10^{-9}$ Nm/rad and a spring constant of 0.2 N/m.

To assess the sensitivity of our device, we have applied an oscillating voltage between the sphere and a nearby metallic plate (as in Fig. 3a) at a frequency $\frac{\omega_1}{2\pi} = 72.2$ Hz. A digital lock-in amplifier equipped with a 24dB roll-off low-pass filter is used to measure the uncertainty in the resulting cantilever oscillation at $2\omega_1$ (144.4 Hz). The result is an RMS noise of $1.2\ \mathrm{pm}/\sqrt{\mathrm{Hz}}$, which indicates that the AC force resolution of our setup is roughly $240\ \mathrm{fN}/\sqrt{\mathrm{Hz}}$. Such a force resolution is more than sufficient for a high precision measurement of the Casimir force.

According to our preliminary results obtained using this device and the method described in the previous section, it appears evident that, as soon as the plate in front of the sphere is set into oscillation, the hydrodynamic interaction between the mechanical beam and the fiber is not negligible and actually hampers the efficiency of the experimental technique. To solve this problem, we are planning to move to larger fiber-top devices (namely, ferrule-top cantilever[31,32]), where the gap between the force sensor and the fiber is much larger ($\simeq 50\ \mu$m). Studies are under way to understand the potentiality of this new instrument. For Casimir force measurements in vacuum, though, the system presented here should work fine.

Acknowledgments

This work was supported by the Netherlands Organisation for Scientific Research (NWO), under the Innovational Research Incentives Scheme VIDI-680-47-209. The authors thank the organizers of QFEXT09 for putting together such an interesting meeting, and the ESF Research Network CASIMIR for providing visit grants. D. I. acknowledges financial support from the European Research Council under the European Community's Seventh Framework Programme (FP7/2007-2013)/ERC Grant Agreement 201739.

One of the authors (DI) would like to express his immense gratitude to Giacomo Bressi, who silently gave fundamental contributions to the Casimir field (and many other topics) for over a decade before a terrible illness brought him away from his family, his friends, and our scientific community.

References

1. F. W. DelRio, M. P. de Boer, J. A. Knapp, E. D. Reedy, P. J. Clews and M. L. Dunn, *Nature Materials* **4**, 629 (2005).
2. H. B. Chan, V. A. Aksyuk, R. N. Kleinman, D. J. Bishop, and F. Capasso, *Science* **291**, 1941 (2001).
3. H. B. Chan, V. A. Aksyuk, R. N. Kleinman, D. J. Bishop, and F. Capasso, *Phys. Rev. Lett.* **87**, 211801 (2001).
4. F. M. Serry, D. Walliser, and G. J. Maclay, *J. Appl. Phys.* **84**, 2501 (1997).
5. F. Capasso, J. N. Munday, D. Iannuzzi, and H. B. Chan, *IEEE J. Select. Topics Quant. Electr.* **13**, 400 (2007).
6. O. Kenneth, I. Klich, A. Mann, and M. Revzen, *Phys. Rev. Lett.* **89**, 033001 (2002).
7. J. N. Munday, D. Iannuzzi, Y. Barash, and F. Capasso, *Phys. Rev.* **A71**, 042102 (2005).
8. P. Ball, *Nature* **447**, 772 (2007).
9. P. Ball, *Nature Materials* **6**, 402 (2007).
10. P. Ball, *Nature Materials* **8**, 705 (2009).
11. A. M. Bevan and D. C. Prieve *Langmuir* **15**, 7925 (1999).
12. J. Israelachvili, *Intermolecular and Surface Forces* (Academic, New York, 1992).
13. J. N. Munday and F. Capasso, *Phys. Rev.* **A75**, 060102(R) (2007).
14. A. A. Feiler, L. Bergstrom, and M. W. Rutland, *Langmuir* **24**, 2274 (2008).
15. J. N. Munday, F. Capasso, and V. A. Parsegian, *Nature* **457**, 170 (2009).
16. S. Lee and W. M. Sigmund, *J. Colloid Interface Sci.* **243**, 365 (2001).
17. S. Lee and W. M. Sigmund, *J. Colloids Surf.* **A204**, 43 (2002).
18. A. Milling, P. Mulvaney and I. Larson, *J. Colloid Interface Sci.* **180**, 460 (1996).
19. H. B. G. Casimir, *Proc. K. Ned. Akad. Wet.* **60**, 793 (1948).
20. S. de Man, K. Heeck, and D. Iannuzzi, *Phys. Rev.* **A79**, 024102 (2009).
21. S. de Man, K. Heeck, R. J. Wijngaarden, *Phys. Rev. Lett.* **103**, 040402 (2009).
22. D. Iannuzzi, S. Deladi, V. J. Gadgil, R. G. P. Sanders, H. Schreuders, and M. C. Elwenspoek, *Appl. Phys. Lett.* **88**, 053501 (2006).
23. G. Jourdan, A. Lambrecht, F. Comin, and J. Chevrier, *Europhys. Lett.* **85**, 31001 (2009).

24. W. J. Kim, M. Brown-Hayes, D. A. R. Dalvit, J. H. Brownell, and R. Onofrio, *Phys. Rev.* **A78**, 020101(R) (2008).
25. S. de Man, K. Heeck, R. J. Wijngaarden, and D. Iannuzzi, submitted to *J. Vac. Sc. Technl. B.*
26. M. Nonnenmacher, M. P. O'Boyle, and H. K. Wickramasinghe, *Appl. Phys. Lett.* **58**, 2921 (1991).
27. V. A. Parsegian, *Van der Waals forces* (Cambridge University Press, New York, 2006).
28. L. M. J. Hocking, *Eng. Math.* **7**, 207 (1973).
29. O. I. Vinogradova. *Langmuir* **11**, 2213 (1995).
30. D. Rugar, H. J. Mamin, and P. Guethner, *Appl. Phys. Lett.* **55**, 2588 (1989).
31. G. Gruca, S. de Man, M. Slaman, J. H. Rector, and D. Iannuzzi, *Proc. SPIE* **7503**, PDP07 (2009).
32. G. Gruca *et al.*, in preparation.

EXPERIMENTAL FEATURES OF THE RECENT LATERAL CASIMIR FORCE MEASUREMENT

HSIANG-CHIH CHIU and U. MOHIDEEN*

Department of Physics and Astronomy, University of California, Riverside, CA 92521
**E-mail: umar.mohideen@ucr.edu*

Some experimental features of the recent lateral Casimir force measurement are discussed. A brief experimental overview, along with the related analysis to obtain calibration constants is presented. The data show the diffraction like coherent scattering from the corrugated surfaces.

Keywords: Lateral Casimir force, diffraction effects.

1. Introduction

The Casimir force [1–4] defined in terms of the modifications of the zero-point photon spectrum due to the presence of boundaries, naturally suggests a strong dependence on the geometry of the boundary. Thus the Casimir force exhibits rich and exotic boundary dependencies, which are not trivial extrapolations following from ideal flat boundaries. It can be attractive or repulsive. Typical examples are perfectly conductive spherical and rectangular shells which are predicted to have a repulsive Casimir force [5, 6]. As the Casimir force exceeds normal electromagnetic and gravitational effects in microelectromechanical systems (MEMS), there is an urgent need to understand and harness these geometry effects.

A consensus has been reached that a complete experimental and theoretical understanding of the role of periodic deviations from the planar geometry needs to be first achieved before attempting experiments on the more fantastic geometry dependences of the Casimir force such as the repulsive force for cubes and spheres. Of the geometry dependences, the lateral Casimir force between uniformly corrugated surfaces is of particular interest [7]. The last five years have seen rapid theoretical advances in the area of some simple geometry dependences [8–14] motivated in part by the demonstration of the lateral Casimir force [15, 16]. The most interesting remains the lateral Casimir force for uniformly corrugated surfaces [17] where the diffraction like coherent scattering effects of the zero-point photons cannot be predicted by simple theories. A key advantage of uniform corrugated surfaces is that exact analytical approaches have now become feasible [17–19].

For real material boundaries, the exact theoretical calculation of the lateral Casimir force was difficult and until recently, approximation methods such as the proximity force approximation (PFA) and the pairwise summation approach (PWS) have been applied. These two methods assume the additivity of the dispersion forces and neglect the diffraction like correlation effect between the zero-point photons scattered from the boundaries [4, 17]. In 2003, a path integral approach was applied to the lateral Casimir force by Emig *et al.* [17] to calculate the role of photon correlation effects for ideal metal boundaries. From these results, it was pointed out that if the lateral Casimir force measurement can be performed with a corrugation period λ which is comparable to the separation distance z, significant deviation of the lateral Casimir force from the additive approach will be observed. In our new experiment, we selected a diffraction grating with a $\lambda = 574.7$ nm to perform the lateral Casimir force measurement [22, 23] so as to observe the deviation from the additive theory and the diffraction-like correlation effects of the zero-point photons. Recently, a modified Scattering Theory (ST) approach has been used to derive the lateral Casimir force between the corrugated sphere and plate [22, 23]. In the ST approach, both the temperature and conductivity correction are considered. Here, we will compare our experimental results with the predictions of PFA and ST.

The first experiment to demonstrate the nontrivial boundary dependence of the normal Casimir force was done by Roy and Mohideen [20] in the additive regime. By comparing the normal Casimir force between an Al coated sphere and a sinusoidal grating, to that between the same sphere and a smooth plate, they showed that the boundary of the surface has to be considered to explain the differences. The ambiguity between the normal and lateral forces resulting in the experimental geometry was not clarified. Chan *et al.* [21] measured the normal Casimir force between a metalized sphere and rectangular silicon trenches in the non-additive regime and showed that the measured Casimir force deviated from the prediction of the additive theory. Since the conductivity corrections to the Casimir force was not included in the theory, the deviation observed in this experiment was approximately 50% less than that expected from ideal metal surfaces.

The lateral Casimir force between a corrugated Au coated sphere and plate covered with sinusoidal corrugations was first demonstrated in the additive regime [15, 16]. The corrugation amplitudes used in this experiment were $A_1 = 59$ nm and $A_2 = 8$ nm. The separation distance in the experiment was around 220 nm which is much less than the corrugation period $\lambda \sim 1200$ nm used in the experiment. The experimental results were found to be in good agreement with a theory using PFA within the experimental error at 95% confidence level [15, 16]. The dielectric properties of Au were considered using simple plasma model with 4th order perturbation theory. The asymmetry of the lateral Casimir force was predicted but not observed due to the small amplitude of the corrugations used [16].

This paper is organized as follows. In section 2, the experimental setup and sample preparation procedures will be discussed. The modification of the cantilever and the *in situ* imprint technique will be reviewed. Section 3 presents the experimental procedure for the lateral Casimir force measurement with an AFM. The measurement of the contact potential, calibration procedure for the force constant of the cantilever, and the data analysis will be presented. Section 4 compares the experimental results with both the PFA and the ST. Section 5 concludes the paper.

2. Experiment Setup and Sample Preparation

The experiment setup of the lateral Casimir force measurement is shown in Fig. 1. It is similar to Ref. [15] but with additional improvements as described in Ref. [22]. A standard atomic force microscope was used and the experiment was performed at a pressure less than 10 mtorr and at room temperature. All the electrical connections and vacuum requirements in this system are the same as described in Ref. [24]. In this section, the modification of the cantilever and the preparation of two aligned corrugated surfaces will be discussed.

2.1. *Modification of the cantilever*

As the lateral Casimir force is of order of pN, the cantilever used in this experiment was modified to enhance its sensitivity. In addition, the extended arm of the cantilever can allow the second sphere to access the central region of the grating surface where there were less surface inhomogenieties. It also decreases the scattering of the laser light used for detecting the cantilever deflection due to the vertical grating edge.

The original cantilever we selected was 320 μm long and had a spring constant close to 0.01 N/m. The cantilever was coated with approximately 40 nm of Al to improve its electrical and thermal conductivity. This is used to prevent cantilever deformation due to the differential thermal expansion

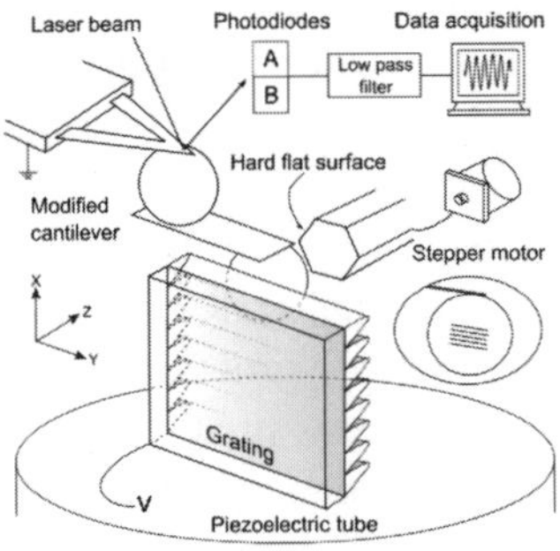

Fig. 1. The experiment setup of the lateral Casimir force measurement. The insert shows the imprinted grating on the second sphere of the cantilever.

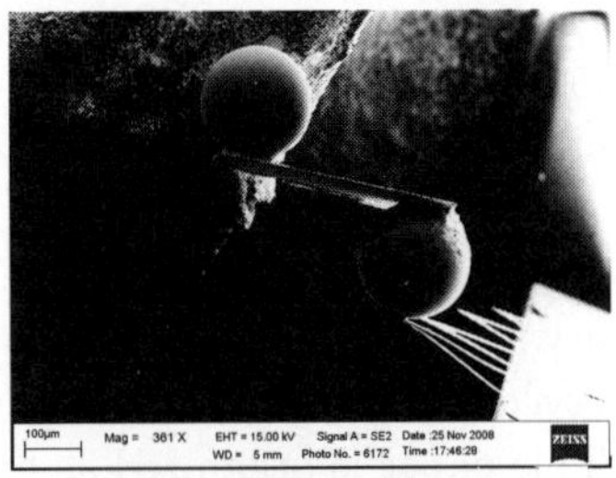

Fig. 2. SEM image of a typical modified cantilever used in the lateral Casimir force measurement. See text for the modification procedure.

when the experiment is performed in vacuum. First, a polystyrene sphere with a diameter of 200 ± 4 μm was attached at the end of the cantilever using conducting silver epoxy. Next, a piece of freshly cleaved mica sheet, 400 μm long, 200 μm wide and few μm thick was attached to the bottom of the first sphere. A second polystyrene sphere of the same size was then attached to the other end of this mica sheet. This second sphere was used for the measurements and its diameter was measured to be $2\text{R} = 194 \pm 0.3$ μm with a calibrated SEM [22]. Next, the whole modified cantilever and spheres were coated with a 10 nm layer of Cr followed by a > 50 nm layer of Au in a thermal evaporator using an oil free vacuum system. The uniformity of all metal coatings was assured by rotating the cantilevers in the thermal evaporator [24]. The SEM image of a typical modified cantilever is shown in Fig. 2. The increased cantilever length will reduce the spring constant of the cantilever by nearly 50% and increase the cantilever sensitivity.

2.2. *Preparation of the corrugated surfaces*

As discussed in the previous section, it takes two aligned uniaxially corrugated surfaces with the same period λ for the observation of the lateral Casimir force. To achieve this, a grating template with sinusoidal uniaxial corrugations was chosen as the first test body. This grating has an average period of 574.7 nm and its corrugation amplitude A_1 was measured by an AFM and found to be homogeneous with a value of 85.4 ± 0.3 nm. The corrugations were covered by 300 nm thick layer of Au. The second corrugated surface was then obtained on the surface of the sphere by an *in situ* imprinting technique. Since this grating template was made of epoxy on a 3 mm thick Pyrex substrate we do not need the additional oxidized aluminum layer to improve the hardness for the imprinting process as described in Refs. [15] and [16].

The grating template usually has to be around 5 mm × 5 mm to be fit into the AFM holder. These gratings were then cleaned with sulfuric-free soap water and agitated in an ultrasonic cleaner for 10 minutes. Next they were rinsed with DI water and pure ethanol, and finally dried with the compressed nitrogen gas. To make sure all the moisture was removed, these gratings were left in a vacuum chamber for at least 12 hours before being mounted in the AFM. The gratings were then examined with an optical microscope to be free of surface damage. This grating was mounted on a L-shaped vertical holder on top of the AFM piezo with conductive silver epoxy. For electric contact, a thin wire was attached to the grating bottom as shown in Fig. 1.

2.3. *The in situ imprint procedure*

The only way to achieve nanoscale aligned corrugations with the same period on two surfaces and keep them separated for distances of 100 nm is by the developed *in situ* imprinting procedure. After the cantilever was in position, a CCD camera was used to change the relative positions between the second sphere of the modified cantilever and the grating surface. First, the second sphere was brought in contact with the grating surface. A hard flat stylus controlled by a stepper motor was approached to the other side of the sphere until they were in contact. Then the z-piezo was moved and the sphere was sandwiched between the grating and the end of the hard flat surface of the stylus, leading to imprint of the corrugations onto the sphere surface. To achieve a deeper imprint amplitude, the stepper motor was used to further press the sphere onto the template grating. Special care was needed at this step. If the applied pressure from the stylus surface was too much, the resulting rotation of the sphere will cause the whole cantilever to break. After the imprint process, the stylus was withdrawn using the stepper motor and the template grating was moved back by reversing the voltage on the piezo. Next, the sphere was moved horizontally to a different position on the grating template for measurement as the amplitude of the template grating might have changed locally after the imprint. The amplitude of the imprinted grating was measured with an AFM after the experiment. An AFM image of the imprinted area on the sphere is shown in Fig. 3. In this Figure, lighter colors show higher points. The sphericity of the imprinted area can be observed. The average imprinted amplitude was

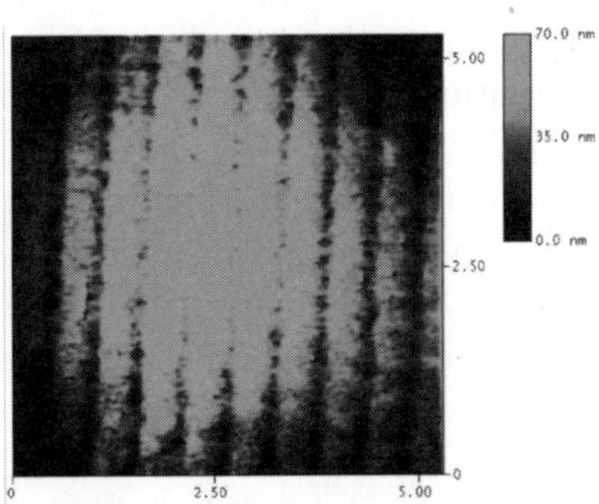

Fig. 3. The AFM image of the imprinted grating on the surface of the second sphere. The average imprinted amplitude is 13.7 ± 0.4 nm. The image also shows the sphericity of the imprinted area.

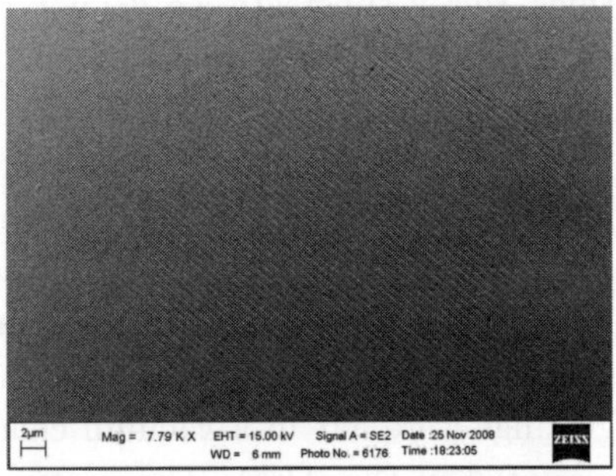

Fig. 4. The SEM of the imprinted grating on the sphere surface. It clearly shows that the imprinted area is larger than $30\ \mu\text{m} \times 30\ \mu\text{m}$.

also measured with an AFM to be 13.7 ± 0.4 nm. A SEM image obtained from the same imprinted area on the sphere is shown in Fig. 4. It shows that the imprinted area is larger than $30\ \mu\text{m} \times 30\ \mu\text{m}$.

3. Experiment Procedure for Measuring the Lateral Casimir Force

After the two aligned corrugated surfaces were obtained, the lateral force measurement was performed. The advantage of using an AFM for a force measurement is the precise control of the separation distance between the two corrugated surfaces. This can be achieved with the piezoelectric scanner which is allowed to move independently in 3 directions as shown in Fig. 1. In this Figure, the x-piezo movement was required to introduce a phase difference between two corrugated surfaces, while the z-piezo movement was used to change the separation distance between them. The minimum voltage that can be applied to the z-piezo is 0.15 V which corresponds to a change in separation distance of 3.6 nm. This provides precise control of the separation distance between two corrugations compared to Ref. [15]. The piezo movements with applied voltages in both x and z directions were calibrated with an optical interferometer [25].

3.1. *The measurement of the contact potential*

As with normal Casimir force measurements, the residual electrostatic force between the two surfaces has to be compensated and hence the potential difference between them has to be measured. Although we can obtain small contact potential between our sample surfaces with our sample preparation procedure, the separation independence of the contact potential has to be confirmed to prevent systematic errors.

The contact potential V_0 between two corrugated surfaces was first measured through the normal electrostatic force between them when the sphere was about 1 μm away from the surface of the template grating. The normal electrostatic force equation between the corrugated sphere and plate is given by

$$\begin{aligned} F_{nor}^{elec}(z,\phi) = k_{tor}S^{nor} &= -2\pi R\varepsilon_0(V-V_0)^2\left[\frac{R}{2z}\,\frac{1}{\sqrt{1-\beta^2}} + c_0 + c_1\,\frac{z}{R}\right. \\ &\quad + c_2\,\frac{z^2(2+\beta^2)}{2R^2} + c_3\,\frac{z^3(2+3\beta^2)}{2R^3} \\ &\quad + c_4\,\frac{z^4(8+24\beta^2+3\beta^4)}{8R^4} + c_5\,\frac{z^5(8+40\beta^2+15\beta^4)}{8R^5} \\ &\quad \left. + c_6\,\frac{z^6(16+120\beta^2+90\beta^4+5\beta^6)}{16R^6}\right], \end{aligned} \tag{1}$$

where

$$\begin{gathered} c_0 = -1.18260\,, \quad c_1 = 22.2375\,, \quad c_2 = -571.366\,, \\ c_3 = 9592.45\,, \quad c_4 = -90200.5\,, \quad c_5 = 383084\,, \quad c_6 = -300357 \end{gathered} \tag{2}$$

and

$$\beta(z,\phi) = \sqrt{\frac{A_1^2 + A_2^2 - 2A_1A_2\cos(\phi)}{z^2}}\,. \tag{3}$$

When a voltage V was applied to the grating template, the electrostatic force will cause the torsional rotation of the modified cantilever. The corresponding deflection S^{nor} of the cantilever can be recorded with the photodiodes of the AFM. The voltages and the deflection signals can thus be used to calculate V_0 using Eq. (1). The measurement of S^{nor} was repeated for eight different voltages between -0.52 to 0.47 V and the mean value was $V_0 = -39.6 \pm 1.6$ mV.

To confirm V_0 is separation-independent, a different measurement was done when the separation distance between two surfaces was only 127.3 nm ($z_0 + 10$ nm). The parabolic dependence between the lateral electrostatic force and the applied voltage V in Eq. (4) was used. The lateral electrostatic force between the sphere-plate corrugations is:

$$\begin{aligned} F_{lat}^{elec}(z,\phi) &= \pi^2\varepsilon_0(V-V_0)^2R\,\frac{A_1A_2}{z^2\lambda}\sin(\phi) \\ &\quad \times \left[\frac{2}{\sqrt{1-\beta^2}\left(1+\sqrt{1-\beta^2}\right)} - 2\sum_{i=1}^{6} ic_i\left(\frac{z}{R}\right)^{i+1} Y_i\right], \end{aligned} \tag{4}$$

where,

$$Y_1 = Y_2 = 1\,, \quad Y_3 = 1 + \frac{1}{4}\beta^2\,, \quad Y_4 = 1 + \frac{3}{4}\beta^2\,,$$
$$Y_5 = 1 + \frac{4}{3}\beta^2 + \frac{1}{8}\beta^4\,, \quad Y_6 = 1 + 10\beta^2 + \frac{5}{8}\beta^4 \tag{5}$$

and the coefficients c_m are defined in Eq. (2).

Several different voltages were applied to the vertical template grating and the corresponding deflection signals S^{lat} of the cantilever were measured. The parabolic dependence of S^{lat} with V was plotted. The value of V_0 was found from the minimum of the parabola to be -39.4 mV. This value is consistent with the previous measurement of the normal electrostatic force between the corrugations. This confirms that the contact potential in this experiment is independent of separation. There was some concern about the precision of the normal electrostatic force, derived using PFA. So we independently verified its precision using a numerical solution of the Poisson equation by finite element analysis. A commercial software package COMSOL Multiphysics [26] was used. The values predicted by Eq. (1) were found to agree with that done numerically within an error of 2.8%.

3.2. *Correction of the non-zero tilt of the vertical template grating*

The grating template was vertically mounted using an L-shaped holder. Thus there will necessarily be a non-zero tilt of the template grating with respect to the x-axis. During the experiment as the template grating was moved along the x-axis to bring about the phase change between the two corrugations, this tilt will lead to changes in the separation distance between the corrugated surfaces. This non-zero tilt can be corrected using the lateral electrostatic force between the two corrugations. The sphere was positioned around a distance of 142.3 nm ($z_0 + 25$ nm) from the vertical grating. Phase changes between the two corrugations were introduced with the x-piezo. The cantilever deflection was measured as a function of this phase change. The change of separation distance due to the non-zero tilt, leads to a steady change of the amplitude of the lateral force with increasing phase as shown in Fig. 5(a).

To correct the non-zero tilt of the vertical grating, a voltage synchronous with that changing the phase was supplied to the z-piezo to counter the change in the separation distance. An adjustable voltage divider was used to adapt the phase changing voltage applied to the x-piezo. The output of

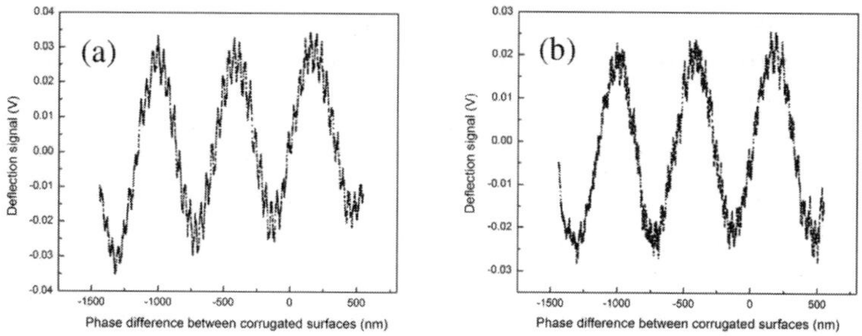

Fig. 5. (a) Lateral electrostatic force signal obtained without correction of non-zero tilt of the template grating. The signal amplitude decreases steadily with increasing phase. (b) Lateral electrostatic signal obtained with the tilt corrected, where the amplitudes of the lateral electrostatic force become similar.

the voltage divider was applied to the z-piezo to change the separation distance synchronous with the change in phase. The amplitude of the lateral electrostatic force was monitored with an oscilloscope as a function of the phase. The voltage output of the voltage divider was adjusted till the amplitude of the lateral electrostatic force became independent of the phase as shown in Fig. 5(b). This corrects the non-zero tilt of the vertical grating and keeps the separation distance between the two corrugated surfaces constant while phase between them is varied.

3.3. *Measurement of the lateral Casimir force*

The lateral Casimir force was next measured. A compensating voltage equal to the measured residual potential difference between the two corrugations of −39.6 mV was applied to the grating template. The lateral Casimir force was measured at 8 different separation distances from contact of $z_0 + 3.96$ nm to $z_0 + 70.86$ nm. The sequence of steps is discussed in Refs. [22] and [23]. The signal to noise ratio was poor at larger separation distances. The deflection signals of the cantilever due to the lateral Casimir force was sent through a low pass filter with a time constant set to 30 ms and recorded as a function of phase difference ϕ at 8192 points corresponding to phase changing steps of 0.4 nm. At the largest separation, 30 scans of force signals obtained with x-piezo moving at 0.103 Hz were averaged and recorded. The number of scans to be averaged was reduced when the two surfaces were closer and signals became stronger. The mechanical drift of the separation distance was independently measured to be 0.14 nm/min, through the measured change in the points of contact of the two corrugations after definite intervals of time.

3.4. *Measurement of the lateral electrostatic force and the determination of k_{ben} and the separation distance at contact z_0*

The lateral electrostatic force was measured to calibrate the deflection signal through the calibration constant k_{ben} of the cantilever and measure the exact separation between the corrugated surfaces through a measurement of the separation on contact z_0. Two different voltages, 0.141 V and 0.101 V, were applied to the grating template for the measurement of the total lateral force (electrostatic + Casimir). This measurement was done at several different separation distances between the two corrugated surfaces. The measurements were repeated twice at each applied voltage.

3.5. *Measurement of the force calibration constant k_{ben} and the separation distance at contact z_0*

To obtain the force calibration constant k_{ben} of the cantilever and separation distance at contact z_0 by fitting the electrostatic force to Eq. (4), the lateral Casimir force has to be subtracted from the total lateral force to obtain the pure lateral electrostatic force. The lateral Casimir force as a function of the phase between two surfaces is asymmetric, whereas the lateral electrostatic force is sinusoidal. Thus the maximum of the lateral Casimir force will not occur at the same phase as that of the lateral electrostatic force. To find the amplitude of the lateral electrostatic force which will occur for a phase difference of $\phi = \pi/2, 5\pi/2 \ldots$. The value of the lateral Casimir force has to be subtracted from that of the total lateral force at the same phase. To accomplish this we fit the deflection data of the lateral Casimir force to a function of the form $\Sigma A_k \sin(k\phi)$ with $k = 1, 2, \ldots, 5$ and its value at $\phi = \pi/2, 5\pi/2 \ldots$ was determined. The values of the lateral Casimir force so determined for different separations are shown in Fig. 6(a). The amplitudes of the total force at the same phase is also shown in the figure. As the total lateral force and lateral Casimir force were determined independently at different times, the separation distances in the two measurements are not identical. Therefore, we need to find the value of the lateral Casimir force signal at the separation distance where the total force signal was measured by interpolation. Then we subtracted the lateral Casimir force from the total force. The obtained lateral electrostatic force signal is shown in Fig. 6(b).

By fitting the lateral electrostatic force signal to Eq. (4), we can obtain the force calibration constant of the cantilever $k_{ben} = 1.27 \pm 0.06$ nN/V and the separation distance at contact $z_0 = 117.3 \pm 2.9$ nm.

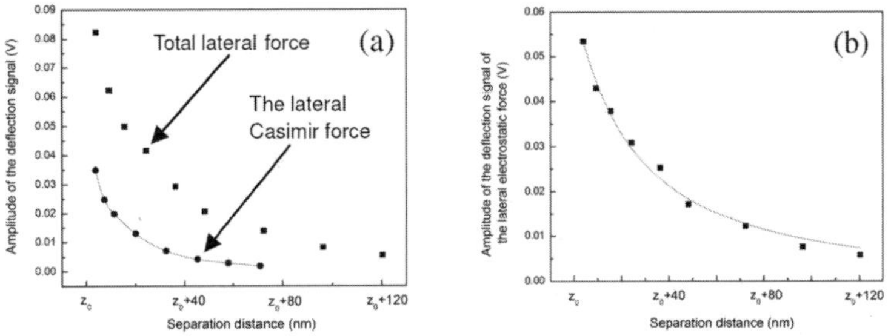

Fig. 6. (a) The amplitude of the deflection signal obtained from the total lateral force and that of the lateral Casimir force at the corresponding phase $= \pi/2$. The cubic spline fit used in the interpolation of the lateral Casimir force values is shown by the solid line. (b) The deflection signal of the lateral electrostatic force after subtraction of the lateral Casimir from the signal of total lateral force. The solid line is the best fit to Eq. (4) to obtain k_{ben} and z_0.

4. Results

The above measurement of k_{ben} can be used to convert the deflection signal to a force. Similarly the measurement of z_0 can be used to convert the relative separation in Fig. 6 to absolute values. The resulting measured lateral Casimir force is shown as dots as a function of the phase shift between the two corrugated surfaces in Fig. 7. This lateral Casimir force was measured at a separation distance $z = 124.7$ nm for corrugations with amplitude $A_1 = 85.4 \pm 0.3$ nm and $A_2 = 13.7 \pm 0.4$ nm. There are no fitting parameters between the theory and experimental results in this figure. Similar results were obtained at other separations. The solid line is the lateral Casimir force calculated using scattering theory. The dielectric properties of Au were described by the generalized plasma-like permittivity with the plasma frequency equal to 9 eV [4]. The asymmetry of the lateral Casimir force can be clearly seen in Fig. 7. This asymmetry occurs because of the larger corrugation amplitudes and shorter separation distances at which the measurements were done. The asymmetry can be quantified by the phase shift of the maximum force amplitude from the midpoint of the two adjacent minima. In Fig. 7, the average value of this phase shift is $0.12 \pm 0.02\lambda$.

Next, we compare the average maximum lateral Casimir force at 8 different separation distances from $z = 124.7$ nm to $z = 188.1$ nm with that calculated using scattering theory (solid line) and the PFA (dashed line) in Fig. 8. The data are shown as crosses with total error bars (random and systematic) at the 95% confidence level. The measured lateral Casimir force is in excellent agreement with the scattering theory. The disagreement with

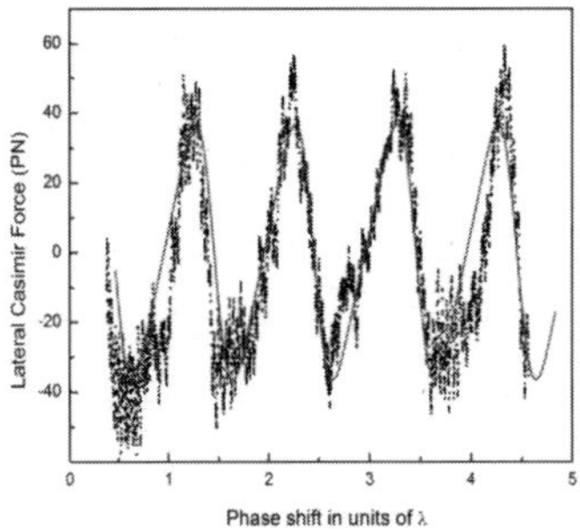

Fig. 7. The measured lateral Casimir force for a sepration between the corrugations of $z = 124.7$ nm is shown as dots as a function of the phase shift between the corrugations. The prediction of the scattering theory at the same separation distance is shown by the solid line. There is no fitting parameter between the experiment and the theory.

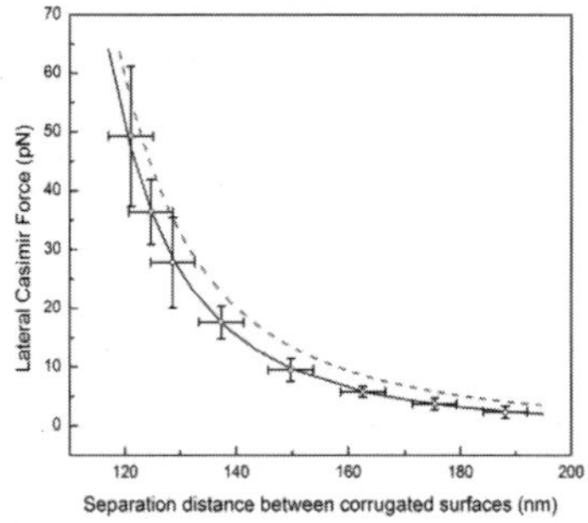

Fig. 8. The measured lateral Casimir force at different sepration distances along with the error bars is shown as crosses. They are compared with the scattering theory (solid line) and the PFA (dashed line). The error bars are indicated at a 95% confidence level.

PFA clearly points to the importance of diffraction-like correlation effects of the zero-point photons.

5. Conclusion

In this paper, some experimental features of the lateral Casimir force were presented. Procedures for measuring the drift and correcting the tilt in the mounting of the corrugations were discussed. An overview of the data analysis for the determination of the calibration constants and the absolute separation between the corrugated surfaces was presented. The data of the measured lateral Casimir force along with the errors were shown and compared to two different theories, one of which, the scattering theory includes the role of the coherent scattering of the photons and the other the PFA ignores the correlation effects. The data was shown to agree with the former and deviate from the latter. The lateral Casimir force has been proposed as the basis for realizing frictionless translation with nano rack and pinion structures for future nano devices [27–29]. This investigation will help in their realization.

Acknowledgments

This work was supported by the NSF Grant No. PHY0653657 (measurement) and DOE Grant No. DE-FG02-04ER46131 (calculation and analysis). The authors wish to thank G. L. Klimchitskaya, V. N. Marachevsky and V. M. Mostepanenko for their help with the theory and analysis. U. M. was also supported by DARPA Grant No. N66001-09-1-2069.

References

1. H. B. G. Casimir, Proc. K. Ned. Akad. Wet. B **51**, 793 (1948)
2. P. W. Milloni, "The quantum vacuum: An introduction to quantum electrodynamics", Academic Press (1994)
3. K. A. Milton, "The Casimir effect", World Scientific Publishing Company (2001)
4. M. Bordag *et al.*, "Advances in the Casimir Effect", Oxford University Press, Oxford, (2009)
5. T. H. Boyer, Phys. Rev. A **9**, 2078 (1974)
6. W. Lukosz, Physica (Utr.) **56**, 109 (1971)
7. R. Golestanian and M. Kardar, Phys. Rev. Lett. **78**, 3421 (1997)
8. H. Gies and K. Klingmüller, Phys. Rev. Lett. **96**, 220401 (2006)
9. T. Emig *et al.*, Phys. Rev. Lett. **96**, 080403 (2006)
10. O. Kenneth and I. Klich, Phys. Rev. Lett. **97**, 160401 (2006)
11. R. L. Jaffe and A. Scardicchio, Phys. Rev. Lett. **92**, 070402 (2004)
12. R. Büscher and T. Emig, Phys. Rev. Lett. **94**, 133901 (2005)
13. S. Reynaud, P. A. M. Neto and A. Lambrecht, J. Phys. A: Math. Theor. **41**, 164004 (2008)
14. R. Golestanian, Phys. Rev. A **80**, 012519 (2009)
15. F. Chen *et al.*, Phys. Rev. Lett. **88**, 101801 (2002)
16. F. Chen *et al.*, Phys. Rev. A **66**, 032113 (2002)
17. T. Emig *et al.*, Phys. Rev. A **67**, 022114 (2003)
18. S. Reynaud, P. A. M. Neto and A. Lambrecht, J. Phys A: Math. Theor. **41**, 164004 (2008)
19. A. Lambrecht and V. N. Marachevsky, Phys. Rev. Lett. **101**, 160403 (2008); Int. J. Mod. Phys. A **24**, 1789 (2009)
20. A. Roy and U. Mohideen, Phys. Rev. Lett. **82**, 4380 (1999)
21. H. B. Chan *et al.*, Phys. Rev. Lett. **101**, 030401 (2008)
22. H.-C. Chiu *et al.*, Phys. Rev. B **80**, 121402(R) (2009)
23. H.-C. Chiu *et al.*, Phys. Rev. B **81**, 115471 (2010)
24. H.-C. Chiu *et al.*, J. Phys. A: Math. Theor. **41**, 164022 (2008)
25. F. Chen and U. Mohideen, Rev. Sci. Instrum. **72**, 3100 (2001)
26. COMSOL Multiphysics, www.comsol.com
27. A. Ashourvan, M. Miri and R. Golestanian, Phys. Rev. Lett. **98**, 140801 (2007)
28. A. Ashourvan, M. Miri and R. Golestanian, Phys. Rev. E **75**, 040103(R) (2007)
29. T. Emig, Phys. Rev. Lett. **98**, 160801 (2007)

LATERAL CASIMIR-POLDER FORCE MEASUREMENT USING BOSE EINSTEIN CONDENSATES

G. A. MORENO

CONICET and Departamento de Física, FCEN, Universidad de Buenos Aires, Ciudad Universitaria, 1428 Buenos Aires, Argentina
E-mail: morenog@df.uba.ar

In this paper we explore the possibility of using Bose Einstein condensates to measure the lateral Casimir-Polder interaction between an atom and a corrugated surface. Among the possible scenarios we calculate the effect of the corrugated surface on the Bose Einstein condensate low energy spectrum and show how it could be measured via Bragg spectroscopy.

Keywords: BEC; Casimir-Polder force.

1. Introduction

Cold atomic gases have been shown to be suitable probes to measure the normal Casimir-Polder (CP) interaction[1] between an atom and a surface. In this work we investigate the possibilities of using Bose-Einstein condensates (BECs) to measure the lateral component of the Casimir-Polder force in presence of a corrugated surface. The main idea involved in the method we propose is that the corrugated surface will modify qualitatively the low energy spectrum of the BEC. This modified spectrum would reveal the information necessary to reconstruct the CP lateral potential through its Fourier transform. In the following sections we will see how the surface modifies the spectrum of a very elongated condensate and calculate the effect that would be observed in a typical experimental situation.

2. BEC near a Corrugated Surface

Let us consider a tightly confined BEC near a corrugated surface, as shown in Figure 1. Due to the fact that CP interactions are relatively weak in the μm scale, the effect of the surface can be included as a small perturbation in the Gross-Pitaevskii equation which describes the mean-field dynamics

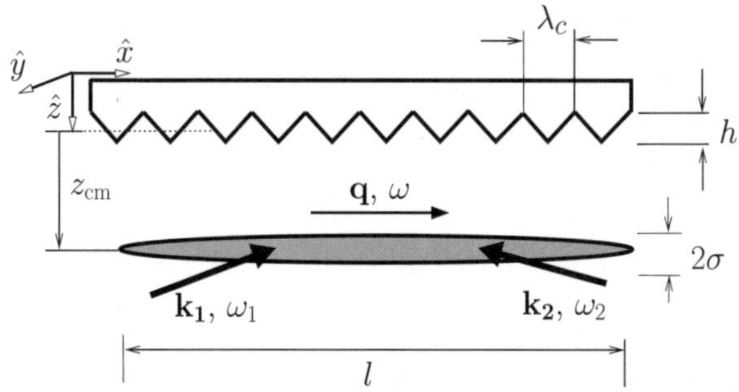

Fig. 1. Set-up for probing Casimir atom-surface interactions by measuring the energy spectrum via Bragg spectroscopy. The energy spectrum of an elongated BEC trapped parallel to a corrugated surface is modified by the lateral CP atom-surface interaction energy.

of the cold atomic cloud,[2] namely:

$$i\hbar\partial_t\varphi = -(\hbar^2/2m)\nabla^2\varphi + [U_N(z) + U_L(x,z)]\varphi \\ +(m/2)(\omega_r^2 r^2 + \omega_x^2 x^2)\varphi + g|\varphi|^2\varphi, \tag{1}$$

where φ is the condensate wavefunction, m is the atomic mass, $g = 4\pi\hbar^2 a/m$, a is the s-wave scattering length, and ω_r (ω_x) is the radial (axial) trapping frequency, $\omega_r \gg \omega_x$. This is a 3D equation but if the radial confinement is tight enough the radial dynamics will be frozen and the condensate wavefunction can be written in the form: $\phi_0(x,t)f_0(r)$, with $f_0(r)$ being the lowest eigenfunction of the operator $-(\hbar^2/2m)\Delta_r + m\omega_r^2 r^2/2$. Thus we can find an effective GP equation for $\phi_0(x,t)$ projecting (1) onto $f_0(r)$ (a more rigorous analysis shows that $\mu - \hbar\omega_r \ll 8\hbar\omega$ must be fulfilled in order to decouple this radial mode from the others, μ being the chemical potential of the BEC[3]). Before doing such projection we expand the potential around a suitable point, namely we expand:

$$\underbrace{\frac{1}{2}m(\omega_z^2(z-z_1)^2 + \omega_y^2 y^2 + \omega_x^2 x^2)}_{trapping} + \underbrace{U(z) + U_L(x,z)}_{Casimir-Polder}, \tag{2}$$

around a point z_0 chosen in such a way that:

$$m\omega_z^2(z_0 - z_1) + \left.\frac{\partial U(z)}{\partial z}\right|_{z=z_0} = 0\ . \tag{3}$$

This yields (order by order):

$$\begin{aligned}
&\frac{1}{2}m\left(\omega_z^2(z_0-z_1)^2+\omega_y^2y^2+\omega_x^2x^2\right)+U(z_0)+U_L(x,z_0)+\\
&\frac{\partial U_L(x,z_0)}{\partial z}(z-z_0)+\\
&\frac{1}{2}m\omega^2(z-z_0)^2+\mathcal{O}(\partial_z^2U_L(z,x)(z-z_0)^2)+\\
&\mathcal{O}(\partial_z^3U(z)(z-z_0)^3)+\mathcal{O}(\partial_z^3U_L(z,x)(z-z_0)^3)+\\
&\mathcal{O}(\partial_z^4U(z)(z-z_0)^4)+\mathcal{O}(\partial_z^4U_L(z,x)(z-z_0)^4)+\dots, \qquad (4)
\end{aligned}$$

where:

$$\omega^2=\omega_z^2+\frac{1}{m}\frac{\partial^2U(z)}{\partial z^2}\bigg|_{z=z_0} \qquad (5)$$

Now we are ready to project equation (1) onto $f_0(r)$. To obtain the effective potential for the 1D problem we integrate equation (4) with a Gaussian weight $\propto e^{-r^2/\sigma^2}$ where $\sigma^2=\hbar/m\omega^3$ (remember $r=\sqrt{y^2+(z-z_0)^2}$). This results in the following 1D potential:

$$U_{eff}(x)=U_0+U_L(x,z_0)+\mathcal{O}\left(\left(\frac{\sigma}{z_0}\right)^2U_L(x,z_0)\right)+\mathcal{O}\left(\left(\frac{\sigma}{z_0}\right)^4U(z_0)\right). \qquad (6)$$

For simplicity, in the previous equation we have taken $\omega_y=\omega$, in order for the cross-section of the condensate to be rotational symmetric. Other choices of ω_y will result in less symmetric cross-section profiles. The nonlinear term in (1) can be easily integrated yielding also a cubic term but with coupling constant $g_{\text{eff}}=g/2\pi\sigma^2$. Therefore, the BEC mean-field dynamics can be described by

$$\begin{aligned}
i\hbar\partial_t\varphi_0(x,t)=&-(\hbar^2/2m)\frac{d^2\varphi_0(x,t)}{dx^2}+U_L(x,z_0)\varphi_0(x,t)\\
&+(m/2)(\omega_x^2x^2)\varphi_0(x,t)+g_{\text{eff}}|\varphi_0(x,t)|^2\varphi_0(x,t). \qquad (7)
\end{aligned}$$

This is the main equation we should consider for the description of the system near the corrugated surface.

3. The Modified Energy Spectrum

We now proceed to calculate the spectrum of excitations, which we expect to be modified due to the presence of $U_L(x,z_0)$. As a first approximation we can neglect the axial confinement because we are assuming $\omega\gg\omega_x$ and $l\gg\lambda_c$, so that the BEC sees a local potential along the x direction just due

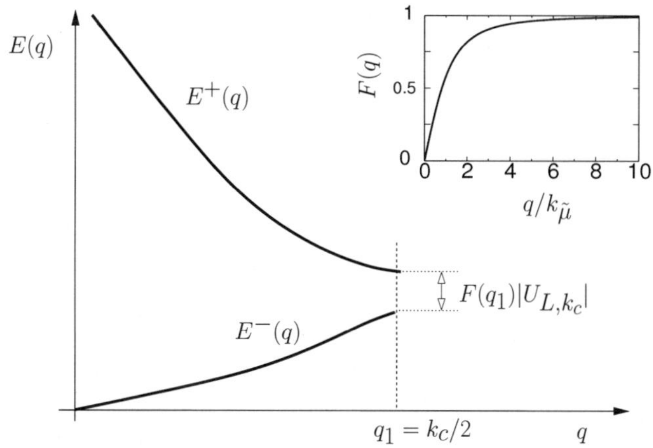

Fig. 2. Modified energy spectrum of an elongated BEC trapped parallel to a surface in the presence of a weak periodic lateral Casimir atom-surface interaction The inset shows the function $F(q)$ that modulates the energy gaps $\Delta E_{q_n} = F(q_n)|U_{L,nk_c}|$.

to the corrugated surface. We can compute the modification to the energy Bogoliubov spectrum of an homogeneous BEC using first order perturbation theory. Solving equation (7) perturbatively one finds that the spectrum is divided into Bloch-bands and has gaps when the wave-number difference in a plane wave expansion is commensurate with the CP potential, the gap reads:

$$\Delta E_{q_n} = |U_{L,nk_c}| \times F(q_n) \; ; \; F(q) = T_q/E_q^{(0)}, \tag{8}$$

where $E_q^{(0)} = E_{\mathrm{B}}(q) = \sqrt{T_q(2\tilde{\mu} + T_q)}$, $T_q = \hbar^2 q^2/2m$ and $\tilde{\mu}$ is the chemical potential. The Fourier coefficient U_{L,nk_c} gives the order of magnitude of the gap and is modulated by the dimensionless suppression factor $F(q)$ (note that in the non-interacting case this factor is one, recovering the usual result for single particle states in a periodic potential). In Fig. 2 we show the modified spectrum and the gap due to the presence of the corrugated surface. In the following section we will evaluate explicitly this effect for a typical experimental situation based on first order approximations for the lateral CP interaction.[4]

4. Measurement of the Spectrum and Numerical Estimates

The energy spectrum of the BEC can be measured using Bragg spectroscopy.[5] In this technique the momentum transferred by two laser beams

depends on two parameters, the detuning ω and difference in the wavenumbers q of the lasers. The momentum transferred to the BEC can be measured by time of flight and it can be shown[3] that the rate of momentum transferred is resonant when the condition $E(q) = \hbar\omega$ is fulfilled. Thus this would give us the necessary information to reconstruct the Fourier coefficients of the lateral CP potential, because once the spectrum is known eq. (8) yields the Fourier coefficients of the lateral CP interaction.
As an example, let us evaluate the effect for a typical experimental situation using a BEC of 10^4 atoms confined to a radius of 0.2μm at 3μm from a perfect metallic surface with 1μm corrugation amplitude and 10μm corrugation period. Under this situation the gap is $\hbar 0.1$Hz centered at $\hbar 485$Hz (see[3] for details). For a larger signal shorter distances should be considered. For example, for $z = 0.7\mu$m, $\lambda_c = 4\mu$m, and $h = 50$nm (see Fig. 1), one obtains a gap of 25Hz centered at $E = 1.2$kHz. This effect is close to sensitivities reported in previous experiments[5] but, at present, the minimum distance to the surface is about 2μm. Measuring this effect seems challenging with present day technology. However, we expect that future experimental improvements will make such a measurement feasible, opening a new window on the physics of the interaction between surfaces and coherent matter.

5. Conclusions

We have analysed the possibility of using BECs to measure the effect of the lateral Casimir-Polder interaction between an atom and a corrugated surface. We have shown how the low energy spectrum of an elongated BEC is modified by the presence of the surface and estimated the effect of a Bragg spectroscopy-based measurement of the Fourier transform of the CP lateral potential.

References

1. D. M. Harber, J. M. Obrecht, J. M. McGuirk, and E. A. Cornell, Phys. Rev. A **72**, 033610 (2005); J. M. Obrecht, R. J. Wild, M. Antezza, L. P. Pitaevskii, S. Stringari, and E. A. Cornell, Phys. Rev. Lett. **98**, 063201 (2007).
2. L. P. Pitaevskii and S. Stringari, *Bose-Einstein Condensation* (Oxford University Press, Oxford, 2003).
3. G. A. Moreno, D. A. R. Dalvit, E. Calzetta. `ArXiv: 0904.0238`.
4. D. A. R. Dalvit, P. A. Maia Neto, A. Lambrecht, and S. Reynaud, Phys. Rev. Lett. **100**, 040405 (2008); J. Phys. A: Math. Theo. **41**, 164028 (2008).
5. J. Steinhauer, R. Ozeri, N. Katz, and N. Davidson, Phys. Rev. Lett. **88**, 120407 (2002); J. Steinhauer, N. Katz, R. Ozeri, N. Davidson, C. Tozzo, and F. Dalfovo, Phys. Rev. Lett. **90**, 060404 (2003).

REPULSIVE CASIMIR AND VAN DER WAALS FORCES: FROM MEASUREMENTS TO FUTURE TECHNOLOGIES

J. N. MUNDAY

Thomas J. Watson Laboratories of Applied Physics, California Institute of Technology, Pasadena, CA 91125, USA

FEDERICO CAPASSO

School of Engineering and Applied Sciences, Harvard University, Cambridge, MA 02138, USA

By engineering the boundary conditions of electromagnetic fields between material interfaces, one can dramatically change the Casimir-Lifshitz force between surfaces as a result of the modified zero-point energy density of the system. Repulsive interactions between macroscopic bodies occur when their dielectric responses obey a particular inequality, as pointed out by Dzyaloshinskii, Lifshitz, and Pitaevskii. We discuss experimental verification of this behavior as well as a description of how this can be used to develop a scheme for quantum levitation. Based on these concepts, we discuss the possible development of a new class of devices based on ultra-low static friction and the ability to sort objects based on their dielectric functions.

Keywords: Casimir; van der Waals; Lifshitz; Levitation.

1. Introduction

The confinement of electromagnetic fields between material surfaces can result in a force between the latter due to quantum fluctuations of the former, which has many interesting features. First, this force represents a macroscopic manifestation of the quantum nature of the vacuum and is measurable using current experimental techniques. Second, a closer examination of several phenomena in nature shows strong evidence that adhesion, friction, wetting and stiction are fundamentally a result of these quantum fluctuations. Third, with the continued miniaturization of devices to the nanoscale, the ability to engineer the vacuum fluctuations between bodies may pave the way for improved device architectures, assembly methods, or functionalities. In this contribution, we will briefly discuss recent measurements of both long- and short-range repulsive forces, measurement schemes for future experiments, and technological opportunities that take advantage of the ability to modify these forces resulting from the confinement of vacuum fluctuations.

2. The Casimir-Lifshitz Force

The general expression for the force between two semi-infinite plates separated by a third medium as a result of the quantum fluctuations of the electromagnetic fields was first derived by Dzyaloshinskii, Lifshitz, and Pitaevskii [1]. In this formulation, the force between two uncharged surfaces, composed of either metals or dielectrics, is derived using the fluctuation-dissipation theorem. Because measurements of the Casimir force are usually compared to this generalized theory rather than the special case of ideal metals developed by Casimir, we refer to the resulting force as the Casimir-Lifshitz force.

Lifshitz's theory has various limiting forms depending on the materials involved and their separations. At very small separations (typically less than a few nm), Lifshitz's theory provides a complete description of the non-retarded van der Waals force. At larger separations, retardation effects give rise to a long-range interaction that in the case of two ideal metals in vacuum reduces to Casimir's result. Thus, both the Casimir force and the van der Waals force are of quantum electrodynamical (QED) origin, but the key physical difference is that in the Casimir case, the retarded nature of the interaction due to the finite speed of light cannot be neglected, as in the van der Waals limit. Retardation effects are actually dominant and lead to a change in the power law of the force with distance [2]. This is true for all materials (metals or dielectrics) when the propagation time of light between the bodies is greater than the inverse characteristic frequency of the materials [2], which for metals is the plasma frequency. The complete theory for macroscopic bodies is valid for any distance between the surfaces and includes, in a consistent way, both limits [1].

3. Origin of Repulsive Forces

As was demonstrated by Dzyaloshinskii, Lifshitz, and Pitaevskii in their seminal paper, the sign of the force depends on the dielectric properties of materials involved [1]. Two plates made out of the same material will always attract, regardless of the choice of the intermediate material (typically a fluid or vacuum); however, between slabs of different materials (here labeled

1 and 2) the force becomes repulsive by suitably choosing the intermediate liquid (labeled 3). Thus, by proper choice of materials, the Casimir-Lifshitz force between slabs 1 and 2 can be either attractive or repulsive. Specifically, the condition for repulsion is:

$$\epsilon_1(i\xi) > \epsilon_3(i\xi) > \epsilon_2(i\xi). \tag{1}$$

Here the dielectric functions ϵ_1, ϵ_2, and ϵ_3 of the materials (Fig. 1) are evaluated at imaginary frequencies [3]. Because they vary with frequency, it is conceivable that inequality [Eq. 1] may be satisfied for some frequencies and not for others. For various separations between the slabs, different frequencies will contribute with different strengths, which can lead to a change in the sign of the force as a function of separation (see for example Ref. [4]).

Fig. 1. Geometry of the setup. Two plates separated by an intermediate fluid.

In order to qualitatively understand the origin of these repulsive forces, we consider the following toy model (see Fig. 2) for the microscopic interaction of the bodies [5]. To first order, the force between the latter is dominated by the pair-wise summation of the van der Waals forces between all the constituent molecules. This additivity is a good approximation for rarefied media; however, the force between two molecules is affected in general by the presence of a third. Hamaker first used this approach in extending the calculations of London to the short-range interaction (i.e. the non-retarded van der Waals force) between bodies and in particular to those immersed in a fluid [6]. Using the previous subscript notation for the three materials and their constituent molecules and suitably choosing them so that their polarizabilities satisfy the inequality $\alpha_1 > \alpha_3 > \alpha_2$, we find the forces between the individual molecules, which are proportional to the product of the polarizabilites integrated over all imaginary frequencies, will obey: $F_{13} > F_{12} > F_{23}$ (Fig. 2). Thus, it is energetically more favorable for molecule 3 to be near molecule 1 than it is for molecule 2 to be near molecule 1. As more molecules of the same species are added to the system, molecules of type 3 will be strongly attracted to those of type 1, resulting in an increased separation for molecules of type 2 from those of type 1. In this way, Hamaker showed, that repulsive forces between two different materials immersed in a liquid are possible by calculating the total interaction energy between the bodies and the fluid as the separation between the bodies is varied. His calculations however were non-rigorous since they neglected non-additivity and retardation effects. When these are included, long-range repulsion between two bodies (materials 1 and 2) separated by a third (material 3) is predicted when their relative dielectric functions obey [Eq. 1]. Note that when the fluid has the largest dielectric function, the cohesive van der Waals interaction within the fluid will results in an attraction between its molecules that is larger than that between the molecules of the fluid and the plate, which leads to an attractive force between the two plates.

Several examples of material systems that obey [Eq. 1] exist in nature. One of the earliest triumphs of Lifshitz's theory was the quantitative explanation of the thickening of a superfluid helium film on the walls of a container [1, 7]. For that system, the dielectric function of liquid is intermediate between that of the container and the surrounding vapor. Thus, it is energetically more favorable for the liquid to be between the vapor and the container, and the liquid climbs the wall. Of course the fact that the superfluid has also zero viscosity means that the fluid can easily spread and even flow out of a container depending on its height. Many other examples of this QED repulsion exist in the realm of wetting or non-wetting of a surface by a fluid, which to a first approximation, is governed by the same inequality for the dielectric functions.

4. Measurements of Repulsive Forces

Few material systems, consisting only of solids separated by a liquid, obey the inequality [Eq. 1] over a large frequency range; however, over the past decade there have been a limited number of force measurements for such systems. In this section we will briefly describe the commonalities and differences between these measurements. Although many methods have been developed to study surface forces [8, 9, 26, 27], the atomic force microscope (AFM) is one of the most widely used methods due to the versatility of material surfaces and surrounding environments that can be studied.

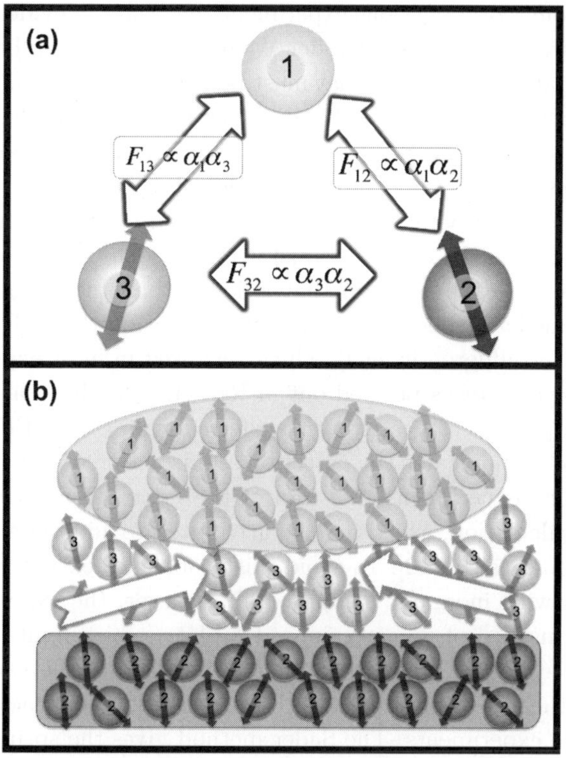

Fig. 2. Toy model of repulsive Casimir-Lifshitz forces. Repulsive forces can exist between two materials, schematically represented as an ensemble of molecules separated by a third, typically a liquid, with specific optical properties. (a) Three individual molecules will all experience attractive interactions. (b) For a collection of molecules, with $\alpha_1 > \alpha_3 > \alpha_2$, it is energetically more favorable for the molecules with the largest polarizabilities (α_1 and α_3 for this example) to be close, resulting in an increased separation between molecules of type 1 and type 2. For a condensed system, the net interaction between material 1 and material 2 is repulsive if the corresponding dielectric functions satisfy $\epsilon_1 > \epsilon_3 > \epsilon_2$, as consequence of the similar inequality between polarizabilities. Note that all the α's and the ϵ's need to be evaluated at imaginary frequencies (see text).

Shortly after the development of the AFM [9], Ducker *et al.* performed one of the first measurements of the force between a colloidal sphere attached to a cantilever and a surface using AFM [10]. To our knowledge, all measurements of repulsive quantum forces to date (in both the retarded and non-retarded regimes) have used a similar setup; however, a variety of calibration methods and experimental techniques were used to probe different distance ranges with different materials.

Prior to our work, pervious experiments have shown evidence for short-range repulsive forces in the van der Waals regime [11–16]; however, there are many experimental issues that must be considered that, as our analysis below shows, were not adequately addressed in many of these experiments. For separations of a few nm or less, liquid orientation, solvation, and hydration forces become important and should be considered, which are not an issue at larger separations. Surface charging effects are important for all distance ranges. In order to satisfy [Eq. 1], one of the solid materials must have a dielectric function that is lower than the dielectric function of the intermediate fluid. One common choice for this solid material is PTFE (polytetrafluoroethylene), which was used in most experiments [12, 14–16]; however, as was pointed out in Ref. [12], residual carboxyl groups and other impurities can easily be transferred from the PTFE to the other surface, which complicates the detection and isolation of the van der Waals force. In a few experiments, the sign of the force did not agree with the theoretical calculation, which may be attributed to additional electrostatic force contributions [11, 12]. To avoid this problem, Meurk *et al.* performed experiments with inorganic samples [13]; however, the experimental configuration consisted of a sharp tip and a plate, which limited the surface separations to below 2 nm. For the determination of the cantilever force constant, either the Sader method [17] or the Cleveland method [18] was used in these experiments. The Sader method gives the spring constant of a cantilever based on the geometry of the cantilever and its resonance frequency, and the Cleveland method uses the resonance frequency shift of a cantilever upon the addition of masses to determine the spring constant. These methods lead to an additional 10-20% error in the determination of the force [19], which could be greatly reduced if a calibration method is performed that uses a known force for the calibration [20–22]. Finally, the determination of the absolute distance was often found by performing a fit of the experimental data to the presumed power law of the van der Waals force [12, 14–16]. Thus, the absolute surface separation could only be determined if one assumed that the measured force was only the van der Waals force and that it was described precisely by a $1/d^2$ force law.

In our recent experiment [23], we measured the long-range repulsive Casimir-Lifshitz force using (a) an improved force and distance calibration scheme, (b) methods to determine and reduce spurious electrostatic forces, and (c) spheres and cantilevers that would allow detection of weak forces at large surface separations. To calibrate the cantilever force constant and the surface separation at contact, a known force, the hydrodynamic force, was applied between the sphere and the plate [20–23]. This allows for in situ calibration and only assumes that the unknown force to be measured is independent of velocity. No assumption about the distance dependence of the force is made to determine the absolute surface separation. To ensure accuracy in the relative distance between the sphere and the plate, a linear variable differential transformer (Asylum Research MFP-3D) is used to control the piezo column, which advances the sphere toward the plate. This reduces hysteresis and nonlinearities inherent in piezoelectric transducers. To minimize electrostatic forces, we chose to use inorganic samples to avoid contamination by charge bearing groups often found with polymers. Further, we performed several investigations to ensure that electrostatic forces were negligible by removing stray charges and fields typically present near the apparatus [22] and performing electrostatic force microscopy on the samples [23–25]. Lifshitz's equation was computed using available dielectric data and corrections for surface roughness (as measured on both the sphere and the plate) to allow for an independent comparison of the theory and the experiment without any fitting parameters. This allowed us to conclude that theory and experiment are consistent within their numerical uncertainties and experimental errors, respectively.

5. Future Measurement Directions and Technological Opportunities

Although the AFM has been the instrument of choice for many surface force measurements, particularly for large surface separations, there is vast literature on other experimental techniques that may be of interest for measuring long-range surface forces in fluids. Techniques include the surface force apparatus (SFA) [8], total internal reflection microscopy (TIRM) [26], video tracking of colloids [27], and their various modifications. We will not discuss these schemes in detail but rather mention that these setups are capable of measuring forces either between macroscopic bodies [8] or between freely moving or confined particles above a surface [26, 27]. The latter offers the opportunity of observing quantum levitation by the Casimir-Lifshitz force without the support of a cantilever.

The ability to modify the Casimir-Lifshitz force opens the door to the possibility of engineering the potential energy landscape for particles based purely on their dielectric functions. Figure 3 shows an example of this behavior. With the appropriate choice of fluid, repulsive forces will occur for asymmetric configurations (Au-SiO2 in this case), while attractive forces will occur for symmetric configurations (Au-Au or SiO2-SiO2). By pattern-

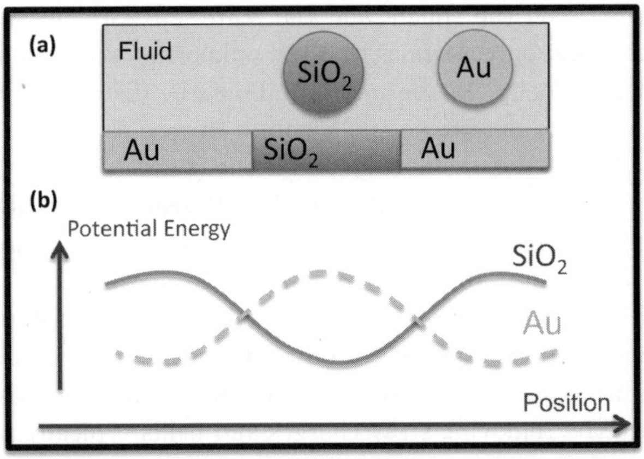

Fig. 3. Schematic of a method aimed at engineering the potential energy landscape of quantum fluctuations. (a) Two spheres made of different materials immersed in a fluid above a plate, which is also composed of two materials. The fluid and the two materials satisfy the inequality of dielectric functions discussed in the text [Eq. 1] (b) Schematic of the potential energy for both the Au (dashed) and the SiO2 (solid) spheres as a function of position for a fixed height above the plate. Each sphere experiences a different minimum energy configuration.

ing a plate with these two different materials, one can study both non-additivity effects and the assembly and sorting of particles based solely on their dielectric functions. Similar sorting and aggregation effects have been observed in the thermodynamic Casimir effect, which is related to classical density fluctuations [28]. The ability to achieve both attractive and repulsive Casimir-Lifshitz forces could be of significant interest technologically as well. One technique might be to develop ultra-sensitive force and torque sensors by counterbalancing gravity to levitate an object immersed in fluid above a surface without disturbing electric or magnetic interactions. Based on this idea, we proposed several devices that would be free to rotate or translate with virtually no static friction [29, 30]. Recent friction measurements have shown that ultralow friction can be obtained in such a configuration [16]. While dynamical damping due to viscosity will put limits on how quickly such a device can respond, in principle even the smallest translations or rotations can be detected on longer time scales. Thus, force and torque sensors could be developed that surpass those currently used. Other recent proposals include the ability to tune chemical reactions [31] and the self-assembly of colloidal scale devices [32, 33] based, at least partially, on manipulating the Casimir-Lifshitz forces.

6. Conclusion

Conclusive experiments have been discussed that demonstrate that the sign of the Casimir-Lifshitz force can be changed by an appropriate choice of materials. Recent proposals and measurements suggest that the role of the Casimir effect may be important for future technologies based on engineering the boundary conditions imposed on the ever-fluctuating electromagnetic fields.

References

1. I. E. Dzyaloshinskii, E. M. Lifshitz, and L. P. Pitaevskii, *Advances in Physics* **10**, 165 (1961).
2. P. W. Milonni, *The Quantum Vacuum: An Introduction to Quantum Electrodynamics* (Academic, San Diego, 1993).
3. Note that $\epsilon(i\xi)$ corresponds to the continuation of $\omega(i\xi)$ in the complex plane and physically represents the material's response to exponentially increasing fields rather than oscillatory ones. $\epsilon(i\xi)$ is real and decrease monotonically to unity as ϵ tends to infinity. For a discussion of these points see L. D. Landau, E. M. Lifshitz, and L. P. Pitaevskii, *Electrodynamics of Continuous Media*, Elsevier, New York, 1984.
4. J. N. Munday et al., *Physical Review A* **71**, 042102 (2005).
5. We would like to acknowledge Lev Pitaevskii and Larry Ford for alerting us of a similar explanation using the optical analogy of the bouyancy force.
6. H. C. Hamaker, *Physica* **4**, 1058 (1937).
7. E. S. Sabisky and C. H. Anderson, *Physical Review A* **7**, 790 (1973).
8. J. N. Israelachvili and G. E. Adams, *Journal of the Chemical Society, Faraday Transactions I* **74**, 975 (1978).
9. G. Binnig, C. F. Quate, and C. Gerber, *Physical Review Letters* **56**, 930 (1986).
10. W. A. Ducker, T. J. Senden, and R. M. Pashley, *Nature* **353**, 239 (1991).
11. J. L. Hutter and J. Bechhoefer, *Journal of Applied Physics* **73**, 4123 (1993).
12. A. Milling, P. Mulvaney, and I. Larson, *Journal of Colloid and Interface Science* **180**, 460 (1996).
13. A. Meurk, P. F. Luckham, and L. Bergstrom, *Langmuir* **13**, 3896 (1997).
14. S. Lee and W. M. Sigmund, *Journal of Colloid and Interface Science* **243**, 365 (2001).
15. S. W. Lee and W. M. Sigmund, *Colloids and Surfaces A: Physicochemical and Engineering Aspects* **204**, 43 (2002).
16. A. Feiler, M. A. Plunkett, and M. W. Rutland, *Langmuir* **24**, 2274 (2008).
17. J. E. Sader *et al.*, *Review of Scientific Instruments* **66**, 3789 (1995).
18. J. P. Cleveland *et al.*, *Review of Scientific Instruments* **64**, 403 (1993).
19. C. T. Gibson, G. S. Watson, and S. Myhra, *Scanning* **19**, 564 (1997).
20. V. S. J. Craig and C. Neto, *Langmuir* **17**, 6018 (2001).
21. J. N. Munday and F. Capasso, *Physical Review A* **75**, 060102 (2007).
22. J. N. Munday *et al.*, *Physical Review A* **78**, 032109 (2008).

23. J. N. Munday, F. Capasso, and V. A. Parsegian, *Nature* **457**, 170 (2009).
24. B. D. Terris *et al.*, *Journal of Vacuum Science and Technology A* **8**, 374 (1990).
25. C. Guillemot *et al.*, *Europhysics Letters* **58**, 566 (2002).
26. D. C. Prieve and N. A. Frej, *Langmuir* **6**, 396 (1990).
27. J. C. Crocker and D. G. Grier, *Journal of Colloid and Interface Science* **179**, 298 (1996).
28. F. Soyka *et al.*, *Physical Review Letters* **101**, 208301 (2008).
29. D. Iannuzzi, J. Munday, and F. Capasso, *Ultra-low friction configuration.* US Patent Application US20070066494 (filed, 19 September 2005).
30. F. Capasso *et al.*, *IEEE Journal of Selected Topics in Quantum Electronics* **13**, 400 (2007).
31. D. P. Sheehan, *The Journal of Chemical Physics* **131**, 104706 (2009).
32. Y. K. Cho *et al.*, *Advanced Materials* **17**, 379 (2007).
33. K. J. M. Bishop *et al.*, *Small* **5**, 1600 (2009).

MACROSCOPIC QUANTUM VACUUM AND MICROSCOPIC GRAVITATION

ROBERTO ONOFRIO

Dipartimento di Fisica "Galileo Galilei", Università di Padova, Padova 35131, Italy
Department of Physics and Astronomy, Dartmouth College, Hanover, NH 03766, USA
ITAMP, Harvard-Smithsonian Center for Astrophysics, Cambridge, MA 02138, USA
E-mail: onofrior@gmail.com

Macroscopic quantum vacuum and modern theories of gravitation share the strong interplay between geometry and physical phenomena. We review selected issues related to the accuracy of the measurement of Casimir forces with particular emphasis on the implications for the search of non-Newtonian gravitational forces in the micrometer range. We then discuss the interplay of the Higgs particle with gravitation, arguing that spectroscopic shifts in atomic transitions due to the modifications of the vacuum expectation value of the Higgs field in regions with strong curvature of space-time may be of conceptual and observational relevance.

Keywords: Casimir Forces; Non-Newtonian Gravitation; Higgs Particle; Compact Astrophysical Objects.

1. Introduction

Casimir forces have been extensively studied in the last decade in a variety of experimental configurations [1, 2]. At the level of demonstrating the existence of this peculiar manifestation of quantum vacuum at the macroscopic level, the situation seems rather settled. A demonstration of an effect which is expected since it is based on *the* prevailing theory, *i.e.* renormalized quantum electrodynamics, without predictions originating from alternative approaches [3], seems rather uneventful and does not necessarily require an in-depth analysis of the data. Debate, however, is still going on to assess the precision and the accuracy of the measurements, since this has an impact on the potential discovery of new forces - predicted by a *variety* of models - with coupling comparable or larger than gravitation, and characterized by an interaction range at or below the micrometer scale [4–6]. In this framework, some previously unidentified systematic effects have been recently evidenced. If not properly taken into account in the data analysis and the experiment-theory comparison, neglecting these effects may translate into significant systematic errors. In this contribution we briefly review the status of the experimental and theoretical knowledge in regard to such systematic effects, and the impact of the latter on the current limits to non-Newtonian gravitational forces. A complementary tool to extend our knowledge of quantum effects in the macroworld is available by considering peculiar quantum effects in strong external fields. In this framework we discuss the possibility to study observable effects due to the influence of curved space-time on the vacuum expectation value of the Higgs field. It turns out that the mass of the electron and, to a smaller extent, of baryons, should be changed in the presence of a strong gravitational field. This in turn gives rise to spectroscopic shifts in atomic spectroscopy of a novel nature with respect to the usual Doppler, gravitational, or cosmological shifts, also raising conceptual puzzles about the assignment of masses via the Higgs mechanism.

2. Casimir force experiments and non-Newtonian limits in the micrometer range

All experiments performed on Casimir forces between conducting bodies since the first attempts by Spaarnay [7] have faced the difficulty of dealing with a voltage present even when the two surfaces are shorted. This potential, called contact or residual potential, is thought to arise from the different Volta potential of the conductors used for the electrical connections, with typical values in the $1-100$ mV range. Apart from being superimposed to the expected Casimir force signal, its presence precludes the possibility to reach small gaps as it will cause, without taking proper precautions, earlier contact between the two surfaces. This is particularly relevant for experiments, such as the one using AFM cantilevers, involving resonators with low stiffness. It is then customary to compensate the contact potential using an external bias voltage kept constant in the entire range of explored distances. However, while performing electrostatic calibrations of an apparatus aimed at measuring the Casimir force in the cylinder-plane configuration, we have found evidence for a distance-dependent contact potential. Intrigued by this finding, we have studied the simpler sphere-plane configuration, since the latter is free from possible issues related to the parallelism and border effects potentially important in the cylinder-plane case. Even in the sphere-plane configuration, we have found a dependence of the contact potential upon distance, as well as an anomalous scaling of the electrostatic force with distance [8]. While this second anomaly does not seem of universal character [9], the dependence of the contact potential on distance has been confirmed by both reanalyzing former experiments (as discussed in [10]), and performing new experiments [9, 11]. The presence of a distance-dependent contact potential creates an electrostatic force unevenly compensated at the various explored distances, unless on-line [9] or off-line [10] compensation techniques are adopted.

The situation remains far from being settled, since there are also several experimental results interpreted as if there is no distance dependence of the contact potential, within the precision of the measurement [12, 13]. This issue is of the outmost importance to assess the best upper limits on Yukawa-like forces using molecular and Casimir forces [14, 15], as well as claims of validation of one specific model to include the finite conductivity and finite temperature correction [16]. Indeed, both the upper limits on Yukawa forces and the validation of thermal corrections rely on the two experimental apparata in which no systematic dependence of the contact potential on distance is observed (see [17] for details). The effect observed in the other experiments is of the order of 4-10 mV in a range of distances of few μm. This voltage should be compared with the equivalent voltage corresponding to the Casimir force at a given distance [18]. When this is done,

Table 1. Summary of relevant formulas for the ideal Casimir force and the Coulomb force in the cases of the sphere-plane, cylinder-plane, and parallel plane geometries. In the first row, the Casimir force is expressed by regrouping the various numerical factors and variables in a common fashion, with numerical values first, then the product $\hbar c$ expected in any Casimir force formula, then the geometrical dependence. The second row represents, in a similar arrangement, the Coulomb force exterted between the various surfaces. In the third row the equivalent Casimir voltage, *i.e.* the bias voltage required to simulate the Casimir force at a given distance d, is reported. In the last row the concrete value of the equivalent Casimir voltage is reported in the case of a typical gap distance of 1 μm. Notice the universality of the equivalent Casimir voltage formulas, with just different numerical factors giving rise to a difference of less than a factor two between the two extreme geometries.

	Sphere-Plane	Cylinder-Plane	Parallel Planes
Casimir	$\left(\frac{\pi^3}{360}\right)\hbar c\left(\frac{R}{d^3}\right)$	$\left(\frac{\pi^3}{384\sqrt{2}}\right)\hbar c\left(\frac{La^{1/2}}{d^{7/2}}\right)$	$\left(\frac{\pi^2}{240}\right)\hbar c\left(\frac{S}{d^4}\right)$
Coulomb	$\pi\epsilon_0\left(\frac{R}{d}\right)V^2$	$\left(\frac{\pi\epsilon_0}{2\sqrt{2}}\right)\left(\frac{La^{1/2}}{d^{3/2}}\right)V^2$	$\left(\frac{\epsilon_0}{2}\right)\left(\frac{S}{d^2}\right)V^2$
$V_{\rm Cas}^{\rm eq}$(d)	$\left(\frac{\pi^2}{360}\right)^{1/2}\left(\frac{\hbar c}{\epsilon_0}\right)^{1/2}\frac{1}{d}$	$\left(\frac{\pi^2}{120}\right)^{1/2}\left(\frac{\hbar c}{\epsilon_0}\right)^{1/2}\frac{1}{d}$	$\left(\frac{\pi^2}{192}\right)^{1/2}\left(\frac{\hbar c}{\epsilon_0}\right)^{1/2}\frac{1}{d}$
$V_{\rm Cas}^{\rm eq}$ (1 μm)	9.85 mV	13.5 mV	17.1 mV

a simple and universal formula, apart from a numerical factor, emerges for the equivalent voltage of the three geometries. As seen in Table 1, the equivalent voltage of the Casimir force at 1 μm is in the 10-20 mV range depending on the specific geometry, and this implies that the expected signal at 1 μm for the Casimir force may be comparable, within one order of magnitude, to the difference between the signals coming from the contact potentials at the two extremes of the explored range of distances.

The current setting is clearly unsatisfactory to the purpose of providing reliable and accurate limits to non-Newtonian forces, regardless of the extent of the systematic effect. Indeed, even without carrying out extensive assessments of the precision of the experiments, we can identify three possible scenarios. If this previously unidentified systematic effect is larger than the quoted systematic error in the experiments used to give the best limits on Yukawa forces, then limits should be revised taking into account this systematic source of error. If it is instead smaller than the quoted systematic error in the same experiments, then one needs to understand whether these experiments, unable to evidence such an effect unlike all others, are actually suitable to provide the best limits to non-Newtonian gravitation. Finally, there is the possibility that specific environmental factors varying from experiment to experiment, such as the geometrical quality and chemical contamination of the surfaces, the radius of curvature of the sphere, the vacuum level, specific electrostatic setting around the apparatus, and temperature control for instance, may explain the different observations. Evidently more experimental and phenomenological work in this direction will be required to provide a consistent framework.

A second issue only addressed quite recently is the validity of the proximity force approximation (PFA) [19] for Yukawa or, in general, volumetric forces. Since the PFA was conceived to deal with proximity forces, *i.e.* forces acting among entities in proximity of each other, it is not *a priori* understood what is their range of validity, and the level of accuracy, whenever PFA is applied to forces acting among entities in the bulk. A simplified form of PFA based on a virtual mapping between the actual sphere-plane configuration used in the experiments and an effective parallel plane configuration has been recently shown to differ from the general expression for the PFA to be used in the case of volumetric forces [20]. This latter form has been in turn shown to coincide with the exact force between two bodies, provided one is an indefinite plane [21]. In the same paper, the sensitivity of the simplified PFA approximation to an unphysical parameter used in the PFA mapping between the sphere-plane and the parallel plane case, the thickness of the hypothetical slab corresponding to the sphere, denoted D_2 in [20] and [21], has been also studied. Taking into account the dependence on this parameter, limits derived through the simplified PFA become quickly unreliable as the Compton wavelength of the Yukawa force approaches the micrometer range, *i.e.* the upper range of distances in which precision Casimir force measurements have been performed. Then the validity of the PFA in the case of volumetric forces of Yukawa type is limited not only by the usual constraint of distance being much smaller than the radius of curvature of the sphere, $a \ll R$, but also by Yukawa range $\lambda \ll R, D_2$. This limits its conceptual validity to the case of nearly pointlike interactions and, above all, to a region of distances smaller than the one in which actual experiments are performed. It seems therefore critical, as precision and accuracy of the experiments will continue to improve, to avoid using the PFA. The exact expression for the Yukawa force in the sphere-plane geometry, already available since a decade [22], has been used recently to improve the limits in the micrometer range using a torsional balance [23, 24].

3. Higgs shifts in astrophysical environments

One of the most important predictions of the standard model of particle physics is the existence of the only fundamental scalar particle held responsible for the spontaneous symmetry breaking of the electroweak sector, providing mass to the intermediate vector bosons $W^{\pm}$ and Z^0 and to all fundamental fermionic matter fields. The Higgs particle is a critical milestone of the standard model, and its discovery and detailed study is the primary focus of research undergoing at high energy accelerators such as Fermilab and, in the close future, at the Large Hadron Collider at CERN. The Higgs particle, if giving mass to all the constituents of matter and provided that it satisfies the equivalence principle, should also play a crucial role in gravitational phenomena. In particular, the Higgs vacuum expectation value in a region of strongly curved space-time should differ from the one in flat space-time. This should give rise to different values for the mass of particles such as electrons and protons, and then to shifts of energy levels of their bound states of spectroscopic relevance. New wavelength shifts in the emission or absorption spectra are then predicted, which could be detected by proper subtraction of the usual Doppler, gravitational, and cosmological shifts.

Quantum field theory in curved space-time has been studied for decades both for non-interacting and interacting fields [25]. The Lagrangian density for an interacting scalar field with parameters μ and λ in a generic curved space-time is written as:

$$L = \frac{1}{2} g^{\mu\nu} \partial_\mu \phi \partial_\nu \phi - \frac{1}{2}(\mu^2 + \xi R)\phi^2 - \frac{\lambda}{4}\phi^4. \tag{1}$$

In the presence of a curved space-time there is an added term to the Lagrangian density with respect to the case of a flat space-time, where ξ measures the coupling between the Higgs field ϕ and the curvature scalar R. In the *minimal* coupling scenario, we should have $\xi = 0$, which however is unnatural if we believe that the standard model at some energy will merge with gravitation, since it precludes any possible crosstalk between the two sectors. Moreover, it does not manifest proper renormalization group behavior, as outlined in [25] and, in a composite scalar model, in [26]. Alternatively, we consider as a working and/or reasonable hypothesis the other possibility of a *conformal* coupling, $\xi = 1/6$. In the spontaneously broken phase, the Higgs field develops a vacuum expectation value obtained by minimizing the effective potential. In the flat space-time this yields a value $v_0 = \sqrt{-\mu^2/\lambda}$, and the masses of the elementary particles are all directly proportional to v via the Yukawa coefficients y_i of the fermion-Higgs Lagrangian density term, $m_i = y_i v_0/\sqrt{2}$. In the presence of curved space-time, the effective coefficient of the Higgs field $\mu^2 \mapsto \mu^2 + \xi R$ and the vacuum expectation value of the Higgs field will become space-time dependent through the curvature scalar as:

$$v = \sqrt{-\frac{\mu^2 + \xi R}{\lambda}} \simeq v_0 \left(1 + \frac{\xi R}{2\mu^2}\right), \tag{2}$$

where the last expression holds for a weak curvature. Notice that the vacuum expectation value is increased by the presence of a curved space-time corresponding to $R > 0$ in the minimal coupling scenario. In the case of elementary particles such as the electrons, provided that the Yukawa couplings are constants yet to be determined - as commonly believed, from algebraic/group theoretic arguments of an underlying fundamental theory incorporating the standard model - the mass m_e will be simply changed proportionally to the Higgs vacuum expectation value, so that in the case of the electron $\delta m_e = y_e(v - v_0)/\sqrt{2} \simeq y_e \xi R/(2^{3/2}\mu^2)$.

The situation for composite particles such as protons and neutrons is more involved. We assume that their masses are made of a flavor-dependent contribution proportional to the masses of the three valence quarks determined by the Higgs coupling, and a color-symmetrical term only dependent on the quark-quark and quark-gluon interaction, *i.e.* proportional to

$\Lambda_{QCD} \simeq 300$ MeV. The latter dominates for lighter, relativistic quarks such as the up and down quarks constituting the valence component of protons and neutrons. Then, due to the universality of the QCD coupling constant for different flavors and for all gluons exchange, the QCD-related mass term will be the same for protons and neutrons, and therefore we will have:

$$m_p = (2y_u + y_d)v/\sqrt{2} + m_{QCD}, \quad m_n = (y_u + 2y_d)v/\sqrt{2} + m_{QCD}. \quad (3)$$

For a generic atom of atomic number Z and atomic mass A we obtain:

$$M(A, Z) = \frac{1}{\sqrt{2}}[y_u(Z + A) + y_d(2A - Z)]v + Am_{QCD}, \quad (4)$$

where we have neglected, to first approximation, the electron mass, the electron-nucleus binding energy and the nucleon binding energy. The purely QCD-dependent mass term should be independent on the curvature of space-time, since otherwise the gluon could acquire a mass giving rise to the explicit breaking of the color symmetry. This is analogous to the case of the other unbroken symmetry of the standard model, $U(1)_{em}$ leading to the electromagnetic charge being conserved even in a generic curved space-time.

The possibility to detect Higgs shifts in atomic and molecular spectroscopy relies on the fact that electronic transitions depend primarily on the mass of the electron, whereas molecular transitions due to vibrational or rotational degrees of freedom depend upon the mass of the nuclei. While the electron mass is directly proportional to the Yukawa coupling, the mass of the nuclei is mainly due to the mass of its proton and neutron constituents, which in turn depends mainly on the color binding energy. We therefore expect that molecular transitions will not be affected by the Higgs shifts at leading order, unlike electronic transitions, not even in the most sensitive case of pyramidal molecules such as ammonia, for which tunneling provides enhanced sensitivity to changes in masses of the nuclei for the inversion lines. It is difficult to detect electronic, vibrational, and rotational transitions in the same region of space from the same species for a gas at thermal equilibrium, due to the very different energy scale difference required to produce these excitations. A comparative analysis of wavelength shifts from different species seems then necessary. This also enables to disentangle the putative Higgs shift contribution from the Doppler shift and the purely gravitational shift. The Doppler shift should be the same for molecules belonging to the same comoving cloud, while the wavelength shift expected from general relativity will act universally on all particles. Thus, unlike the Higgs shift, the latter will not discriminate between fundamental particles and interactions binding energies, *i.e.*, between the electron mass and the main contribution to the proton mass due to QCD. A promising object to look for possible Higgs shifts is the Galactic center, where a compact object with an estimated mass of $M \simeq 2.6 \times 10^6$ solar masses, Sagittarius A* [27–31], is present and supposed to be a black hole with a Schwarzschild radius of $R_s = 2GM/c^2 \simeq 4 \times 10^9$ m. Unfortunately, the curvature scalar is zero for a Schwarzschild black hole, but the lack of detailed knowledge of this compact object cannot rule out its possible description in terms of a Kerr black hole or a naked singularity. The latest possibility is promising since it may provide effective curvatures comparable to the Higgs Compton wavelength, although the very existence of naked singularities is a hotly debated theme in general relativity. As benchmarks from the observational viewpoint, with a 1 pc resolution survey it is now possible to obtain spectra of atoms or molecules at a distance of $r \simeq 2 \times 10^{16}$ m from the Galactic center, and recent surveys of ammonia [32, 33] have a spectral sensitivity corresponding to a Doppler shift of about 2-3 km/s^{-1}, *i.e.* $\delta\nu/\nu \simeq 10^{-5}$.

In the case of atomic hydrogen spectroscopy, the spectral lines depend directly on the reduced mass $\mu = m_e m_p/(m_e + m_p)$ and ultimately, due to the large mass ratio m_p/m_e, on the electron mass. Observation of atomic lines from the Galactic center is difficult due to the strong absorption at optical wavelengths, and therefore one should focus on the high-precision monitoring of the 21-cm neutral hydrogen line, still dependent on the electron to proton mass ratio. This implies measuring the 21-cm neutral hydrogen line from interstellar clouds near the Galactic Center, or from stars with highly eccentric orbits. In the latter case, clear signatures might be available by looking at the temporal variability of the 21 cm hyperfine line along the star orbit.

Finally, we want to point out that in the standard model the masses of fundamental particles have a different treatment with respect to the mass coming from interaction among themselves. If the equivalence principle holds, the gravitational mass of the electrons constituting a test body will change if the Higgs field is coupled to curvature, while the nucleons will continue to keep, to leading order, the usual gravitational charge. This is in striking contrast to the case of general relativity where all sources of energy contribute without any distinctive feature to the emergence of space-time, originating at least an unappealing contrast in the way masses are considered in Higgs physics and gravitation. Stability issues of back-reaction on the metric may also arise depending on the relative signs of ξ and R, potentially originating catastrophic growths of the local curvature and of the mass of the test particles.

4. Conclusions

We have briefly discussed two critical issues in the interplay between macroscopic quantum vacuum and microscopic gravitation, namely systematic effects in Casimir force measurements affecting our knowledge of possible non-Newtonian forces of gravitational origin, and the influence of strong space-time curvature on the generation of mass via the Higgs field. By smearing out the two traditional regions of applicability of quantum physics and gravitation, the microworld and the macroworld respectively, one could gain important insights on their mutual compatibility, as already attempted in [34–39] for the relationship between Casimir energy and curved space-time, and in [40] concerning the validity of the equivalence principle for macroscopically distinguishable quantum states. All this may be considered part of a program aimed at narrowing the gap between quantum vacuum and the standard model of elementary particle physics and its proposed extensions, as recently outlined in [41].

Acknowledgments

I would like to thank the QFEXT09 organizers for the kind invitation to such a stimulating and successful meeting. I also acknowledge partial support from the Julian Schwinger Foundation through grant JSF 08070000 on Astrophysics of Quantum Vacuum.

References

1. M. Bordag, U. Mohideen, and V. M. Mostepanenko, *Phys. Rep.* **353**, 1 (2001).
2. S. K. Lamoreaux, *Rep. Progr. Phys.* **68**, 201 (2005).
3. For a different viewpoint see: C. R. Hagen, *Eur. Phys. J. C* **19**, 677 (2001).
4. S. Dimopoulos and G. F. Giudice, *Phys. Lett. B* **379**, 105 (1996).
5. Y. Fujii, *Nature* **234**, 5 (1971); *Ann. Phys.* **69**, 494 (1972); *Phys. Rev. D* **9** 874 (1974); *Int. J. Mod. Phys. A* **6**, 3505 (1991).
6. E. Fischbach and C. L. Talmadge, *The Search for Non-Newtonian Gravity* (AIP/Springer-Verlag, New York, 1999).
7. M. J. Sparnaay, *Physica* **24**, 751 (1948).
8. W. J. Kim, M. Brown-Hayes, D. A. R. Dalvit, J. H. Brownell, and R. Onofrio, *Phys. Rev. A* **78**, 020101(R) (2008).
9. S. de Man, K. Heeck, and D. Iannuzzi, *Phys. Rev. A* **79**, 024102 (2009).
10. W. J. Kim, M. Brown-Hayes, D. A. R. Dalvit, J. H. Brownell, and R. Onofrio, *J. Phys. Conf. Ser.* **161**, 012104 (2009).
11. W. J. Kim, A. O. Sushkov, D. A. R. Dalvit, and S. K. Lamoreaux, *Phys. Rev. Lett.* **103**, 060401 (2009).
12. H.-C. Chiu, C.-C. Chang, R. Castillo-Garza, F. Chen, and U. Mohideen, *J. Phys. A* **41**, 164022 (2008).

13. R. S. Decca, E. Fischbach, G. L. Klimchitskaya, D. E. Krause, D. López, U. Mohideen, and V. M. Mostepanenko, *Phys. Rev. A* **79**, 026101 (2009).
14. V. A. Kuzmin, I. I. Tkachev, and M. E. Shaposhnikov, *Pis'ma Zh. Eksp. Teor. Fiz.* **36**, 49 (1982) [*JETP Lett.* **36**, 59 (1982)].
15. V. M. Mostepanenko and I. Yu. Sokolov, *Phys. Lett. A* **125**, 405 (1987).
16. R. S. Decca, D. López, E. Fischbach, G. L. Klimchitskaya, D. E. Krause, and V. M. Mostepanenko, *Ann. Phys. (N.Y.)* **318**, 37 (2005).
17. G. L. Klimchitskaya, U. Mohideen, and V. M. Mostepanenko, *Rev. Mod. Phys.* **81**, 1827 (2009).
18. R. Onofrio and G. Carugno, *Phys. Lett. A* **198**, 365 (2005).
19. B. V. Derjaguin and I.I. Abrikosova, *Sov. Phys. JETP* **3**, 819 (1957).
20. R. S. Decca, E. Fischbach, G. L. Klimchitskaya, D. E. Krause, D. López, and V. M. Mostepanenko, *Phys. Rev. D* **79**, 124021 (2009).
21. D. A. R. Dalvit and R. Onofrio, *Phys. Rev. D* **80**, 064025 (2009).
22. M. Bordag, B. Geyer, G. L. Klimchitskaya, and V. M. Mostepanenko, *Phys. Rev. D* **58**, 075003 (1998).
23. M. Masuda, M. Sakaki, and A. Araya, *Class. Quantum Grav.* **24**, 3965 (2007).
24. M. Masuda and M. Sasaki, *Phys. Rev. Lett.* **102**, 171101 (2009).
25. N. D. Birrell and P. C. W. Davies, *Quantum Fields in Curved Space*, (Cambridge University Press, 1982).
26. C. T. Hill and D. S. Salopek, *Ann. Phys. (N.Y.)* **213**, 21 (1992).
27. M. Morris and E. Serabyn, *Annu. Rev. Astron. Astrophys.* **34**, 645 (1996).
28. A. Eckart and R. Genzel, *Mon. Not. R. Astron. Soc.* **284**, 576 (1997).
29. R. Genzel, A. Eckart, T. Ott, and F. Eisenhauer, *Mon. Not. R. Astron. Soc.* **291**, 219 (1997).
30. A. M. Ghez, B. L. Klein, M. Morris, and E. E. Blecklin, *Astrophys. J.* **509**, 678 (1998).
31. M. J. Reid, A. C. S. Readhead, R. C. Vermeulen, and R. N. Treuhaft, *Astrophys. J.* **524**, 816 (1999).
32. C. Henkel, N. Jethava, A. Kraus, K. M. Menten, C. L. Carilli, M. Grasshoff,D. Lubowich, and M. J. Reid, *Astron. Astrophys.* **440**, 893 (2005).
33. T. L. Wilson, C. Henkel, and S. Höttemeister, *Astron. Astrophys.* **460**, 533 (2006).
34. M. Karim, A. H. Bokhari, and B. J. Ahmedov, *Class. Quantum Grav.* **17**, 2459 (2000).
35. F. Sorge, *Class. Quantum Grav.* **22**, 5109 (2005).
36. S. A. Fulling, K. A. Milton, P. Parashar, A. Romeo, K. V. Shajesh, and J. Wagner, *Phys. Rev. D* **76**, 025004 (2007).
37. K. A. Milton, P. Parashar, K. V. Shajesh, and J. Wagner, *J. Phys. A* **40**, 10395 (2007).
38. K. V. Shajeah, K. A. Milton, P. Parashar, and J. A. Wagner, *J. Phys. A* **41**, 164058 (2008).
39. K. A. Milton, S. A. Fulling, P. Parashar, A. Romeo, K. V. Shajesh, and J. A. Wagner, *J. Phys. A* **41**, 164052 (2008).
40. L. Viola and R. Onofrio, *Phys. Rev. D* **55**, 455 (1997).
41. H. Gies, *J. Phys. A: Math. Theor.* **41**, 164039 (2008).

THE PVLAS EXPERIMENT AND ITS RESULTS: PROBING THE QUANTUM VACUUM WITH POLARIZED LIGHT AND MAGNETIC FIELDS

F. DELLA VALLE and E. MILOTTI

Dip. di Fisica and INFN, Sez. di Trieste, Via Valerio, 4 - 35127 Trieste, Italy

G. DI DOMENICO and G. ZAVATTINI

Dip. di Fisica and INFN, Sez. di Ferrara, Via Saragat, 2 - 44100 Ferrara, Italy

U. GASTALDI, R. PENGO and G. RUOSO*

INFN, Laboratori Nazionali di Legnaro, Viale dell'Università, 2 - 35020 Legnaro, Italy
* *e-mail: ruoso@lnl.infn.it*

The experiment PVLAS studies the optical properties of the vacuum, that behaves much like a material medium when it is permeated by an external (electric or magnetic) field. Using a strong superconducting magnet and a very sensitive ellipsometer we have searched for modifications of the index of refraction of the vacuum due to the presence of a magnetic field. A birefringence is predicted to arise because of the vacuum fluctuations of the electromagnetic field, and a similar effect (dichroism) could be due to the presence of yet undiscovered low mass particles interacting with two photons.

At present PVLAS has set the best existing limits on such processes, its sensitivity being limited by external noise sources which have now been accounted for. A completely redesigned prototype apparatus is now under construction: it is based on rotating permanent magnets and an ellipsometer employing an ultra stable Fabry-Perot resonator. A 50-fold improvement in the sensitivity of the ellipsometer has now been achieved, and we hope to improve our best limits when the magnets system will be installed.

Keywords: Ellipsometry; Quantum Vacuum; Quantum Electrodynamics.

1. Introduction

Vacuum magnetic birefringence and elastic light-light scattering have been predicted many years ago. Both effects are associated with electron-positron vacuum fluctuations and can be calculated in the framework of the Euler-Heisenberg-Weisskopf (EHW) effective Lagrangian correction L_{EHW} [1,2].

Calculation is performed for photon energies well below the electron mass and for fields much smaller than their critical values, $B \ll B_{\rm crit} = m_e^2c^2/e\hbar = 4.4 \cdot 10^9$ T, $E \ll E_{\rm crit} = m_e^2c^3/e\hbar = 1.3 \cdot 10^{18}$ V/m (S.I. units). Writing the total index of refraction as $n_{\rm tot} = n_{\rm r} - \imath n_{\rm i}$ one finds that the birefringence $\Delta n_{\rm r}$ induced by a magnetic field perpendicular to the propagation direction is [2,3]

$$\Delta n_{\rm r}^{\rm (EHW)} = 3A_eB^2 \tag{1}$$

where the parameter A_e describing the non linearity is given by (S.I. units)

$$A_e = \frac{2}{45\mu_0}\frac{\alpha^2\lambda\!\!\!^{-3}_e}{m_ec^2} = \left(\frac{\alpha}{45\pi}\right)\frac{1}{2B_{\rm crit}^2} = 1.32 \cdot 10^{-24}\ {\rm T}^{-2} \tag{2}$$

with $\lambda\!\!\!^{-}_e = \frac{\hbar}{m_ec}$ being the Compton wavelength of the electron, $\alpha = e^2/(\hbar c 4\pi\epsilon_0)$ the fine structure constant, m_e the electron mass, c the speed of light in vacuum and μ_0 the magnetic permeability of vacuum. A more general expression can be derived in the framework of post-Maxwellian (pM) description of non linear electrodynamics [4].

In the presence of a birefringence, the measurable quantity is the induced ellipticity on a linearly polarized light beam

$$\psi = \frac{\pi\Delta n_{\rm r}L_{\rm eff}}{\lambda}\sin 2\vartheta \tag{3}$$

where $L_{\rm eff}$ is the effective path length within the magnetic field region, ϑ is the angle between the polarization direction and the magnetic field and λ is the wavelength of the light. For $\lambda = 1064$ nm, $B = 2.3$ T and $L_{\rm eff} = 500$ km (present laboratory values) the induced ellipticity is $\psi_{\rm EHW} = 3.1 \cdot 10^{-11}$.

Applying the Euler-Heisenberg-Weisskopf (EHW) effective Lagrangian density also to photon-photon scattering of linearly polarized photons, one can relate the light-by-light cross section to the non linear parameter A_e, hence to the birefringence [5]:

$$\sigma_{\gamma\gamma}^{\rm (EHW)}(E_\gamma) = \frac{1}{45^2}\frac{973}{5\pi}\alpha^4\left(\frac{E_\gamma}{m_ec^2}\right)^6\lambda\!\!\!^{-2}_e = \frac{973\mu_0^2}{20\pi}\frac{E_\gamma^6}{\hbar^4c^4}A_e^2 \tag{4}$$

For $E_\gamma = 1.17$ eV (1064 nm wavelength) this cross section is exceedingly small: $\sigma_{\gamma\gamma}^{\rm (EHW)}(E_\gamma = 1.17\ {\rm eV}) = 1.9 \cdot 10^{-65}\ {\rm cm}^2$. other hypothetical situation which would cause both magnetically induced birefringence and dichroism is the existence of light neutral bosons which couple to two photons [6], called axion-like particles (ALP). In our experimental configuration (see below) the coupling of the pseudoscalar (scalar) particle to the two fields will occur only if the photon electric field and external magnetic fields are parallel (perpendicular). These hypothetical interactions will lead to both dichroism and birefringence. The former effect is due to a real production of bosons whereas the latter effect is due to its virtual production. The induced birefringence and dichroism will lead to an ellipticity $\psi_{\rm ALP}$ and a rotation $\epsilon_{\rm ALP}$, respectively, (in natural Heavyside-Lorentz units)

$$\epsilon_{\rm ALP} = \left(\frac{BL}{4M_{a,s}}\right)^2 N \left(\frac{\sin x}{x}\right)^2 \tag{5}$$

$$\psi_{\rm ALP} = \frac{\omega B^2 L}{2M_{a,s}^2 m_{a,s}^2} N \left(1 - \frac{\sin 2x}{2x}\right) \tag{6}$$

where, in vacuum, $x = \frac{Lm_{a,s}^2}{4\omega}$, ω is the photon energy, L is the magnetic field length and N is the number of passes through the magnetic field region.

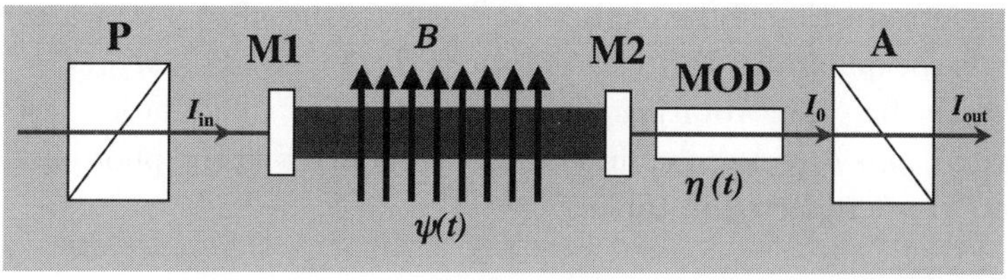

Fig. 1. Scheme of a sensitive ellipsometer based on a Fabry-Perot cavity.

2. Current experimental set-up

In the past we carried out repeated measurement runs with an apparatus based on a rotating superconducting dipole magnet in the Legnaro Laboratories of INFN [7,8], and now we are developing an improved prototype based on permanent magnets. The ellipsometer is based on a Fabry-Perot cavity (FP) with rotating permanent dipole magnets. The magnetic field is perpendicular to the laser beam used to probe the magnetized vacuum. A schematic drawing of the sensitive ellipsometer is shown in figure 1.

A polarizer **P** defines the linear polarization of the incoming laser beam. Two mirrors **M1** and **M2** define a Fabry-Perot optical cavity with very high finesse $\mathcal{F}$. Part of the region between the mirrors is filled by the magnetic field and the observed ellipticity Ψ is the ellipticity ψ acquired in a single pass multiplied by a factor $\frac{2\mathcal{F}}{\pi}$: $\Psi = \psi\frac{2\mathcal{F}}{\pi}$. After mirror **M2** a photoelastic modulator **MOD** adds a known time-dependent ellipticity $\eta(t)$ to the already acquired total ellipticity Ψ. After the modulator an analyzer **A**, rotated at 90° with respect to **P**, selects the electric field component perpendicular to the input polarization. The transmitted light is then detected by a low noise photodiode. See ref. [8] for a detailed discussion.

For higher sensitivity Ψ is also made time dependent by rotating the magnetic field. Experimentally, spurious noise sources are also present: slowly varying ellipticities $\alpha(t)$ are present in all optical elements; polarizers are non ideal and have extinction ratio σ^2. The total measured intensity at the output of the ellipsometer can therefore be written as [8]

$$I_{\rm out}(t) \simeq I_0 \left[\sigma^2 + \eta(t)^2 + \alpha(t)^2 + 2\eta(t)\Psi(t)\sin 2\vartheta(t) + 2\eta(t)\alpha(t)\right] \qquad (7)$$

where $\vartheta(t)$ represents the angle between the magnetic field and the input polarization direction. In the PVLAS experiment the magnetic field rotates with angular frequency $\Omega_{\rm Mag}$. Due to the B^2 dependence of the induced birefringence, the induced ellipticity will have an angular frequency of $2\Omega_{\rm Mag}$.

The introduction of the modulator linearizes the effect $\Psi(t)$ which would otherwise be quadratic. With the ellipticity $\Psi(t) = \Psi_0 \cos(2\Omega_{\rm Mag}t + 2\theta_{\rm Mag})$ generated by the rotating magnet and the modulation $\eta(t) = \eta_0 \cos(\omega_{\rm Mod}t + \theta_{\rm Mod})$ the main Fourier components in the photodiode current are those reported in table 1.

Table 1. Amplitudes and phases of the frequency components of the signal after the analyzer **A**.

Frequency	Fourier component	Intensity/I_0	Phase
DC	$I_{\rm DC}$	$\sigma^2 + \alpha^2_{\rm DC} + \eta_0^2/2$	$-$
$\omega_{\rm Mod}$	$I_{\omega_{\rm Mod}}$	$2\alpha_{\rm DC}\eta_0$	$\theta_{\rm Mod}$
$\omega_{\rm Mod} \pm 2\Omega_{\rm Mag}$	$I_{\omega_{\rm Mod}\pm 2\Omega_{\rm Mag}}$	$\eta_0 \frac{2\mathcal{F}}{\pi} \psi_0$	$\theta_{\rm Mod} \pm 2\theta_{\rm Mag}$
$2\omega_{\rm Mod}$	$I_{2\omega_{\rm Mod}}$	$\eta_0^2/2$	$2\theta_{\rm Mod}$

Following the schematic layout of figure 1, a new compact 50 cm long benchtop ellipsometer has been assembled at the Department of Physics of the University of Ferrara, Italy, to study the noise sources present in ellipsometers based on Fabry-Perot cavities. The mechanical origin of this noise in the original PVLAS setup in Legnaro, Italy, has been suspected for a long time. Indeed the sensitivity reached with the Legnaro setup, for long integration times, was $S_{\rm PVLAS} \approx 10^{-6}\,{\rm Hz}^{-1/2}$ about a factor 100 above the calculated shot noise limit. The new setup, therefore, was designed to be used both with and without seismic isolation; the basic seismic isolation uses pneumatic legs, which act as a low-pass filter with a cutoff frequency of about 5 Hz. At first, for a less critical setup, we used a cavity with finesse about $\mathcal{F} = 3000$. After achieving the best possible result, we mounted higher finesse mirrors and reached a value of $\mathcal{F} = 414000$. We also achieved an input coupling of 75% with a total transmission of 25%. In the 50 cm long benchtop ellipsometer the optical elements, including the polarizer, cavity mirrors, modulator and analyzer, were mounted on a single breadboard in a vacuum chamber. The distance between the mirrors of the Fabry-Perot cavity was about 50 cm. The polarizer **P**, entrance mirror **M1**, and analyzer **A** were mounted on motorized rotating mounts for polarization alignement and to achieve the best possible extinction ratio with the cavity inserted. We used the standard Pound-Drever-Hall method to lock the laser to the cavity but with the laser itself being used as the optical phase modulator [9]. The laser was a 200 mW Innolight Nd:YAG 1064 nm laser.

3. Results and perspectives

The best limits on induced magnetic birefringence and dichroism can be extracted from the ellipticity and rotation measurements performed with the PVLAS apparatus. The results are summarized in Table 2. Interpreting the

Table 2. Best present limits @ 95% c.l. on magnetically induced birefringence and dichroism of vacuum set by the PVLAS experiment.

Mode	$\lambda = 1064$ nm, $L_{\rm eff} = 45$ km	$\lambda = 532$ nm, $L_{\rm eff} = 23$ km
Ellipticity	$1.4 \cdot 10^{-8}$ @ 2.3 T $\Delta n_{\rm r} < 6.6 \cdot 10^{-20}$ $A_{\rm e} < 4.2 \cdot 10^{-21}$ T^{-2}	$1.4 \cdot 10^{-8}$ @ 2.3 T $\Delta n_{\rm r} < 6.3 \cdot 10^{-20}$ $A_{\rm e} < 4.0 \cdot 10^{-21}$ T^{-2}
Rotation	$1.2 \cdot 10^{-8}$ rad @ 5.0 T $\Delta\kappa < 5.3 \cdot 10^{-15}$ cm^{-1} $\Delta n_{\rm i} < 9.0 \cdot 10^{-20}$	$1.0 \cdot 10^{-8}$ rad @ 2.3 T $\Delta\kappa < 8.7 \cdot 10^{-15}$ cm^{-1} $\Delta n_{\rm i} < 15 \cdot 10^{-20}$

ellipticity measurements as limits on the EHW non linear electrodynamic parameter $A_{\rm e}$, from equations (1) and (3) one finds [8]

$$A_{\rm e}^{(1064{\rm nm})} < 4.2 \cdot 10^{-21}\ {\rm T}^{-2};\ A_{\rm e}^{(532{\rm nm})} < 4.0 \cdot 10^{-21}\ {\rm T}^{-2} \tag{8}$$

Interpreting the rotation limit measured at $\lambda = 1064$ nm as a limit on the existence of axion-like particles, from equation (5) one can set a model independent lower limit on the ALP coupling constant to two photons for masses $m_{\rm a,s} < 1$ meV: $M_{\rm a,s} > 2.4 \cdot 10^6$ GeV [7].

With the new 50 cm long test ellipsometer mounted on a seismically isolated optical bench a record ellipticity sensitivity of $S_{50{\rm cm}} = 3 \cdot 10^{-8}\,{\rm Hz}^{-1/2}$ above 5 Hz was obtained even with a finesse $\mathcal{F} = 414000$ (see figure 2 left). The seismic isolation was active above about 5 Hz with a vibration reduction factor of 20–50 above 10 Hz. Translated to birefringence sensitivity assuming a 50 cm long interaction region one finds $\Delta n_{50\text{cm}} = 7.7 \cdot 10^{-20}\,\text{Hz}^{-1/2}$. Figure 2 shows the magnetic field length needed to reach a unitary signal-to-noise ratio for the EHW magnetically induced birefringence, as a function of ellipsometer sensitivity, with finesse $\mathcal{F} = 414000$ and an integration time of $T = 10^6$ s. We believe that with a 2 meter-long, 2.3 T magnet, the first direct measurement of this vacuum polarization effect should finally become possible.

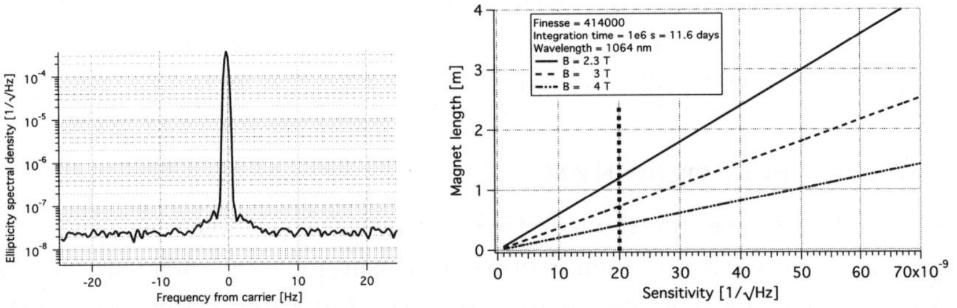

Fig. 2. Left: Ellipticity spectral density around the modulator carrier frequency. Right: Necessary magnetic field length for reaching the EHW magnetically induced birefringence as a function of sensitivity having fixed the finesse to $\mathcal{F} = 414000$ and with an integration time of $T = 10^6$ s.

References

1. H. Euler and B. Kochel, Naturwiss. **23** (1935) 246; W. Heisenberg and H. Euler, Z. Phys. **98** (1936) 718; V.S. Weisskopf, Kgl. Danske Vid. Sels., Math.-fys. Medd. **14** (1936) 6; J. Schwinger, Phys. Rev. **82** (1951) 664.
2. R. Baier and P. Breitenlohner, Acta Phys. Austriaca **25** (1967) 212; R. Baier and P. Breitenlohner, Nuovo Cimento **47** (1967) 261; S.L. Adler, Ann. Phys. **67** (1971) 559; Z. Bialynicka-Birula and I. Bialynicki-Birula, Phys. Rev. D **2** (1970) 2341.
3. E. Iacopini and E. Zavattini, Phys. Lett. B **85** (1979) 151.
4. V.I. Denisov *et al.*, Phys. Rev. D **69** (2004) 066008; G. Zavattini and E. Calloni, Eur. Phys. J. C **62** (2009) 459.
5. B. De Tollis, Nuovo Cimento **35** (1965) 1182; B. De Tollis, Nuovo Cimento **32**, (1964) 757; R. Karplus *et al.*, Phys. Rev. **83** (1951) 776; D.A. Dicus *et al.*, Phys. Rev. D **57** (1998) 2443; D. Bernard *et al.*, Eur. Phys. J. D **10** (2000) 141; F. Moulin *et al.*, Z. Phys. C **72** (1996) 607.
6. L. Maiani *et al.*, Phys. Lett. B**175** (1986) 359; P. Sikivie, Phys. Rev. Lett. **51** (1983) 1415; M. Gasperini, Phys. Rev. Lett. **59** (1987) 396.
7. E. Zavattini *et al.*, Phys. Rev. D **77** (2008) 032006.
8. M. Bregant *et al.*, Phys. Rev. D **78** (2008) 032006.
9. G. Cantatore *et al.*, Rev. Sci. Instrum. **66** (1995) 2785.

THE THERMAL CASIMIR EFFECT FOR CONDUCTING PLATES AND THE BOHR-VAN LEEUWEN THEOREM

GIUSEPPE BIMONTE

Dipartimento di Scienze Fisiche, Università di Napoli Federico II, Complesso Universitario MSA Via Cintia, Napoli, I-80126, Italy and INFN, Sezione di Napoli, Italy
E-mail: bimonte@na.infn.it

We examine recent prescriptions for estimating the thermal Casimir force between two metallic plates from the point of view of their consistency with the Bohr-van Leeuwen theorem of classical statistical physics. We find that prescriptions including the effect of ohmic dissipation satisfy the theorem, while prescriptions that neglect ohmic dissipation do not.

Keywords: Casimir, thermal, statistical.

1. Introduction

Much attention has been devoted recently to the study of the influence of temperature on the Casimir force [1]. In this contribution, we shall focuss our attention on the thermal Casimir force for two metallic, non-magnetic plates, a problem that involves subtle theoretical difficulties not yet understood as we write. We recall that the thermal Casimir pressure $P(a,T)$ between two plane-parallel dielectric plates in vacuum at distance a is provided by the following formula due to Lifshitz:

$$P(a,T) = -\frac{k_B T}{2\pi^2} \sum_{l \geq 0} \left(1 - \frac{1}{2}\delta_{l,0}\right) \int d^2\mathbf{k}_\perp q_l \sum_{\alpha=\mathrm{TE,TM}} \left(\frac{e^{2aq_l}}{r_\alpha^2(i\xi_l, k_\perp)} - 1\right)^{-1}, \tag{1}$$

where $\mathbf{k}_\perp$ is the on-plane wave-vector, $\xi_l = 2\pi k_B T l/\hbar$ are the Matsubara frequencies, $q_l = \sqrt{k_\perp^2 + \xi_l^2/c^2}$, and $r_\alpha(i\xi_l, k_\perp)$ are the reflection coefficient of the slabs, evaluated at imaginary frequencies $\omega_l = i\,\xi_l$. In the case of conducting plates, the object of controversy is the correct magnitude of the terms with $l = 0$, whose evaluation requires making a prescription for the reflection coefficients $r_\alpha(0, k_\perp)$ of the slabs at zero-frequency (the terms with $l > 0$ pose no problems, and can be evaluated very accurately using optical data of the material constituting the plates). We shall only consider in what follows normal (i.e. non superconducting) non-magnetic good conductors, like gold (poor conductors have also recently attracted much attention [2], but we shall not consider them here). For good conductors $r_{\rm TM}(0, k_\perp)$ is obviously equal to one, but the correct magnitude of $r_{\rm TE}(0, k_\perp)$ is controversial. The alternative prescriptions that have been proposed for $r_{\rm TE}(0, k_\perp)$ can be dubbed as the Drude and the plasma prescription, respectively. On one hand, the Drude prescription maintains that

$$r_{\rm TE}(0, k_\perp)|_{\rm Dr} = 0\ , \tag{2}$$

which represents the limiting value for zero-frequency of the TE Fresnel reflection coefficient, once the Drude model of a ohmic conductor is adopted. On the contrary, the plasma prescription neglects altogether ohmic dissipation and takes for $r_{\rm TE}(0, k_\perp)$ the following non-vanishing value:

$$r_{\rm TE}(0, \mathbf{k}_\perp)|_{\rm pl} = \frac{\sqrt{\Omega_P^2/c^2 + k_\perp^2} - k_\perp}{\sqrt{\Omega_P^2/c^2 + k_\perp^2} + k_\perp}\ , \tag{3}$$

where Ω_P is the plasma frequency of the conductor, as obtained from optical data at IR frequencies. There is no room here to discuss in detail the reasons in favor or against either prescription, and we address the reader to Ref. [1].

In order to discriminate between the Drude and the plasma prescriptions, we recently proposed [3] to test their consistency with a well known result from classical statistical physics, i.e. the Bohr- van Leeuwen theorem [4]. To justify this approach, we note on one hand that the Casimir effect is an equilibrium phenomenon, and therefore the theoretical models used for the plates should be consistent with the dictates of statistical physics. It is interesting to note, in this regard, that statistical physics does indeed imply very general constraints, known as Onsager's reciprocity relations, on the possible form of the 2×2 reflection matrix of a homogeneous possibly anisotropic surface, that for example rule out certain phenomenological models of chiral materials [5]. On the other hand, to justify recourse to a theorem of classical statistical physics, we observe that, differently from the $l > 0$ terms, the troublesome $l = 0$ terms of Lifshitz formula have essentially a classical character, since they do not explicitly involve Planck's constant. Obviously this remark applies only to normal metals, and indeed the Bohr- van Leeuwen theorem was originally derived to explain their weak magnetic properties. Obviously, this is not the case for magnetic or superconducting materials, whose response functions depend on quantum effects that disappear in the classical limit.

2. The Bohr-van Leeuwen theorem

The Bohr-van Leeuwen theorem originated early in the 20th century, in an attempt to explain the absence of strong diamagnetism in normal conductors placed in an external magnetic field [4]. By slightly generalizing its initial content, that referred to a material placed in a static external magnetic field, we can state the theorem as saying that in classical systems at thermal equilibrium *matter decouples from transverse em fields.* To prove it, consider the *classical microscopic* Hamiltonian for a system of N charged particles interacting with the em field, in the Coulomb gauge ($\nabla \cdot \mathbf{A} = 0$):

$$H = \sum_{i=1}^{N} \frac{1}{2m_i} \left[\mathbf{P}_i - \frac{e_i}{c} (\mathbf{A}(\mathbf{r}_i) + \mathbf{A}_{\text{ext}}(\mathbf{r}_i)) \right]^2 + \sum_{i<j} e_i e_j v(|\mathbf{r}_i - \mathbf{r}_j|) + \sum_{i=1}^{N} V^{(\text{walls})}(\mathbf{r}_i) + H_0^{(\text{rad})} , \tag{4}$$

where $v(|\mathbf{r}_i - \mathbf{r}_j|)$ is the Coulomb electrostatic potential, $V^{(\text{walls})}(\mathbf{r})$ is a confining potential mimicking the walls of the particles' containers, $H_0^{(\text{rad})}$ is the free Hamiltonian for the fluctuating em field $\mathbf{A}$ and we possibly allow for the presence of an external static magnetic field with potential $\mathbf{A}_{\text{ext}}$. Consider now the classical partition function of the system $Z = \int d\mu_{\text{part}} \int d\mu_{\text{rad}} \exp(-\beta H)$, where $d\mu_{\text{part}}$ and $d\mu_{\text{rad}}$ are the phase-space canonical measures for the particles and the em field, respectively. It is then straightforward to check that under the canonical transformation:

$$\mathbf{P}_i \to \mathbf{P}'_i = \mathbf{P}_i - \frac{e_i}{c} (\mathbf{A}(\mathbf{r}_i) + \mathbf{A}_{\text{ext}}(\mathbf{r}_i)) \tag{5}$$

Z factorizes into the product

$$Z = Z^{(\text{part})} \times Z_0^{(\text{rad})} , \tag{6}$$

where $Z_0^{(\text{rad})} = \int d\mu_{\text{rad}} \exp(-\beta H_0^{(\text{rad})})$ is the partition function for the free em field in empty space, and $Z^{(\text{part})} = \int d\mu_{\text{part}} \exp(-\beta H^{(\text{part})})$, where $H^{(\text{part})} = \sum_{i=1}^{N} \mathbf{P}_i^2/(2m_i) + \sum_{i<j} e_i e_j v(|\mathbf{r}_i - \mathbf{r}_j|)$. The factorization property Eq. (6) shows that at *thermal equilibrium* the particles and the transverse em field are completely decoupled. This decoupling property has two important physical consequences: on one hand, it implies that an external magnetic field $\mathbf{A}_{\text{ext}}$ does not affect the thermodynamic properties of

matter, and on the other hand it implies that the presence of matter has no influence on thermal averages of the fluctuating em field. The former implication provides the original content of the Bohr-van Leeuwen theorem [4], while the second implication is the one we need for the purposes of Casimir physics, as we discuss below.

3. The Bohr-van Leewuen theorem and the Casimir effect

In the previous Section, we have shown that in classical statistical physics, the thermal averages of the transverse em field are independent of matter. As we now explain, this result has important consequences for the thermal Casimir effect. We consider the simple case of two identical plane-parallel homogeneous and isotropic slabs with complex permittivity $\epsilon(\omega)$, lying in the (x, y) plane, separated by an empty gap of width a. As it is well known, the Casimir pressure $P(a, T)$ between the slabs is equal to the (renormalized) quantum thermal average $\langle T_{zz} \rangle$ of the component T_{zz} of the Maxwell stress tensor, evaluated at any point in the empty gap. It is possible to split $\langle T_{zz} \rangle$ as

$$\langle T_{zz} \rangle = \langle T_{\| zz} \rangle + \langle T_{\perp zz} \rangle \,, \tag{7}$$

where $\langle T_{\| zz} \rangle$ and $\langle T_{\perp zz} \rangle$ represent the contributions from the longitudinal and the transverse em field, respectively. It is shown in Ref. [3] that $\langle T_{\| zz} \rangle$ and $\langle T_{\perp zz} \rangle$ have the expressions:

$$\langle T_{\| zz} \rangle = \frac{1}{\pi^2} \int_0^\infty \frac{d\omega}{\omega} E_\beta(\omega) \int dk_\perp \, k_\perp^2 \, \mathrm{Im} \left[\left(1 - \frac{e^{2k_\perp d}}{\bar{r}^2(\omega)} \right)^{-1} \right] , \tag{8}$$

$$\langle T_{\perp zz} \rangle = -\frac{2}{\pi} \int_0^\infty \frac{d\omega}{\omega} E_\beta(\omega) \, \mathrm{Im} \left[\mathcal{T}_\perp(\omega) \right] , \tag{9}$$

$$\mathcal{T}_\perp(\omega) = \frac{1}{2\pi} \int_0^\infty dk_\perp k_\perp \left\{ q \sum_{\alpha=\mathrm{TE,TM}} \left(\frac{e^{2qd}}{r_\alpha^2(\omega, k_\perp)} - 1 \right)^{-1} - k_\perp \left(\frac{e^{2k_\perp d}}{\bar{r}^2(\omega)} - 1 \right)^{-1} \right\} . \tag{10}$$

Here $q = \sqrt{k_\perp^2 - \omega^2/c^2}$, $\bar{r} = (\epsilon(\omega) - 1)/(\epsilon(\omega) + 1)$ and $E_\beta(\omega) = \hbar\omega/2 \coth[\hbar\omega/(2k_B T)]$. Consider the transverse contribution $\langle T_{\perp zz} \rangle$: since, as shown in the previous Section, the classical statistical averages of the transverse em field are independent of matter, it follows that $\langle T_{\perp zz} \rangle$ must vanish in the classical limit. For $\hbar \to 0$, $E_\beta(\omega) \to k_B T$, and then it can be taken outside the integral on the r.h.s. of Eq. (9). After rotating the ω domain of integration from the real positive axis to the imaginary positive axis (such a rotation is permitted by the analyticity properties of the complex permittivity $\epsilon(\omega)$ of causal media), one finds [3]

$$\lim_{\hbar \to 0} \langle T_{\perp zz} \rangle = - k_B T \lim_{\omega \to 0} \mathcal{T}_{\perp}(\omega) \,. \tag{11}$$

It can be easily verified that for vanishing frequency the TM contribution to $\mathcal{T}_{\perp}(\omega)$ always cancels against the second term between the curly brackets on the r.h.s. of Eq. (10), leaving us with:

$$\lim_{\hbar \to 0} \langle T_{\perp zz} \rangle = -\frac{k_B T}{2\pi} \int_0^{\infty} dk_{\perp} \, k_{\perp}^2 \left(\frac{e^{2k_{\perp} d}}{r_{\rm TE}^2(0, k_{\perp})} - 1 \right)^{-1} \,. \tag{12}$$

The Bohr-van Leeuwen theorem requires that the quantity on the r.h.s. vanishes for all separations, and this is only possible if

$$r_{\rm TE}(0, k_{\perp}) = 0 \,. \tag{13}$$

The conclusion is that the Drude prescription Eq. (2) is consistent with the theorem, while the plasma prescription Eq. (3) is not.

4. Concluding remarks

The Bohr-van Leeuwen theorem of classical statistical physics states that in the classical limit the transverse em field decouples from matter. We have used this theorem as a criterion to discriminate between the so-called Drude and plasma prescriptions that have been recently proposed to evaluate the thermal Casimir force between two metallic non-magnetic plates. We have shown that the Drude prescription is consistent with this theorem, while the plasma prescription is not. The results derived in this paper do not apply to magnetic or superconducting materials, because the properties of such materials arise from quantum effects that disappear in the classical limit.

Acknowledgments

The author acknowledges financial support by the European Science Foundation (ESF) within the activity 'New Trends and Applications of the Casimir Effect' (www.casimir-network.com).

References

1. G.L. Klimchitskaya, U. Mohideen and V.M. Mostepanenko, arXiv:0902.4022.
2. L.P. Pitaevskii, Phys. Rev. Lett. **101**, 163202 (2008).
3. G. Bimonte, Phys. Rev. A **79**, 042107 (2009).
4. H. J., van Leeuwen, J. Phys. Radium **2**, 362 (1921); J. H. Van Vleck, *The Theory of Electric and Magnetic Susceptibilities* (Clarendon Press, Oxford, 1932).
5. G. Bimonte and E. Santamato, Phys. Rev. A **76**, 013810 (2007).

ELECTROMAGNETIC CASIMIR EFFECT IN WEDGE GEOMETRY AND THE ENERGY-MOMENTUM TENSOR IN MEDIA

I. BREVIK* and S. Å. ELLINGSEN†

Department of Energy and Process Engineering,
Norwegian University of Science and Technology,
N-7491 Trondheim, Norway
** E-mail: iver.h.brevik@ntnu.no*
† E-mail: simen.a.ellingsen@ntnu.no

K. A. MILTON

Oklahoma Center for High Energy Physics and
H.L. Dodge Department of Physics and Astronomy,
The University of Oklahoma, Norman, OK 73019, USA
E-mail: milton@nhn.ou.edu

The wedge geometry closed by a circular-cylindrical arc is a nontrivial generalization of the cylinder, which may have various applications. If the radial boundaries are not perfect conductors, the angular eigenvalues are only implicitly determined. When the speed of light is the same on both sides of the wedge, the Casimir energy is finite, unlike the case of a perfect conductor, where there is a divergence associated with the corners where the radial planes meet the circular arc. We advance the study of this system by reporting results on the temperature dependence for the conducting situation. We also discuss the appropriate choice of the electromagnetic energy-momentum tensor.

1. Introduction

Casimir theory for the wedge geometry continues to attract interest. The reasons for this are many-faceted – probably the most important one being that the material boundaries are plane, thus avoiding some of the formal divergences that so often plague calculations in the presence of curved boundaries. The wedge geometry moreover implies a formalism closely related to that of cylindrical geometry, and actually also to that of cosmic string theory. Finally, the wedge geometry is a convenient testing ground for experimental tests of Casimir-Polder forces.

The Casimir energy and stress in a wedge geometry was approached already in the 1970s.[1,2] Various embodiments of the wedge with perfectly conducting walls were treated by Brevik and co-workers[3–5] and others.[6,7] More recently a wedge intercut by a cylindrical shell was considered by Nesterenko and co-workers.[8,9] Local Casimir stresses were considered by Saharian and co-workers.[10–12] The interaction of an atom with a wedge was studied experimentally by Sukenik *et al.*[13] The theory of that interaction was worked out by Barton[14] and others.[15–18] The semitransparent wedge has very recently been considered by Milton, Wagner, and Kirsten.[19] The closely related case of circular symmetry has been treated in several papers, dealing with a perfectly conducting circular boundary,[20–22] as well as the case of a dielectric circular boundary.[24–27]

The typical wedge geometry is sketched in figure 1a. The planes are situated at $\theta = 0$ and $\theta = \alpha$. We shall assume in the present paper that the interior of the wedge is filled with an isotropic medium of spatially constant and nondispersive refractive index $n = \sqrt{\varepsilon\mu}$.

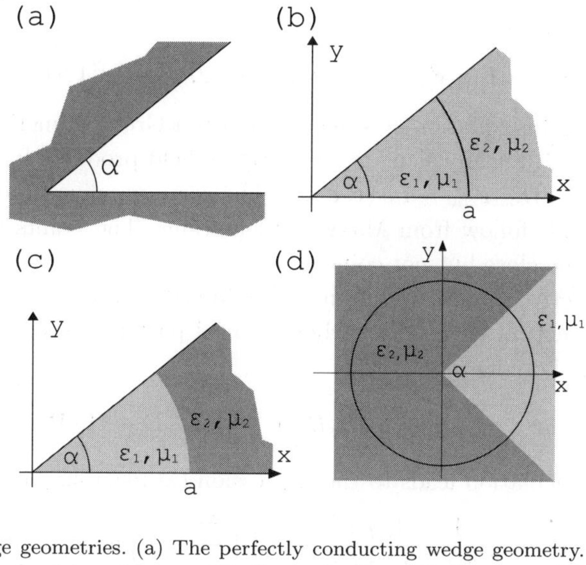

Fig. 1. Wedge geometries. (a) The perfectly conducting wedge geometry. (b) The geometry of a wedge intercut by a perfectly conducting cylindrical arc. (c) Wedge with magnetodielectric arc. (d) Diaphanous wedge in a perfectly conducting cylindrical shell.

In the simplest version of the wedge model, the planes are taken to be perfectly conducting. Various modifications of this simple wedge model can be envisaged. In sections 3 and 4 below we consider two generalizations of the simple wedge geometry, called Wedge I and II, in which the interior region is closed by a circular boundary thus implying an eigenvalue problem for the photon frequencies. The Wedge II model treated in section 4, in particular, removes the strict perfect boundary condition of the radial walls. The material of these two sections is based on two recent papers.[28,29] New developments are a closer examination of the behavior at finite temperature. As an introductory step, we delineate in the next section the essentials of classic Casimir theory for the perfectly conducting wedge. In section 6 we relate the macroscopic electromagnetic theory for the dielectric wedge region to the more general question about which electromagnetic energy-momentum tensor is to be preferred in media. This one-hundred-year-old question has actually attracted considerable interest recently.

2. Extracts from the Classic Theory for the Perfectly Conducting Wedge

The geometry is shown in Fig. 1a and was considered in a number of publications.[1–5] The governing equation for the Fourier transform of Green's function $\mathbf{\Gamma}(x, x')$ is

$$\nabla \times \nabla \times \mathbf{\Gamma}(\mathbf{r}, \mathbf{r}', \omega) - \varepsilon\mu\omega^2 \mathbf{\Gamma}(\mathbf{r}, \mathbf{r}', \omega) = -\mu\omega^2 \mathbf{1}\delta(\mathbf{r} - \mathbf{r}'). \tag{1}$$

After solving this equation in terms of the scalar Green's functions $F_m(r, r')$ and $G_m(r, r')$[5] we can calculate the effective field products for the electric fields as $i\langle E_i(\mathbf{r})E_k(\mathbf{r}')\rangle_\omega = \Gamma_{ik}(\mathbf{r}, \mathbf{r}', \omega)$. The corresponding products for the magnetic fields follow from Maxwell's equations. The points $\mathbf{r}$ and $\mathbf{r}'$ are assumed to be close but not coincident.

The effective products can now be inserted in the electromagnetic energy-momentum tensor $S_{\mu\nu}$, whose spatial part in classical notation is, in Heaviside-Lorentz units,

$$S_{ik} = -E_i D_k - H_i B_k + \frac{1}{2}\delta_{ik}(\mathbf{E} \cdot \mathbf{D} + \mathbf{H} \cdot \mathbf{B}). \tag{2}$$

A lengthy calculation leads to the expression (Minkowski metric assumed)

$$\langle S_{\mu\nu}(\mathbf{r})\rangle = \frac{1}{720\pi^2 n}\frac{1}{r^4}\left(\frac{\pi^2}{\alpha^2} + 11\right)\left(\frac{\pi^2}{\alpha^2} - 1\right)\mathrm{diag}(1, -3, 1, 1), \tag{3}$$

where we have subtracted off the term $\langle S_{\mu\nu}^{\alpha=\pi}(\mathbf{r})\rangle$ corresponding to a plane sheet. The components are ordered as $\langle S_{\mu\nu}\rangle = \langle S_{rr}, S_{\theta\theta}, S_{zz}, -w\rangle$, where w is the energy density.

The expression (3) refers to zero temperature. It is worth noticing that in the limit $\alpha \to 0, r \to \infty$ such that αr becomes the separation between parallel plates, the expression agrees with Barton.[14] Moreover, the expression agrees with that of a cosmic string if the string's deficit angle $8\pi GM$ is identified with $2\pi - 2\alpha$.[28]

3. Wedge I: Perfectly Conducting Walls; Circular Boundary at $r = a$

The geometry is shown in Fig. 1b and 1c. The walls are perfectly conducting as before, while we assume now that there is as boundary a circular arc with radius a. The wedge thus has an interior region $r < a$ (region 1) where the material parameters are ε_1 and μ_1, and an exterior region $r > a$ (region 2) with analogous parameters ε_2 and μ_2. The materials are assumed to be nondispersive. We assume the diaphanous (isorefractive) condition $\varepsilon_1\mu_1 = \varepsilon_2\mu_2 = n^2$. The transverse wave numbers $k_\perp$ in the two regions are accordingly the same, $k_\perp^2 = n^2\omega^2 - k_z^2$. It is convenient to introduce the symbol $p = \pi/\alpha$, and also $\lambda_\nu(x) = (I_\nu(x)K_\nu(x))'$,where I_ν and K_ν are modified Bessel functions.

3.1. *The boundary $r = a$ being perfectly conducting*

This is the simplest case. Detailed expansions of the electric and magnetic fields are given in Ref. 28. In region 1 there are two independent polarizations, one TM polarization where the mode eigenvalues are determined by $J_{mp}(k_\perp a) = 0$ with $m = 1, 2, 3, ...$, and one TE polarization where the eigenvalues are determined by $J'_{mp}(k_\perp a) = 0$. One azimuthally symmetric TE mode exists, corresponding to $m = 0$, but there is no such TM mode.

In region 2 the TM polarization yields $H_{mp}^{(1)}(k_\perp a) = 0$, $m = 1, 2, 3, ..$, whereas the TE polarization yields $H_{mp}^{(1)'}(k_\perp a) = 0$, $m = 0, 1, 2, \ldots$. Summing over all modes and making use of the argument principle, we arrive at the following expression for the total zero-point energy

$$\tilde{\mathcal{E}} = \frac{1}{4\pi n a^2} \sum_{m=0}^{\infty}{}' \int_0^\infty dx\, x \ln[1 - x^2 \lambda_{mp}^2], \tag{4}$$

the prime meaning that the mode $m = 0$ is counted with half weight. This is the boundary-induced contribution to the zero-point energy. If the boundary $r = a$ were removed and either the interior or the exterior medium were to fill the whole region, we would get $\tilde{\mathcal{E}} = 0$. Moreover, we have omitted a zero-mode divergence caused by the sharp corners where the arc meets the wedge. If $p = 1$, Eq. (4) is one-half that for a conducting circular cylinder.

3.2. *Dielectric boundary at r = a*

The most important change compared with the previous subsection is that regions 1 and 2 become coupled via electromagnetic boundary conditions. Assuming $n_1 = n_2$ we find the following simple expression for the zero-point energy

$$\tilde{\mathcal{E}} = \frac{1}{4\pi n a^2} \sum_{m=0}^{\infty}{}' \int_0^\infty dx\, x \ln[1 - \xi^2 x^2 \lambda_{mp}^2], \tag{5}$$

with

$$\xi = \frac{\varepsilon_2 - \varepsilon_1}{\varepsilon_2 + \varepsilon_1}. \tag{6}$$

The conducting case is obtained by setting $\xi = 1$. The case $n_1 \neq n_2$ is more complicated, but for weak-coupling, $|\varepsilon_1 - \varepsilon_2| \ll 1$, a self-energy can still be extracted[28] by generalizing the work done on dielectric cylinders.[24,25]

The expressions (4) and (5) are still not in general finite. A finite self-energy can be extracted from this formula by a method of zeta function regularisation,[23] generalizing the standard formal result for a circular cylinder ($p = 1$).[20] Further details and numerical results are reported in Ref. 28.

4. Wedge II: Diaphanous (Isorefractive) Wedge in Cylindrical Shell

Consider the geometry of Fig. 1d wherein a diaphanous magnetodielectric wedge ($n_1 = n_2$) inside a perfectly conducting cylindrical shell of radius a is considered. More details were published in Ref. 29. The sum over orders of the Bessel function partial waves is now not simply equidistant values $\nu = mp$ as before but the zeros of the dispersion function for $\nu = i\eta$:

$$D(i\eta; \alpha) = \sinh^2 \eta\pi - \xi^2 \sinh^2 \eta(\pi - \alpha). \tag{7}$$

The reflection coefficient ξ was defined in Eq. (6). In the absence of any wedge this becomes $D_0(i\eta) = \sinh^2 \eta\pi$. The energy of the diaphanous wedge enclosed by a perfectly conducting cylindrical shell is thus found as

$$\tilde{\mathcal{E}} = \frac{1}{16\pi^3 i} \int_{-\infty}^{\infty} dk \int_{-\infty}^{\infty} d\zeta\zeta \int_{-\infty}^{\infty} d\eta \left(\frac{D'}{D} - \frac{D_0'}{D_0}\right) \frac{d}{d\zeta} \ln[1 - x^2 \lambda_{i\eta}^2(x)]. \tag{8}$$

In the non-dispersive case where ξ is independent of ζ this may be simplified by means of partial integration w.r.t. ζ,] introduction of polar

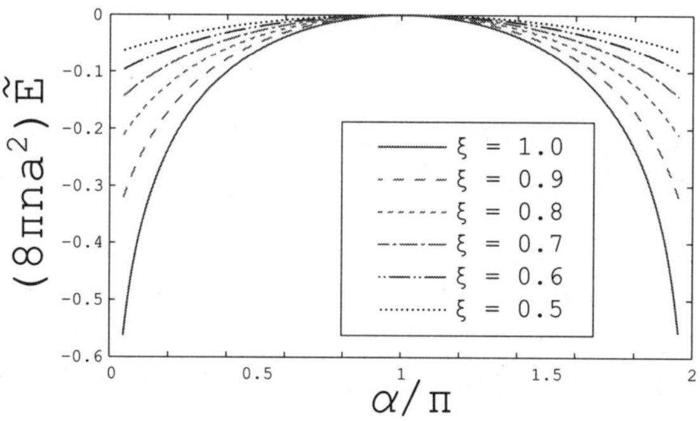

Fig. 2. The energy per length of the wedge, divided by a reference energy $(8\pi na^2)^{-1}$, as a function of opening angle α for different values of the reflection coefficient ξ.

coordinates, and use of symmetry properties, to

$$\tilde{\mathcal{E}} = -\frac{1}{4\pi^2 na^2}\int_0^\infty d\eta \left(\frac{D'}{D} - \frac{D_0'}{D_0}\right)\int_0^\infty dx\, x\,\mathrm{Im}\ln[1 - x^2\lambda_{i\eta}^2(x)]. \qquad (9)$$

Since $\mathrm{Re}\{x^2\lambda_{i\eta}^2\} \le 1$ we may use for numerical purposes

$$\mathrm{Im}\ln(1 - x^2\lambda_{i\eta}^2) = -\arctan\frac{x^2\mathrm{Im}\lambda_{i\eta}^2}{1 - x^2\mathrm{Re}\lambda_{i\eta}^2}.$$

A numerical evaluation of this energy expression was performed in Ref. 29 and the details of the procedure will not be iterated here. The energy is plotted as a function of the opening angle α for different values of the reflection coefficient $|\xi|$ in Fig. 2.

5. Considerations of Finite Temperature

We present for the first time some considerations on the Casimir energy of a closed perfectly conducting wedge such as in figure 1b when $T > 0$. (For earlier work on the cylinder at high temperatures see Ref. 30.) By letting $p = 1$ in the following, these considerations automatically apply to the case of the perfectly conducting cylinder. We consider the finite part of the energy, given for zero T in equation (4). As is customary, finite temperature implies a compactification of the imaginary time axis so that the integral over imaginary frequencies $i\zeta$ becomes a sum over Matsubara frequencies $\zeta_j = 2\pi jT$:

$$\int_0^\infty d\zeta f(\zeta) \to 2\pi T\sum_{j=0}^{\infty}{}' f(\zeta_j). \qquad (10)$$

A little calculation shows that the expression (4) may then be written $\tilde{\mathcal{E}} = \sum_{m=0}^{\infty\prime} \mathcal{E}_m$ where

$$\mathcal{E}_m = \frac{T}{\pi a}\sum_{j=0}^{\infty}{}' \int_{j\tau}^{\infty} \frac{dx\, x}{\sqrt{x^2 - j^2\tau^2}} \ln[1 - x^2\lambda_{mp}^2(x)] = \frac{T}{\pi a}\sum_{j=0}^{\infty}{}' e_{m,j}. \quad (11)$$

The nondimensional temperature is $\tau = 2\pi naT$. As previously encountered at zero temperature the simple expression for $\tilde{\mathcal{E}}$ is divergent and the challenge is to regularize it. We follow a scheme closely analogous to that of Ref. 23 (c.f. appendix A of Ref. 28) using zeta functions by subtracting and adding the asymptotic behaviour of the integrand,

$$\ln[1 - x^2\lambda_{mp}^2(x)] \sim -\frac{x^4}{4(m^2p^2 + x^2)^3}, \quad m, x \to \infty. \quad (12)$$

This asymptotic behaviour is responsible for divergences both for $j \to \infty$ and $m \to \infty$. The work on this problem is still in progress and details of the calculations and further discussion will be published elsewhere.[31]

Adding and subtracting the asymptotic behaviour we write

$$\mathcal{E}_m = \frac{T}{\pi a}\sum_{j=0}^{\infty}{}' \left[\tilde{e}_{\underline{m},j} - \frac{1}{4}\int_{j\tau}^{\infty} \frac{dx\, x^5}{\sqrt{x^2 - j^2\tau^2}(\underline{m}^2p^2 + x^2)^3}\right], \quad (13)$$

using compact notation $\underline{m}$ defined as $\underline{m}(m) = m$ for $m \geq 1$ and $\underline{m}(0) = 1/p$. The symbol $\tilde{e}$ implies that the leading asymptotic term (12) has been subtracted from the integrand, with $\underline{m}$ replacing m. The double sum resulting from the integral in (13) is formally divergent but may be regularised by use of the Chowla-Selberg formula.[32] After some calculation we obtain the following result

$$\tilde{\mathcal{E}} = \frac{T}{\pi a}\sum_{m,j=0}^{\infty}{}' \tilde{e}_{\underline{m},j} - \frac{\tau(3 - 3\tau\partial_\tau + \tau^2\partial_\tau^2)[1 + 2\mathcal{K}(\tau) + \mathcal{S}(\tau, p)]}{512\pi na^2}; \quad (14a)$$

$$\mathcal{K}(\tau) = \sum_{j=1}^{\infty} \frac{j\tau - \sqrt{1 + j^2\tau^2}}{j\tau\sqrt{1 + j^2\tau^2}}, \quad (14b)$$

$$\mathcal{S}(\tau, p) = \frac{2}{p}[\gamma - \ln(4\pi p/\tau)] + \frac{8}{p}\sum_{l=1}^{\infty} \sigma_0(l) K_0(2\pi l\tau/p), \quad (14c)$$

where γ is Euler's constant, $\sigma_0(l)$ is the number of positive divisors of l, K_0 is the modified Bessel function of the second kind, and $\partial_\tau = \partial/\partial\tau$. Notably, $\mathcal{S}$ obeys the symmetry relation $\mathcal{S}(\tau, p) = \mathcal{S}(p, \tau)$. One may show that this result reduces to the zero temperature expression[28] when $\tau \to 0$,[31] and we have verified that when $p = 1$ the coefficients of the two leading-order terms as $\tau \to \infty$, of order τ and $\tau \ln \tau$, equal twice those found previously for the cylindrical shell in Ref. 30 as they should.

6. On the Electromagnetic Energy Momentum Tensor in Media

The expression (2) for the spatial part of the energy-momentum tensor (equal to minus the Maxwell stress tensor T_{ik} according to Møller[33]), is common for the Minkowski and Abraham tensor alternatives as long as the medium is isotropic. The extraction of the "correct" form of the tensor has however been discussed for a long time. Thus it is to be noted that

- the expression is different from that of Einstein and Laub (1908);[34]
- it is different from that of Peierls (1976);[35]
- and it is different from that of Raabe and Welsch (2005);[36] cf. also the comment[37] of Brevik and Ellingsen. A survey up to 1979 is given by Brevik.[38]

When combined with the Minkowski momentum density $\mathbf{g}^M = \mathbf{D} \times \mathbf{B}$ one obtains the Minkowski energy-momentum tensor $S_{\mu\nu}^M$ whose covariant form can be written as[33]

$$S_{\mu\nu}^M = F_{\mu\alpha} H_{\nu\alpha} - \frac{1}{4}\delta_{\mu\nu} F_{\alpha\beta} H_{\alpha\beta}. \tag{15}$$

Here $F_{4k} = iE_k, H_{4k} = iD_k, F_{ik} = B_l$ (cycl), $H_{ik} = H_l$ (cycl). This energy-momentum tensor has several attractive properties: it has zero divergence for a pure radiation field, the four components of momentum and energy forming a four-vector; it is a convenient expression for field quantization (for instance, the quantization for a radiation field in a nondispersive medium can be found via a mapping technique leading one from vacuum to a medium[39]), and it has the peculiar property of being a space-like expression leading to negative photon energies in certain coordinate systems, strikingly found in connection with the Cherenkov effect in the emitter's rest frame.

One might think that it should be relatively easy to test the Minkowski tensor by measuring electromagnetic forces in optics. Actually, this is not so easy as most experiments measure only the surface force density $\mathbf{f} = -\frac{1}{2}E^2\nabla\varepsilon$, acting on surfaces. Let us give some examples:

The classic experiment of Ashkin and Dziedzic[40] showed how a narrow light beam incident on a free liquid surface acts by giving rise to an outward pull. A related experiment is that of Zhang and Chang,[41] demonstrating the oscillations of a water droplet by a laser pulse. The optical stretcher experiment of Guck *et al.*[42] is of the same kind, as is the series of two-fluid experiments of Delville *et al.*[43] near the critical point. The important factor in these experiments is simply the surface force, not electromagnetic momentum. And the recent fiber experiment of She *et al.*[44] belongs in our opinion to the same category.[45] The experiment of Campbell *et al.*[46] is however different in nature, as it is one of the very few experiments being able to test the Minkowski momentum directly.

Under stationary conditions, where the high-frequency forces average out when averaged over a period, the Minkowski tensor is able to describe all experiments that we are aware of. And this gives support to our expression (2) for the stress tensor.

Acknowledgments

The work of KAM was supported by the US National Science Foundation under Grant No. PHY-0554926 and by the US Department of Energy under Grants Nos. DE-FG02-04ER41305 and DE-FG02-04ER-46140.

References

1. J.S. Dowker and G. Kennedy, *J. Phys. A* **11**, 895 (1978).
2. D. Deutsch and P. Candelas, *Phys. Rev. D* **20**, 3063 (1979).
3. I. Brevik and M. Lygren, *Ann. Phys.* **251**, 157 (1996).
4. I. Brevik, M. Lygren, and V. Marachevsky, *Ann. Phys.* **267**, 134 (1998).
5. I. Brevik and K. Pettersen, *Ann. Phys.* **291**, 267 (2001).
6. V.V. Nesterenko et al., *Ann. Phys.* **298**, 403 (2002).
7. H. Razmi and S.M. Modarresti, *Int. J. Mod. Phys.* **44**, 229 (2005).
8. V.V. Nesterenko et al., *J. Math. Phys.* **42**, 1974 (2001).
9. V.V. Nesterenko et al., *Class. Quant. Grav.* **20**, 431 (2003).
10. A.H. Rezaeian and A.A. Saharian, *Class. Quant. Grav.* **19**, 3625 (2002).
11. A.A. Saharian, *Eur. Phys. J. C* **52**, 721 (2007).
12. A.A. Saharian, Casimir densities for wedge-shaped boundaries, in *The Casimir Effect and Cosmology*, (Tomsk State Pedagogical University Press, Tomsk, Russia, 2008), `arXiv: 0810.5207`.
13. C.I. Sukenik et al., *Phys. Rev. Lett.* **70**, 560 (1993).
14. G. Barton, *Proc. R. Soc. London* **410**, 175 (1987).
15. S.C. Skipsey et al., *Opt. Commun.* **254**, 262 (2005).
16. S.C. Skipsey et al., *Phys. Rev. A* **73**, 011803(R) (2006).
17. T.N.C. Mendes et al., *J. Phys. A* **41**, 164029 (2008).
18. F.S.S. Rosa et al., *Phys. Rev. A* **78**, 012105 (2008).
19. K.A. Milton, J. Wagner, and K. Kirsten, arXiv:0911.1123, *Phys. Rev. D*, in press.
20. L.L. DeRaad, Jr and K.A. Milton, *Ann. Phys.* **136**, 229 (1981).
21. P. Gosdzinsky and A. Romeo, *Phys. Lett. B* **441**, 265 (1998).

22. G. Lambiase et al., *J. Math. Phys.* **40**, 6254 (1999).
23. K.A. Milton et al., *Phys. Rev. D* **59**, 105009 (1999).
24. I. Cavero-Peláez and K.A. Milton, *Ann. Phys.* **320**, 108 (2005).
25. I. Cavero-Peláez and K.A. Milton, *J. Phys. A* **39**, 6225 (2006).
26. A. Romeo and K. A. Milton, *Phys. Lett. B* **621**, 309 (2005).
27. I. Brevik and A. Romeo, *Phys. Scripta* **76**, 48 (2007).
28. I. Brevik, S.Å. Ellingsen, and K.A. Milton, *Phys. Rev. E* **79**, 041120 (2009).
29. S.Å. Ellingsen, I. Brevik, and K.A. Milton, *Phys. Rev. E* **80**, 021125 (2009).
30. M. Bordag, V.V. Nesterenko, and I.G. Pirozhenko, *Phys. Rev. D* **65**, 045011 (2002); *Nucl. Phys. B* **104**, 228 (2002).
31. S.Å. Ellingsen, I. Brevik, and K.A. Milton, in preparation.
32. E. Elizalde, *Ten Physical Applications of Spectral Zeta Functions* (Springer, Berlin, 1995), p.83.
33. C. Møller, *The Theory of Relativity* (Clarendon Press, Oxford, 1972).
34. A. Einstein and J. Laub, *Ann. Physik* **26**, 541 (1908).
35. R. Peierls, *Proc. Roy. Soc. A* **347**, 475 (1976).
36. C. Raabe and D.-G. Welsch, *Phys. Rev. A* **71**, 013814 (2005).
37. I. Brevik and S. A. Ellingsen, *Phys. Rev. A* **79**, 027801 (2009); cf. also Reply *ibid.* (in press).
38. I. Brevik, *Phys. Rep.* **52**, 133 (1979).
39. I. Brevik and B. Lautrup, *Mat. Fys. Medd. Dan. Vid. Selsk.* **38**, No. 1 (1970).
40. A. Ashkin and J.M. Dziedzic, *Phys. Rev. Lett.* **30**, 139 (1979).
41. J.Z. Zhang and R.K. Chang, *Opt. Lett.* **13**, 916 (1988).
42. J. Guck *et al.*, *Biophys. J.* **81**, 767 (2001).
43. J.P. Delville *et al.*, *Trends in Electro-Optic Research* (Nova Sciences, New York, 2006), pp. 1-58.
44. W. She *et al.*, *Phys. Rev. Lett.* **101**, 243601 (2008).
45. I. Brevik, *Phys. Rev. Lett.* **103**, 219301 (2009); cf. also Reply *ibid.* **103**, 219302 (2009).
46. G.K. Campbell *et al.*, *Phys. Rev. Lett.* **94**, 170403 (2005).
47. R.V. Jones and J.C.S. Richards, *Proc. Roy. Soc. London Ser. A* **221**, 480 (1954).

CASIMIR-POLDER POTENTIAL IN THERMAL NON-EQUILIBRIUM

S. Å. ELLINGSEN

Department of Energy and Process Engineering, Norwegian University of Science and Technology, N-7491 Trondheim, Norway

Y. SHERKUNOV

Department of Physics, University of Warwick, Coventry CV4 7AL, United Kingdom

S. Y. BUHMANN and S. SCHEEL

Quantum Optics and Laser Science, Blackett Laboratory, Imperial College London, Prince Consort Road, London SW7 2AZ, United Kingdom

Different non-equilibrium situations have recently been considered when studying the thermal Casimir–Polder interaction with a body. We show that the Keldysh Green function method provides a very general common framework for such studies where non-equilibrium of either the atom or the body with the environment can be accounted for. We apply the results to the case of ground state polar molecules out of equilibrium with their environment, observing several striking effects. We consider thermal Casimir–Polder potentials in planar configurations, and new results for a molecule in a cylindrical cavity are reported, showing similar characteristic behaviour as found in planar geometry.

1. Introduction

Casimir–Polder (CP) or retarded Van der Waals forces[1] is the name given to electromagnetic dispersion forces between electrically neutral, but polarisable particles (atoms, molecules) and macroscopic objects. In the present paper we discuss CP potentials on particles when the atom–body–environment system is not in thermal equilibrium. There exists a rich literature on CP forces in thermal equilibrium (cf. the citations in Ref. 2), and non-equilibrium systems of an excited atom inside planar structures have been studied for a long time.[3,4] The latter works showed that an atom excited to an energy higher than thermal energies will have a spatially oscillating component in both force and heating rate.

Two different types of non-equilibrium have recently been studied. The first study concerned atoms next to a substrate whose temperature differs from the environment temperature[5,6] where the present radiation was assumed to be practically unable to excite the atoms. It was found that the imbalance between environment and substrate temperature may lead to strong force components whose sign depends on which of the two temperatures is greater. This prediction was subsequently confirmed in an experiment by the group of Cornell.[7]

The second study considered particles in an equilibrium thermal background of bodies and environment at uniform temperature, but prepared in an arbitrary superposition of internal eigenstates, and thus not necessarily at equilibrium with this background[8] (cf. the similar results reported in Ref. 9; for an extension to non-uniform temperature environments, see Ref. 10). It was demonstrated that even ground state particles are subject to an oscillating force component in the presence of macroscopic bodies at nonzero temperature. While utterly unobservable for atoms (which are essentially in their ground state when thermalized at room temperature), there could be some hope of observing and even using this effect for molecules,[11,12] which have excited states of very low energy.

A similar investigation was recently made[13] by applying Keldysh theory to the system of two atoms prepared in an arbitrary state, in the presence of an external (thermal) electromagnetic field. In the first half of this article, we discuss the correspondence of the Keldysh formalism with both of the above non-equilibrium theories (Sec. 2). In the second half (Sec. 3), we apply the results for a particle in a uniform-temperature environment to planar and cylindrical cavities.

2. Correspondence of Thermal Non-equilibrium Theories

The recent studies by Antezza, Pitaevskii and Stringari[5] (APS) on the one hand and Buhmann and Scheel[8] (BS) on the other hand both deal with non-equilibrium situations. However, the theory of APS is concerned with effects due to a thermal disequilibrium between different parts of the environment. The theory of BS describes an in some sense opposite situation where the focus is on a disequilibrium between the atom in a non-thermalized state and its environment which is itself in thermal equilibrium with uniform temperature.

In this section we demonstrate how the two situations may be bridged through a non-equilibrium theory of CP forces based on the Keldysh Green function method[13] by showing that the known results for the BS and APS configurations can both be obtained within this formalism. We restrict our derivation to the interaction between a ground-state particle and a non-magnetic medium embedded in an external (thermal) electromagnetic field, neglecting the non-equilibrium dynamics. A more general situation will be considered elsewhere.

The CP potential of a polarisable ground-state particle at position $\mathbf{r}$ next to a non-magnetic body embedded in an electromagnetic field can be calculated with the help of the Keldysh Green function technique. We calculate the density matrix of the atom using the Keldysh method.[14] Resticting our considerations to the potential at initial time, we may neglect terms associated with atomic non-equilibrium dynamics. The shift of the atomc ground-state due to the interaction with the electromagnetic field can then be found in terms of the photon Green functions (for details, see Ref. 13]:

$$U(\mathbf{r}) = -\frac{1}{2\pi}\int_0^\infty d\omega\,\omega^2 \mathrm{Tr}\mathrm{Im}\left\{\boldsymbol{\alpha}(\omega)\cdot\left[\hbar\mu_0 \mathbf{G}^{(1)}(\mathbf{r},\mathbf{r},\omega) - \boldsymbol{\rho}(\mathbf{r},\mathbf{r},\omega)\right]\right\}, \quad (1)$$

where $\boldsymbol{\alpha}(\omega)$ is the ground-state polarizability of the atom and $\mathbf{G}^{(1)}(\mathbf{r},\mathbf{r},\omega)$ is the scattering part of the Green tensor of the electromagnetic field, $\mathbf{G}(\mathbf{r},\mathbf{r}',\omega)$, which in turn is the unique solution to

$$\left[\boldsymbol{\nabla}\times\boldsymbol{\nabla}\times -\frac{\omega^2}{c^2}\,\varepsilon(\mathbf{r},\omega)\right]\mathbf{G}(\mathbf{r},\mathbf{r}',\omega) = \boldsymbol{\delta}(\mathbf{r}-\mathbf{r}') \quad (2)$$

together with the boundary condition at infinity. Finally, $\boldsymbol{\rho}(\mathbf{r},t;\mathbf{r}',t') = -i\langle\hat{\mathbf{A}}(\mathbf{r}',t')\hat{\mathbf{A}}(\mathbf{r},t)\rangle^{\mathrm{T}}$, where $\hat{\mathbf{A}}$ is the vector potential of the external electromagnetic field. Note that the Keldysh result offers a clear distinction between the influence of the zero-point fluctuations of the field, the first term in Eq. (1) and the contributions from external fields as contained in the second term.

Let us first compare the Keldysh method with the BS calculations.[8] If the environment containing the body and the external field is at global thermal equilirium with a uniform temperature T_E, the density matrix $\boldsymbol{\rho}$ can be calculated with the help of the fluctuation-dissipation theorem:

$$\boldsymbol{\rho}(\mathbf{r},\mathbf{r}',\omega) = -2i\hbar\mu_0 N(\omega,T_E)\mathrm{Im}\mathbf{G}(\mathbf{r},\mathbf{r}',\omega), \quad (3)$$

where $N(\omega,T_E) = [\exp(\hbar\omega/k_{\mathrm{B}}T_E)-1]^{-1}$ denotes the thermal photon numbers. Substituting this into Eq. (1) and discarding the position-independent contribution associated with the bulk Green tensor ($\mathbf{G} \mapsto \mathbf{G}^{(1)}$), we find:

$$\begin{aligned} U(\mathbf{r}) =& -\frac{\hbar\mu_0}{2\pi}\int_0^\infty d\omega\,\omega^2[2N(\omega,T_E)+1]\mathrm{Tr}\mathrm{Im}\left[\boldsymbol{\alpha}(\omega)\cdot\mathbf{G}^{(1)}(\mathbf{r},\mathbf{r},\omega)\right] \\ &+\frac{\hbar\mu_0}{\pi}\int_0^\infty d\omega\,\omega^2 N(\omega,T_E)\mathrm{Tr}\left[\mathrm{Im}\boldsymbol{\alpha}(\omega)\cdot\mathrm{Re}\mathbf{G}^{(1)}(\mathbf{r},\mathbf{r},\omega)\right]. \end{aligned} \quad (4)$$

The first term in Eq. (4) can be cast into an alternative form by writing $\mathrm{Im}z = (z - z^*)/(2i)$, using the identities $\mathbf{G}^*(\omega) = \mathbf{G}(-\omega)$, $\boldsymbol{\alpha}^*(\omega) = \boldsymbol{\alpha}(-\omega)$ and making the substitution $\omega \mapsto -\omega$. The emerging integral over the entire real frequency axis can be completed to a closed contour by adding a vanishing integral over an infinite semi-circle in the upper half of the complex frequency plane. Evaluating the contour integral via Cauchy's theorem, we are left with the contributions from the poles $i\xi_m$ of $[2N(\omega,T_E)+1]$, viz.

$$\begin{aligned} U(\mathbf{r}) =& \mu_0 k_\mathrm{B} T \sum_{m=0}^{\infty}{}' \xi_m^2 \mathrm{Tr}\left[\boldsymbol{\alpha}(i\xi_m)\cdot\mathbf{G}^{(1)}(\mathbf{r},\mathbf{r},i\xi_m)\right] \\ &+\frac{\hbar\mu_0}{\pi}\int_0^\infty d\omega\,\omega^2 N(\omega,T_E)\mathrm{Tr}\left[\mathrm{Im}\boldsymbol{\alpha}(\omega)\cdot\mathrm{Re}\mathbf{G}^{(1)}(\mathbf{r},\mathbf{r},\omega)\right], \end{aligned} \quad (5)$$

with $\xi_m = 2\pi m k_\mathrm{B} T/\hbar$. The prime at the Matsubara sum indicates that the $m = 0$ term carries half-weight. For an isotropic particle, the ground-state polarizability in the perturbative limit can be given as

$$\boldsymbol{\alpha}(\omega) = \lim_{\epsilon\to 0}\frac{1}{3\hbar}\sum_k\left[\frac{|\mathbf{d}_{0k}|^2}{\omega+\omega_k+i\epsilon} - \frac{|\mathbf{d}_{0k}|^2}{\omega-\omega_k+i\epsilon}\right]\mathbf{I} \quad (6)$$

[$\omega_k = (E_k - E_0)/\hbar$ are transition frequencies; $\mathbf{d}_{0k}$ are electric dipole matrix elements; $\mathbf{I}$ is the unit tensor]. Using $\lim_{\epsilon\to 0} 1/(x+i\epsilon) = \mathcal{P}/x - i\pi\delta(x)$ (with $\mathcal{P}$ principal value), the thermal CP potential of an isotropic ground-state atom is given by

$$\begin{aligned} U(\mathbf{r}) =& \mu_0 k_\mathrm{B} T \sum_{m=0}^{\infty}{}' \xi_m^2 \alpha(i\xi_m)\mathrm{Tr}\mathbf{G}^{(1)}(\mathbf{r},\mathbf{r},i\xi_m) \\ &+\frac{\mu_0}{3}\sum_k |\mathbf{d}_{0k}|^2\,\omega_k^2 N(\omega_k,T_E)\mathrm{Tr}\mathrm{Re}\mathbf{G}^{(1)}(\mathbf{r},\mathbf{r},\omega_k). \end{aligned} \quad (7)$$

The corresponding force $\mathbf{F}(\mathbf{r}) = -\boldsymbol{\nabla}U(\mathbf{r})$ agrees exactly with the BS result for the force at initial time on a ground-state atom in the perturbative limit, cf. Eq. (25) of Ref. 8. Cf. this reference for details of the derivation and the dynamical and nonperturbative generalisation of this result for arbitrary (incoherent) initial-state preparation of the atom. The first term of Eq. (7) is the non-resonant force, while the second term is a resonant term due to absorption processes, which are pure non-equilibrium effects. Note that such processes will unavoidably heat the atom, so that our result is only valid at initial time when the atom is still in its ground state. When the atom has become thermalized, the resonant term vanishes and the expression reduces to a pure Matsubara-type sum similar to the non-resonant expression.

In the APS case,[5] the body is assumed to be held at a uniform temperature T_S different from that of the environment, T_E, such that the total system is in a stationary non-equilibrium. The atom is assumed to be at zero temperature in the sense that the thermal energies $k_{\mathrm{B}}T_S, k_{\mathrm{B}}T_E$ are much smaller than the energies $\hbar\omega_k$ necessary to excite the atom. This means that terms $N(\omega, T_E)\mathrm{Im}\boldsymbol{\alpha}(\omega)$, which are proportional to the occupation numbers of photons at the atomic frequencies, cf. Eq. (6), can be neglected. The density matrix $\boldsymbol{\rho}$ obeys the kinetic-like equation emerging in the Keldysh method:[15]

$$\begin{aligned}\boldsymbol{\rho}(\mathbf{r},t;\mathbf{r}',t') &= \boldsymbol{\rho}_0(\mathbf{r},t,\mathbf{r}',t') + \frac{\mu_0}{\hbar}\int d^3r_1\int dt_1\int d^3r_2\int dt_2\,[\boldsymbol{\rho}_0(\mathbf{r},t;\mathbf{r}_1,t_1)\\ &\quad\cdot\boldsymbol{\Pi}_A(\mathbf{r}_1,t_1;\mathbf{r}_2,t_2)\cdot\mathbf{G}_A(\mathbf{r}_2,t_2;\mathbf{r}',t')\\ &\quad+\mathbf{G}_0(\mathbf{r},t;\mathbf{r}_1,t_1)\cdot\boldsymbol{\Pi}_R(\mathbf{r}_1,t_1;\mathbf{r}_2,t_2)\cdot\boldsymbol{\rho}(\mathbf{r}_2,t_2;\mathbf{r}',t')\\ &\quad-\mathbf{G}_0(\mathbf{r},t;\mathbf{r}_1,t_1)\cdot\boldsymbol{\Pi}_{12}(\mathbf{r}_1,t_1;\mathbf{r}_2,t_2)\cdot\mathbf{G}_A(\mathbf{r}_2,t_2;\mathbf{r}',t')/(\mu_0\hbar)];\end{aligned}\tag{8}$$

$$\begin{aligned}\mathbf{G}(\mathbf{r},t;\mathbf{r}',t') &= \mathbf{G}_0(\mathbf{r},t;\mathbf{r}',t') + \frac{\mu_0}{\hbar}\int d^3r_1\int dt_1\int d^3r_2\int dt_2\,\mathbf{G}_0(\mathbf{r},t;\mathbf{r}_1,t_1)\\ &\quad\cdot\boldsymbol{\Pi}_R(\mathbf{r}_1,t_1;\mathbf{r}_2,t_2)\cdot\mathbf{G}(\mathbf{r}_2,t_2;\mathbf{r}',t').\end{aligned}\tag{9}$$

Here, the index 0 corresponds to free fields; $\mathbf{G}_A = \mathbf{G}^\dagger$ is the advanced Green tensor; and $\boldsymbol{\Pi}$ are polarisation operators defined in terms of fluctuating currents $\boldsymbol{\Pi}_{12}(\mathbf{r},t;\mathbf{r}',t') = -i\langle\hat{\mathbf{j}}(\mathbf{r}',t')\hat{\mathbf{j}}(\mathbf{r},t)\rangle^{\mathrm{T}}$, $\boldsymbol{\Pi}_{21}(\mathbf{r},t;\mathbf{r}',t') = -i\langle\hat{\mathbf{j}}(\mathbf{r},t)\hat{\mathbf{j}}(\mathbf{r}',t')\rangle$, and $\boldsymbol{\Pi}_R(\mathbf{r},t;\mathbf{r}',t') = [\boldsymbol{\Pi}_{12}(\mathbf{r},t;\mathbf{r}',t') - \boldsymbol{\Pi}_{21}(\mathbf{r},t;\mathbf{r}',t')]\Theta(t-t')$, $\boldsymbol{\Pi}_A = \boldsymbol{\Pi}_R^\dagger$, where Θ is the Heaviside function. If the body was in equilibrium with the environment, the solution of Eq. (8) would be given by Eq. (3). To account for effects due to the temperature difference between the body and the environment, we now solve Eq. (8) iteratively starting with the equilibrium solution (3). The first approximation reads:

$$\begin{aligned}\boldsymbol{\rho}^{(1)}(\mathbf{r},\mathbf{r}',\omega) &= -2i\hbar\mu_0 N(T_E,\omega)\mathrm{Im}\mathbf{G}(\mathbf{r},\mathbf{r}',\omega) - \mu_0^2\int d^3r_1 d^3r_2\,\mathbf{G}_0(\mathbf{r},\mathbf{r}_1,\omega)\\ &\quad\cdot\{N(T_E,\omega)\boldsymbol{\Pi}_{21}(\mathbf{r}_1,\mathbf{r}_2,\omega) - [N(T_E,\omega)+1]\boldsymbol{\Pi}_{12}(\mathbf{r}_1,\mathbf{r}_2,\omega)\}\cdot\mathbf{G}^*(\mathbf{r}_2,\mathbf{r}',\omega).\end{aligned}\tag{10}$$

Here, we have used Eq. (9), the free-field density matrix $\boldsymbol{\rho}_0(\mathbf{r},\mathbf{r}',\omega) = -2i\hbar\mu_0 N(\omega,T_E)\mathrm{Im}\mathbf{G}_0(\mathbf{r},\mathbf{r}',\omega)$ and the relation[15] $\boldsymbol{\Pi}_R - \boldsymbol{\Pi}_A = \boldsymbol{\Pi}_{12} - \boldsymbol{\Pi}_{21}$

for the polarization operators. Resubstituting Eq. (10) into the r.h.s. of Eq. (8), we obtain a result of the form (10) with $\mathbf{G}_0 \mapsto \mathbf{G}_0 + \mathbf{G}_0 \cdot \boldsymbol{\Pi}_R \cdot \mathbf{G}_0$. Repeating the iterations many times and performing the sum via a Born series $\mathbf{G} = \mathbf{G}_0 + \mathbf{G}_0 \cdot \boldsymbol{\Pi}_R \cdot \mathbf{G}_0 + \mathbf{G}_0 \cdot \boldsymbol{\Pi}_R \cdot \mathbf{G}_0 \cdot \boldsymbol{\Pi}_R \cdot \mathbf{G}_0 + \ldots$, we arrive at

$$\boldsymbol{\rho}(\mathbf{r}, \mathbf{r}', \omega) = -\hbar\mu_0 \left[2iN(T_E, \omega)\mathrm{Im}\mathbf{G}(\mathbf{r}, \mathbf{r}', \omega) + \int d^3r_1 d^3r_2 \, \mathbf{G}(\mathbf{r}, \mathbf{r}_1, \omega) \right.$$
$$\left. \cdot \{N(T_E, \omega)\boldsymbol{\Pi}_{21}(\mathbf{r}_1, \mathbf{r}_2, \omega) - [N(T_E, \omega) + 1]\boldsymbol{\Pi}_{12}(\mathbf{r}_1, \mathbf{r}_2, \omega)\} \cdot \mathbf{G}^*(\mathbf{r}_2, \mathbf{r}', \omega) \right]. \tag{11}$$

As stated, the body is assumed to be at local thermal equilibrium with temperature T_S. This allows us to implement the fluctuation–dissipation theorem to calculate the polarisation operators:

$$\boldsymbol{\Pi}_{12}(\omega, \mathbf{r}, \mathbf{r}') = -\frac{i\hbar\varepsilon_0}{2\pi} \omega^2 N(\omega, T_S) \mathrm{Im}\varepsilon(\mathbf{r}, \omega) \boldsymbol{\delta}(\mathbf{r} - \mathbf{r}'); \tag{12a}$$

$$\boldsymbol{\Pi}_{21}(\omega, \mathbf{r}, \mathbf{r}') = -\frac{i\hbar\varepsilon_0}{2\pi} \omega^2 [N(\omega, T_S) + 1] \mathrm{Im}\varepsilon(\mathbf{r}, \omega) \boldsymbol{\delta}(\mathbf{r} - \mathbf{r}'). \tag{12b}$$

Inserting Eqs. (12) and (11) into Eq. (1), we find:

$$U(\mathbf{r}) = -\frac{\hbar\mu_0}{2\pi} \int_0^\infty d\omega \, \omega^2 [2N(\omega, T_E) + 1] \mathrm{ImTr} \left[\boldsymbol{\alpha}(\omega) \cdot \mathbf{G}^{(1)}(\mathbf{r}, \mathbf{r}, \omega) \right]$$
$$+ \mathrm{Re} \frac{\hbar\mu_0}{(2\pi)^2} \int_0^\infty d\omega \, \frac{\omega^4}{c^2} \int d^3s \, \mathrm{Im}\varepsilon(\mathbf{s}, \omega) \{N(\omega, T_E)[N(\omega, T_S) + 1]$$
$$-[N(\omega, T_E) + 1]N(\omega, T_S)\} \, \mathrm{Tr} \left[\boldsymbol{\alpha}(\omega) \cdot \mathbf{G}(\mathbf{r}, \mathbf{s}, \omega) \cdot \mathbf{G}^*(\mathbf{s}, \mathbf{r}, \omega) \right]. \tag{13}$$

The first term of Eq. (13) describes the equilibrium CP force, cf. Eq. (4) above, the second term corresponds to the case when the medium is not in equilibrium with the field. If the medium is a homogeneous body of permittivity $\varepsilon(\omega)$ occupying a volume V_S, then the latter leads to just the APS result for the non-equilibrium force $\mathbf{F}_{\mathrm{neq}}(\mathbf{r}) = -\boldsymbol{\nabla} U_{\mathrm{neq}}(\mathbf{r})$ [see Eqs. (7) and (9) of Ref. 5]:

$$\mathbf{F}_{\mathrm{neq}}(\mathbf{r}) = \frac{\hbar\mu_0}{2\pi^2} \int_0^\infty d\omega \, \frac{\omega^4}{c^2} \int_{V_S} d^3s \left[\frac{1}{e^{\hbar\omega/k_{\mathrm{B}}T_S} - 1} - \frac{1}{e^{\hbar\omega/k_{\mathrm{B}}T_E} - 1} \right]$$
$$\times \mathrm{Im}\varepsilon(\omega) \left\{ \boldsymbol{\nabla}' \mathrm{ReTr} \left[\boldsymbol{\alpha}(\omega) \cdot \mathbf{G}(\mathbf{r}, \mathbf{s}, \omega) \cdot \mathbf{G}^*(\mathbf{s}, \mathbf{r}', \omega) \right] \right\}_{\mathbf{r}' = \mathbf{r}}. \tag{14}$$

3. Thermal CP Potential on a Particle in Uniform Temperature Environment

To illustrate the non-equilibrium effects, we apply the general theory of the previous section to specific scenarios. We will concentrate on a non-equilibrium between the particle and the electromagnetic field and consider the potential (7) of a ground-state particle in an environment of uniform temperature.

3.1. *Planar systems*

The trace of the Green tensor at a distance z to the right of a half-space with reflection coefficients r_s, r_p for s, p polarization is[16]

$$\mathrm{Tr}\mathbf{G}(\mathbf{r},\mathbf{r},\omega) = \frac{i}{4\pi}\int_0^\infty \frac{q\,dq}{\beta}\left[\frac{2\beta^2c^2}{\omega^2}r_p - \sum_{\sigma=s,p} r_\sigma\right]e^{2i\beta z} \tag{15}$$

with $\beta = \sqrt{\omega^2/c^2 - q^2}$. The integral over transverse momentum q naturally separates into a propagating part $q < \omega/c$ and an evanescent part $q > \omega/c$.

We use Eq. (15) to calculate the force on a ground state LiH molecule outside a gold half-space at $T = 300$K. The result is striking (Fig. 1a): the evanescent part almost exactly cancels the non-resonant part, and the propagating part is spatially oscillating and dominates in the retarded regime.

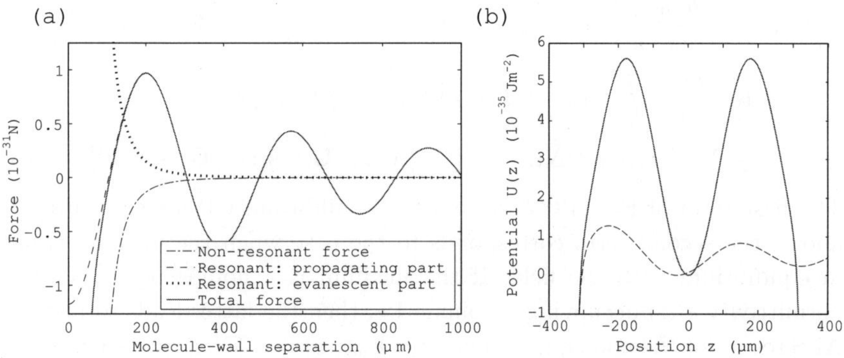

Fig. 1. (a) Components of the thermal CP force $F_z = -\partial U/(\partial z)$ on a ground state LiH molecule outside a gold half space at 300K. (b) Enhanced potential from cavity of width $a = \lambda_{k0}$ (solid line). The dashed line indicates the potential outside a single half-space.

Unfortunately, the spatially oscillating propagating force component is very weak outside a half-space. We have investigated a scheme to enhance the amplitude of the oscillating potential by fine-tuning the width of a planar cavity to exactly one wavelength of light which resonates with the dominating molecular transition.[12] Figure 1b shows that the scheme works in principle, but the enhancement factor thus achieved is not enough to bring the oscillations into a regime which is likely to be observable. The reason for this is primarily that the enhancement factor scales with the logarithm of the Q-factor of the cavity (shown analytically and numerically in Ref. 12) which strongly limits the potentiality of such a scheme.

3.2. *Cylindrical cavity*

Another candidate geometry is a molecule situated inside a cylindrical cavity of radius R. At certain specific radii, resonances like that for the planar cavity are expected to occur, where the two-dimensional mode confinement holds the promise that resonant force components may be much stronger than in the planar case. As a first step towards exploiting this effect to realise strong thermal CP forces, we investigate the spatial profile of resonant forces inside the cylindrical cavity. Work on this problem is continuing.

The trace of the Green tensor for points $\mathbf{r}' = \mathbf{r} = (\rho, \theta, z)$ inside a cylindrical vacuum cavity in an unbounded non-magnetic medium of permittivity ε may, after much simplification, be written as[17,18]

$$\begin{aligned}\mathrm{Tr}\mathbf{G}(\rho,\rho;\omega) =& \frac{ik}{2\pi}\int_0^\infty dt \sum_{n=0}^{\infty}{}' \Big\{(r_M + t^2 r_N) \\ &\times \left[\frac{n^2}{\phi^2 x^2} J_n^2(\phi x) + J_n'^2(\phi x)\right] + r_N \frac{x^2}{g^2} J_n^2(\phi x)\Big\}. \qquad (16)\end{aligned}$$

Here, $x = g\sqrt{1-t^2}$, $g = kR$ and $k = \omega/c$. The dimensionless radial co-ordinate is $\phi = \rho/R$ and the integration variable t is the dimensionless momentum component along the cylinder axis, h, relative to k.† The reflection coefficients read $r_{M,N} = -[H_n^{(1)}(x)/J_n(x)]\tilde{r}_{M,N}$ with $\tilde{r}_{M,N} = (A + B_{M,N})/(A + B_D)$ and

$$A = n^2[x^6 - (2x_1^2 + g^2)x^4 + (2g^2 + x_1^2)x_1^2 x^2 - g^2 x_1^4], \qquad (17a)$$

$$B_M = g^2 x_1^2 x^2[\varepsilon \tilde{h}_1^2 x^2 - (\tilde{h}_1 \tilde{j}_2 + \varepsilon \tilde{h}_1 \tilde{h}_2) x_1 x + \tilde{h}_2 \tilde{j}_2 x_1^2], \qquad (17b)$$

$$B_N = g^2 x_1^2 x^2[\varepsilon \tilde{h}_1^2 x^2 - (\varepsilon \tilde{h}_1 \tilde{j}_2 + \tilde{h}_1 \tilde{h}_2) x_1 x + \tilde{h}_2 \tilde{j}_2 x_1^2], \qquad (17c)$$

$$B_D = g^2 x_1^2 x^2[\varepsilon \tilde{h}_1^2 x^2 - (\varepsilon + 1) \tilde{h}_1 \tilde{j}_2 x_1 x + \tilde{j}_2^2 x_1^2] \qquad (17d)$$

$[x_1 = g\sqrt{\varepsilon - t^2};\ x_2 = x;\ \tilde{j}_j = J_n'(x_j)/J_n(x_j);\ \tilde{h}_j = H_n^{(1)\prime}(x_j)/H_n^{(1)}(x_j)]$.

In the perfectly conducting limit $|\varepsilon| \to \infty$ (at nonzero ω) we find

$$r_M \to -\frac{H_n^{(1)\prime}(x)}{J_n'(x)}; \qquad r_N \to -\frac{H_n^{(1)}(x)}{J_n(x)}. \qquad (18)$$

†For the non-resonant term the dimensionless substitution variable $t = h/k$ cannot be used since $k = 0$ for the zeroth Matsubara term. We use $\tilde{t} = hR = gt$ in this case.

For the perfectly conducting cylinder, thus, the resonant radii for radiation of a given frequency are $R_{ni}(\omega)$ and $R'_{ni}(\omega)$ with $R^{(\prime)}_{ni}(\omega) = cj^{(\prime)}_{ni}/\omega$ wherein j_{ni} and j'_{ni} are the ith zero of $J_n(x)$ and $J'_n(x)$, respectively. When $\varepsilon < \infty$ the resonances move away from these values. Further details and analysis will be reported elsewhere.

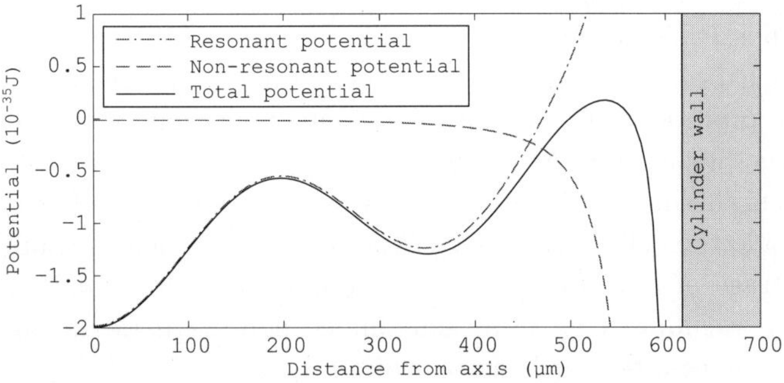

Fig. 2. The thermal CP potential on a LiH molecule in an infinitely thick cylindrical gold cavity at $T = 300\text{K}$. The radius is $R = 1.5R_{11} \approx 618\mu\text{m}$.

We plot the potential for the example of $R = 1.5R_{11}$ for ground-state LiH in Fig. 2. For technical reasons, in the cylindrical geometry splitting the resonant potential into propagating and evanescent parts is no longer natural and straightforward. As a benchmark we have checked that the potential outside a plane is regained close to the cylinder boundary in the limit of large cylinder radius $R \gg 2\pi/k_{10}$.

Figure 2 shows clearly that the peculiar traits observed for the resonant potential outside a half-space, depicted in Fig. 1a, are present also in the cylindrical cavity as one would expect. The resonant potential once again almost cancels the non-resonant term close to the surface giving a resulting attractive force in the near zone which is dramatically reduced compared to the non-resonant term alone. As before the retarded regime is dominated by oscillating behaviour. However, we also observe qualitative differences from the planar case, e.g., the potential minimum closest to the wall is no longer the deepest one as is the case in a planar cavity.[12]

4. Summary

We have demonstrated that the recent complementary theories for the thermal CP force between a ground-state particle and a body at non-equilibrium between either the particle or the body and the electromagnetic field may both be obtained using the Keldysh formalism. Applying the results to planar and cylindrical geometries of uniform temperature, we have found that even a ground-state particle is subject to resonant force components which in both geometries strongly cancel the well-known resonant force in the nonretarded regime and lead to spatially oscillating forces for retarded distances. The latter are expected to be much enhanced for the cylindrical cavity, opening the perspective towards molecule guiding with thermal photons.

Acknowledgments

This work was financially supported partially by the UK Engineering and Physical Sciences Research Council. We acknowledge financial support by the European Science Foundation (ESF) within the activity 'New Trends and Applications of the Casimir Effect' (`www.casimir-network.com`) and useful discussions with Dr. Mauro Antezza and Dr. Carsten Henkel. S.Y.B. is grateful for support by the Alexander von Humboldt foundation. We finally thank Professor Kimball Milton and the local organisers for an excellent conference.

References

1. H.B.G. Casimir and D. Polder, *Phys.Rev.* **73**, 360 (1948).
2. S.Y. Buhmann and D.-G. Welsch, *Prog.Quantum Electron.* **31**, 51 (2007).
3. G. Barton, *Proc.R.Soc. London A* **320**, 251 (1970); **367**, 117 (1979); **410**, 141 (1987).
4. W. Jhe, *Phys.Rev. A* **43**, 5795 (1991); **44**, 5932 (1991).
5. M. Antezza et al., *Phys.Rev.Lett.* **95**, 113202 (2005).
6. M. Antezza et al., *Phys.Rev. A* **77**, 022901 (2008).
7. J.M. Obrecht et al., *Phys.Rev.Lett.* **98**, 063201 (2007).
8. S.Y. Buhmann and S. Scheel, *Phys.Rev.Lett.* **100**, 253201 (2008).
9. M.-P. Gorza and M. Ducloy, *Euro.Phys.J. D* **40**, 343 (2006).
10. S.Y. Buhmann and S. Scheel, *Phys.Scripta T* **135**, 014013 (2009).
11. S.Å. Ellingsen, S.Y. Buhmann, and S. Scheel, *Phys.Rev. A* **79**, 052903 (2009).
12. S.Å. Ellingsen, S.Y. Buhmann, and S. Scheel, *Phys.Rev. A* **80**, 022901 (2009).
13. Y. Sherkunov, *Phys.Rev. A* **79**, 032101 (2009).
14. L.V. Keldysh, *Sov.Phys. JETP* **20**, 1018 (1965).
15. E.M. Lifshitz and L.P. Pitaevskii, *Course of Theoretical Physics, Physical Kinetics* (Pergamon Press, Oxford, 1981).
16. M.S. Tomaš, *Phys.Rev. A* **51**, 2545 (1995).
17. L.-W. Li et al., *J. Electromag. Waves Appl.* **14**, 961 (2000).
18. S. Scheel and S.Y. Buhmann, *Acta Phys. Slov.* **58**, 675 (2008).

GEOMETRY-TEMPERATURE INTERPLAY IN THE CASIMIR EFFECT

H. GIES

Theoretisch-Physikalisches Institut, Universität Jena, D-07743 Jena, Germany
and
Helmholtz Institute Jena, D-07743 Jena, Germany

A. WEBER

Institute for Theoretical Physics, Heidelberg University,
D-69120 Heidelberg, Germany

We discuss Casimir phenomena which are dominated by long-range fluctuations. A prime example is given by "geothermal" Casimir phenomena where thermal fluctuations in open Casimir geometries can induce significantly enhanced thermal corrections. We illustrate the underlying mechanism with the aid of the inclined-plates configuration, giving rise to enhanced power-law temperature dependences compared to the parallel-plates case. In limiting cases, we find numerical evidence even for fractional power laws induced by long-range fluctuations. We demonstrate that thermal energy densities for open geometries are typically distributed over length scales of $1/T$. As an important consequence, approximation methods for thermal corrections based on local energy-density estimates such as the proximity-force approximation are expected to become unreliable even at small surface separations.

Keywords: Casimir effect; finite-temperature field theory; worldline approach.

1. Introduction

Beyond its many attributes and genuine properties, the Casimir effect[1–3] is also a phenomenon that can be dominated by long-range fluctuations. At first sight, this statement may seem surprising as many standard Casimir examples do not show manifest signatures of a long-range fluctuation phenomenon (such as, e.g., critical phenomena). Even though massless fluctuations of the radiation field naturally support long-range correlations, the geometry of the Casimir configuration typically provides a length scale beyond which fluctuations become less relevant. For instance, the length scale of fluctuations associated with the classic Casimir effect between parallel plates is certainly set by the plate separation a, serving effectively as a long-range cutoff.

Another seeming counter-example for the above-given statement is the sphere-plate configuration at small separation distances a, which is experimentally highly relevant.[4] Here, the Casimir interaction energy behaves as $E \sim R/a^2$ for $a/R \to 0$, where R is the sphere radius. It thus exhibits a simple power law with integer coefficients that follow from a geometric scaling analysis known as the proximity force approximation (PFA).[5] (The validity of this approximation in the asymptotic regime has been confirmed by analytical as well as numerical methods of quantum field theory,[6–10] see also Ref. 11 for a solution at larger distances.) Similar geometric analyses work equally well for, say, electrostatic forces in the same configuration which are not related to any long-range fluctuation phenomenon.

In the present work, we argue that these examples do not reveal the long-range nature of the Casimir effect, as the corresponding interaction energies are dominated by rather localized energy densities. E.g., for the sphere-plate case, the dominant contribution to the Casimir force arises from the region between the surfaces at closest separation. By contrast, we present Casimir phenomena in the following where the energy density is distributed over a wide range of scales, such that the potential long-range nature of the Casimir effect becomes most prominent.

An example of this class of phenomena is the nontrivial interplay of finite-temperature and geometry dependences of the Casimir effect. As first conjectured in Ref. 12, the thermal modifications of the Casimir effect can differ qualitatively for different geometries. This is because the thermal corrections arise from thermal excitations of the fluctuation spectrum, which in turn depends strongly on the geometry. First analytical as well as numerical evidence of this "geothermal" interplay has been provided in Ref. 13 by applying the worldline formalism to a perpendicular-plates configuration. A detailed study of this phenomenon for the more general inclined-plates case has been performed in Ref. 14, the results of which will be used as a quantitative example for our arguments in the following.

The purpose of the present contribution is to develop the general physics picture underlying the geothermal Casimir phenomena. In Sect. 2, we discuss the origin and various general perspectives on the interplay between geometry and finite temperature. Section 3 briefly summarizes the worldline method which is a powerful tool to analyze this interplay quantitatively. Several examples will be discussed in Sect. 4, where we also present new results for the thermal force density of specific open geometries. Conclusions are given in Sect. 5.

2. Origin of Geothermal Casimir Phenomena

The origin of a nontrivial interplay between geometry and temperature in the Casimir effect can be understood in simple terms. Consider the classical parallel-plate case: as the wavelengths of the fluctuations orthogonal to the plates have to be commensurate with the distance a between the plates, this corresponding relevant part of the spectrum $\omega_n^2(a) = p_\parallel^2 + (\pi n/a)^2$ has a gap of wave number $k_{\rm gap} = \pi/a$. As is obvious, e.g., from the partition function $Z = \sum_n \exp(-\omega_n(a)/T)$, the gapped modes are exponentially suppressed at small temperatures $T \ll k_{\rm gap}$. In $D = 4$ spacetime dimensions, the integration over the parallel modes $p_\parallel$ converts this exponential dependence into the low-temperature power law T^4 for the parallel-plate Casimir force. The corresponding thermal contribution to the free interaction energy (apart from a distance-independent term) is

$$\Delta E_\parallel(T) = \frac{\pi^2}{90} A\, a\, T^4, \quad aT \ll 1, \tag{1}$$

where A denotes the plate's area.[a] The above-given argument for a suppression of thermal contributions applies to all geometries with a gap in the relevant part of the spectrum (e.g. concentric cylinders or spheres, Casimir pistons, etc.). These geometries are called *closed.*

By contrast, *open* geometries with a gapless relevant part of the spectrum have no such suppression of thermal contributions. Any small value of the temperature can always excite the low-lying modes in the spectrum. Therefore, we expect a generically stronger thermal contribution $\sim (aT)^\alpha$ with $0 < \alpha < 4$.

Another argument for the fundamental difference between open and closed geometries and thermal corrections is the following: Eq. (1) can also

---

[a] As the thermal contribution to the force resulting from a derivative of Eq. (1) is independent of the distance, a real experiment may need to take the boundaries of the apparatus or other nearby surfaces into account. This is especially required in the extreme low-temperature limit when T times the separation to any of these auxiliary surfaces is small; e.g., for thermal forces in piston-like geometries that explicitly account for apparatus boundaries, see Ref. 15. On the other hand, if the length scale provided by $1/T$ is much smaller than any of these auxiliary distances to the apparatus boundaries, the force estimate resulting from Eq. (1) is expected to be reasonably accurate for $aT \ll 1$. In the remainder of this article, the low-temperature limit therefore always refers to $aT \ll 1$, but with $1/T$ being still smaller than any other distance to further auxiliary surfaces.

be written as $\Delta E_{\|}(T) = V\,\epsilon_{\mathrm{SB}}$, where $V = aA$ is the volume between the parallel plates, and ϵ_{SB} is the Stefan-Boltzmann free energy density of the radiation field. Hence, we can understand the low-temperature correction in the parallel-plate case as an excluded volume effect: the thermal modes of the radiation field at low temperatures do not fit in between the plates, and therefore the corresponding volume does not contribute to the total thermal free energy. By contrast, open geometries by construction cannot be associated with any (unambiguously defined) excluded volume, such that significant deviations from a T^4 behavior can be expected.

These considerations immediately point to the possibility that the thermal part of the low-temperature Casimir effect can be dominated by long-range fluctuations. This is because a temperature much lower than the inverse distance, $aT \ll 1$, sets a new length scale which can be much larger as the plate distance as well as any other length scale of the geometry (such as a sphere radius). In closed geometries, this length scale is effectively cut off by the gap in the spectrum, implying the parametric suppression of thermal effects. In open geometries, this length scale sets a relevant scale that can, for instance, reflect the spatial extent of the distribution of the thermal energy density. The total thermal energy thus can receive dominant contributions from long-range modes corresponding to significantly extended thermal energy distributions.

An important consequence can already be anticipated at this point: approximation methods that are based on local considerations will generically fail to predict the correct low-temperature correction in open geometries. An example is given by the PFA which is based on the assumption that the Casimir energy can be estimated by integrating over local parallel-plates energy densities. Whereas this approximation may or may not work at zero temperature depending on the geometric details of the configuration, it is even conceptually questionable at finite temperature, as open geometries should not be approximated by closed-geometry building blocks. Quantitatively, such a procedure is expected to fail, as local energy-density approximations will not be able to capture the contributions from larger length scales induced by long-range modes.

The temperature-geometry interplay is not an academic problem: experimentally important configurations such as the sphere-plate or the cylinder-plate geometry belong to this class of open geometries, but thermal corrections have so far been approximated by the PFA. Whether or not a potentially significant geothermal interplay may exist in the relevant parameter range $aT \sim 0.01 \ldots 0.1$ is a technically challenging quantitative problem.

The considerations so far have concentrated on the low-temperature limit. In fact, the high-temperature limit exhibits a universal linear dependence on the temperature for a simple reason. At high temperature in the imaginary-time formalism, only the zeroth Matsubara mode can contribute as all higher modes acquire thermal masses $\sim \pi T$ and hence are largely suppressed. The zeroth Matsubara mode has no temperature dependence at all, such that the only temperature dependence arises from the measure of the fluctuation trace which is linear in T. A less technical argument with the same result can be based on the underlying Bose-Einstein distribution governing the bosonic thermal fluctuations of the radiation field. This distribution increases as $\sim T$ in the high-temperature limit, inducing this linear temperature dependence directly in the free energy. The properties of the geometry only enter the prefactor in the high-temperature limit. Universal features of thermal Casimir energies with an emphasis on the high-temperature limit have been systematically studied in Refs. 16–19.

3. Worldline Method for the Casimir Effect

Let us briefly review the worldline approach[20] to the Casimir effect as it is needed for the present line of argument. For the remainder of the paper, we consider a fluctuating massless scalar field satisfying Dirichlet boundary conditions on the Casimir surfaces. For details of the formalism, see Refs. 7,14,21. Consider a configuration Σ consisting of two static surfaces Σ_1 and Σ_2. The worldline representation of the Casimir interaction energy reads

$$E_{\mathrm{c}} = -\frac{1}{2(4\pi)^{D/2}} \int_0^\infty \frac{\mathrm{d}\mathcal{T}}{\mathcal{T}^{1+D/2}} \int \mathrm{d}^d x_{\mathrm{CM}} \, \langle \Theta_\Sigma[\mathbf{x}(\tau)] \rangle \,, \tag{2}$$

where $D = d+1$ denotes the spacetime dimensions. The worldline functional obeys $\Theta_\Sigma[\mathbf{x}] = 1$ if a worldline $\mathbf{x}(\tau)$ intersects both surfaces $\Sigma = \Sigma_1 \cup \Sigma_2$, and is zero otherwise.

The expectation value in Eq. (2) is taken with respect to an ensemble of d-dimensional closed Gaußian worldlines with center of mass $\mathbf{x}_{\mathrm{CM}}$,

$$\langle \dots \rangle = \frac{\int_{\mathbf{x}_{\mathrm{CM}}} \mathcal{D}\mathbf{x} \dots e^{-\frac{1}{4}\int_0^{\mathcal{T}} \mathrm{d}\tau \, \dot{\mathbf{x}}^2(\tau)}}{\int_{\mathbf{x}_{\mathrm{CM}}} \mathcal{D}\mathbf{x} \, e^{-\frac{1}{4}\int_0^{\mathcal{T}} \mathrm{d}\tau \, \dot{\mathbf{x}}^2(\tau)}}. \tag{3}$$

Equation (2) expresses the fact that all worldlines intersecting both surfaces do not satisfy Dirichlet boundary conditions on both surfaces. They are removed from the ensemble of allowed fluctuations by the Θ functional and thus contribute to the negative Casimir interaction energy. The auxiliary

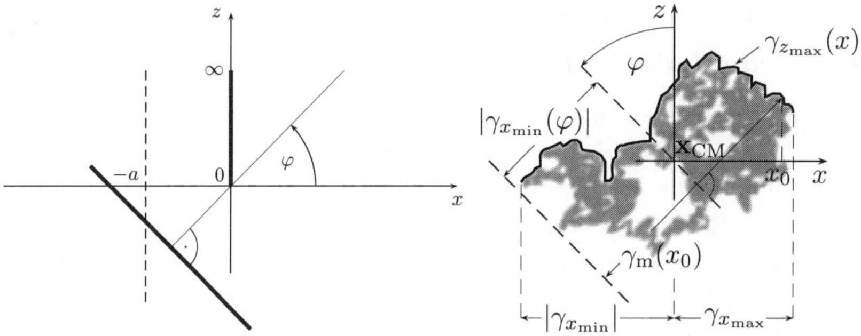

Fig. 1. Left panel: sketch of the inclined-plates configuration. The infinite plate (dashed line) is rotated in the x, z plane by an angle φ. As special cases, $\varphi = 0$ corresponds to the configuration of one semi-infinite plate parallel to an infinite plate (1si configuration), whereas $\varphi = \pi/2$ yields the perpendicular-plates configuration. Right panel: all relevant information for the evaluation of the Casimir energy (6) of inclined plates is encoded in the function $\gamma_{\mathrm{m}}(x)$, which has to be integrated from $\gamma_{x_{\min}}$ to $\gamma_{x_{\max}}$.

propertime parameter $\mathcal{T}$ scales the extent of a worldline by a factor of $\sqrt{\mathcal{T}}$. Large $\mathcal{T}$ correspond to long-range, small $\mathcal{T}$ to short-range fluctuations.

Finite temperature $T = 1/\beta$ in the Matsubara formalism is equivalent to a compactified Euclidean time on the interval $[0, \beta]$. The worldlines now live on $S^1 \times \mathbb{R}^d$ and can carry a winding number. Summing over all winding numbers, the Casimir free energy (2) becomes

$$E_{\mathrm{c}} = -\frac{1}{2(4\pi)^{D/2}} \int_0^\infty \frac{\mathrm{d}\mathcal{T}}{\mathcal{T}^{1+D/2}} \left(1 + 2\sum_{n=1}^{\infty} e^{-\frac{n^2\beta^2}{4\mathcal{T}}} \right) \int \mathrm{d}^d x_{\mathrm{CM}} \left\langle \Theta_\Sigma[\mathbf{x}(\tau)] \right\rangle . \quad (4)$$

The finite-temperature worldline formalism for static configurations thus boils down to an additional winding-number prefactor in front of the worldline expectation value. Notice that the winding-number sum is directly related to the standard Matsubara sum by a Poisson resummation.

Finally, it is advantageous to rescale the worldlines such that the velocity distribution becomes independent of $\mathcal{T}$,

$$\boldsymbol{\gamma}(t) := \frac{1}{\sqrt{\mathcal{T}}} \mathbf{x}(\mathcal{T} t) \quad \rightarrow \quad e^{-\frac{1}{4}\int_0^{\mathcal{T}} \dot{\mathbf{x}}^2 \mathrm{d}\tau} = e^{-\frac{1}{4}\int_0^1 \dot{\boldsymbol{\gamma}}^2 \mathrm{d}t}, \quad (5)$$

where $\dot{\boldsymbol{\gamma}} = \mathrm{d}\boldsymbol{\gamma}(t)/\mathrm{d}t$. The worldline integrals can be evaluated also numerically by Monte Carlo methods in a straightforward manner. Various efficient ab initio algorithms for generating discretized worldlines with Gaußian velocity distribution have been developed, see, e.g., Refs. 7, 22.

4. Geothermal Phenomena for Inclined Casimir Plates

4.1. *Inclined plates at zero temperature* $T = 0$

The inclined-plates (i.p.) configuration turns out to be an ideal work horse to study geothermal phenomena, in particular, the transition from open to closed geometries and the role of long-range fluctuations. It was studied in detail in Ref. 14 for general D. Very recently, results for inclined plates have been obtained for the electromagnetic case at zero temperature using scattering theory.[23] Here, we summarize our results for $D = 4$ and provide more details on the geometry-temperature interplay.

The inclined-plates configuration consists of a perfectly thin semi-infinite plate above an infinite plate at an angle φ, see Fig. 1. The semi-infinite plate has an edge of length L_y. The area of the infinite plate is A. The limit L_y, $A \to \infty$ is implicitly understood. Let a be the minimal distance between the plates. Evaluating the Θ functional for this configuration[14] in (2), the Casimir energy can be written as

$$\frac{E_{\mathrm{c}}^{\mathrm{i.p.},\varphi}}{L_y} = -\frac{\csc(\varphi)}{96\,\pi^2\,a^2}\left\langle \int_{\gamma_{x_{\min}}}^{\gamma_{x_{\max}}} \mathrm{d}x\ \gamma_{\mathrm{m}}^3(x) \right\rangle, \tag{6}$$

where

$$\gamma_{\mathrm{m}}(x) \equiv x\cos(\varphi) + \sin(\varphi)\gamma_{z_{\max}}(x) - \gamma_{x_{\min}}(\varphi), \tag{7}$$

$$\gamma_{x_{\min}}(\varphi) \equiv \min_t\left(\gamma_x(t)\cos(\varphi) + \gamma_z(t)\sin(\varphi)\right). \tag{8}$$

Here, the Casimir energy has been related to simple geometrical properties of the worldlines: $\gamma_{x_{\min}}(\varphi)$ measures the extent of the worldline in the negative x direction of a coordinate system rotated by the angle φ, and $\gamma_{z_{\max}}(x)$ denotes the x-dependent envelope of the worldline in positive z direction, see Fig. 1. Equation (6) is shown as a function of φ in Fig. 2.

For $\varphi = \pi/2$ we rediscover the perpendicular plates result[24,25] as a special case. For $\varphi = 0$, the integral in Eq. (6) can be done analytically resulting in $\langle\lambda^4/4\rangle = \pi^4/30$. Together with the φ-dependent prefactor, Eq. (6) diverges as $\varphi \to 0$ as it should. This is because Eq. (6) corresponds to the energy per unit *edge length*, whereas for $\varphi \to 0$ the Casimir energy becomes proportional to the *area* of the semi-infinite plate.

The result for the parallel limit $\varphi \to 0$ has to arise from the general inclined-plates formula Eq. (6), but involves a subtle limiting process which was performed in Ref. 14. For this 1si case, the total Casimir interaction energy decomposes into,[24,25]

$$E_{\mathrm{c}}^{\mathrm{1si}} = E_{\mathrm{c}}^{\mathrm{1si},\|} + E_{\mathrm{c}}^{\mathrm{1si,edge}}, \tag{9}$$

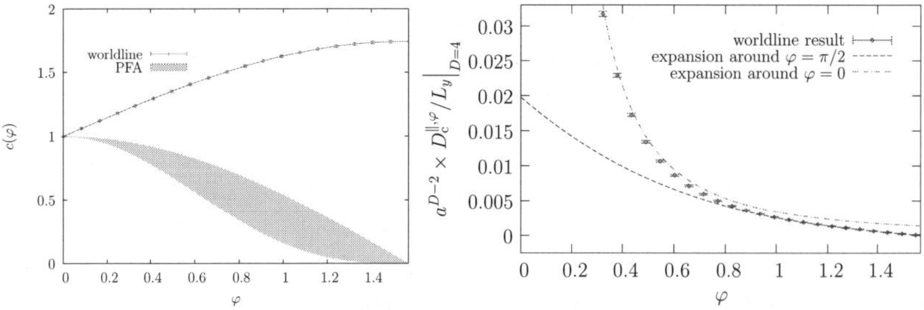

Fig. 2. Left panel: normalized Casimir energy per edge length of the inclined-plates configuration, $c(\varphi) = -2880E_c^{\text{i.p.},\varphi}a^2\sin(\varphi)/\pi^2 L_y$. The blue area corresponds to the PFA prediction: for the upper bound, we have used the infinite plate as a basis, for the lower bound the semi infinite plate, yielding $\cos(\varphi)$ and $\cos^3(\varphi)$ as bounding curves, respectively. Right panel: normalized Casimir torque per edge length (Eq. 10) and its expansion around $\varphi = \pi/2$ (Eq. (11)) and $\varphi = 0$ (Eq. (12)), respectively.

where $E_c^{1\text{si},\|}/A$ is the usual Casimir energy per unit area of two parallel plates, with A now being the area of the semi-infinite plate. The subleading edge energy $E_c^{1\text{si,edge}}$ arises solely due to the presence of the edge and is proportional to the length of the latter. For finite plates, the edge effect contributes to the Casimir force, effectively increasing the plates' area.[14,24]

The Casimir torque $D_c^{\text{i.p.},\varphi}$ can easily be obtained from

$$D_c^{\text{i.p.},\varphi} = \frac{\mathrm{d}E_c^{\text{i.p.},\varphi}}{\mathrm{d}\varphi}. \tag{10}$$

Fitting the numerical data for the torque to an odd polynomial in the vicinity of $\varphi \to \pi/2$, we obtain (see Fig. 2)

$$\frac{D_c^{\text{i.p.},\varphi\to\pi/2}a^2}{L_y} \approx 0.00329\left(\frac{\pi}{2} - \varphi\right) + 0.0038\left(\frac{\pi}{2} - \varphi\right)^3. \tag{11}$$

For the other limit $\varphi \to 0$, the Casimir torque diverges. The leading order,[14]

$$D_c^{\text{i.p.},\varphi\to 0} \cong \frac{L_y\pi^2}{2880a^2\varphi^2} \approx 0.00343\frac{L_y}{a^2\varphi^2}, \tag{12}$$

is an excellent approximation to Eq. (10) for φ not too close to $\pi/2$. The divergent Casimir torque per length can be converted into a finite torque per unit area which leads to the classical result for the torque,

$$D_c^{\|,\varphi\to 0} = \frac{AL_z\pi^2}{960a^4} \approx 0.0103\frac{AL_z}{a^4}. \tag{13}$$

Here, A and L_z denote the semi-infinite plate's area and the extent in z direction, respectively.

A new characteristic contribution emerges from the edge effect Eq. (9) which effectively changes the shape of the upper plate seen by worldlines, as the upper plate appears to be higher near and at the edge itself. This leads to a contribution which works against the standard torque (13). The correction to Eq. (13) emerging from the edge effect then is[14] $\approx -(0.003660 \pm 0.000038) L_y/a^2$.

4.2. *Inclined plates at finite temperature*

Decomposing the Casimir free energy at finite temperature $T = 1/\beta$ into its zero-temperature part $E_c(0)$ and finite-temperature correction $\Delta E_c(T)$,

$$E_c(T) = E_c(0) + \Delta E_c(T), \tag{14}$$

is straightforward in the worldline picture by using the relation (4). The finite-temperature correction is purely driven by the worldlines with nonzero winding number, whereas the complicated geometry-dependent part of the calculation remains the same for zero or finite temperature. The same statement holds for the Casimir force $F_c(T) = F_c(0) + \Delta F_c(T)$.

In the following, we concentrate on the low-temperature limit, $aT \ll 1$. Full expressions for arbitrary temperature can be found in Ref. 14. From dimensional analysis of Eq. (14), we would naively expect the Casimir free energy to be of the form

$$E_c(T) = E_c(0)(1 + c_1 aT + c_2(aT)^2 + c_3(aT)^3 + \dots). \tag{15}$$

No negative exponents should appear in (15) since the thermal part of the free energy has to disappear as $T \to 0$. Generically, the $T = 0$ Casimir energy $E_c(0)$ diverges for surfaces approaching contact $a \to 0$. From Eq. (15), we would naively expect the same for the thermal correction. If, however, sufficiently many of the first c_i's in Eq. (15) vanish, then the thermal part of the Casimir energy will be well behaved in a without a divergence for $a \to 0$.

This is indeed the case for parallel plates ($c_1 = c_2 = 0$, and $E_c(0) \sim 1/a^3$) and for inclined plates ($c_1 = 0$, and $E_c(0) \sim 1/a^2$). As a consequence, an extreme simplification arises: the thermal contribution in the low-temperature limit can be obtained by first taking the formal limit $a = 0$ (only in the thermal contribution, of course).

In the following, we argue that there is no divergence in the thermal contribution in the limit $a \to 0$ for general geometries: Imagine a fancy geometry. The a-divergent part arises from the regions near the points (or lines or surfaces) of contact as $a \to 0$. The surfaces in these regions by construction bend away from each other. The thermal contribution can now be made larger by flattening the surfaces in the contact region. Let us now substitute these regions by broader parallel plates. Then, the local thermal contribution to the Casimir energy of the original configuration will be clearly smaller than the finite thermal contribution of parallel plates. As the latter does not lead to divergences for $a \to 0$, there can also be no divergence for the general curved case arising from the contact regions. Of course, infinite geometries may still experience an infinite thermal force, as it is the case for two infinitely extended parallel plates, but the local thermal contribution to the force density will be finite.

Another distinct feature of low-temperature effects is the spread of the thermal force density over regions of size $\sim 1/T$ even for very small separations a. These effects are, of course, relevant for open geometries such as a sphere and cylinder above a plate.[26] But it can also be demonstrated by calculating the thermal force density $\Delta f_{\rm c}(r,T) = f_{\rm c}(r,T) - f_{\rm c}(r,0)$ for two perpendicular plates at a distance $a = 0$ as a function of the coordinate r on the infinite surface measuring the distance from the edge (i.e., the contact point). The result can easily be obtained fully analytically on the worldline,[26] yielding

$$\Delta f_{\rm c}(r,T) = -\frac{T^4 L_y}{2\pi^2}\left[\frac{\pi^4}{45} - \sum_{n=1}^{\infty}\left\langle \frac{2e^{-(n\lambda_1/2rT)^2}}{n^4} + \frac{\lambda_1^2 e^{-(n\lambda_1/2rT)^2}}{2T^2 n^2 r^2}\right\rangle\right], \quad (16)$$

where λ_1 measures the extent of half a unit worldline, i.e., the distance measured in x direction from the left end to the center of mass. Rescaling the radial coordinate $r \to \lambda_1 r$ per worldline, the following rescaled force density leads to the same force upon integration over r,

$$\Delta \tilde{f}_{\rm c}(r,T) = -\frac{T^4 L_y \sqrt{\pi}}{4\pi^2}\left[\frac{\pi^4}{45} - \sum_{n=1}^{\infty}\left(\frac{2e^{-(n/2rT)^2}}{n^4} + \frac{e^{-(n/2rT)^2}}{2T^2 n^2 r^2}\right)\right], \quad (17)$$

where we have used $\langle\lambda_1\rangle = \sqrt{\pi}/2$. (Equations (16) and (17) possibly differ by a total derivative, but both provide for a reasonable thermal force density.) Upon integration, we obtain the thermal force of the perpendicular plates at $a = 0$, $\Delta F_{\rm c}(T) = -\zeta(3)\, L_y T^3/4\pi$ in agreement with Ref. 13. In the limit $a \to 0$, the configuration has a scale invariance, which is reflected in the fact that Eqs. (16, 17) remain unchanged under $T \to T\alpha$, $r \to r/\alpha$ and $\Delta\tilde{f}_{\rm c} \to \Delta\tilde{f}_{\rm c}/\alpha^4$ for arbitrary α. That means that evaluating (16, 17) for say $T = 1$ is sufficient to infer its form at all other T. Equation (17) is shown for $T = 1$ in Fig. 3. The inflection point of each term in the n sum is at $r_0 = n/T\sqrt{10}$. For $r < 1/T\sqrt{10}$ the force density stays nearly

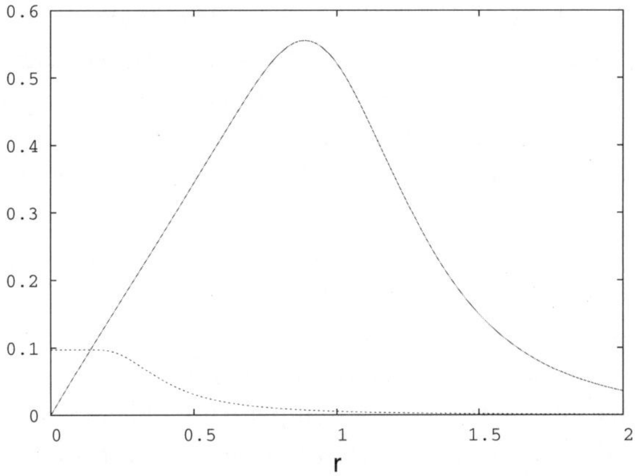

Fig. 3. The negative rescaled thermal force density Eq. (17) for perpendicular plates (dashed blue line) and the negative radial thermal force density for a sphere of radius $R = 1$ above an infinite plate (solid red line) for $T = 1$. The thermal force density in the sphere-plate case has a maximum of $\approx 2\pi \times \pi^2/90$, where the factor 2π arises from the cylindrical measure. Note that a considerable fraction of the force density lies outside the sphere which only extends to $r = 1$. As temperature drops, the maximum moves monotonously to the right.

constant (and is equal to the first term in (16)) and rapidly goes to zero for $r > 1/T\sqrt{10}$. From this, we draw the important conclusion, that the region of constant force density can be made arbitrarily large in r direction by choosing sufficiently low T.

Similar important consequences arise for temperature effects in other geometries. For example, the radial force density of a sphere above a plate exhibits a maximum due to the cylindric measure factor r, see Fig. 3. Although this force density is not scale invariant due to the additional dimensionful scale R (sphere radius), its maximum will nevertheless move away from the sphere as the temperature drops. No local approximate tools such as the PFA will be able to predict the correct thermal force. The fact that the force density is not scale invariant leads to different temperature behaviors for $T < 1/R$ and $T > 1/R$ even in the limit $a \to 0$.[26] For very recent results on thermal corrections in the sphere-plane geometry, see Ref. 27

Let us now compare the low-temperature limit of the Casimir free energy of two inclined plates with that of two parallel plates. For $(aT) \to 0$, the correction $\Delta E_{\rm c}^{\|}$ to the well-known parallel-plates energy reads

$$\frac{\Delta E_{\rm c}^{\|}(aT \to 0)}{A} = -\frac{\zeta(3)T^3}{4\pi} + \frac{\pi^2 aT^4}{90} \approx -0.0957T^3 + 0.110aT^4. \qquad (18)$$

Note that only the T^4 term contributes to the force. The thermal correction to the inclined-plates energy is much more sensitive to temperature,

$$\frac{\Delta E_{\mathrm{c}}^{\mathrm{i.p.},\varphi}(aT \to 0)}{L_y} = -\frac{c_{\varphi,T_0} T^2}{24\sin(\varphi)} + \frac{\zeta(3) a T^3}{4\pi \sin(\varphi)}, \tag{19}$$

where c_{φ,T_0} was calculated numerically in Ref. 14 as a function of φ. Only the second term, which is a purely analytical result, contributes to the force. Equation (19) is the generalization of a result for perpendicular plates, $\varphi = \pi/2$, see Ref. 13.

Equation (19), being an energy per edge length, diverges as $\varphi \to 0$. As in Eq. (9), it has to be replaced by the energy of a semi-infinite plate above a parallel one, $E_{\mathrm{c}}^{\mathrm{1si}}(T) = E_{\mathrm{c}}^{\mathrm{1si,edge}}(T) + E_{\mathrm{c}}^{\mathrm{1si},\|}(T)$. The thermal part of $E_{\mathrm{c}}^{\mathrm{1si},\|}(T)$ is as in Eq. (18), where A is the area of the semi-infinite plate. The leading thermal correction to the edge effect $\Delta E_{\mathrm{c}}^{\mathrm{1si,edge}}(T)$ reads

$$\frac{\Delta E_{\mathrm{c}}^{\mathrm{1si,edge}}(T)}{L_y} = -\frac{c_{\varphi,T_0} T^2}{24} + 0.063 a^{1.74} T^{3.74}. \tag{20}$$

We find that the low-temperature regime of the 1si edge effect is well described by a non-integer power law, $\Delta F_{\mathrm{c}}^{\mathrm{1si,edge}} \sim T^{3.74}$, where the fractional exponent arises from the geometry-temperature interplay in this open geometry. Of course, our numerical analysis cannot guarantee to determine the true asymptotic behavior in the limit $aT \to 0$, but our data in the low-temperature domain $0.01 \lesssim aT \lesssim 0.4$ are well fitted by the non-integer scaling also at higher fit orders.[14] This result is very reminiscent to non-integer exponents known from critical phenomena. In both cases, this result arises from fluctuation contributions on all length scales, clearly revealing the long-range nature of both phenomena. Note that the leading temperature exponent of the 1si geometry is between the parallel-plates exponent 4 and the inclined-plates exponent 3, reflecting the fact that thermal properties of the 1si geometry lie between those of the parallel and inclined plates.

The long-range nature of Casimir phenomena becomes also visible at the thermal correction to the torque. This is immediately transparent from Eq. (19). Whereas the a-independent first term of Eq. (19) does not contribute to the force, both terms in Eq. (19) contribute to the low-temperature limit of the Casimir torque, the thermal contribution being $\mathrm{d}\Delta E_{\mathrm{c}}^{\mathrm{i.p.},\varphi}(T)/\mathrm{d}\varphi$. Concentrating on the limit $aT \to 0$ for small deviations from the perpendicular-plates case, $\varphi = \pi/2 - \delta\varphi$, an expansion to first order in $\delta\varphi$ yields:

$$\Delta D_{\mathrm{c}}^{\mathrm{i.p.},\varphi=\pi/2-\delta\varphi}(aT \to 0)/L_y = (0.0716 - 0.0957 aT)\, T^2 \delta\varphi. \tag{21}$$

In the validity regime of the low-temperature expansion, $aT \ll 1$, the positive first term is always dominant, hence the perpendicular-plates case remains a repulsive fixed point. Most importantly, we would like to stress that the quadratic dependence of the torque on the temperature $\sim T^2$ ($\sim T^{D-2}$ in the general case) for the inclined-plates configuration represents the strongest temperature dependence of all observables discussed here.

5. Conclusions

A nontrivial interplay between finite temperature fluctuations and the geometry of a configuration can give rise to a variety of qualitatively different thermal corrections to Casimir phenomena. This effect becomes most pronounced in geometries where the relevant part of the spectrum is gapless. In these so-called open geometries, any small value of the temperature can excite low-lying thermal modes, giving rise to thermal corrections. By contrast, a gap in the relevant part of the spectrum of closed geometries suppresses thermal excitations at low temperature.

In the present work, we have developed the general picture underlying these geothermal Casimir phenomena. Open geometries, for instance, support a stronger influence of long-range fluctuations on thermal Casimir phenomena. We have illustrated the underlying mechanisms with the aid of the inclined-plates configuration and also presented first results for the experimentally important sphere-plate configuration.

Furthermore, we have presented a general argument that low-temperature corrections to Casimir forces become much more easily accessible by taking the (formal) contact limit $a \to 0$ (only for the thermal contributions), as thermal corrections remain well behaved in this limit. Whereas the existence of this limit is well known for parallel plates, we have argued that the same result holds for general geometries. The existence of this limit is also a reason why thermal corrections, for instance, in the perpendicular-plate case can be determined analytically. We expect that this observation will be useful for many other geometries as well. This should lead to practical simplifications also in other field theory approaches such as functional-integral approaches,[28,29] scattering theory,[30–34] and mode summation.[35]

This particular geothermal interplay which we have observed in the context of the Casimir effect is certainly not restricted to Casimir physics. The crucial ingredients are a gapless fluctuation spectrum (though small gaps may not necessarily exert a strong quantitative influence) in a spatially inhomogeneous background. We expect that similar phenomena can occur for the thermal response of a system with an inhomogeneous condensate and an (almost) gapless fluctuation spectrum.

We conclude with the remark the geothermal interplay is only one out of several highly nontrivial interferences between deviations from the ideal Casimir limit. For instance, the interplay between dielectric material properties and finite temperature[36] is still a subject of intense theoretical investigations and has created a long-standing controversy.[37–41] Also the interplay between dielectric properties and geometry has been shown to lead to significant deviations from ideal curvature effects as well.[42] Whereas we have concentrated on Casimir forces and interaction energies between disconnected surfaces, also Casimir free energies of single bodies can exhibit a sophisticated temperature dependence, as has recently been analyzed for wedges and cylindrical shells.[43] All this exemplifies that a profound understanding of the Casimir effect requires a thorough quantum field theoretic basis.

Acknowledgments

We would like to thank Kim Milton and all organizers of QFEXT09 for creating such a stimulating and productive conference atmosphere. This work was supported by the National Science. We have benefited from activities within the ESF Research Network CASIMIR and acknowledge support from the Landesgraduiertenförderung Baden-Württemberg, the Heidelberg Graduate School of Fundamental Physics (AW), and from the DFG grant Gi328/5-1 and SFB-TR18 (HG).

References

1. H.B.G. Casimir, Kon. Ned. Akad. Wetensch. Proc. **51**, 793 (1948).
2. M. Bordag, U. Mohideen and V. M. Mostepanenko, Phys. Rept. **353**, 1 (2001); R. Onofrio, New J. Phys. **8**, 237 (2006) [arXiv:hep-ph/0612234]; S. Y. Buhmann and D. G. Welsch, Prog. Quant. Electron. **31**, 51 (2007) [arXiv:quant-ph/0608118].
3. K. A. Milton, "The Casimir effect: Physical manifestations of zero-point energy," *River Edge, USA: World Scientific (2001).*
4. S. K. Lamoreaux, Phys. Rev. Lett. **78**, 5 (1997); U. Mohideen and A. Roy, Phys. Rev. Lett. **81**, 4549 (1998); H.B. Chan *et al.*, Science 291, 1941 (2001); R.S. Decca *et al.*, Phys. Rev. D **68**, 116003 (2003); Phys. Rev. Lett. **94**, 240401 (2005).
5. B.V. Derjaguin, I.I. Abrikosova, E.M. Lifshitz, Q.Rev. **10**, 295 (1956); J. Blocki, J. Randrup, W.J. Swiatecki, C.F. Tsang, Ann. Phys. (N.Y.) **105**, 427 (1977).

6. M. Schaden and L. Spruch, Phys. Rev. A **58**, 935 (1998); Phys. Rev. Lett. **84** 459 (2000)
7. H. Gies, K. Langfeld and L. Moyaerts, JHEP **0306**, 018 (2003); arXiv:hep-th/0311168.
8. A. Scardicchio and R. L. Jaffe, Nucl. Phys. B **704**, 552 (2005); Phys. Rev. Lett. **92**, 070402 (2004).
9. H. Gies and K. Klingmuller, Phys. Rev. Lett. **96**, 220401 (2006) [arXiv:quant-ph/0601094].
10. M. Bordag, Phys. Rev. D **73**, 125018 (2006); Phys. Rev. D **75**, 065003 (2007).
11. A. Bulgac, P. Magierski and A. Wirzba, Phys. Rev. D **73**, 025007 (2006) [arXiv:hep-th/0511056]; A. Wirzba, A. Bulgac and P. Magierski, J. Phys. A **39** (2006) 6815 [arXiv:quant-ph/0511057].
12. A. Scardicchio and R. L. Jaffe, Nucl. Phys. B **743** (2006) 249 [arXiv:quant-ph/0507042].
13. H. Gies and K. Klingmuller, J. Phys. A **41**, 164042 (2008).
14. A. Weber and H. Gies, Phys. Rev. D **80**, 065033 (2009) [arXiv:0906.2313 [hep-th]].
15. K. Kirsten and S. A. Fulling, Phys. Rev. D **79**, 065019 (2009) [arXiv:0901.1902 [hep-th]].
16. R. Balian and B. Duplantier, Annals Phys. **112**, 165 (1978).
17. I. Klich, J. Feinberg, A. Mann and M. Revzen, Phys. Rev. D **62**, 045017 (2000) [arXiv:hep-th/0001019].
18. V. V. Nesterenko, G. Lambiase and G. Scarpetta, Phys. Rev. D **64**, 025013 (2001) [arXiv:hep-th/0006121].
19. M. Bordag, V. V. Nesterenko and I. G. Pirozhenko, Phys. Rev. D **65**, 045011 (2002) [arXiv:hep-th/0107024].
20. Z. Bern and D.A. Kosower, Nucl. Phys. **B362**, 389 (1991); **B379**, 451 (1992); M.J. Strassler, Nucl. Phys. **B385**, 145 (1992); M. G. Schmidt and C. Schubert, Phys. Lett. B **318**, 438 (1993) [arXiv:hep-th/9309055]; H. Gies and K. Langfeld, Nucl. Phys. B **613**, 353 (2001); for a review, see C. Schubert, Phys. Rept. **355**, 73 (2001).
21. H. Gies and K. Klingmuller, Phys. Rev. D **74**, 045002 (2006) [arXiv:quant-ph/0605141].
22. H. Gies, J. Sanchez-Guillen and R. A. Vazquez, JHEP **0508**, 067 (2005) [arXiv:hep-th/0505275].
23. N. Graham, A. Shpunt, T. Emig, S. J. Rahi, R. L. Jaffe and M. Kardar, arXiv:0910.4649 [quant-ph].
24. H. Gies and K. Klingmuller, Phys. Rev. Lett. **97**, 220405 (2006) [arXiv:quant-ph/0606235].
25. K. Klingmuller, PhD Dissertation, Heidelberg U., URN: urn:nbn:de:bsz:16-opus-78464, URL: http://www.ub.uni-heidelberg.de/archiv/7846 ˜(2007).
26. H. Gies and A. Weber, in preparation.
27. A. Canaguier-Durand, P.A. Maia Neto, A. Lambrecht, S. Reynaud, arXiv:0911.0913 [quant-ph] (2009).
28. M. Bordag, D. Robaschik and E. Wieczorek, Annals Phys. **165**, 192 (1985).
29. T. Emig, A. Hanke and M. Kardar, Phys. Rev. Lett. **87** (2001) 260402.

30. T. Emig, R. L. Jaffe, M. Kardar and A. Scardicchio, Phys. Rev. Lett. **96** (2006) 080403.
31. O. Kenneth and I. Klich, Phys. Rev. Lett. **97**, 160401 (2006); arXiv:0707.4017.
32. T. Emig, N. Graham, R. L. Jaffe and M. Kardar, Phys. Rev. Lett. **99**, 170403 (2007) [arXiv:0707.1862]; arXiv:0710.3084.
33. R. B. Rodrigues, P. A. Maia Neto, A. Lambrecht and S. Reynaud, Phys. Rev. Lett. **96**, 100402 (2006) [arXiv:quant-ph/0603120]; Phys. Rev. A **75**, 062108 (2007).
34. K. A. Milton and J. Wagner, Phys. Rev. D **77**, 045005 (2008) [arXiv:0711.0774 [hep-th]]; J. Phys. A **41**, 155402 (2008) [arXiv:0712.3811 [hep-th]]; K. A. Milton, P. Parashar and J. Wagner, arXiv:0806.2880 [hep-th].
35. F. D. Mazzitelli, D. A. R. Dalvit and F. C. Lombardo, New J. Phys. **8**, 240 (2006); D. A. R. Dalvit, F. C. Lombardo, F. D. Mazzitelli and R. Onofrio, Phys. Rev. A **74**, 020101 (2006).
36. M. Boström and Bo E. Sernelius, Phys. Rev. Lett. **84**, 4757 (2000).
37. V. M. Mostepanenko *et al.*, J. Phys. A **39**, 6589 (2006) [arXiv:quant-ph/0512134].
38. I. Brevik, S. A. Ellingsen and K. A. Milton, New J. Phys. **8**, 236 (2006) [arXiv:quant-ph/0605005].
39. G. Bimonte, Phys. Rev. A **79**, 042107 (2009) [arXiv:0903.0951 [quant-ph]].
40. G.-L. Ingold, A. Lambrecht, S. Reynaud, Phys. Rev. E **80**, 041113 (2009) [arXiv:0905.3608 [quant-ph]].
41. F. Intravaia and C. Henkel, Phys. Rev. Lett. **103**, 130405 (2009) [arXiv:0903.4771 [quant-ph]].
42. A. Canaguier-Durand, P. A. Maia Neto, I. Cavero-Pelaez, A. Lambrecht and S. Reynaud, Phys. Rev. Lett. **102**, 230404 (2009) [arXiv:0901.2647 [quant-ph]].
43. I. Brevik, S. A. Ellingsen and K. A. Milton, arXiv:0911.2688 [hep-th].

THERMAL EFFECTS IN THE MAGNETIC CASIMIR-POLDER INTERACTION

H. HAAKH*, F. INTRAVAIA†, C. HENKEL

Institut für Physik und Astronomie, Universität Potsdam,
Karl-Liebknecht-Straße 24/25, 14476 Potsdam, Germany
**E-mail: haakh@uni-potsdam.de*
www.quantum.physik.uni-potsdam.de

We investigate the magnetic dipole coupling between a metallic surface and an atom in a thermal state, ground state and excited hyperfine state. This interaction results in a repulsive correction and – unlike the electrical dipole contribution – depends sensitively on the Ohmic dissipation in the material.

Keywords: Casimir-Polder interaction, temperature, Ohmic dissipation.

1. Introduction

Exact knowledge of the Casimir-Polder interaction between an atom and a conducting surface is rapidly becoming important in modern microtrap experiments (atom chips), whose stability is limited by an attractive potential well at sub-micron distances. We investigate here the magnetic dipole contribution to the atom-surface interaction and find it to be repulsive for atoms prepared in certain states. The magnetic Casimir-Polder potential differs greatly from its electrical counterpart, on which previous research has concentrated. Above all, the much smaller transition frequencies lead to a stronger dependence on temperature and the Ohmic dissipation in the material. This results in a strong suppression at distances above the thermal wavelength which is absent in the case of the plasma and, more generally, in superconductors.

The fundamental quantity calculated in this work is the free energy of the magnetic dipole interaction between an atom and a planar surface. For

†Present address: Theoretical Division, MSB213, Los Alamos National Laboratory, Los Alamos, NM87545, USA.

an atom prepared in a state $|a\rangle$ the free energy is given by the expression[1,2]

$$\mathcal{F}(L,T) = -k_B T \sum_{n=0}^{\infty}{}' \beta_{ij}^a(\mathrm{i}\xi_n)\mathcal{H}_{ji}(L,\mathrm{i}\xi_n) + \sum_b n(\omega_{ba})\mu_i^{ab}\mu_j^{ba}\mathrm{Re}[\mathcal{H}_{ji}(L,\omega_{ba})]\,, \tag{1}$$

where $\xi_n = 2\pi n k_B T/\hbar$ are the Matsubara frequencies and the term $n = 0$ comes with a weight $\frac{1}{2}$ in the primed sum. The second (resonant) term involves the mean thermal photon number $n(\omega)$ and the magnetic transition dipole matrix $\boldsymbol{\mu}$. The state-specific polarizability is[1,3]

$$\beta_{ij}^a(\omega) = \sum_b \frac{\mu_i^{ab}\mu_j^{ba}}{\hbar}\frac{2\omega_{ba}}{\omega_{ba}^2 - (\omega + \mathrm{i}0^+)^2}\,, \tag{2}$$

while $\boldsymbol{\mathcal{H}}$ is the magnetic Green tensor

$$\boldsymbol{\mathcal{H}}(L,\omega) = \frac{\mu_0}{8\pi}\int_0^\infty k dk\, \kappa \left[\left(r_\mathrm{s} + \frac{\omega^2}{c^2\kappa^2}r_\mathrm{p}\right)[\hat{\boldsymbol{x}}\hat{\boldsymbol{x}} + \hat{\boldsymbol{y}}\hat{\boldsymbol{y}}] + 2\frac{k^2}{\kappa^2}r_\mathrm{s}\hat{\boldsymbol{z}}\hat{\boldsymbol{z}}\right] e^{-2\kappa L}\,. \tag{3}$$

Here, μ_0 is the vacuum permeability, k is the in plane wave vector, $\hat{\boldsymbol{x}}\hat{\boldsymbol{x}}$, $\hat{\boldsymbol{y}}\hat{\boldsymbol{y}}$, $\hat{\boldsymbol{z}}\hat{\boldsymbol{z}}$ are the cartesian dyadic products, and $\kappa = [k^2 - (\omega + \mathrm{i}0^+)^2/c^2]^{1/2}$. The Fresnel reflection amplitudes[4] $r^{\mathrm{s,p}}(k,\omega)$ are taken here for a local, isotropic and nonmagnetic ($\mu(\omega) = 1$) medium. All information about the optical properties and the Ohmic dissipation in the surface is then encoded in the dielectric function $\varepsilon(\omega)$. We will consider a *Drude metal*[4] $\varepsilon_{\mathrm{Dr}}(\omega) = 1 - \omega_p^2/[\omega(\omega + \mathrm{i}\gamma)]$ with plasma frequency ω_p and dissipation rate $\gamma > 0$, independent of temperature. In the dissipationless *plasma model*, we get $\varepsilon_{\mathrm{pl}}(\omega)$ by setting $\gamma = 0$. It can be read as the limiting case of a *superconductor*[5] well below the transition temperature. For a more thorough discussion of Casimir(-Polder) effects in superconductors, see Refs.6,7.

2. Casimir-Polder potential in global equilibrium

For an atom in a thermal state, the polarizability Eq.(2) must be averaged over thermal state occupation numbers. For a two-level system with transition frequency Ω_m, this yields

$$\boldsymbol{\beta}^T(\omega) = \tanh\left(\frac{\hbar\Omega_m}{2k_BT}\right)\boldsymbol{\beta}^g(\omega)\,, \tag{4}$$

in terms of the ground state polarizability [Eq. (2) with $a = g$]. In thermal equilibrium, Eq.(1) reduces to

$$\mathcal{F}(L,T) = -k_B T \sum_{n=0}^{\infty}{}' \beta_{ij}^T(\mathrm{i}\xi_n)\mathcal{H}_{ji}(L,\mathrm{i}\xi_n). \tag{5}$$

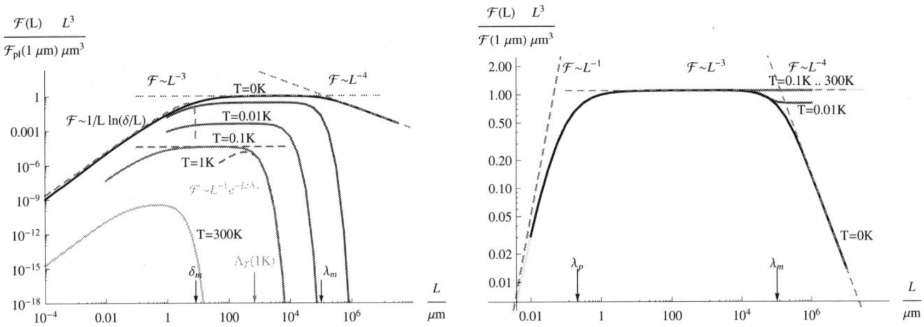

Fig. 1. Magnetic Casimir-Polder free energy for a thermalized two-level atom near a Drude metal (left) and a plasma (right). Parameters are $\omega_p = 8.9 \times 10^{15}\text{s}^{-1}, \gamma = 0.01\omega_p, \Omega_m = 3 \times 10^9\text{s}^{-1} \approx 23\,\text{mK}$. The energy is scaled to $\mathcal{F}_{\text{pl}}(1\,\mu\text{m}, 0\,\text{K}) = 9.79 \times 10^{-37}$ J. Dashed asymptotes: see Ref. 7.

Both $\boldsymbol{\beta}^T(\text{i}\xi)$ and $\boldsymbol{\mathcal{H}}(L, \text{i}\xi)$ are real expressions for $\xi > 0$. We assume a static magnetic dipole aligned perpendicular to the surface, not unrealistic in magnetic traps, and have $\beta^T_{xx} = \beta^T_{yy}$ and $\beta^T_{zz} = 0$.

The free energies for a Drude metal and a plasma are shown in Fig. 1. A striking difference occurs at large distances: the Drude metal is transparent to static magnetic fields $[\boldsymbol{\mathcal{H}}(L, 0) = 0]$, and the zeroth term of the Matsubara sum vanishes. The free energy is then dominated by the first term, which decays exponentially for $L > \Lambda_T = \hbar c/4\pi k_B T$. We call this the *thermal decoupling* of the magnetic dipole.

A plasma (superconductor) shields static magnetic fields (Meißner-Ochsenfeld effect, $\boldsymbol{\mathcal{H}}(L, 0) \neq 0$), leading to an enhanced interaction energy at nonzero temperature. The linear dependence on T of the zeroth Matsubara term in Eq. (5) cancels with the thermal polarizability (4) for $\hbar\omega_{ab} \ll k_B T$. Only around this (quite low) temperature, there are any thermal effects.[7] The close coincidence between the thermal and the $T = 0$ potentials at small distances can be understood from the low-frequency behaviour of the Green tensor, see Ref. 7 where also the dashed asymptotes of Fig. 1 are discussed.

3. Casimir-Polder potential for nonthermal states

Many experimental settings involve atoms prepared in a special state rather than in thermal equilibrium with the environment. We consider a two-level atom prepared in the ground state as an example for an ultracold gas. The assumption $k_B T \gg \hbar|\omega_{ba}|$ is quite realistic for most temperatures, and now

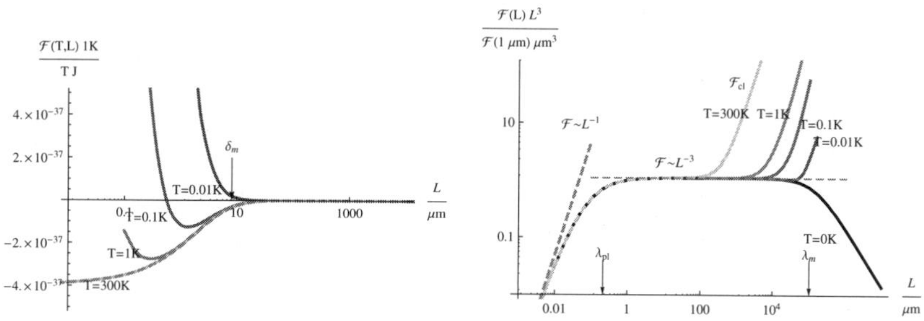

Fig. 2. Magnetic Casimir-Polder free energy in the atomic ground state (left: Drude metal, right: plasma). Note the different scaling in the left plot. Parameters as before.

the resonant part of Eq. (1) contributes, too, and the free energy becomes

$$\begin{aligned}\mathcal{F}(L,g,T) &\approx -2k_BT\sum_{n\geq 1}\beta^g(\mathrm{i}\xi_n)\mathcal{H}_{xx}(L,\mathrm{i}\xi_n) \\ &+ k_BT\beta^g(0)\left\{\mathrm{Re}[\mathcal{H}_{xx}(L,\Omega_m)] - \mathcal{H}_{xx}(L,0)\right\} \ .\end{aligned} \tag{6}$$

At the temperatures considered, the first line (without the $n = 0$ term) is nearly identical for the plasma and Drude model. Anyway, the second line is not. In the Drude model $\mathcal{H}(L,0) = 0$, but the remaining resonant part is significant in the non-retarded regime. It actually changes the sign of the Casimir-Polder potential already at short distances, as soon as $k_BT \gtrsim \hbar\Omega_m$, see Fig. 2 (left). This leads to an *attractive* potential well of approximately 0.02 pK [environmental temperature $T = 1\,\mathrm{K}$] at distances below 1 μm, possibly accessible to quantum reflection experiments.[8] If we consider an atom prepared in an excited state (e.g. another hyperfine state), the interaction changes sign globally, because of the transition energies $\hbar\omega_{ab} < 0$ in the polarizability (2).

In contrast, the magnetic coupling to a plasma is entirely repulsive at small distances (Fig. 2, right). Here, the second line of Eq.(6) nearly vanishes because $\mathcal{H}_{xx}(L,\omega)$ is approximately independent of frequency, at least in the non-retarded regime. Hence, the zeroth Matsubara term is removed from the Casimir-Polder potential and the next order in the expansion of the occupation number $n(\Omega_m) \approx k_BT/\hbar\Omega_m - \frac{1}{2}$ gives the leading contribution to the resonant term. In the non-retarded regime this restores the $T = 0$ behaviour (cf. right panels of Figs. 1 and 2). At larger distances (retarded regime) the resonant term becomes dominant and oscillates with distance. A scheme to enhance these oscillations for the electric Casimir-Polder interaction has been recently proposed, using rovibrational states of polar molecules.[9]

4. Discussion

At experimentally relevant temperatures, the magnetic Casimir-Polder interaction shows much richer effects than its electric counterpart. This is because the electric dipole coupling is dominated by electric fields which in all conductors are screened by surface charges at large scales. This masks any difference between a normal metal and a superconductor (or the plasma model). Differently, the boundary conditions for magnetic fields, relevant for the magnetic dipole, depend on surface currents and hence on Ohmic dissipation. We have seen that including dissipation in the surface response leads to the thermal decoupling of the atom at distances beyond the thermal wavelength. For atoms prepared in a nonthermal state, the balance between repulsive and attractive contributions can produce local extrema, whose sign is controlled by the atomic (hyperfine) state. Experimental tests of the Casimir-Polder interaction may also answer remaining open questions on the temperature dependence of the Casimir interaction, offering the advantage of a well-defined system that can be handled with high precision.

Acknowledgement

We acknowledge financial support by the European Science Foundation (ESF) within the activity 'New Trends and Applications of the Casimir Effect', by the German-Israeli Foundation for Scientific Research and Development (GIF) and by the Alexander von Humboldt foundation.

References

1. J. M. Wylie and J. E. Sipe, *Phys. Rev. A* **32**, 2030 (1985).
2. M. Gorza and M. Ducloy, *Eur. Phys. J. D* **40**, 343 (2006).
3. A. McLachlan, *Proc. Roy. Soc. (London) A.* **271**, 387 (1963).
4. J. D. Jackson, *Classical electrodynamics* (Wiley, New York, 1975).
5. J. Schrieffer, *Theory of Superconductivity* (Perseus Books, 1999).
6. G. Bimonte, *Phys. Rev. A* **78**, 062101 (2008).
7. H. Haakh, F. Intravaia, C. Henkel, S. Spagnolo, R. Passante, B. Power and F. Sols, submitted to *Phys. Rev. A*, preprint arXiv:0910.3133 (2009).
8. V. Druzhinina and M. DeKieviet, *Phys. Rev. Lett.* **91**, 193202 (2003).
9. S. Ellingsen, S. Buhmann and S. Scheel, *Phys. Rev. A* **80**, 22901(2009).

MODE CONTRIBUTIONS TO THE CASIMIR EFFECT

F. INTRAVAIA*§ and C. HENKEL

Institut für Physik und Astronomie, Universität Potsdam,
Karl-Liebknecht-Str. 24/25, 14476 Potsdam, Germany
**E-mail: francesco.intravaia@qipc.org*
www.quantum.physik.uni-potsdam.de

Applying a sum-over-modes approach to the Casimir interaction between two plates with finite conductivity, we isolate and study the contributions of surface plasmons and Foucault (eddy current) modes. We show in particular that for the TE-polarization eddy currents provide a repulsive force that cancels, at high temperatures, the Casimir free energy calculated with the plasma model

Keywords: Mode contributions, surface plasmons, eddy currents.

1. Introduction

Intense theoretical effort is currently devoted to the understanding of the Casimir effect for real experimental setups. This involves the impact of temperature, finite conductivity, engineered materials, and may identify routes to *design* the final Casimir pressure. Almost all analyses rely on the Lifshitz formula[1,2] where the physical properties of the material are encoded in the scattering amplitudes (i.e., reflection coefficients in planar geometries). Their evaluation at imaginary frequencies obscures, however, how the material objects modify the modes of the electromagnetic field. A 'sum over modes' approach is nevertheless possible, even if the eigenfrequencies ω_m are complex (due to material absorption, for example). For two objects at distance L the Casimir energy at zero temperature can be written as[3]

$$E = \frac{\hbar}{2} \sum_{p,\mathbf{k}}{}' \operatorname{Re}\Big[\sum_m \big(\omega_m - \frac{2\mathrm{i}\omega_m}{\pi}\ln\frac{\omega_m}{\Lambda}\big)\Big]_\infty^L, \qquad \operatorname{Im}\Big[\sum_{p,\mathbf{k},m}{}' \omega_m\Big]_\infty^L = 0 \quad (1)$$

where the prime indicates that purely imaginary eigenfrequencies are weighted with $1/2$. Eq.(1) generalizes Casimir's formula for the vacuum en-

§Present address: Theoretical Division, MS B213 Los Alamos National Laboratory, Los Alamos NM 87545, U. S. A.

ergy between two perfect reflectors[4] and is valid for generic (causal) mirrors with arbitrary thickness. Note that one does not simply take real parts of the complex eigenfrequencies, as suggested some time ago[5] (see also Ref.6). The logarithmic correction in Eq.(1) is consistent with the 'system+bath' paradigm that describes the thermodynamics of quantum dissipative systems.[7] In this context, the frequency scale Λ is interpreted as the cutoff frequency of the bath spectral density. The Casimir energy does not depend on this constant because of the sum rule in (1).

The sum-over-modes approach provides an 'anatomic view' of the Casimir effect where contributions from different modes are clearly identified. This is useful to understand unusual behaviours and may suggest new ways to taylor the Casimir force.[8–10] In the following, we illustrate Eq.(1) with the help of a few examples.

2. Dissipative Plasmons at short distance

One of the most interesting contributions to the Casimir force originates from surface modes bound to the vacuum/medium interface.[11] These modes have a dispersion relation that splits in two branches, $\omega = \Omega_\pm(k)$, as two surfaces are approached. Substituting these frequencies in Eq.(1), we get a plasmonic contribution to the Casimir energy (A: surface area)

$$E_{\text{pl}} = \frac{\hbar A}{2} \int \frac{k\text{d}k}{2\pi} \text{Re} \Big[\sum_{i=\pm} \big(\Omega_i(k) - \frac{2\text{i}\Omega_i(k)}{\pi} \ln \frac{\Omega_i(k)}{\Lambda} \big) \Big]_\infty^L \tag{2}$$

Consider the case of two metals at a distance smaller than the plasma wavelength $\lambda_{\text{pl}} = 2\pi c/\omega_{\text{pl}}$. We are then in the quasi-electrostatic regime, and the surface plasmon modes are given by[12] (red and blue points in Fig.1)

$$\Omega_\pm = \sqrt{\omega_\pm^2 - \frac{\gamma^2}{4}} - \text{i}\frac{\gamma}{2}, \qquad \omega_\pm^2 = \frac{\omega_{\text{pl}}^2}{2} \left(1 \pm e^{-kL}\right) \tag{3}$$

where γ is the damping rate in a Drude description of the metal. One can easily check that the sum rule in Eq.(1) is automatically satisfied. To leading order in $\gamma \ll \omega_{\text{pl}}$ (good conductors) Eq.(2) yields

$$E_{\text{pl}} \approx -\frac{\pi^2 \hbar c A}{720 L^3} \frac{3}{2} \left(\alpha \frac{L}{\lambda_{\text{pl}}} - \frac{15\zeta(3)}{\pi^4} \frac{\gamma L}{c} \right), \qquad \alpha = 1.193\ldots \tag{4}$$

where $\zeta(3) \approx 1.202$ is a Zeta function. This corresponds exactly to the total Casimir force calculated in Ref.13, including the dissipative correction. In fact, in this short distance limit, the Casimir energy is completely dominated by the plasmonic contribution.[13–15] Eq.(2) is valid also beyond the

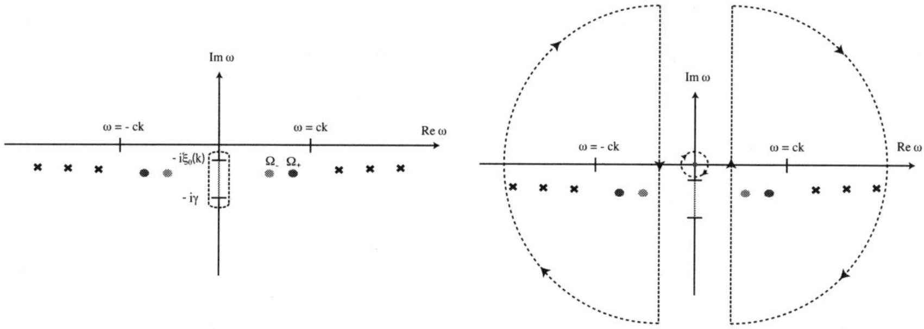

Fig. 1. (Left) Complex eigenfrequencies in the parallel plate geometry, for a fixed wavevector k (not to scale). Red and blue points: dissipative surface plasmons. Red line: bulk continuum of eddy currents. Black crosses: propagating modes in the cavity between the plates. (Right) A counter-clockwise path around the eddy current continuum is equivalent to a clockwise path around the whole complex plane, encircling all other modes.

good conductor limit, however, and could be used, e.g., to analyze semiconductors where surface plasmons appear in a different frequency range and can have much stronger damping.

3. Eddy currents

As a second example, consider the contribution from eddy current modes. They are connected with low-frequency currents that satisfy a diffusion equation in the conducting metal[16] and are completely absent within the lossless description of the so-called plasma model.[2] We have analyzed these modes recently[10] and constructed from the 'system+bath' paradigm their quantum thermodynamics. They behave like free Brownian particles, since the eigenfrequencies of bulk eddy currents are purely imaginary $\omega_m = -\mathrm{i}\xi_m$ ($\xi_m > 0$). From Eq.(1), we get the Casimir energy

$$E_{\text{eddy}} = -\sum_{p,\mathbf{k}} \Big[\sum_m \frac{\hbar \xi_m}{2\pi} \ln \frac{\xi_m}{\Lambda} \Big]_\infty^L \tag{5}$$

For these modes alone, the sum rule [Eq.(1)] is not satisfied, and the eddy current contribution to the Casimir energy depends on the cutoff Λ. This is also well-known from quantum Brownian motion where bath modes up to Λ are entangled to the particle.

Mathematically, eddy currents form a mode continuum that can be identified in the complex frequency plane from the branch cut of the root $k_m = \sqrt{\epsilon(\omega)\omega^2/c^2 - k^2}$ which describes the propagation of the electromagnetic field inside the medium. For a Drude metal, the cut is located between

$\omega_m = -\mathrm{i}\xi_0(\mathbf{k}) \approx -\mathrm{i}Dk^2$ (for $k \ll \omega_{\mathrm{pl}}/c$) and $\omega_m = -\mathrm{i}\gamma$ (see Fig. 1), where $D = \gamma(\lambda_{\mathrm{pl}}/2\pi)^2$ is the electromagnetic diffusion constant. We get the L-dependent change in the mode density along the branch cut by applying the logarithmic argument theorem to the Green function of the electromagnetic field. Using the contour sketched in Fig.1(left), it is possible to show that Eq.(5) can be written as

$$E_{\mathrm{eddy}} = \int_0^\infty \frac{\mathrm{d}\xi}{\pi} \sum_{p,\mathbf{k}} \partial_\xi \Big(\frac{\hbar\xi}{2\pi} \ln \frac{\xi}{\Lambda} \Big) \mathrm{Im} \ln \left[1 - r_p^2(-\mathrm{i}\xi - 0^+) e^{-2\kappa L} \right], \qquad (6)$$

with $\kappa = \sqrt{\xi^2 + k^2}$ and r_p the reflection coefficient of the mirrors in polarization $p = \mathrm{TE}, \mathrm{TM}$. This gives rise to a repulsive Casimir force (Fig.1 of Ref. 10), provided Λ is sufficiently large, e.g., $\Lambda \geq \gamma$.

The structure of Eq.(6) allows for an immediate translation to the high-temperature (classical) limit. Replace the zero-point energy with the classical free energy per mode, $k_B T \ln(\hbar\xi/k_B T)$, and get

$$\mathcal{F}_{\mathrm{eddy}} \approx - \int_0^\infty \frac{d\xi}{\pi} \sum_{p,\mathbf{k}} \frac{k_B T}{\xi} \mathrm{Im} \ln \left[1 - r_p^2(-\mathrm{i}\xi - 0^+) e^{-2\kappa L} \right], \qquad (7)$$

(A more rigorous proof follows from the representation for the free energy given in Ref. 10.) Eq. (7) is thus the result of the logarithmic argument theorem applied to the high-temperature limit of the free energy. Now the contour around the eddy current continuum can also be interpreted as a contour encircling the whole complex plane, i.e., the surface plasmon and propagating modes [Fig. 1(right)]. This is particularly interesting in the TE-polarization because there are no surface plasmons, and the residue at $\omega = 0$ vanishes $[r_{\mathrm{TE}}^2(\omega \to 0) = 0]$. This means that eddy currents and propagating modes give, up to a sign, the same Casimir energy at high temperature (or large distance). Since propagating modes are only slightly affected by conduction on the metal (i.e., they behave similarly in the Drude and plasma models), we find the simple relation

$$\mathcal{F}_{\mathrm{eddy}}^{\mathrm{TE}} \approx -\mathcal{F}_{\mathrm{C}}^{\mathrm{TE}}(\mathrm{pl.m.}), \qquad \gamma/\omega_p \ll 1 \qquad (8)$$

where $\mathcal{F}_{\mathrm{C}}^{\mathrm{TE}}(\mathrm{pl.m.})$ is the Casimir free energy at high temperature calculated within the plasma model.[2] In the Drude model, the two contributions are present and cancel each other when they are both in the high-temperature regime (which happens at different distances, see Fig.4 of Ref.10).

A different scenario occurs in the TM-polarization. The residue at $\omega = 0$ does not vanish and corresponds exactly to the high-temperature limit of the plasma model.[2] Indeed, we have checked that eddy currents give only a very small contribution.

4. Conclusions

Using a mode-summation approach, we have isolated and analyzed the contribution of two classes of modes to the Casimir effect, allowing for complex eigenfrequencies of the electromagnetic field. A previous result for the short-distance limit between good conductors[13] has been generalized to any conductivity and distance by considering coupled surface plasmonic modes (for the lossless case, see Refs.8,9). We also considered eddy currents which are overdamped or diffusive modes in the bulk of a Drude metal, and showed that they contribute a repulsive Casimir interaction, in agreement with Ref.10. At high temperature and for a good conductor, we found in a simple way that their free energy in the TE-polarization differs only slightly from the Casimir free energy within a dissipationless description (the plasma model), but is of the opposite sign. In this way, eddy currents nearly cancel out the attractive Casimir interaction from propagating modes. This explains the strong difference between the Drude and plasma models for the temperature correction of the electromagnetic Casimir effect.[2]

Acknowledgments

We acknowledge financial support by the European Science Foundation within the activity 'New Trends and Applications of the Casimir Effect' (www.casimir-network.com). F.I. acknowledges financial support by the Alexander von Humboldt Foundation.

References

1. E. Lifshitz, *Sov. Phys. JETP (USA)* **2**, p. 73 (1956).
2. G. L. Klimchitskaya, U. Mohideen and V. M. Mostepanenko, *Rev. Mod. Phys.* (2010 in press), arXiv:0902.4022v1.
3. F. Intravaia and C. Henkel, *J. Phys. A* **41**, 164018 (2008).
4. H. Casimir, *Proc. kon. Ned. Ak. Wet.* **51**, p. 793 (1948).
5. D. Langbein, *Phys. Rev. B* **2**, p. 3371 (1970).
6. B. E. Sernelius, *Phys. Rev. B* **74**, 233103 (2006).
7. U. Weiss, *Quantum Dissipative Systems* (World Scientific, Singapore 2008).
8. F. Intravaia and A. Lambrecht, *Phys. Rev. Lett.* **94**, 110404 (2005).
9. F. Intravaia, C. Henkel and A. Lambrecht, *Phys. Rev. A* **76**, 033820 (2007).
10. F. Intravaia and C. Henkel, *Phys. Rev. Lett.* **103**, 130405 (2009).
11. G. Barton, *Rep. Prog. Phys.* **42**, p. 963 (1979).
12. E. N. Economou, *Phys. Rev.* **182**, p. 539 (1969).
13. C. Henkel, K. Joulain, J.-P. Mulet and J.-J. Greffet, *Phys. Rev. A* **69**, 023808 (2004).
14. N. V. Kampen, B. Nijboer and K. Schram, *Phys. Lett. A* **26**, p. 307 (1968).
15. E. Gerlach, *Phys. Rev. B* **4**, p. 393 (1971).
16. J. Jackson, *Classical Electrodynamics* (Wiley & Sons, New York, 1975).

THERMAL CASIMIR FORCE BETWEEN MAGNETIC MATERIALS

G. L. KLIMCHITSKAYA,[1,2] B. GEYER[2] and V. M. MOSTEPANENKO[3,2]

[1] *North-West Technical University, Millionnaya St. 5, St.Petersburg, 191065, Russia*
[2] *Institute for Theoretical Physics, Leipzig University, Postfach 100920, D-04009, Leipzig, Germany*
[3] *Noncommercial Partnership "Scientific Instruments", Tverskaya St. 11, Moscow, 103905, Russia*

We investigate the Casimir pressure between two parallel plates made of magnetic materials at nonzero temperature. It is shown that for real magnetodielectric materials only the magnetic properties of ferromagnets can influence the Casimir pressure. This influence is accomplished through the contribution of the zero-frequency term of the Lifshitz formula. The possibility of the Casimir repulsion through the vacuum gap is analyzed depending on the model used for the description of the dielectric properties of the metal plates.

Keywords: Magnetic materials; Lifshitz formual; plasma model; Drude model.

1. Introduction

The Casimir effect which results in a force acting between two parallel electrically neutral material plates separated with a gap of width a finds many prospective applications ranging from fundamental physics to nanotechnology.[1] For real dissimilar material plates described by the dielectric permittivities $\varepsilon^{(1,2)}(\omega)$ and magnetic permeabilities $\mu^{(1,2)}(\omega)$ at temperature T in thermal equilibrium the generalized Lifshitz formula for the Casimir (van der Waals) pressure takes the form[2–5]

$$P(a,T) = -\frac{k_BT}{\pi}\sum_{l=0}^{\infty}{}' \int_0^{\infty} q_l k_\perp dk_\perp \left\{\left[\frac{e^{2aq_l}}{r_{\rm TM}^{(1)}(i\xi_l,k_\perp)r_{\rm TM}^{(2)}(i\xi_l,k_\perp)}-1\right]^{-1} + \left[\frac{e^{2aq_l}}{r_{\rm TE}^{(1)}(i\xi_l,k_\perp)r_{\rm TE}^{(2)}(i\xi_l,k_\perp)}-1\right]^{-1}\right\}. \quad (1)$$

Here, k_B is the Boltzmann constant, prime adds a multiple one half to the term with $l = 0$, $k_\perp$ is the projection of the wave vector on the plane of the plates, and $\xi_l = 2\pi k_B T l/\hbar$ with $l = 0, 1, 2, \ldots$ are the Matsubara frequencies. The reflection coefficients for the transverse magnetic and transverse electric polarizatons of the electromagnetic field are given by

$$r_{\rm TM}^{(n)}(i\xi_l, k_\perp) = \frac{\varepsilon_l^{(n)} q_l - k_l^{(n)}}{\varepsilon_l^{(n)} q_l + k_l^{(n)}}, \qquad r_{\rm TM}^{(n)}(i\xi_l, k_\perp) = \frac{\mu_l^{(n)} q_l - k_l^{(n)}}{\mu_l^{(n)} q_l + k_l^{(n)}}, \tag{2}$$

where $\varepsilon_l^{(n)} \equiv \varepsilon^{(n)}(i\xi_l)$, $\mu_l^{(n)} \equiv \mu^{(n)}(i\xi_l)$, the index $n = 1, 2$ numerates the plates and

$$q_l^2 = k_\perp^2 + \frac{\xi_l^2}{c^2}, \qquad {k_l^{(n)}}^2 = k_\perp^2 + \varepsilon_l^{(n)} \mu_l^{(n)} \frac{\xi_l^2}{c^2}. \tag{3}$$

As was noticed long ago,[6] “In the majority of cases, the contribution to the van der Waals interaction due to the magnetic properties of real materials is extremely small.” A large contribution (including the Casimir repulsion through a vacuum gap for some range of parameters) was found[7] using the approximation of frequency-independent ε and μ. Later, however, it was shown[8] that for real materials μ is nearly equal to unity in the range of frequencies which gives a major contribution to the Casimir pressure. This problem was reconsidered[4] at both zero and nonzero temperature for one metallic and one magnetodielectric plate using the description of a metal by means of the Drude model and of magnetodielectric by a simplified model of the Drude-Lorentz type. At $T \neq 0$ the Casimir force was found to be always attractive.

In this paper we investigate the thermal Casimir pressure between plates made of ferromagnetic metal, ferromagnetic dielectric and nonmagnetic metal taking into account realistic dependences of ε and μ on the frequency and using different approaches to the theory of the thermal Casimir force suggested in the literature. We demonstrate how the use of different approaches influences the Casimir pressure and find when the Casimir repulsion through a vacuum gap is feasible. In Sec. 2 we provide a brief review of magnetic properties. Sec. 3 deals with ferromagnetic metals and Sec. 4 with ferromagnetic dielectrics. In Sec. 5 we consider the behavior of the Casimir pressure in the vicinity of Curie temperature. Sec. 6 contains our conclusions.

2. Review of Magnetic Properties

The magnetic permeability along the imaginary frequency axis is represented in the form

$$\mu(i\xi) = 1 + 4\pi\chi(i\xi), \tag{4}$$

where $\chi(i\xi)$ is the magnetic susceptibility. The magnitude of $\chi(i\xi)$ decreases monotonously when ξ increases. All materials possess diamagnetic polarization for which[9–11] $\chi(0) < 0$, $\mu(0) < 1$ and $|\mu(0)-1| \sim 10^{-5}$. Diamagnets (such materials as, for instance, Au, Si, Cu and Ag) do not possess any other type of magnetic polarization. For them one can put $\mu_l = 1$, $l = 0, 1, 2, \ldots$ in computations using (1) so that magnetic properties of diamagnets do not influence the Casimir force.

Some materials also possess paramagnetic polarization (in a broad sense) which is larger in magnitude than the diamagnetic one and leads to[11] $\chi(0) > 0$, $\mu(0) > 1$. Paramagnets (in a narrow sense) are materials with $\mu(0) > 1$ if the interaction of magnetic moments of their constituent particles is neglibibly small. Paramagnets may consist of microparticles which are paramagnetic magnetizable but have no intrinsic magnetic moment (the Van Vleck polarization paramagnetism[12]) and of microparticles possessing a permanent magnetic moment (the orientational paramagnetism[9–12]). For all paramagnets in a narrow sense it is true that $\chi(0) < 10^{-4}$ and one can put $\mu_l = 1$ for all l. This conclusion is unchanged for all paramagnets in a broad sense (with the single exception of ferromagnets) because $\chi(0)$ remains as small as mentioned above and takes only a slightly larger values in the vicinity of $T = 0$ even at temperatures below the critical temperature $T_{\rm cr}$ of the magnetic phase transitions[9–11,13–15] (for different materials $T_{\rm cr}$ varies from a few K to more than thousand K).

For the subdivision of paramagnetic materials called ferromagnets it is true that $\mu(0) \gg 1$ at $T < T_{\rm cr}$ (in this case $T_{\rm cr}$ is referred to as the Curie temperature, $T_{\rm cr} \equiv T_C$). There is a lot of ferromagnetic materials, both metals and dielectrics.[16] The rate of decrease of $\mu(i\xi)$ for ferromagnets depends on their electric resistance. Thus, for ferromagnetic metals and dielectrics $\mu(i\xi)$ becomes approximately equal to unity for ξ above 10^4 and $10^9\,$Hz, respectively. Keeping in mind that the first Matsubara frequency $\xi_1 \sim 10^{14}\,$Hz at $T = 300\,$K we arrive at the conclusion that ferromagnets can affect the Casimir force between macroscopic bodies only through the contribution of the zero-frequency term of the Lifshitz formula (1). In all terms of this formula with $l \geq 1$ one can put $\mu_l = 1$. Note that below we do not consider so-called hard ferromagnetic materials possessing a spontaneous magnetization because the magnetic interaction between the plates made of such materials far exceeds any conceivable Casimir force. The subject of our interset is the soft ferromagnetic materials which do not possess a spontaneous magnetization. It is well known that the magnetic permeability of ferromagnets depends on the applied magnetic field.[9–11] Since in the Casimir interaction the mean applied field is equal to zero, below we consider the so-called *initial* (zero field) permeability, i.e., $\mu = \mu(\boldsymbol{H} = 0)$.

3. Ferromagnetic Metals

First we consider the case when both Casimir plates are made of common ferromagnetic metal Co, Cd, Fe or Ni. The dielectric properties of a metal are described by the Drude[17,18] or the plasma[19,20] model approaches, i.e., using the dielectric functions of the form

$$\varepsilon_D(i\xi) = 1 + \frac{\omega_p^2}{\xi(\xi+\gamma)}, \qquad \varepsilon_D(i\xi) = 1 + \frac{\omega_p^2}{\xi^2}, \tag{5}$$

where ω_p is the plasma frequency, γ is the relaxation parameter. By considering different models proposed in the literature we aim to determine whether or not the magnetic properties influence the magnitude of the Casimir pressure and is it possible to experimentally distinguish between alternative theoretical predictions. For all $l \geq 1$ we put $\mu_l = 1$.

For two similar plates Eq. (2) leads to the following reflection coefficients at $\xi = 0$ if the Drude and plasma models are used

$$r_{\mathrm{TM},D}(0,k_\perp) = r_{\mathrm{TM},p}(0,k_\perp) = 1, \quad r_{\mathrm{TE},D}(0,k_\perp) = \frac{\mu(0)-1}{\mu(0)+1} \equiv r_\mu,$$
$$r_{\mathrm{TE},p}(0,k_\perp) = \frac{\mu(0)ck_\perp - [c^2k_\perp^2 + \mu(0)\omega_p^2]^{1/2}}{\mu(0)ck_\perp + [c^2k_\perp^2 + \mu(0)\omega_p^2]^{1/2}}. \tag{6}$$

In the limiting case of large separations (high T) only the zero-frequency term in Eq. (1) contributes to the Casimir pressure and all calculations can be performed analytically. When the Drude model is used, the result is

$$P_D(a,T) = -\frac{k_BT}{8\pi a^3}\left[\zeta(3) + \mathrm{Li}_3(r_\mu^2)\right], \tag{7}$$

where $\zeta(z)$ is the Riemann zeta function and $\mathrm{Li}_n(z)$ is polylogarithm function. Under the conditions $\mu(0) \gg 1$ (valid for ferromagnetic metals) and $\mu(0) = 1$ (valid for nonmagnetic metals) Eq. (7) leads to

$$P_{D,\mathrm{fm}}(a,T) = -\frac{k_BT}{4\pi a^3}\zeta(3), \qquad P_{D,\mathrm{nm}}(a,T) = -\frac{k_BT}{8\pi a^3}\zeta(3), \tag{8}$$

respectively. As can be seen from Eq. (8), if the Drude model is used, the account of magnetic properties of ferromagnetic metals doubles the magnitude of the Casimir pressure at large separations. If, however, the plasma model is used at large separations under the condition $\sqrt{\mu(0)}\delta_0/a \ll 1$, Eq. (1) results in

$$P_{p,\mathrm{fm}}(a,T) = -\frac{k_B T}{4\pi a^3}\zeta(3)\left[1 - 3\sqrt{\mu(0)}\frac{\delta_0}{a}\right], \tag{9}$$

where $\delta_0 = c/\omega_p$ is the skin depth. The same expression, but with $\mu(0) = 1$, is obtained for nonmagnetic metals described by the plasma model. At $T = 300\,\mathrm{K}$ Eqs.(7)–(9) are applicable for $a > 6\,\mu$m.

From an experimental point of view the most interesting region is from $a = 0.5\,\mu$m to $a = 1\,\mu$m (for $a < 0.5\,\mu$m the contribution of the zero-frequency term and, thus, of magnetic properties is not large enough). We have performed numerical computations of the Casimir pressure, Eq. (1), in the region from 0.5 to $6\,\mu$m for Co with parameters[21,22] $\omega_{p,\mathrm{Co}} = 3.97\,\mathrm{eV}$, $\gamma_{\mathrm{Co}} = 0.036\,\mathrm{eV}$ and $\mu_{\mathrm{Co}}(0) = 70$. In Fig. 1 we plot the ratio of the Casimir pressure P between two Co plates at $T = 300\,\mathrm{K}$ to $P_0 = -\pi^2\hbar c/(240a^4)$ computed using (a) the Drude model and (b) the plasma model. The solid lines take into account the magnetic properties and the dashed lines are computed with magnetic properties neglected. Note that the solid line in Fig. 1(a) is almost coincident with the dashed line in Fig. 1(b). At small separations $a < 1\,\mu$m, the difference between the dashed line in Fig. 1(a) and the solid line in Fig. 1(b) is also not observable in the limits of the experimental precision. Thus the experiments on an indirect measurement

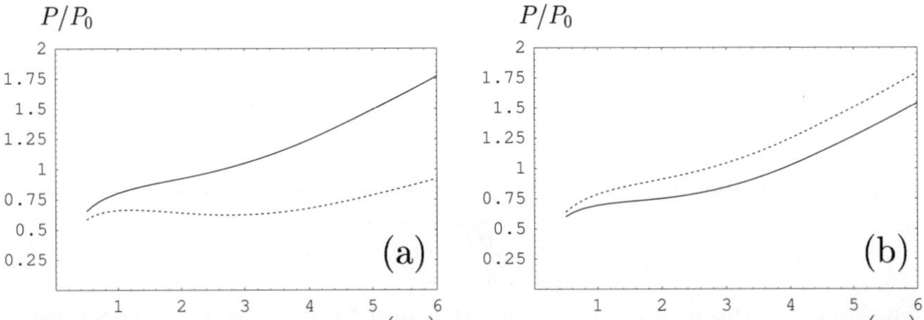

Fig. 1. The relative Casimir pressure as a function of separation in the configuration of two parallel Co plates with inclusion of magnetic properties (the solid lines) and with magnetic properties neglected (the dashed lines). Computations are performed with the dielectric permittivity (a) of the Drude model and (b) of the plasma model.

of the Casimir pressure by means of a micromechanical ocsillator[23] at separations of about 500–600 nm can allow us to choose one of the following situations.

1. The experimental data are in favour of the solid line in Fig. 1(a) and the dashed line in Fig. 1(b). This means that either the magnetic properties affect the Casimir pressure and metals should be described by the Drude model or the magnetic properties do not affect the Casimir pressure and metals should be described by the plasma model.

2. The experimental data are in favour of the dashed line in Fig. 1(a) and the solid line in Fig. 1(b). In this case either the magnetic properties affect the Casimir pressure and metals should be described by the plasma model or the magnetic properties do not affect the Casimir pressure and metals should be described by the Drude model.

Now let one plate be made of a ferromagnetic metal ($n = 1$) and the other of a nonmagnetic metal ($n = 2$). Here, in the limit of large separations one obtains

$$P_D(a,T) = -\frac{k_B T}{8\pi a^3}\zeta(3) \tag{10}$$

if the Drude model is used. For the plasma model under conditions $\sqrt{\mu(0)}\delta_{01}/a \ll 1$ and $\delta_{02}/a \ll 1$ it follows

$$P_p(a,T) = -\frac{k_B T}{4\pi a^3}\zeta(3)\left[1 - \frac{3(\sqrt{\mu(0)}\delta_{01} + \delta_{02})}{2a}\right]. \tag{11}$$

Note than when the Drude model is used $r_{\rm TE}^{(2)}(0,k_\perp) = 0$ and, thus, the magnetic properties of a ferromagnetic plate entering only through $r_{\rm TE}^{(1)}(0,k_\perp)$ do not influence the result. The results of the numerical computations for the Co plate interacting with the Au plate ($\omega_{p,\rm Au} = 9.0\,$eV, $\gamma_{\rm Au} = 0.035\,$eV) in the case when the plasma model is used are shown in Fig. 2(a). It is seen that here the inclusion of the magnetic properties (the solid line) decreases the magnitude of the Casimir pressure. The influence of the magnetic properties is, however, very moderate and can be observed only in the experiment on measuring the difference Casimir pressure above a patterned plate[24] one section of which is made of Co and the other of Au. Such an experiment allows one to choose between the two alternatives in each of the situations described above. This will provide a complete experimental answer to questions whether the magnetic properties influence the Casimir force and what dielectric model should be used in the Lifshitz theory to describe real metals.

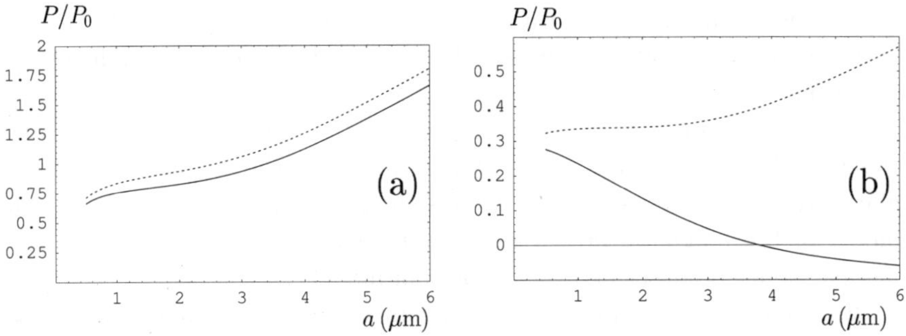

Fig. 2. The relative Casimir pressure as a function of separation in the configuration of two parallel plates, one made of Au and the other of (a) Co and (b) ferromagnetic dielectric with inclusion of magnetic properties (the solid lines) and with magnetic properties neglected (the dashed lines). Computations are performed using the plasma model for the dielectric permittivity of metals.

4. Ferromagnetic Dielectrics

Ferromagnetic dielectrics are materials that, while displaying physical properties characteristic of dielectrics, show ferromagnetic behavior under the influence of an external magnetic field.[25] Such materials are widely used in different magneto-optical devices. As an example, we consider a composite material of polystyrene with a volume fraction f of ferromagnetic metal nanoparticles in the mixture. The permittivity of such a material can be presented in the form[26]

$$\varepsilon_{\text{fd}}(i\xi) = \varepsilon_{\text{d}}(i\xi)\left(1 + \frac{3f}{1-f}\right), \tag{12}$$

where ε_{d} is the permittivity of polystyrene.[27]

We have performed computations of the Casimir pressure for two parallel plates one of which is made of ferromagnetic dielectric [$f = 0.25$, $\varepsilon_{\text{fd}}(0) = 5.12$, $\mu(0) = 25$] and the other of Au described by the plasma model. Recall that if Au is described by the Drude model the magnetic properties do not influence the Casimir pressure as explained in Sec. 3. The computational results for P/P_0 as a function of a are presented in Fig. 2(b) where the solid line takes the magnetic properties into account and the dashed line neglects them. As can be seen in Fig. 2(b), magnetic properties have an important influence on the Casimir pressure and even lead to the change of sign of the force (from attraction to repulsion).

This important conclusion can be confirmed analytically in the limiting case of large a. If the metallic properties of Au plate are described by the

Drude model, one obtains

$$P_D(a,T) = -\frac{k_BT}{8\pi a^3}\mathrm{Li}_3(r_\varepsilon), \qquad r_\varepsilon \equiv \frac{\varepsilon_{\mathrm{fd}}-1}{\varepsilon_{\mathrm{fd}}+1}. \tag{13}$$

This does not depend on the magnetic properties. If, however, the plasma model is used, then, under the condition $\delta_{02}/a \ll 1$, one arrives at

$$P_p(a,T) = -\frac{k_BT}{8\pi a^3}\left[\mathrm{Li}_3(r_\varepsilon) + \mathrm{Li}_3(-r_\mu)\left(1 - 3\frac{\delta_{02}}{a}\right)\right]. \tag{14}$$

The expression on the right-hand side of Eq. (14) is positive and the respective Casimir force is repulsive if the following condition is satisfied:

$$\mathrm{Li}_3(r_\varepsilon) < \left|\mathrm{Li}_3(-r_\mu)\left(1 - 3\frac{\delta_{02}}{a}\right)\right|. \tag{15}$$

This condition is easily satisfied for real materials.

5. Vicinity of the Curie Temperature

At the Curie temperature T_C specific for each material, ferromagnets undergo a magnetic phase transition.[14,16] At higher temperature they lose ferromagnetic properties and become paramagnets in the narrow sense. Thus, for Fe, Co, Ni and Gd the Curie temperature is equal to 1043 K, 1388 K, 627 K and 293 K, respectively.[28] Here, we consider the behavior of the Casimir pressure under the magnetic phase thansition which occurs with the increase of T in the configuration of two similar plates made of Gd. The Drude parameters of Gd are equal to[29] $\omega_{p,\mathrm{Gd}} = 9.1\,\mathrm{eV}$, $\gamma_{\mathrm{Gd}} = 0.58\,\mathrm{eV}$.

Computations of the Casimir pressure between two parallel plates made of Gd in the vicinity of the Curie temperature require respective values of $\mu(0)$ for Gd at $T < T_C$ [at $T > T_C$, $\mu_{\mathrm{Gd}}(0) = 1$ to high accuracy]. In Fig. 3(a) the magnetic permeability of Gd is shown as a function of temperature in the region from 280 K to 300 K on the basis of the experimental data.[30] The Casimir pressure as a function of temperature was computed at the separation $a = 500\,\mathrm{nm}$ between the plates using Eq. (1). The computational results obtained using the Drude and the plasma models are shown in Fig. 3(b) by the pairs of lines 1 and 2, respectively. In each pair the solid line takes into account the magnetic properties and the dashed line is computed with these properties disregarded. As can be seen from Fig. 3b, experiments on the magnetic phase transition can also be used to determine the influence of magnetic properties on the Casimir force and as a test for different models of the dielectric properties of metals.

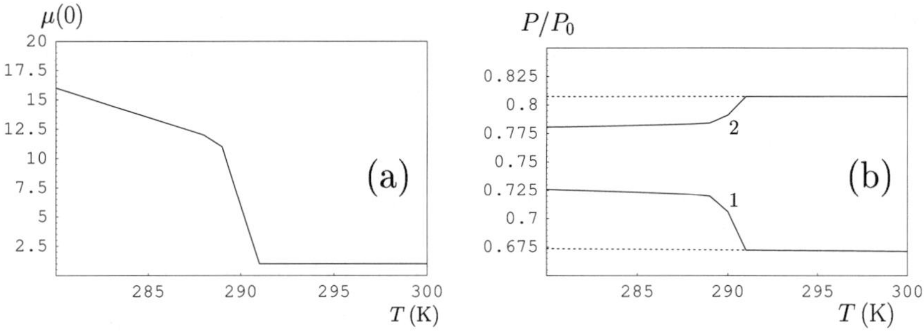

Fig. 3. (a) The static magnetic permeability of Gd at the magnetic phase transition as a function of temperature. (b) The relative Casimir pressure as a function of temperature in the configuration of two parallel Gd plates at the separation $a = 0.5\,\mu$m. The solid and dashed lines include and neglect the magnetic properties, respectively. The pairs of lines marked 1 and 2 indicate the respective computational results obtained using the Drude and the plasma models.

6. Conclusions

The investigation of the influence of magnetic properties on the Casimir force performed above leads to the following conclusions.

1. Of all the real materials, only ferromagnets might affect the Casimir force.
2. At all feasible temperatures the possible influence of ferromagnets on the Casimir force occurs solely through the contribution of the zero-frequency term in the Lifshitz formula.
3. In the framework of the Lifshitz theory the Casimir repulsion of two macroscopic bodies separated by a vacuum gap arises for only the case when one body is made of ferromagnetic dielectric and the other is metallic. In doing so the metal is described by the plasma model.
4. Modern experimental techniques present good opportunities to check whether the magnetic properties of the plate material influence the Casimir force. Experiments with magnetic bodies allow independent test of the plasma and Drude model approaches to the description of the dielectric properties of metals.

Acknowledgments

The authors are grateful to the Deutsche Forschungsgemeinschaft Grant No. GE 696/9–1 for partial financial support.

References

1. M. Bordag, G. L. Klimchitskaya, U. Mohideen and V. M. Mostepanenko, *Advances in the Casimir Effect* (Oxford University Press, Oxford, 2009).
2. P. Richmond and B. W. Ninham, *J. Phys. C: Solid St. Phys.* **4**, 1988 (1971).
3. S. Y. Buhmann, D.-G. Welsch and T. Kampf, *Phys. Rev. A* **72**, 032112 (2005).
4. M. S. Tomaš, *Phys. Lett. A* **342**, 381 (2005).
5. S. J. Rahi, T. Emig, N. Graham, R. L. Jaffe and M. Kardar, *Phys. Rev. D* **80**, 085021 (2009).
6. Yu. S. Barash and V. L. Ginzburg, *Sov. Phys. Usp.* **18**, 305 (1975).
7. O. Kenneth, I. Klich, A. Mann and M. Revzen, *Phys. Rev. Lett.* **89**, 033001 (2002).
8. D. Iannuzzi and F. Capasso, *Phys. Rev. Lett.* **91**, 029101 (2003).
9. P. W. Selwood, *Magnetochemistry* (Interscience Publ., New York, 1956).
10. A. H. Morrish, *The Physical Principles of Magnetism* (J. Wiley, New York, 1965).
11. S. V. Vonsovskii, *Magnetism* (J. Wiley, New York, 1974).
12. J. H. Van Vleck, *The Theory of Electric and Magnetic Susceptibilities* (Oxford University Press, Oxford, 1932).
13. C. M. Hurd, *Contemp. Phys.* **23**, 480 (1982).
14. *Magnetic Phase Transitions*, ed. M. Ausloos (Springer, Berlin, 1983).
15. B. Barbara, D. Gignoux and C. Vettier, *Lectures on Modern Magnetism* (Springer, Berlin, 1988).
16. A. Goldman, *Handbook of Modern Ferromagnetic Materials* (Springer, New York, 1999).
17. M. Boström and B. E. Sernelius, *Phys. Rev. Lett.* **84**, 4757 (2000).
18. I. Brevik, J. B. Aarseth, J. S. Høye and K. A. Milton, *Phys. Rev. E* **71**, 056101 (2005).
19. M. Bordag, B. Geyer, G. L. Klimchitskaya and V. M. Mostepanenko, *Phys. Rev. Lett.* **85**, 503 (2000).
20. V. B. Bezerra, G. L. Klimchitskaya, V. M. Mostepanenko and C. Romero, *Phys. Rev. A* **69**, 022119 (2004).
21. M. A. Ordal, R. J. Bell, R. W. Alexander Jr., L. L. Long and M. R. Querry, *Appl. Otp.* **24**, 4493 (1985).
22. *Ferromagnetic Materials*, ed. E. P. Wohlfarth (North-Holland, Amsterdam, 1980).
23. R. S. Decca, D. López, E. Fischbach, G. L. Klimchitskaya, D. E. Krause and V. M. Mostepanenko, *Eur. Phys. J. C* **51**, 963 (2007).
24. R. Castillo-Garza, C.-C. Chang, D. Jimenez, G. L. Klimchitskaya, V. M. Mostepanenko and U. Mohideen, *Phys. Rev. A* **75**, 062114 (2007).
25. S. A. Chambers, *Surf. Sci. Reports* **61**, 345 (2006).
26. L. Lewin, *J. Inst. Elect. Eng. (part C)* **94**, 64 (1947).
27. V. A. Parsegian and G. H. Weiss, *J. Coll. Interface Sci.* **81**, 285 (1981).
28. P. Heller, *Rep. Progr. Phys.* **30**, 731 (1967).
29. *Handbook of Optical Constants of Solids III*, ed. E. D. Palik (Academic, New York, 1998).
30. J. M. D. Coey, V. Skumryev and K. Gallagher, *Nature* **401**, 35 (1999).

THE CASIMIR EFFECT AND THE FOUNDATIONS OF STATISTICAL PHYSICS

V. M. MOSTEPANENKO[1,2] and G. L. KLIMCHITSKAYA[1,3]

[1] *Institute for Theoretical Physics, Leipzig University, Postfach 100920, D-04009, Leipzig, Germany*
[2] *Noncommercial Partnership "Scientific Instruments", Tverskaya St. 11, Moscow, 103905, Russia*
[3] *North-West Technical University, Millionnaya St. 5, St. Petersburg, 191065, Russia*

The Lifshitz theory and its modifications are discussed with respect to the Nernst heat theorem and the experimental data of several recent experiments. An analysis of all available information leads to the conclusion that some concepts of statistical physics might need reconsideration.

Keywords: Lifshitz theory; Casimir force; Nernst heat theorem.

1. Introduction

In the last few years there has been an explosion of interest in the Casimir effect[1] which has resulted in new precise experiments, elaboration of powerful theoretical methods and in suggestions of prospective applications (a modern overview of the subject can be found in Ref. 2). Coincident with many developments of a conclusive character, starting from 2000 there were also controversial discussions in the literature on the nature and size of the thermal effects in the Lifshitz theory of the Casimir force.[2,3] Boström and Sernelius[4] were the first who predicted the existence of large thermal corrections to the Casimir force between two plane parallel metallic plates described by the Drude model spaced at separation of a few hundred nanometers. Bordag et al.[5] argued that such corrections are nonphysical and suggested to calculate the thermal Casimir force using the dielectric permittivity of the plasma model (for the latter purpose the plasma model was also used in Ref. 6). Later both approaches were further developed in Refs. 7,8 and 9,10, respectively. A step of paramount importance was made by the experiments of Decca et al.[11,12] which excluded the existence of large thermal corrections predicted by the Drude model at almost 100% confidence level. Related experiments for semiconductor[13] and dielectric[14] materials leading to similar conclusions[15] were subsequently performed. On the one hand, thermodynamic arguments based on the Nernst heat theorem favored the plasma model approach for metals[10] and neglect of the dc conductivity for dielectrics.[16] On the other hand, statistical physics applied in the so-called classical limit was in support of the Drude model.[17] The situation was so extraordinary that it was even suggested[18–20] to modify the Lifshitz theory providing the fundamental description of both the van der Waals and Casimir forces between real materials. For this purpose the standard reflection coefficients were replaced with their generalizations taking into account the screening effects and diffusion currents. It was shown, however, that the modified theory still violates the Nernst heat theorem[21–26] and is in contradiction with the experimental data. These conclusions were disputed[27–29] by the authors of the modified theory.

Keeping in mind that controversial discussion on this subject has lasted for already ten years and consensus is not yet achieved, it seems pertinent to collect and analyze all the proposed arguments. Such an analysis seems to be especially useful because there were discussions in the previous literature which appear one sided by dealing with only selected facts and disregarding others. By taking into account all known facts in a fair manner (i.e., by assuming that published experimental and theoretical results are correct if we cannot indicate any specific mistake invalidating them), we arrive at the conclusion that some of the concepts of statistical physics commonly used for the theoretical description of the interaction of fluctuating fields with matter need to be reconsidered.

The structure of this paper is as follows. In Sec. 2 we briefly discuss the Nernst theorem in the Lifshitz theory. Sec. 3 is devoted to the same subject in application to the proposed modifications of the Lifshitz theory. In Sec. 4 we consider what the experiments say and if they are reliable. Sec. 5 considers what statistical physics says. Sec. 6 contains our conclusions.

2. The Lifshitz Theory and the Nernst Heat Theorem

The Lifshitz theory provides an expression for the free energy $\mathcal{F}(a,T)$ of the fluctuating electromagnetic field interacting with two thick uncharged plates (semispaces) separated by a gap of width a per unit area of plates. It is supposed that this system is in thermal equilibrium at temperature T. Material of the plates is described by the dielectric permittivity $\varepsilon(\omega)$ depending only on the frequency. Under these conditions $\mathcal{F}(a,T)$

is expressed in terms of the Fresnel reflection coefficients $r_{\rm TM,TE}(i\xi_l, k_\perp)$ for the transverse magnetic (TM) and electric (TE) polarizations of the electromagnetic field calculated at the imaginary Matsubara frequencies $\xi_l = 2\pi k_B T l/\hbar$, where k_B is the Boltzmann constant, $l = 0, 1, 2, \ldots$, and $\boldsymbol{k}_\perp = (k_x, k_y)$ is the projection of the wave vector on the plane of the plates.[2]

For materials with no free charge carriers (insulators) $\varepsilon(i\xi)$ can be represented in the oscillator form

$$\varepsilon(i\xi) = 1 + \sum_{j=1}^{K} \frac{g_j}{\omega_j^2 + \xi^2 + \gamma_j \xi}, \tag{1}$$

where $\omega_j \neq 0$ are the oscillator frequencies and $\varepsilon_0 \equiv \varepsilon(0) < \infty$. Electrons in metals are usually described by the Drude or plasma models

$$\varepsilon_D(i\xi) = 1 + \frac{\omega_p^2}{\xi(\xi+\gamma)}, \qquad \varepsilon_p(i\xi) = 1 + \frac{\omega_p^2}{\xi^2}, \tag{2}$$

where ω_p is the plasma frequency, γ is the relaxation parameter.

It was suggested[9,10] to use the Nernst heat theorem as a test of applicability of different models of ε in the Lifshitz theory. The entropy of the system under consideration (the two plates interacting with the fluctuating field) per unit area of plates is finite and can be calculated as

$$S_{\rm syst}(a,T) = -\frac{\partial \mathcal{F}(a,T)}{\partial T} - \frac{\partial \mathcal{F}_n(T)}{\partial T} \equiv S(a,T) + S_n(T). \tag{3}$$

Here, $S(a,T)$ is the separation-dependent part of the entropy related to the interaction between the fluctuating field and the plates, and $\mathcal{F}_n$ (S_n) are the parts of the free energy (entropy) of the system which do not depend on a. The quantities $\mathcal{F}_n$ (S_n) are related to the noninteracting case (specifically, they contain the large free energy and entropy of remote plates) and do not contribute to the Casimir force.

There are different formulations of the third law of thermodynamics (the Nernst heat theorem) in the literature (some of them are discussed in Ref. [30]). Below throughout the text we use only the standard formulation from textbooks which is the following.[31,32] When $T \to 0$, the entropy of an equilibrium system [in our case $AS_{\rm syst}(a,T)$ where A is the area of the plates] goes to a finite limit $S_{\rm syst,0}$ which does not depend on volume, pressure, density or other thermodynamic parameters of the system. According to quantum statistical physics, we get[31,32]

$$S_{\rm syst,0} = k_B \ln W_0, \tag{4}$$

where W_0 is an integer number describing the degree of degeneracy of the ground state of the system. If the ground state is nondegenerate, $W_0 = 1$, one has $S_{\rm syst,0} = 0$. The latter, however, is not necessary to satisfy the Nernst theorem as formulated above. It is important only that $S_{\rm syst,0}$ does not depend on the continuous thermodynamic parameters, specifically, on a. Keeping in mind that $S(a,T)$ is the part of the entropy depending on a, the necessary requirement for the satisfaction of the Nernst theorem is that $S(a,T) \to 0$ when $T \to 0$. If $S(a,T)$ goes to some function of a, $f(a)$, when $T \to 0$, the Nernst theorem is violated because $f(a)$ cannot be compensated by the a-independent limit of the quantity $S_n(T)$.

When ε_D of the Drude model (2) is substituted into the Lifshitz formula for metals with perfect crystal lattices, we get[10]

$$S(a,0) = S_D(a,0) = -\frac{k_B\zeta(3)}{16\pi a^2}\left[1 - 4\frac{\delta_0}{a} + 12\left(\frac{\delta_0}{a}\right)^2 - \cdots\right] < 0, \qquad (5)$$

where $\delta_0 = c/\omega_p$ is the skin depth and $\zeta(z)$ is the Riemann zeta function. A metal with perfect crystal lattice is a truely equilibrium system. Thus in this case we deal with the violation of the Nernst heat theorem. It was argued in the literature[27] that with the decrease of T the frequency region of the anomalous skin effect, where local description by means of $\varepsilon_D(\omega)$ is inapplicable, extends to low frequencies. This objection, however, does not solve the problem. First, for any low T there exists some narrow region of small frequencies $[0, \omega_0]$ where local desctiption by means of $\varepsilon_D(\omega)$ is applicable. Then the result (5) remains valid because it originates from the zero-frequency term of the Lifshitz formula. Second, the Drude model at low T was used for the interpolation between the regions of the normal skin effect and infrared optics in the classical theories by Bloch, Grüneisen and Debye.[33] Although such a model approach does not provide an exact description of real metals due to the existence of the anomalous skin effect, it seems strange that it leads to the violation of the Nernst theorem when used in combination with the Lifshitz formula. Note that for metals with impurities the Lifshitz formula combined with the Drude model satisfies the Nernst theorem.[8,34] This is a step forward in the resolution of the problem but does not solve it because the introduction of impurities might result in a violation of the thermal equilibrium which for sure takes place for perfect crystal lattices (see a discussion[35]). At the same time the substitution of ε_p of the plasma model (2) into the Lifshitz formula leads to $S_p(a,0) = 0$. For insulators it was shown[16] that $S(a,T)$ calculated with the dielectric permittivity (1) goes to zero when T vanishes. If, however, the dc conductivity

$\sigma_0(T)$ is taken into account,

$$\varepsilon_{\rm dc}(i\xi,T)=\varepsilon(i\xi)+4\pi\frac{\sigma_0(T)}{\xi}, \tag{6}$$

it results in the violation of the Nernst theorem

$$S(a,0)=S_{\rm dc}(a,0)=\frac{k_B}{16\pi a^2}\left[\zeta(3)-{\rm Li}_3(r_0^2)\right]>0. \tag{7}$$

Here, ${\rm Li}_3(z)$ is the polylogarithm function and $r_0=(\varepsilon_0-1)/(\varepsilon_0+1)$.

To avoid the violation of the Nernst theorem and contradictions with the experimental data (see Sec. 4) in numerous applications of the Lifshitz theory, the following phenomenological prescription was proposed.[36,37] When applying the Lifshitz theory to metals, conduction electrons should be described by the plasma model. In the application of this theory to dielectrics, dc conductivity should be omitted. Keeping in mind that all materials can be divided into metals (whose conductivity is not equal to zero at $T=0$) and dielectrics (whose conductivity vanishes when $T\to 0$), this prescription can be considered as universally applicable. In some sense it is not new because metals were often described in the literature by means of the plasma model[6,38] and the dc conductivity of dielectrics was almost always omitted.[14] It was generally believed, however, that with account of relaxation properties of conduction electrons (i.e., using the Drude model) and of the dc conductivity of dielectrics slightly more exact results would be obtained. The new fact recognized in the last few years is that the inclusion of these features leads to drastically different calculational results which are in conflict with thermodynamics and contradict the experimental data. This fact invites reconsideration of the Lifshitz theory and careful analysis of all assumptions laid in its foundation.

3. The Nernst Heat Theorem in the Modifications of the Lifshitz Theory

The most general modification[19] leaves the formalism of the Lifshitz theory unchanged but replaces the Fresnel reflection coefficients, $r_{\rm TM,TE}(i\xi_l,k_\perp)$ with the modified ones, $\tilde{r}_{\rm TM,TE}(i\xi_l,k_\perp)$, which take into account both the drift and diffusion currents by means of the Boltzmann transport equation. The modified reflection coefficients depend on a new parameter κ which has the physical meaning of an inverse screening radius. It is equal to $\kappa_{\rm DH}$ or $\kappa_{\rm TF}$ for Debye-Hückel and Thomas-Fermi screening radia applicable for the Maxwell-Boltzmann and Fermi-Dirac statistics, respectively. For dielectrics ($\kappa=\kappa_{\rm DH}$) at $\xi=0$ the coefficient $\tilde{r}_{\rm TM}(0,k_\perp)$ was first obtained in Ref. 18.

The modified reflection coefficients $\tilde{r}_{\rm TM,TE}$ were also phenomenologically expressed[20] in terms of $\boldsymbol{k}$-dependent dielectric permittivities in the random phase approximation (recall that in the presence of a gap between semispaces the translation invariance in space is violated and the nonlocal dielectric permittivity ε_z depending on $\boldsymbol{k}$ does not exist as a rigorous mathematical concept[39]).

For metals with perfect crystal lattice by using $\kappa = \kappa_{\rm TF}$ it was shown[22,23] that the modified entropy $\tilde{S}(a,0) = S_D(a,0) < 0$, as can be seen from Eq. (5). Thus, in this case the modification of the Lifshitz theory proposed[18–20] suffers from the same thermodynamic difficulty as the standard Drude model. For dielectric materials ($\kappa = \kappa_{DH}$) the situation turned out to be more involved. Under the condition that the density of charge carriers $n(T) \to 0$ more quickly than $T^{1+\alpha}$ with $\alpha > 0$ (this is the case for intrinsic semiconductors) it was shown[21,24] that the modified entropy $\tilde{S}(a,0) = 0$, i.e., the Nernst theorem is satisfied. In the two Comments[25,26] it was stressed, however, that for dielecric materials not satisfying this condition (for instance, for doped semiconductors with $n < n_{\rm cr}$, semimetals of dielectric type and solids with ionic conductivity) the modified Lifshitz theory violates the Nernst heat theorem. In this case it holds $\tilde{S}(a,0) = S_{dc}(a,0) > 0$ where S_{dc} is defined in Eq. (7). For dielectric materials under consideration $n(T)$ does not go to zero with vanishing T and conductivity vanishes with temperature due to the vanishing mobility of charge carriers.

The result that the modifications of the Lifshitz theory are in disagreement with thermodynamics was disputed in the literature. Thus, it was claimed[29] that the approach of Ref. 19 satisfies the Nernst theorem for all dielectrics. However, in the respective proof it was assumed that $n(T) \to 0$ when $T \to 0$. The above-mentioned dielectric materials for which this is not the case were not discussed. Reply[27] claimed that the materials leading to conflicts with thermodynamics in the modified Lifshitz theory are amorphous glass-like bodies which are out of equilibrium state and have a big entropy at $T = 0$. The Nernst theorem is not valid for such bodies. It is true that glass-like bodies must not satisfy the Nernst theorem. The arguments in the Reply[27] are, however, somewhat contradictory. The point is that we consider not the entropy $S_n(T)$ of the plate made of a glass-like material (SiO_2 for instance[14]), but the entropy of the interaction with the fluctuating field $S(a,T)$ (see Sec. 2). If the fluctuating field is in equilibrium with the plate (as is assumed in Ref. 14), one can apply the Lifshitz theory. In this case, however, in accordance with the Nernst theorem, $S(a,T)$ must vanish when T vanishes. In fact the input data for the Lifshitz formula are the values of $\varepsilon(i\xi)$ which are quite similar for the amorphous and polycryctal SiO_2. The Lifshitz formula is applicable when the fluctuating field is in equilibrium with the material of the plate. This formula is incapable of distinguishing between the cases when the plate material is in equilibrium or out of equilibrium. Reply[27] does not also provide a response concerning the existence of crystallic materials (semimetals of the dielectric type, for instance) leading to the violation of the Nernst theorem in the proposed modifications of the Lifshitz theory.

Both Replies[27,28] cast doubts on the fact that there are dielectric materials for which $n(T)$ does not go to zero when $T \to 0$ with a reference to the measurements[40] for SiO_2 performed in the region from 433 K to 473 K. Such high-T results seem to be irrelevant to the problem under consideration. Independent measurements of all three parameters, conductivity, n and mobility, demonstrate[41] that "mobility has the dominant influence upon the conductivity-temperature dependence." As was recently confirmed,[42] "On long time scales the 'mobile' ion density must be the total ion concentration. This 'long run' may be years or more, and ions trapped for so long are for all practical purposes immobile. Nevertheless, unless there are infinite barriers in the solid, which is unphysical, in the very long run all ions are equivalent." Thus, for ionic conductors (like amorphous SiO_2) n does not vanish when $T \to 0$. The same conclusion holds for compensated semiconductors of the dielectric type. If the density of donor atoms n_d is larger than the density of acceptor atoms n_a, the density of charge carriers at low T, $n_d - n_a$, remains constant.[43] One more example is provided by semimetals of the dielectric type which are crystal materials with a regular structure. For these materials the Fermi energy is at a band where the density of states is not equal to zero. The number of charge carriers near the Fermi surface is fixed and determined by the structure of the crystal lattice. For both compensated semiconductors and semimetals of dielectric type conductivity vanishes due to vanishing mobility.[44,45] All the above testifies that the problem of thermodynamic inconsistency of the proposed modifications of the Lifshitz theory deserves serious attention.

4. What Experiments Say and Is It Reliable

It was widely discussed in the literature that the measurement data of the experiments with a micromechanical oscillator[11,12] exclude the use of the Drude model for the calculation of the thermal Casimir force between metals but are consistent with the use of the plasma model. The experiments with an atomic force microscope[13] and Bose-Einstein condensate[14] are inconsistent with the inclusion of the dc conductivity of a dielectric plate but consistent with the theory omitting this conductivity. These results are related to the standard Lifshitz theory. They are obtained at a 99.9% and 95% confidence levels with respect to experiments of Refs. 12,13 and at a 70% confidence level for the experiment of Ref. 14.

Just after the modifications of the Lifshitz theory were proposed, the obtained theoretical results (which are almost coincident for all three variants of the modified theory) were compared with the experimental data. For metals, it was found[22,23,26] that the experimental data[12] exclude the modified Lifshitz theory at a 99.9% confidence level. For dielectrics, the data of the experiment[13] exclude the predictions of the modified theory at a 70% confidence level.[21,24,25] It was found also that the data of the experiment[14] determined at a 70% confidence level are not precise enough and do not permit to make a conclusive comparison with theory. The point is that it is consistent with both the standard Lifshitz theory with dc conductivity excluded and with the modified Lifshitz theory.

Note that it was claimed[20] that the experimental data[13] can hardly distinguish between the standard Lifshitz theory with omitted dc conductivity of dielectric Si and the modified theory. This claim is based on a complete misunderstanding of statistical procedures used for the comparison between experiment and theory. Thus, in Fig. 1a of Ref. 20 the experimental data are shown with errors determined at a 70% confidence level, but the width of the theoretical band related to the modified theory was calculated at a 95% confidence level (i.e., artificially widened in order to make theory consistent with the data). Such a comparison is evidently irregular. In the Erratum,[20] instead of plotting the theoretical band at a 70% confidence level, the experimental errors were increased by calculating them at a 95% confidence level. This is, however, meaningless because the data[13] are not of sufficient precision for the conclusive comparison with the modified Lifshitz theory at a 95% confidence level.[21] If a comparison at the 70% confidence level would be made, the result[21] on the exclusion of the modified theory is reproduced.

Thus, the Drude model approach and the modified Lifshitz theory are in disagreement with the experimental data. The question arises what is the reliability of these experiments. The experiments under consideration were repeated several times with the same result and the most conservative statistical procedures for the data processing and error analysis have been used. It was claimed, however, that there is an anomalous distance dependence of the gradient of the electric force, used for calibration of the Casimir setup, between an Au plate and an Au spherical lens of 30 mm radius.[46] The respective contact potential was found to be separation-dependent. On this basis it was suggested to perform a reanalysis of the previous experiments mentioned above. These doubts cast on previous experiments with small spheres of about 100 μm radia are not justified. The reason is that the contact potential in the experiments[11,12] was measured to be constant over a wide range of separations and the standard force-distance dependence for the electric force was observed, as predicted by classical electrodynamics. The possible reason for the anomalous dependence observed[46] is deviation of the mechanically polished and ground surface of the centimeter-size radius from a perfect spherical shape.[47] An attempt to avoid this conclusion using the capacitance measurements at large separations[48] was shown to be based on incorrect computations.[49] Because of this, continuing claims that important systematic effects have not been properly taken care of in the electrostatic calibrations in previous experiments, in our opinion, are unfair and cannot be considered as a scientific argument against these experiments. This does not mean that there is no need to look for systematic effects which might be present in previously performed experiments. It would be desirable, however, that such kind investigations were performed in the experimental configurations maximally similar to the original ones and were not based on far-reaching extrapolations.

5. What Statistical Physics Says

Classical statistical physics permits one to calculate the free energy for two remote plates consisting of mobile quantum charges interacting with the quantized electromagnetic field. In doing so, photons and charges are supposed to be in thermal equilibrium at temperature T. The obtained free energy[17] is equal to the one calculated by using the Lifshitz formula combined with the Drude model (i.e., equal to one half of the result valid for ideal metal plates).

Another consequence of statistical physics is the Bohr-van Leeuwen theorem which states that in classical systems at thermal equilibrium matter decouples from the transverse electromagnetic field. Recently it was shown[50] that this theorem is satisfied if and only if at large separations the reflection coefficient $r_{\mathrm{TE}}(0, k_\perp)$ of nonmagnetic materials is equal to zero leading to the same result for the free energy as the Lifshitz formula combined with the Drude model. Thus, in the classical limit (at large separations) the Drude model approach finds support from the source side of statistical physics although it has difficulties with respect to the Nernst theorem and disagrees with the experimental data at short separations.

In this situation it is useful to reformulate the problem in an equivalent way. It was shown[51] that large negative temperature correction arising in the Drude model approach at short separations can be described as the contribution of eddy currents. The absence of this contribution in the measurement data was interpreted in a way that it was somehow reduced.[51] The mechanism of this reduction remains, however, unclear. As a possible resolution of the problem the standard Planck distribution was modified[52] by including a phenomenological parameter D taking into account the "saturation effects". In this way an agreement between the Lifshitz theory combined with the Drude model and experimental data[12–14] was achieved. However, the relative arbitrariness in the value of D remains a problem.

The roots of the controversial situation under consideration might be connected with the use of some basic statements of statistical physics outside of their application region. It is common knowledge that when a physical system deviates from the equilibrium state (for instance, when a semiconductor is placed in an external field) the fluctuation-dissipation theorem is violated. In this respect it is pertinent to recall that both the Lifshitz theory combined with the Drude model and its modifications[18–20] include transport phenomena in an external field and, thus, violate the applicability condition of the fluctuation-dissipation theorem on which they are based. The possibility of such violation is explicitly admitted by the statement[18] that "It is not clear if the fields with the very low frequencies... are in thermal equilibrium with bodies. The problem is worth experimental investigation." In our opinion experiments[12–14] have already solved this problem in the most unambiguous manner.

6. Conclusions

From the foregoing we arrive to the following conclusions.

1) For metals with perfect crystal lattices the Lifshitz theory combined with the Drude model violates the Nernst theorem. The Nernst theorem is satisfied when the relaxation is nonzero at zero temperature, i.e. when impurities are taken into account. The Lifshitz theory including the dc conductivity of dielectrics and modifications of this theory violate the Nernst theorem for wide classes of different materials.

2) The experimental data of several experiments are inconsistent with the Lifshitz theory combined with the Drude model or including the dc conductivity and with the modifications of this theory. Keeping in mind that the Drude relation correctly describes the response of a metal to real (external) electric field, the reason of this inconsistency might be connected with some fundamental differences between real and fluctuating fields.
3) Phenomenologically, contradictions of the Lifshitz theory with both the Nernst theorem and the experimental data disappear if the free charge carriers are described by means of the plasma model in metals and are disregarded in dielectrics. Similar to any phenomenological approach, this one is useful as a practical matter but cannot be offered as an alternative to a complete theoretical description which remains unknown.
4) In our opinion, there are concepts of statistical physics related to the theoretical description of the interaction of classical and quantum fluctuating fields with matter that might need a reconsideration. Opinions on this subject vary and the consensus is not yet achieved.

Acknowledgments

The authors are grateful to the Deutsche Forschungsgemeinschaft Grant No. GE 696/9–1 for partial financial support.

References

1. H. B. G. Casimir, *Proc. K. Ned. Akad. Wet.* **51** 793 (1948).
2. M. Bordag, G. L. Klimchitskaya, U. Mohideen and V. M. Mostepanenko, *Advances in the Casimir Effect* (Oxford University Press, Oxford, 2009).
3. G. L. Klimchitskaya, U. Mohideen and V. M. Mostepanenko, *Rev. Mod. Phys.* **81**, N4 (2009); ArXiv:0902.4022.
4. M. Boström and B. E. Sernelius, *Phys. Rev. Lett.* **84**, 4757 (2000).
5. M. Bordag, B. Geyer, G. L. Klimchitskaya and V. M. Mostepanenko, *Phys. Rev. Lett.* **85**, 503 (2000).
6. C. Genet, A. Lambrecht and S. Reynaud, *Phys. Rev. A* **62**, 012110 (2000).
7. J. S. Høye, I. Brevik, J. B. Aarseth and K. A. Milton, *Phys. Rev. E* **67**, 056116 (2003); I. Brevik, J. B. Aarseth, J. S. Høye and K. A. Milton, *Phys. Rev. E* **71**, 056101 (2005).
8. J. S. Høye, I. Brevik, S. A. Ellingsen and J. B. Aarseth, *Phys. Rev. E* **75**, 051127 (2007).
9. V. B. Bezerra, G. L. Klimchitskaya and V. M. Mostepanenko, *Phys. Rev. A* **66**, 062112 (2002); G. L. Klimchitskaya, U. Mohideen and V. M. Mostepanenko, *J. Phys. A: Math. Theor.* **40**, (F)339 (2007).

10. V. B. Bezerra, G. L. Klimchitskaya, V. M. Mostepanenko and C. Romero, *Phys. Rev. A* **69**, 022119 (2004).
11. R. S. Decca, E. Fischbach, G. L. Klimchitskaya, D. E. Krause, D. López and V. M. Mostepanenko, *Phys. Rev. D* **68**, 116003 (2003); R. S. Decca, D. López, E. Fischbach, G. L. Klimchitskaya, D. E. Krause and V. M. Mostepanenko, *Ann. Phys. (N.Y.)* **318**, 37 (2005).
12. R. S. Decca, D. López, E. Fischbach, G. L. Klimchitskaya, D. E. Krause and V. M. Mostepanenko, *Phys. Rev. D* **75**, 077101 (2007); *Eur. Phys. J. C* **51**, 963 (2007).
13. F. Chen, G. L. Klimchitskaya, V. M. Mostepanenko and U. Mohideen, *Optics Express* **15**, 4823 (2007); *Phys. Rev. B* **76**, 035338 (2007).
14. J. M. Obrecht, R. J. Wild, M. Antezza, L. P. Pitaevskii, S. Stringari and E. A. Cornell, *Phys. Rev. Lett.* **98**, 063201 (2007).
15. G. L. Klimchitskaya and V. M. Mostepanenko, *J. Phys. A: Math. Theor.* **41**, (F)312002 (2008).
16. B. Geyer, G. L. Klimchitskaya and V. M. Mostepanenko, *Phys. Rev. D* **72**, 085009 (2005).
17. P. R. Buenzli and Ph. A. Martin, *Phys. Rev. A* **77**, 011114 (2008); B. Jancovici and L. Šamaj, *Europhys. Lett.* **72**, 35 (2005).
18. L. P. Pitaevskii, *Phys. Rev. Lett.* **101**, 163202 (2008).
19. D. A. R. Dalvit and S. K. Lamoreaux, *Phys. Rev. Lett.* **101**, 163203 (2008).
20. V. B. Svetovoy, *Phys. Rev. Lett.* **101**, 163603 (2008); **102**, (E)219903 (2009).
21. G. L. Klimchitskaya, U. Mohideen and V. M. Mostepanenko, *J. Phys. A: Math. Theor.* **41**, (F)432001 (2008).
22. V. M. Mostepanenko, R. S. Decca, E. Fischbach, B. Geyer, G. L. Klimchitskaya, D. E. Krause, D. López and U. Mohideen, *Int. J. Mod. Phys. A* **24**, 1721 (2009).
23. V. M. Mostepanenko, *J. Phys.: Conf. Series.* **161**, 012003 (2009).
24. G. L. Klimchitskaya, *J. Phys.: Conf. Series.* **161**, 012002 (2009).
25. B. Geyer, G. L. Klimchitskaya, U. Mohideen and V. M. Mostepanenko, *Phys. Rev. Lett.* **102**, 189301 (2009).
26. R. S. Decca, E. Fischbach, B. Geyer, G. L. Klimchitskaya, D. E. Krause, D. López, U. Mohideen and V. M. Mostepanenko, *Phys. Rev. Lett.* **102**, 189303 (2009).
27. L. P. Pitaevskii, *Phys. Rev. Lett.* **102**, 189302 (2009).
28. D. A. R. Dalvit and S. K. Lamoreaux, *Phys. Rev. Lett.* **102**, 189304 (2009).
29. D. A. R. Dalvit and S. K. Lamoreaux, *J. Phys.: Conf. Series.* **161**, 012009 (2009).
30. M. Aizenman and E. H. Lieb, *J. Stat. Phys.* **24**, 279 (1981).
31. D. Kondepugi and I. Prigogine, *Modern Thermodynamics* (John Wiley & Sons, New York, 1998).
32. Yu. B. Rumer and M. S. Ryvkin, *Thermodynamics, Statistical Physics and Kinetics* (Mir, Moscow, 1980).
33. N. W. Ashcroft and N. D. Mermin, *Solid State Physics* (Saunders College, Philadelphia, 1976).

34. I. Brevik, S. A. Ellingsen, J. S. Høye and K. A. Milton, *J. Phys. A: Math. Theor.* **41**, 164017 (2008).
35. G. L. Klimchitskaya and V. M. Mostepanenko, *Phys. Rev. E* **77**, 023101 (2008); J. S. Høye, I. Brevik, S. A. Ellingsen and J. B. Aarseth, *Phys. Rev. E* **77**, 023102 (2008).
36. V. M. Mostepanenko and B. Geyer, *J. Phys. A: Math. Theor.* **41**, 164014 (2008).
37. G. L. Klimchitskaya and B. Geyer, *J. Phys. A: Math. Theor.* **41**, 164032 (2008).
38. J. Schwinger, L. L. DeRaad and K. A. Milton, *Ann. Phys. (N.Y.)* **115**, 1 (1978).
39. G. L. Klimchitskaya and V. M. Mostepanenko, *Phys. Rev. B* **75**, 036101 (2007); B. E. Sernelius, *Phys. Rev. B* **75**, 036102 (2007).
40. M. Tomazawa and D.-W. Shin, *J. Non-Cryst. Solids* **241**, 140 (1998).
41. H. J. Schütt and E. Gerdes, *J. Non-Cryst. Solids* **144**, 14 (1992).
42. J. C. Dure, P. Maass, B. Roling and D. L. Sidebottom, *Rep. Progr. Phys.* **72**, 046501 (2009).
43. K. Seeger, *Semiconductor Physics* (Springer, New York, 1973).
44. N. F. Mott, *Phil. Mag.* **19**, 835 (1969).
45. N. F. Mott, *Metal-Insulator Transitions*, 2nd edition (Taylor and Francis, London, 1990).
46. W. J. Kim, M. Brown-Hayes, D. A. R. Dalvit, J. H. Brownell and R. Onofrio, *Phys. Rev. A* **78**, (R)020101 (2008).
47. R. S. Decca, E. Fischbach, G. L. Klimchitskaya, D. E. Krause, D. López, U. Mohideen and V. M. Mostepanenko, *Phys. Rev. A* **79**, 026101 (2009).
48. W. J. Kim, M. Brown-Hayes, D. A. R. Dalvit, J. H. Brownell and R. Onofrio, *Phys. Rev. A* **79**, 026102 (2009).
49. R. S. Decca, E. Fischbach, B. Geyer, G. L. Klimchitskaya, D. E. Krause, D. López, U. Mohideen and V. M. Mostepanenko, ArXiv:0904.4720.
50. G. Bimonte, *Phys. Rev. A* **79**, 042107 (2009).
51. F. Intravaia and C. Henkel, *Phys. Rev. Lett.* **103**, 130405 (2009).
52. B. E. Sernelius, *Europhys. Lett.* **87**, 14004 (2009); *Phys. Rev. A* **80**, 043828 (2009).

CASIMIR-LIFSHITZ FORCES AND ENTROPY

L. P. PITAEVSKII

CNR INFM-BEC, Department of Physics, University of Trento,
I-38100 Povo, Trento, Italy
Kapitza Institute for Physical Problems,
Kosygina 2, 119334 Moscow, Russia
E-mail:lev@science.unitn.it

It is shown that the violation of the positiveness of the entropy due to the Casimir-Lifshitz interaction claimed in several papers is an artifact related to an improper interpretation of the "Casimir entropy", which actually is a difference of two positive terms. It is explained that at definite condition this "Casimir entropy" must be negative. A direct derivation of the low temperature behavior of the "Casimir entropy" of a metallic surface in conditions of the anomalous skin effect is given and singular temperature dependency of this quantity is discussed. In conclusion a hydrodynamic example of the entropy of a liquid film is considered. It occurs that the entropy of a film of finite thickness and a liquid half-space behave differently at $T \to 0$.

1. Physical Meaning and Properties of the "Casimir Entropy"

Recently much attention has been devoted to the calculation of the entropy related to the Casimir-Lifshitz forces.[1–8] The common point of these papers is the calculation of the so-called "Casimir entropy", which is defined as

$$S_C(T,l) = -\int_l^\infty \left(\frac{\partial f}{\partial T}\right)_l dl \, , \tag{1}$$

where $f(T,l)$ is the force between bodies and l is the distance between bodies. (I am using units such that $k_B = \hbar = 1$. In my notation f is negative for attraction.) It was taken for granted that S_C is the proper definition of the entropy and must possess all properties of the entropy, particularly it must be positive. I will present here considerations that it is not so in general.

It is important to recognize that the Lifshitz theory only gives rigorously the quantity for which it was developed, i. e. the force $f(T,l)$ for $l \gg d$, where d is the interatomic distance.[9,10] It is impossible to calculate in this theory full free energy or entropy in terms of dielectric properties of bodies, because these quantities are mainly defined by short-distances interactions and are divergent in the Lifshitz approximation. According to thermodynamics

$$f = -\left(\frac{\partial F}{\partial l}\right)_T , \tag{2}$$

where F is the free energy. Differentiation with respect to T gives

$$\left(\frac{\partial S}{\partial l}\right)_T = \left(\frac{\partial f}{\partial T}\right)_l . \tag{3}$$

The derivative $(\partial S/\partial l)_T$ is the *only* information about entropy given by the Lifshitz theory. For *regular* bodies (quantum fluids and their mixtures, regular solids) Nernst's theorem demands that $S(T,l) \to 0$ as $T \to 0$. Thus, in this limit

$$\frac{\partial S}{\partial l} = \frac{\partial f}{\partial T} \to 0 . \tag{4}$$

This is the only consequence of the theorem we can derive. Notice that the limit $T \to 0$ must be taken at constant l. Integration gives

$$\Delta S(T,l,l_0) = S(T,l) - S(T,l_0) = \int_{l_0}^{l} \frac{\partial f}{\partial T} dl . \tag{5}$$

Of course, again $\Delta S \to 0$, when $T \to 0$. To my knowledge this condition and condition (4) were not violated in any *proper* calculations for regular bodies, i. e. in calculations where meaningful models were solved correctly. However, ΔS is a *difference* of two positive terms and it is impossible to say anything about its sign. There are also no restrictions on the sign of the specific heat difference which can be obtained from (5):

$$T\frac{\partial \Delta S}{\partial T} = C_V(T,l) - C_V(T,l_0) = T\int_{l_0}^{l} \frac{\partial^2 f}{\partial T^2} dl . \tag{6}$$

The "Casimir entropy" (1) can be obtained from (5) choosing $l_0 = \infty$:

$$S_C(T,l) = \int_{\infty}^{l} \frac{\partial f}{\partial T} dl = S(T,l) - S(T,l=\infty) . \tag{7}$$

In general, it is impossible to say anything about the sign of S_C. However, sometimes it is possible to connect the appearance of a negative S_C

with the temperature dependence of S_C and $\partial S_C/\partial l$ as $T \to 0$. The point is that the temperature dependence of $\partial S_C/\partial l$ is the same as the first term in S_C. However, the situation with the second term is different. The limit $l \to \infty$ results in the appearance of two free surfaces of the bodies. They are new physical objects and their thermodynamic properties can be different from the properties of a finite gap between bodies. If it happens in the limit $T \to 0$ that $|S_C| \gg |\partial S_C/\partial l|$, then the second term in (7) dominates and $S_C < 0$. In this case S_C must obviously be l-independent. On the contrary, the leading at $T \to 0$ l-dependent term is originated from the first term and must be positive at large l.

We can now review the situation in different cases. There is no problem for dielectrics. At low temperatures only the low frequency photons with $\hbar\omega \sim k_B T$ are excited. Because the imaginary part of the dielectric function $\varepsilon''(\omega)$ tends to 0 when $\omega \to 0$, these photons are well defined elementary excitations and their thermodynamic functions behave in a usual way at $T \to 0$.

The situation in conductors is more tricky. It was shown that for metals with impurities, where the conductivity $\sigma(T) \to const$ as $T \to 0$, both $\partial S_C/\partial l$ and S_C tend to zero by the same law.[2–4] Thus there is no any peculiarities again.

For pure metals, where the anomalous skin effect theory must be used, the situation is quite interesting. As it was shown by Svetovoy and Esquivel,[8] in this case $|\partial S_C/\partial l| \propto T$, while $|S_C| \propto T^{2/3}$ (see next section). Thus $|S_C| \gg |\partial S_C/\partial l|$. A calculation gives that $S_C < 0$ and does not depend on l, as must be the case. The leading l-dependent term is proportional to T and positive.

We see that there are no violations of the thermodynamic conditions in the Lifshitz theory of Casimir-Lifshitz forces for conductive media, if the problem is formulated and solved in a correct way.

For *disordered* glass-like media Nernst's theorem *is not valid.* They are not at an equilibrium state at low temperatures due to a very long relaxation time and have a large finite entropy at zero temperature. (See, for example, a recent paper[11] and references therein.) There is no reason why the difference (7) must in this case tend to zero at $T \to 0$. Thus general considerations cannot say anything about the properties of S_C and I will not discuss this case anymore. Notice only that the specific heat difference (6) must tend to zero also for disordered bodies. However, this condition is satisfied in a trivial way in all known examples due to the factor T in the left hand side of the equation.

2. The Surface Entropy of a Metal in Condition of Anomalous Skin Effect

As I noted previously, the presence of the l-independent negative term in the "Casimir entropy" of a metal in conditions of anomalous skin effect implies that the free surface of this metal gives the main contribution to the entropy. It is instructive to examine this mechanism more closely by mens of a direct calculation of the entropy for a metallic half-space. For this problem, the most convenient method is, in my opinion, the method of calculation of the Casimir-Lifshitz contribution to the free energy, which was suggested by Barash and Ginzburg in 1975.[12] This powerful method, unfortunately, did not attract the attention which it surely deserves. (See, however, an interesting example of its use in Ref.[13]) The method is analogous to the one presented in Ref.[10] Differently from the latter, however, it does not demand the calculation of the full Green's function of electromagnetic field, but it only requires a dispersion relation for electromagnetic waves. The dispersion relation can be presented in the form

$$D(q,\omega) = 0. \tag{8}$$

Then the Ginzburg-Barash equation[12] for the contribution of the Van der Waals interactions to the free energy can be written as:

$$\begin{aligned} F(T) &= T\sum_{n=0}^{\infty}{}' \int d^2q \log D(q, i\zeta_n) \\ &= E(0) + \frac{1}{2\pi i}\mathrm{P}\int_{-\infty}^{\infty} \frac{d\omega}{e^{\omega/T}-1}\int d^2q \log D(q,\omega) \\ &= E(0) + \frac{T}{2\pi i}\mathrm{P}\int_{-\infty}^{\infty} \frac{d\xi}{e^{\xi}-1}\int d^2q \log D(q,T\xi), \end{aligned} \tag{9}$$

where q is the wave vector in the plane of the surface and $\zeta_n = 2\pi T n$. Sometimes it is convenient to use the energy E instead of the free energy. Taking into account that $E = F - T(\partial F/\partial T)$, one gets the following contribution to the energy:

$$E(T) = E(0) - \frac{T^2}{2\pi i}\int_{-\infty}^{\infty} \frac{\xi d\xi}{e^{\xi}-1}\int d^2q \frac{D'(q,T\xi)}{D(q,T\xi)}\,, \tag{10}$$

where

$$D' = (\partial D/\partial \omega)\,. \tag{11}$$

In the situation of the anomalous skin effect the dispersion relation must be expressed in terms of the surface impedance $Z(q,\omega)$:

$$D(q,\omega) \propto \sqrt{\omega^2 - q^2/c^2}\,Z(q,\omega) - i\omega/c = 0\,. \tag{12}$$

The impedance for waves of s-polarization has the form

$$Z(q,\omega) = (i\omega/qc)F(\omega\omega_p^2/q^3c^2v_F) \tag{13}$$

where ω_p is the plasma frequency and v_F is the Fermi velocity of electrons in the metal. The function F has been calculated in.[6] However, the goal of this section is to calculate only the temperature dependence of the l-independent term in S_C. Then we need not an explicit expression for F.

Notice first of all that in the calculation of the low temperature behavior of the entropy the values $\omega \sim T/\hbar \to 0$ are always important. Further, it is obvious from (18) and (13) that the values $q \propto \omega^{1/3} \gg \omega$ are also important. Then we can neglect ω in comparison with qc and $D \approx -(\omega/c)F(\omega\omega_p^2/q^3c^2v_F)$. The factor (ω/c) can be omitted and, finally, we find the estimate

$$E(T) - E(0) \propto T^2 \int_{-\infty}^{\infty} \frac{\xi d\xi}{e^\xi - 1} \int \frac{d^2q}{q^3} \Phi\left(\frac{q}{(T\xi)^{1/3}}\right) \propto T^{5/3}\,. \tag{14}$$

This estimate assumes that the integral converges. This can be easily checked. From (14) we finally find that

$$S \propto T^{2/3}\,. \tag{15}$$

It is worth noticing that we succeeded to calculate the leading term in the full entropy of the body only because of its singular behavior with respect to T. The singularity is produced by the contribution of small $q \propto T^{1/3}$ and the corresponding integrals converge. Notice also that the temperature dependence (15) corresponds to the contribution to the entropy of surface waves with dispersion law $\omega = Aq^3$. Of course, under the considered conditions there are no undamped surface waves propagating along the surface. However, the evanescent waves give in this case the same type of contribution.

3. Entropy of a Helium Film on a Solid Surface

In the previous section we have shown that in some conditions the temperature dependence of the entropy for a finite gap between bodies can be different from one of a free surface. This result might look strange. However, in this section we will show that the same situation takes place in a quite simple system - a superfluid film on a surface of a solid body, where the main contribution to the entropy is due to quantized surface waves, so called ripplons.[15]

The surface waves on a surface of superfluid liquid are well-defined elementary excitations and their contribution to the free energy can be calculated in the usual way from the knowledge of the excitations energy spectrum $\omega = \omega(q)$:

$$F = T \int \log\left(1 - e^{-\omega(q)/T}\right) \frac{qdq}{2\pi}$$
$$= -\frac{1}{4\pi} \int \frac{q^2 d\omega}{e^{\omega/T} - 1} \,. \tag{16}$$

Neglecting the effects of the gravity and of the Van der Waals interaction with the substrate, the waves on the surface of a liquid with thickness l has the following dispersion law (see Problem 1 in §62, of Ref. 14):

$$\omega^2(q) = \frac{\alpha q^3}{\rho} \tanh(ql) \,, \tag{17}$$

where α is the surface tension of the liquid at $T = 0$ and ρ is its density. When $l \to \infty$ we obtain the dispersion law for the free surface of a liquid half-space:

$$\omega^2(q) = (\alpha q^3)/\rho \,. \tag{18}$$

Substitution (18) into (16) and integration give (in usual units):[15]

$$F = -\Gamma(7/3)\zeta(7/3)\frac{(k_B T)^{7/3}\rho^{2/3}}{4\pi\hbar^{4/3}\alpha^{2/3}} \,. \tag{19}$$

Correspondingly the entropy behaves as $S \propto T^{4/3}$. The values $q \sim (\rho/\alpha)^{1/3}(k_B T/\hbar)^{2/3}$ give the main contribution to the entropy and the result (19) is valid under the condition $k_B T \gg \hbar\alpha^{1/2}/(l^{3/2}\rho^{1/2})$.

In the opposite limit $ql \ll 1$ the dispersion law is:

$$\omega^2(q) = (\alpha l q^4)/\rho \tag{20}$$

and

$$F = -\frac{\pi}{24}\frac{(k_B T)^2\rho^{1/2}}{\hbar\alpha^{1/2}l^{1/2}} \,. \tag{21}$$

Correspondingly $S \propto T$. Again we find that the temperature dependence of the entropy is different for a free surface and a film of finite thickness. In the hydrodynamic example of surface waves the situation is opposite to the one of the Casimir-Lifshitz forces in section 2: here the film of finite thickness has larger entropy.

Acknowledgments

I thank V. Svetovoy for clarification of important points of previous papers and Yu. Barash, F. Dalfovo, C. Henkel, F. Intravaia and E. Taylor for discussions. I acknowledge financial support by the European Science Foundation (ESF) within the activity "New Trends and Applications of the Casimir Effect" (www.casimir-network.com).

References

1. V. B. Bezerra, G. L. Klimchitskaya, and V. M. Mostepanenko, *Phys. Rev. A* **65** , 052113 (2002).
2. J. S. Høye, I.Brevik, J. B. Aarseth, and K. A. Milton, *Phys. Rev. E* **67**, 056116 (2003).
3. M. Böstrom and Bo E. Sernelius, *Physica A* **339**, 73 (2004).
4. V. B. Bezerra, G. L. Klimchitskaya, V. M. Mostepanenko, and C. Romero, *Phys. Rev. A* **69**, 022119 (2004).
5. Bo E. Sernelius, *Phys. Rev. B* **71**, 235114 (2005).
6. V.B. Svetovoy and R. Esquivel, Phys. *Rev. E* **72**, 036113 (2005).
7. V. M. Mostepanenko et al., *J. Phys. A: Math. Gen.* **39**, 6589 (2006).
8. V. B. Svetovoy and R. Esquivel, *J. Phys. A: Math. Gen.* **39**, 6777 (2006).
9. E. M. Lifshitz, *Sov. Phys. JETP* **2**, 73 (1956).
10. I. E. Dzyaloshinskii, E. M. Lifshitz, and L. P. Pitaevskii, *Advances in Physics* **10**, 165 (1961).
11. S. A. Langer and J. P. Sethna, *Phys. Rev. Lett.* **61**, 570 (1998).
12. Yu. S. Barash and V. L. Ginzburg, *Sov. Phys. Usp.* **18**, 305 (1975).
13. I. Brevik and E. Elizalde, *Phys. Rev. D* **49**, 5319 (1994).
14. L. D. Landau and E. M. Lifshitz, *Fluid Mechanics* (Elsevier, Oxford, 2004).
15. K. R. Atkins, *Can. J. Phys.* **31**, 1165 (1953).

THE THERMAL CASIMIR EFFECT: SATURATION

BO E. SERNELIUS

Division of Theory and Modeling, Department of Physics, Chemistry and Biology, Linköping University, SE-581 83 Linköping, Sweden
E-mail: bos@ifm.liu.se

This article addresses the discrepancies between theoretical and experimental results obtained for the thermal Casimir effect. Here we test the possibility that saturation effects may be the root of the problems. We present graphs that describe the numerical derivations in great detail.

Keywords: Casimir; thermal effects; dissipation; saturation; massless bosons.

1. Introduction

We realized[1] that the correct way to treat the materials in the Casimir experiments is to include finite temperature, finite conductivity and dissipation, all at the same time. Up till then one had started from perfect metals and added corrections, one by one, for the effects just mentioned. Very soon it became evident that this more realistic treatment of the materials led to worse agreement with the experimental results. This divided the Casimir community into two groups: one in favor of the more realistic treatment based on known material properties; one in favor of a theory that agrees better with experiment but, for this to be the case, has to invoke various prescriptions — a phenomenological approach. The general situation is one of great frustration. The consensus in the Casimir community at present is that the correct approach is to use the proper material properties and that both experimentalists and theorists should do their best in finding the root of the problem.

In the present work we test if the cause of all the problems can be saturation effects. The electromagnetic normal modes involved in the Casimir effect are assumed to be independent massless bosons. However, these bosons are formed from electron excitations and the electrons are fermions. This means that the modes are not completely independent. Saturation effects are expected to appear in case a huge number of modes are excited. We test this idea on several different Casimir experiments where theory and experiment disagree. First out is the classical Casimir force measurement between two metal half spaces; here both in the form of the torsion pendulum experiment by Lamoreaux[2] and in the form of the Casimir pressure measurement between a gold sphere and a gold plate as performed by Decca et al.;[3] theory predicts a large negative thermal correction, absent in the high precision experiments. The third experiment is the measurement of the Casimir force between a metal plate and a laser irradiated semiconductor membrane as performed by Chen et al.;[4] the change in force with laser intensity is larger than predicted by theory. The fourth experiment is the measurement of the Casimir force between an atom and a wall in the form of the measurement by Obrecht et al.[5] of the change in oscillation frequency of a ^{87}Rb Bose-Einstein condensate (BEC) trapped to a fused silica wall; the change is smaller than predicted by theory.

The main results from the work is to be presented elsewhere.[6,7] To avoid too much overlap we here make a much briefer description of the formal derivation and expand more on where the cause of the deviations are to be found and what the effects are from our saturation corrections. In Sec. 2 we summarize the formalism; in Sec. 3 we discuss the experiments by Lamoreaux and Decca; Sec. 4 is devoted to the experiment by Chen et al.; in Sec. 5 we discuss the experiment by Obrecht et al.; Sec. 6 contains a brief summary.

2. Basic Formalism

The theory corresponding to all experiments treated here are based on the two plate geometry. Even the Casimir force between an atom and a wall can be obtained from the results of this geometry; one takes the limit when the thickness of one of the plates goes to zero and at the same time lets the material of the thin plate be diluted. In the two plate geometry the interaction energy per unit area, $V(d)$, can at zero temperature be written on the form[8]

$$V(d) = \frac{\hbar}{\Omega} \sum_{\mathbf{k}} \int_0^\infty \frac{d\omega}{2\pi} \ln\left[f(\mathbf{k}, i\omega)\right], \tag{1}$$

where d is the distance between the plates, $\mathbf{k}$ is the two-dimensional wave vector in the plane of the plates, Ω is the area of a plate, and $f(\mathbf{k}, \omega) = 0$ is the condition for an electromagnetic normal mode in the particular geometry. This expression one arrives at using the extended argument principle[8]

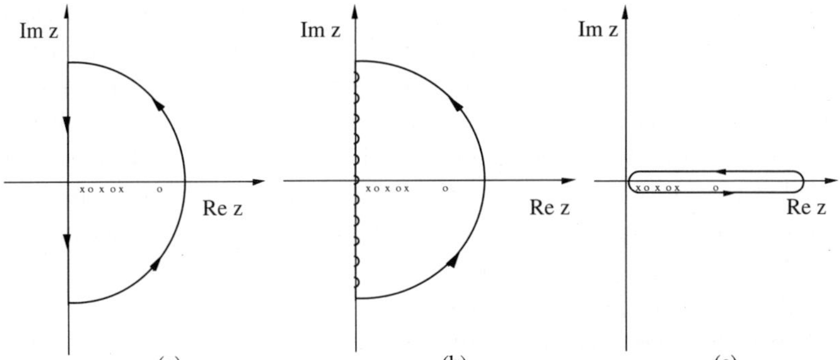

Fig. 1. Integration contours in the complex frequency plane used in connection with the extended argument principle: (a) contour used for zero temperature; (b) contour used for finite temperature; (c) contour that can be used both at zero and finite temperature.

with the contour shown in Fig. 1(a). At finite temperature the integration is replaced by a discrete summation over Matsubara frequencies,

$$V(d) = \frac{1}{\beta\Omega}\sum_{\mathbf{k}}\sum_{\omega_n}{}' \ln\left[f(\mathbf{k}, i\omega_n)\right]; \; \omega_n = \frac{2\pi n}{\hbar\beta}. \tag{2}$$

This expression one arrives at using the contour shown in Fig. 1(b). Alternatively one may use the contour shown in Fig. 1(c) and integrate along the real frequency axis,

$$V(d) = \frac{2\hbar}{\Omega}\sum_{\mathbf{k}} \mathrm{Im}\int_0^\infty \frac{d\omega}{2\pi}\left[n(\omega) + 1/2\right]\ln\left[f(\mathbf{k}, \omega)\right], \tag{3}$$

where $n(\omega) = \left[\exp(\hbar\beta\omega) - 1\right]^{-1}$ is the distribution function for massless bosons. This form can also be used at zero temperature; then the distribution function vanishes. Problems encountered using the real frequency integration is discussed in detail elsewhere.[9] In the two plate geometry there are two groups of normal mode, transverse magnetic (TM) and transverse electric (TE), each with a different mode condition function. The interaction potential is a sum of two terms, $V(d) = V^{TM}(d) + V^{TE}(d)$.

The distribution function in Eq. (3) diverges for zero frequency and it is the low energy modes that cause the problems in all experiments treated here. We shift the distribution function downwards in frequency, so that it never reaches the point of divergence, by adding a damping parameter, D,

$$\tilde{n}(\omega) = \left[\exp(\hbar\beta\omega + D) - 1\right]^{-1}. \tag{4}$$

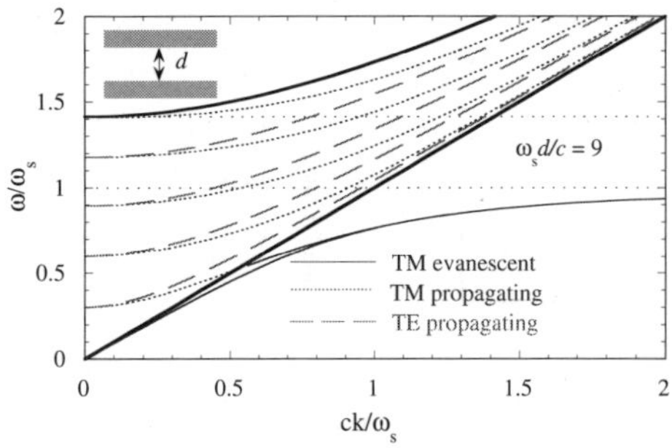

Fig. 2. Dispersion curves for the modes between two gold plates in absence of dissipation. The frequencies are in units of ω_s, the surface plasmon frequency. The solid straight line is the light dispersion curve in vacuum; the dashed (dotted) curves are TE (TM) propagating modes; the thin solid curves are evanescent TM modes; the thick solid curve is the lower boundary for transverse bulk modes in the plates. From Ref. [10].

The discrete frequency summation in Eq. (2) is the result of the poles of the distribution function that all fall on the imaginary axis.[8] Our new distribution function has its poles shifted away from the axis the distance $D/\hbar\beta$ into the left half plane. The new form of the interaction potential is

$$V(d) = \frac{1}{\beta\Omega}\sum_{\mathbf{k}}\sum_{\omega_n}{}' \frac{1}{\pi}\int_{-\infty}^{\infty} d\omega' \frac{(D/\beta)\ln\left[f(k, i\omega')\right]}{(\omega' - \omega_n)^2 + (D/\beta)^2}. \tag{5}$$

Each term in the original summation is replaced by an integral. For small D values it is enough to replace only the zero frequency term.

3. Two Parallel Metal Plates

This geometry is applicable to the torsion pendulum experiment by Lamoreaux[2] and the Casimir pressure measurement performed by Decca et al.[3] The dispersion curves for the electromagnetic normal modes for two gold plates[10] are shown in Fig. 2. This figure is valid in neglect of dissipation in the plate materials. The modes are propagating (evanescent) above and to the left (below and to the right) of the light dispersion curve. The light dispersion curve is the straight diagonal line in the figure; it has slope unity with the chosen scaling of the axes. Note that there are no TE evanescent modes. When the system is allowed to have dissipation there are modes everywhere. Each original mode is replaced by a continuum of modes.[11] Evanescent TE modes appear and the continuum extends all the way down to the momentum axis. These modes are the cause of all the problems with the thermal Casimir force in this geometry. The integrand of Eq. (1) and

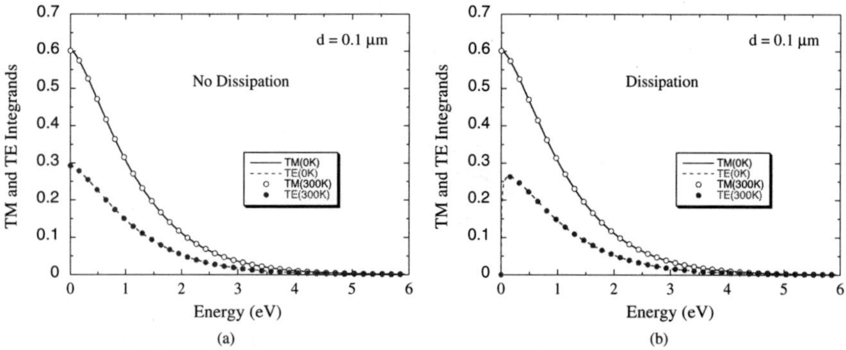

Fig. 3. Integrands of Eq. (1) (curves) and summands of Eq. (2) (circles) in the case of two gold plates separated by $0.1\mu m$.

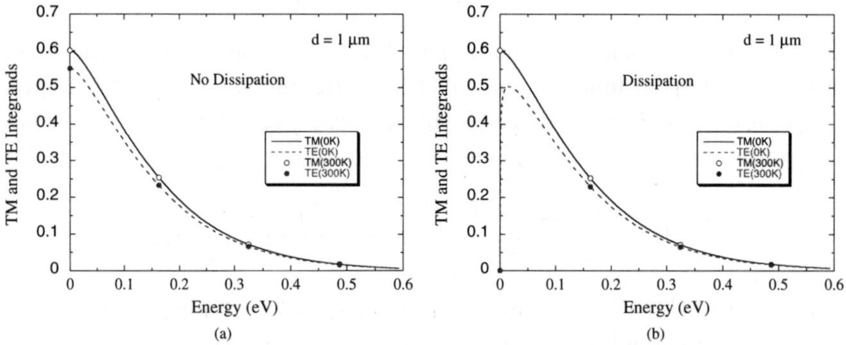

Fig. 4. Same as Fig. 3 but now for $1.0\mu m$.

summand of Eq. (2) in the case of two gold plates separated by $0.1\mu m$ are displayed in Fig. 3. Fig. 3(a) is the result when dissipation is neglected and Fig. 3(b) when dissipation is included. The solid (dashed) curves is the TM (TE) integrand; the open (filled) circles is the TM (TE) summand. We note that the effect of dissipation is to push down the TE integrand toward zero for small energies or frequencies. In Eq. (2) the n = 0 term vanishes for the TE contribution. We further note that for this separation the circles come closely spaced and the summation and integration give very similar results — there are very small temperature effects; the integrand is pushed down very close to the vertical axis — the dissipation has a very small effect on the zero temperature results.

In Fig. 4 we show the results at $1.0\mu m$ separation. Here we note that summation points are much fever and hence the temperature effect is much

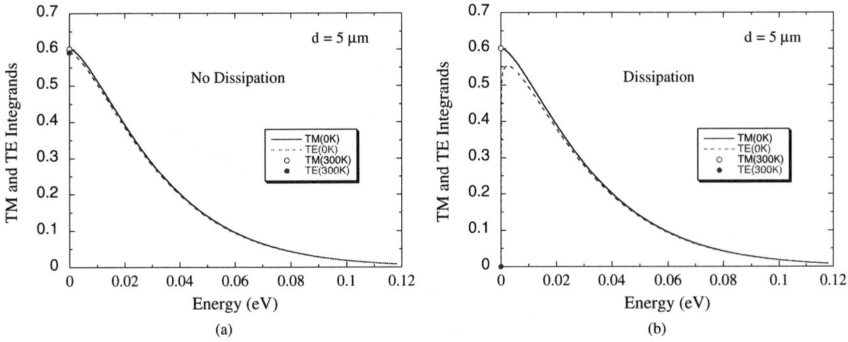

Fig. 5. Same as Fig. 3 but now for $5\mu m$.

larger; the dissipation has still negligible effect on the zero temperature result; on the room temperature result, on the other hand, it has a huge effect — the TE contribution has been reduced by a factor of two. For $5\mu m$ separation, Fig. 5, we note that only one summation point contributes; the dissipation has the effect that at room temperature there is nolonger any TE contribution at all.

In Fig. 6(a) we show the energy correction factor for the two-gold-plate geometry. The energy correction factor is the Casimir energy per unit area divided by the zero temperature Casimir energy per unit area for two perfectly reflecting metal plates, $-\hbar c\pi^2/(720d^3)$. The experimental results[2] obtained by Lamoreaux are shown as squares with error bars. There is a cluster of experimental points near $1\mu m$ separation. Here the deviation between theory and experiment is over 20%. Theory and experiment are in clear disagreement. The experimental result[3] for the normalized Casimir pressure at 295 K is shown as dots in Fig. 6(b). The bars are the endpoints of the experimental error bars. The upper (lower) thick solid curve is the theoretical result for zero temperature (295 K) calculated with Eqs. (1) and (2), respectively. The dielectric function on the imaginary frequency axis was derived from experimental tabulated optical data for gold. We note that the zero temperature result agrees much better with the experimental result. The large negative thermal correction comes entirely from the TE evanescent modes. All curves are normalized with the zero temperature Casimir pressure between two perfectly reflecting metal plates, $\hbar c\pi^2/(240d^4)$. The circles are the results for different damping parameters from using Eq. (3) with the modified distribution function of Eq. (4) in the contribution from the TE evanescent waves. To each set of circles

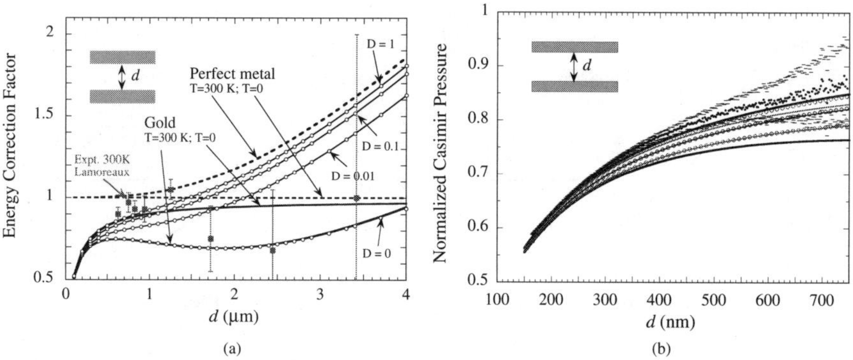

Fig. 6. (a): Energy correction factor for two gold plates. The filled squares with error bars are the Lamoreaux' experimental[2] values from the torsion pendulum experiment. The dashed curves are the perfect metal results. The thick solid curves are the results for real gold plates using Eq. (1) for zero temperature and Eq. (2) for room temperature. The curves with circles are the results from our model calculations with different saturation parameters; (b): Casimir pressure between two gold plates. The experimental result[3] is shown as dots and the endpoints of the error bars are indicated by horizontal bars; the upper (lower) thick solid curve is the traditional theoretical zero (room) temperature result; the circles are the present results, obtained from Eq. (3) with the distribution function for TE evanescent waves modified according to Eq. (4), with damping parameters 0.01, 0.1, and 1.0, respectively, counting from below; the corresponding results obtained by shifting the zero frequency pole into the left of the complex frequency plane are shown as thin solid curves.

corresponds a thin solid curve. This curve is the result of using Eq. (2), where just the zero frequency pole has been moved into the left half of the complex frequency plane and the corresponding term in the summation has been modified according to Eq. (5). We note that for the two lowest set of curves with small damping the two results agree. For very high damping there are deviations. These deviations have two reasons: One is that in the thin solid curves all mode types are affected by the damping; the second is that for strong enough damping more terms in the summation should be modified. In Fig. 7(a) we show the thermal correction to the Casimir energy. The contribution from all four mode types are given separately. These curves have been obtained from Eq. (3) and the Drude dielectric function suited for gold has been used. We note that the TE evanescent waves give a negative contribution. For larger separations the total TE contribution saturates when it has completely eliminated the zero temperature TE contribution. Fig. 7(b) shows how the integrand for the TE evanescent modes is affected by the choice of damping parameter.

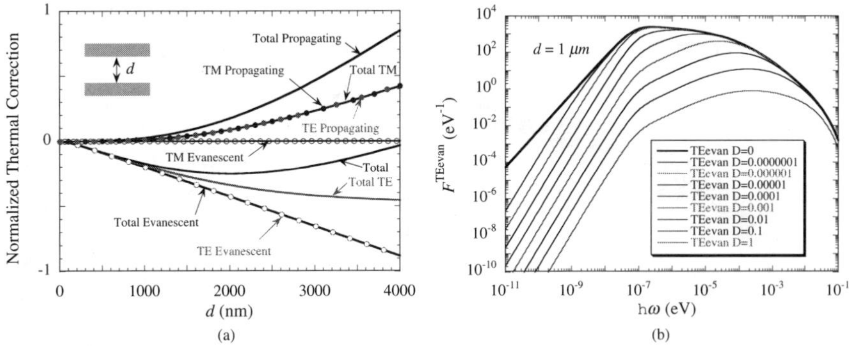

Fig. 7. (a): The thermal contribution to the Casimir energy from the four mode types between two gold plates. The results are also lumped together in various ways. All results are divided by the zero temperature Casimir energy for perfectly reflecting plates; (b): The integrand for the TE evanescent modes for different choices of damping parameter.

4. Two Parallel Plates, One Metallic and One Semiconducting

The third experiment we consider here is the measurement of the Casimir force between a gold plate and a laser irradiated semiconductor membrane as performed by Chen et al.[4] They measured the change in force with the laser irradiation compared to without any irradiation. The idea is to find out how the force varies with carrier concentration in the semiconducting membrane. The results are shown in Fig. 8(a). The open squares with error bars are the experimental result. The dashed curve with open circles is the theoretical result for 300 K. The deviations are clear. In this geometry it is not enough to neglect dissipation to get agreement with experiment; the theoretical results with and without dissipation are very similar. Besides, it is now the TM modes that cause the problems. The solid curve with filled (open) circles is our saturation based result with D equal to 0.01 (0.1). We find that both these curves agree with the experiment within the experimental uncertainty. Here we have used Eq. (2) and just modified the zero frequency contribution according to Eq. (5).

5. Atom Wall Geometry

In the experiment by Obrecht et al.[5] one studied indirectly the force between a rubidium atom and a dielectric substrate. This was done by measuring the collective oscillation frequency of the mechanical dipole mode of a BEC near enough to a dielectric substrate for the force to measurably

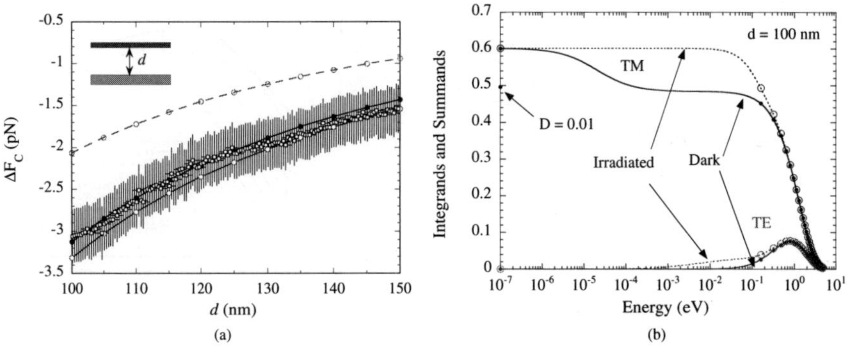

Fig. 8. (a): The change in Casimir force, at 300 K, between a gold sphere and a silicon membrane with and without laser irradiation. The open squares with error bars are the experimental[4] result. The dashed curve with open circles is the theoretical result without saturation effects. The solid curve with filled (open) circles is our present result with D equal to 0.01(0.1); (b): The TM and TE integrands and summands with (dotted) and without (solid) laser irradiation at $100nm$ separation. The arrow to the left shows how the dark sample $n = 0$ term is shifted down at the damping parameter 0.01.

distort the trapping potential. The fractional change in trap frequency is defined as $\gamma_x \equiv (\omega_0 - \omega_x)/\omega_0$ in terms of the unperturbed trap frequency, ω_0, and ω_x, the trap frequency perturbed by the force. In Fig. 9(a) the experimental result[5] is shown as open squares with error bars. The upper (lower) curve is the theoretical result, without saturation, including (neglecting) the conductivity from the few thermal carriers in the silica wall. We see that also here the neglect of the contribution, to the dielectric function of the silica wall, from the very few thermally excited carriers brings the theoretical result into agreement with experiment. In this geometry, just as in the gold-plate silicon-wafer geometry, the TM modes cause the problems and it is not enough to neglect dissipation to get good agreement between theory and experiment. To include saturation effects we have just modified the zero frequency contribution in analogy with Eq. (5). We note that in this experiment it is enough to have a damping parameter as small as 10^{-10} to bring the theoretical result into agreement with experiment.

6. Summary

In summary we have proposed that saturation effects are responsible for the discrepancy between theory and experiment in several quite different Casimir geometries. We have demonstrated that the problems may go away in all cases.

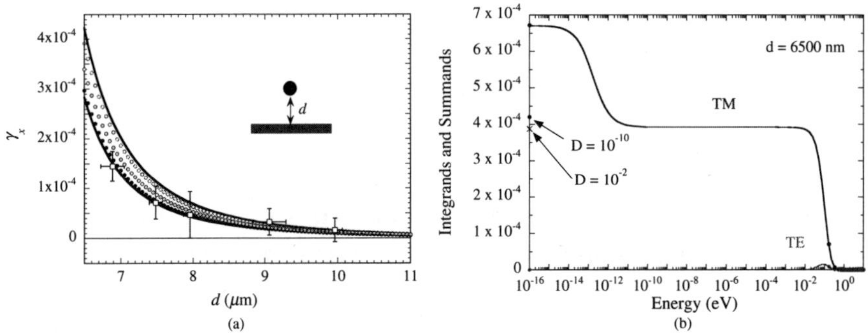

Fig. 9. (a): Fractional change in trap frequency for a Rb atom near a silica wall versus separation. The open squares are the experimental result.[5] The upper (lower) curve is the theoretical result including (neglecting) the conductivity from the few thermal carriers in the silica wall. The circles are our present results for the D values 10^{-10}, 10^{-11}, and 10^{-12}, respectively, counted from below; (b): The TM and TE integrands and summands at 6500nm separation. The arrows to the left shows how the $n = 0$ term is shifted down for the damping parameter 10^{-10} and 10^{-2}, respectively.

Acknowledgments

The research was sponsored by the VR-contract No:70529001 and support from the VR Linné Centre LiLi-NFM and from CTS is gratefully acknowledged.

References

1. M. Boström and Bo E. Sernelius, Phys. Rev. Lett **84**, 4757 (2000).
2. S. K. Lamoreaux, Phys. Rev. Lett. **78**, 5 (1997).
3. R. S. Decca, D. Lopéz, E. Fischbach, G. L. Klimchitskaya, D. E. Krause, and V. M. Mostepanenko, Phys. Rev. D **75**, 077101 (2007).
4. F. Chen, G. L. Klimchitskaya, V. M. Mostepanenko, and U. Mohideen, Phys. Rev. B **76**, 035338 (2007).
5. J. M. Obrecht, R. J. Wild, M. Antezza, L. P. Pitaevskii, S. Stringari, and E. A. Cornell, Phys. Rev. Lett. **98**, 063201 (2007).
6. Bo E. Sernelius, EPL, **87** 14004 (2009).
7. Bo E. Sernelius, Phys. Rev. A, **80**, 043828 (2009).
8. Bo E. Sernelius, *Surface Modes in Physics* (Wiley-VCH, Berlin, 2001).
9. S. A. Ellingsen, Phys. Rev. E **78**, 021120 (2008).
10. Bo E. Sernelius, Phys. Rev. B **71**, 235114 (2005).
11. Bo E. Sernelius, Phys. Rev. B **74**, 233103 (2006).

FINITE TEMPERATURE CASIMIR EFFECT IN THE PRESENCE OF EXTRA DIMENSIONS

L. P. TEO

Faculty of Engineering, University of Nottingham Malaysia Campus, Semenyih, 43500, Selangor, Malaysia
E-mail: LeePeng.Teo@nottingham.edu.my

K. KIRSTEN

Department of Mathematics, Baylor University, Waco, Texas 76798-7328, USA
E-mail: klaus_kirsten@baylor.edu

We consider the finite temperature Casimir force acting on two parallel plates in a closed cylinder with the same cross section of arbitrary shape in the presence of extra dimensions. Dirichlet boundary conditions are imposed on one plate and fractional Neumann conditions with order between zero (Dirichlet) and one (Neumann) are imposed on the other plate. Formulas for the Casimir force show that it is always attractive for Dirichlet boundary conditions, and is always repulsive when the fractional order is larger than 1/2. For some fractional orders less than 1/2, the Casimir force can be either attractive or repulsive depending on the size of the internal manifold and temperature.

Keywords: Finite temperature Casimir effect, Extra dimensions, Fractional Neumann conditions.

1. Introduction

Piston configurations have received considerable interest because they allow for an unambiguous prediction of Casimir forces.[1] The force depends on the boundary conditions imposed, the cross section of the piston and on properties of additional Kaluza-Klein dimensions that might be present.[2–7] In this contribution we consider the same geometrical setting as in refs.,[4,5] but with fractional boundary conditions imposed. This is an extension of ref.[6] to the case where additional dimensions are considered.

2. Finite temperature Casimir effect in spacetimes with extra dimensions

Consider a massive scalar field $\varphi(x)$ of mass m in the background spacetime $M \times \mathcal{N}$, where M is the $(d+1)$-dimensional Minkowski spacetime and $\mathcal{N}$ is a compact n-dimensional internal manifold. We want to find the Casimir force acting between two parallel plates with arbitrary shape in a cylinder of constant simply connected cross section Ω. To interpolate between Dirichlet and Neumann boundary conditions, we consider fractional Neumann conditions

$$\left.\frac{\partial^{\eta_1}}{\partial x^{1\eta_1}}\varphi(x)\right|_{x^1=0}=0, \qquad \left.\frac{\partial^{\eta_2}}{\partial x^{1\eta_2}}\varphi(x)\right|_{x^1=a}=0, \qquad \eta_1,\eta_2\in[0,1],$$

on the plates. It can be shown that the Casimir force only depends on the difference $\eta_1-\eta_2$. Therefore without loss of generality, we fix Dirichlet boundary conditions on the left plate and impose fractional Neumann conditions of order η on the right plate. For concreteness, on the walls of the cylinder Dirichlet boundary conditions are imposed.

Using the piston approach,[1] we first consider the Casimir force acting on a piston moving freely inside a closed cylinder $[0,L]\times\Omega\times\mathcal{N}$, and let the right end of the cylinder go to infinity, i.e., $L\to\infty$. Let $x^1=a$ denotes the position of the piston. The finite temperature Casimir force acting on the piston is given by

$$F_{\rm Cas}^{\rm piston}(a)=-\frac{\partial}{\partial a}\left\{E_{\rm Cas}^{\rm cylinder}(a)+E_{\rm Cas}^{\rm cylinder}(L-a)\right\}, \tag{1}$$

where $E_{\rm Cas}^{\rm cylinder}(a)$ is the finite temperature Casimir energy inside the closed cylinder $[0,a]\times\Omega\times\mathcal{N}$; formally

$$E_{\rm Cas}^{\rm cylinder}(a)=\sum_{k=-\infty}^{\infty}\sum_{j=1}^{\infty}\sum_{l=0}^{\infty}\left\{\frac{1}{2}\omega_{k,j,l}+T\log\left(1-e^{-\omega_{k,j,l}/T}\right)\right\}. \tag{2}$$

The eigenfrequencies of the field $\varphi(x)$ which satisfies Dirichlet boundary conditions on the plate at $x^1=0$ and the wall $[0,a]\times\partial\Omega\times\mathcal{N}$ of the cylinder, and fractional Neumann conditions of order η on the plate at $x^1=a$, are given by

$$\omega_{k,j,l}=\sqrt{\left(\frac{\pi\left(k-\frac{\eta}{2}\right)}{a}\right)^2+\omega_{\Omega,j}^2+\omega_{\mathcal{N},l}^2+m^2}=\sqrt{\left(\frac{\pi\left(k-\frac{\eta}{2}\right)}{a}\right)^2+m_{j,l}^2}.$$

Here, $\omega_{\Omega,j}^2$, $j=1,2,\ldots$, are the eigenvalues for the Laplace operator with Dirichlet boundary conditions on Ω, $\omega_{\mathcal{N},l}^2$, $l=0,1,2,\ldots$, are the eigenvalues of the Laplace operator on $\mathcal{N}^n$, with $\omega_{\mathcal{N},0}^2 = 0$. For generic $\eta \in (0,1)$, $k \in \mathbb{Z}$. For $\eta = 0$ or 1, $k \in \mathbb{N}$. In the following we shall assume $\eta \in (0,1)$; for $\eta = 0$ and $\eta = 1$ the Casimir force will have to be divided by two.

It is worth mentioning that if one considers a cylinder of length $2a$ instead of a and if one imposes twisted boundary conditions

$$\varphi(x^1 = 2a) = e^{-i\pi\eta}\varphi(x^1 = 0), \quad \varphi'(x^1 = 2a) = e^{-i\pi\eta}\varphi'(x^1 = 0)$$

for a complex massive scalar field $\varphi(x)$, one obtains the same set of eigenfrequencies.

Using the zeta regularization method to compute the Casimir energy (2), we find that

$$\begin{aligned} E_{\text{Cas}}^{\text{cylinder}}(a) &= \Lambda_0 + a\Lambda_1 \\ &- T\text{Re} \sum_{k=1}^{\infty} \sum_{j,l} \sum_{p=-\infty}^{\infty} \frac{e^{i\pi k\eta}}{k} \exp\left(-2ka\sqrt{m_{j,l}^2 + (2\pi pT)^2}\right), \end{aligned}$$

where Λ_0 and Λ_1 are terms that are independent of a. $\Lambda_0 + a\Lambda_1$ does not contribute to the Casimir force (1) acting on the piston, and we find by taking the limit $L \to \infty$ in (1) that the Casimir force acting between two parallel plates is

$$F_{\text{Cas}}^{\parallel}(a) = -2T\text{Re} \sum_{j,l} \sum_{p=-\infty}^{\infty} \frac{e^{i\pi\eta}\sqrt{m_{j,l}^2 + (2\pi pT)^2}}{\exp\left(2a\sqrt{m_{j,l}^2 + (2\pi pT)^2}\right) - e^{i\pi\eta}}. \tag{3}$$

This expression shows that in the high temperature limit, the leading term of the Casimir force is linear in T given by the sum of the terms with $p = 0$, which is usually called the classical term. For the low temperature behavior, some computations give (we use the notation $M_{k,p}^2 = k^2a^2 + (p/(2T))^2$)

$$\begin{aligned} F_{\text{Cas}}^{\parallel}(a) = &-\frac{1}{\pi} \sum_{k=1}^{\infty} \sum_{j,l} \cos(\pi k\eta) \left\{ \frac{m_{j,l}}{ka} K_1(2kam_{j,l}) + 2m_{j,l}^2 K_0(2kam_{j,l}) \right\} \\ &+ \frac{2}{\pi} \sum_{k=1}^{\infty} \sum_{p=1}^{\infty} \sum_{j,l} \cos(\pi k\eta) \frac{m_{j,l}}{M_{k,p}} \times \\ &\qquad \left\{ K_1(2m_{j,l}M_{k,p}) - \frac{2a^2k^2 m_{j,l}}{M_{k,p}} K_2(2m_{j,l}M_{k,p}) \right\}. \end{aligned}$$

The first two terms are the zero temperature Casimir force. The sum of the last two terms is the thermal correction. It goes to zero exponentially fast when $T \to 0$.

For a pair of Dirichlet-Dirichlet plates, i.e., $\eta = 0$, (3) (divided by two) shows that the Casimir force is always attractive (negative). For Dirichlet-Neumann plates, i.e., $\eta = 1$, (3) (divided by two) shows that the Casimir force is always repulsive. In fact, this is true for all $\eta \geq 1/2$ since the function

$$f(u) = -\mathrm{Re}\left\{\frac{e^{i\pi\eta}}{u - e^{i\pi\eta}}\right\} = \frac{1 - u\cos(\pi\eta)}{(u - \cos(\pi\eta))^2 + \sin^2(\pi\eta)} \tag{4}$$

is positive for $\eta \geq 1/2$ and $u \geq 0$. In all these cases the force is enhanced by the presence of extra dimensions. For $\eta \in (0, 1/2)$, one can show that when the plate separation is large enough, the Casimir force eventually becomes attractive.

When the plate separation a is much smaller than the size $r = \mathrm{Vol}(\Omega)^{1/(d-1)}$ of the cross section, some computations show that the leading term of the Casimir force density acting on the plates is

$$\begin{aligned}\frac{F^{\|}_{\mathrm{Cas}}(a)}{\mathrm{Vol}\,(\Omega)} &\sim -\frac{4T}{2^d\pi^{\frac{d}{2}}}\sum_{k=1}^{\infty}\sum_{l}\sum_{p=-\infty}^{\infty}\cos(\pi k\eta) \\ &\times\Bigg\{(d-1)\left(\frac{\sqrt{m_l^2 + (2\pi pT)^2}}{ka}\right)^{\frac{d}{2}} \times K_{\frac{d}{2}}\left(2ka\sqrt{m_l^2 + (2\pi pT)^2}\right) \\ &+2\frac{\left(\sqrt{m_l^2 + (2\pi pT)^2}\right)^{\frac{d+2}{2}}}{(ka)^{\frac{d-2}{2}}}K_{\frac{d-2}{2}}\left(2ka\sqrt{m_l^2 + (2\pi pT)^2}\right)\Bigg\},\end{aligned} \tag{5}$$

where $m_l = \sqrt{\omega_{\mathcal{N},l}^2 + m^2}$. In the massless case ($m = 0$), the sum of the terms with $l = p = 0$ should be replaced by

$$-\frac{(d-1)T\,\Gamma\left(\frac{d}{2}\right)}{2^{d-1}\pi^{\frac{d}{2}}\,a^{d_1}}\sum_{k=1}^{\infty}\frac{\cos(\pi k\eta)}{k^d}.$$

A form of (5) more suitable to study the low temperature behavior is

$$\begin{aligned}\frac{F^{\|}_{\mathrm{Cas}}(a)}{\mathrm{Vol}\,(\Omega)} &\sim \frac{2}{2^d\pi^{\frac{d_1+1}{2}}}\sum_{k=1}^{\infty}\sum_{p=-\infty}^{\infty}\sum_{l}\cos(\pi k\eta)\left(\frac{m_l}{M_{k,p}}\right)^{\frac{d+1}{2}} \\ &\times\left\{K_{\frac{d+1}{2}}\left(2m_lM_{k,p}\right) - \frac{2k^2a^2m_l}{M_{k,p}}K_{\frac{d+3}{2}}\left(2m_lM_{k,p}\right)\right\}.\end{aligned} \tag{6}$$

When $m=0$, the sum of the terms with $l=0$ in (6) should be replaced by

$$-\frac{d}{2^d\pi^{\frac{d+1}{2}}}\frac{\Gamma\left(\frac{d+1}{2}\right)}{a^{d+1}}\sum_{k=1}^{\infty}\frac{\cos(\pi k\eta)}{k^{d+1}}-\frac{2\Gamma\left(\frac{d+1}{2}\right)\zeta_R(d+1)}{\pi^{\frac{d+1}{2}}}T^{d+1}$$
$$+\frac{\pi}{2^{\frac{d-2}{2}}}\frac{T^{\frac{d-2}{2}}}{a^3}\sum_{k=-\infty}^{\infty}\sum_{p=1}^{\infty}\left(k-\frac{\eta}{2}\right)^2\left(\frac{|k-\frac{\eta}{2}|}{pa}\right)^{\frac{d-2}{2}}K_{\frac{d-2}{2}}\left(\frac{\pi p\left(|k-\frac{\eta}{2}|\right)}{aT}\right).$$

One can show that when $am\ll 1$, the massive correction to the Casimir force is of order $(am)^2\log(am)^2$. Therefore, (5) and (6) show that when $am\ll 1$ and $R\ll a\ll r$, where $R=\mathrm{Vol}(\mathcal{N})^{1/n}$ is the size of the internal manifold, the nature of the Casimir force is determined by the sign of

$$-\sum_{k=1}^{\infty}\frac{\cos(\pi k\eta)}{k^q}=-B_q(\eta),\qquad 0\leq\eta\leq 1,$$

with $q=d$ if $aT\gg 1$ and $q=d+1$ if $aT\ll 1$. For any q, the function $B_q(\eta)$, $\eta\in[0,1]$ is a monotonically decreasing function and it has a unique zero η_q in $[0,1]$. One can show that $1/3=\eta_1<\eta_2<\ldots<1/2$. Therefore, we find that when $R\ll a\ll r$, for $\eta_d<\eta<\eta_{d+1}$, the Casimir force can change from attractive to repulsive when one increases the temperature. On the other hand, when the plate separation a becomes much smaller than the size of the internal manifold, then the internal manifold should be treated in the same way as the large cross section. In this case, we find that the sign of the Casimir force is determined by $-B_q(\eta)$ with $q=d+n$ if $aT\gg 1$ and with $q=d+n+1$ if $aT\ll 1$. Again, we see that decreasing the distance between the plates such that it is much smaller than the size of the internal manifold can change the Casimir force from repulsive to attractive for some values of η.

References

1. R.M. Cavalcanti, *Casimir force on a piston*, Phys. Rev. D **69** (2004), 065015.
2. M.P. Hertzberg, R.L. Jaffe, M. Kardar and A. Scardicchio, *Attractive Casimir Forces in a Closed Geometry*, Phys. Rev. Lett. **95** (2005), 250402.
3. V. Marachevsky, *Casimir interaction of two plates inside a cylinder*, Phys. Rev. D **75** (2007), 085019.
4. L.P. Teo, *Finite temperature Casimir effect in Kaluza-Klein spacetime*, Nucl. Phys. B **819** (2009), 431–452.
5. K. Kirsten and S.A. Fulling, *Kaluza-Klein models as pistons*, Phys. Rev. D **79** (2009), 065019.
6. S.C. Lim and L.P. Teo, *Repulsive Casimir force from fractional Neumann boundary conditions*, Phys. Lett. B **679** (2009), 130–137.
7. E. Elizalde, S.D. Odintsov and A.A. Saharian, *Repulsive Casimir effect from extra dimensions and Robin boundary conditions: From branes to pistons*, Phys. Rev. D **79** (2009), 065023.

THE ROLE OF MAGNETOPLASMONS IN CASIMIR FORCE CALCULATIONS

R. ESQUIVEL-SIRVENT

Instituto de Física, Universidad Nacional Autónoma de México, Apdo. Postal. 20-364, México D.F. 01000
E-mail: raul@fisica.unam.mx

R. GARCÍA-SERRANO, M. A. PALOMINO-OVANDO

Facultad de Ciencias Fisico-Matemáticas, Universidad Autónoma de Puebla, Apartado Postal 5214, Puebla 72000, México

G. H. COCOLETZI

Instituto de Física, Universidad Autónoma de Puebla, Apartado Postal J-48, Puebla 72570, México

In this paper we review the role of magneto plasmon polaritons in the Casimir force calculations. By applying an external constant magnetic field a strong optical anisotropy is induced on two parallel slabs reducing the reflectivity and thus the Casimir force. As the external magnetic field increases, the Casimir force decreases. Thus, with an an external magnetic field the Casimir force can be controlled.The calculations are done in the Voigt configuration where the magnetic field is parallel to the slabs. In this configuration the reflection coefficients for TE and TM modes do not show mode conversion.

Keywords: Casimir, magnetoplasmons, Voigt.

1. Introduction

The optics of surfaces plays an important role in the calculation of the Lifshitz-Casimir formula.[1] Indeed the Lifshitz formula can be obtained from the sum of surface-polariton modes between two metallic slabs.[2–6] In metallic and semiconducting surfaces the effect of an externally applied magnetic field $\mathbf{B_0}$ leds to the excitation of magnetoplasmon modes. This external magnetic field changes significantly the behavior of the plasma modes and induces an optical anisotropy that is magnetic field dependent.[7] This has the effect of reducing the Casimir force significantly. This reduction on the force has been applied to the problem of pull-in dynamics in micro and nano electrodynamical systems (mems and nems), and as has been shown to increase the detachment length in cantilever mems and nems.[8]

To illustrate this point, consider the Drude model in the presence of the external magnetic field. The equation of motion for the electrons in the material is

$$m\frac{d\mathbf{v}}{dt} = q(\mathbf{E} + \mathbf{v} \times \mathbf{B}) - \frac{m}{\tau}\mathbf{v} \tag{1}$$

where m is the effective mass of the electron, q the charge and τ is the relaxation time. Assuming an harmonic electric field $e^{-i\omega t}$ the current $\mathbf{j} = nq\mathbf{v}$ can be found and thus the conductivity can be calculated[9]

$$\sigma_{ij}(\omega, \mathbf{B}_0) = \frac{nq^2}{\tau^* m}\frac{\delta_{ij} + \omega_c\tau^* e_{ijk}(B_k/B_0) + (wc\tau^*)^2(B_iB_j/B_0^2)}{1 + (\omega_c\tau^*)^2}, \tag{2}$$

where $\tau^* = \tau/(1 - i\omega\tau)$, $w_c = q|\mathbf{B}_0|/mc$ is the cyclotron frequency, and e_{ijk} is the Levi-Civita symbol. The dielectric tensor is obtained from

$$\epsilon_{ij}(\omega, \mathbf{B}_0) = \delta_{ij} + \frac{4\pi i}{\omega}\sigma_{ij}. \tag{3}$$

Clearly if $\mathbf{B}_0 = 0$ we recover the results for the isotropic case.

2. Voigt and Faraday Configurations

For an arbitrary direction of the magnetic filed, the calculation of the dispersion relation of the surface magneto plasmons and of the optical reflectivity is difficult. To simplify the problem, specific directions of the magnetic field have to be chosen.[10] In the so called *Faraday* configuration, the magnetic field is perpendicular to the slab. In this case, there is mode conversion upon reflection from the slab. This is, if a TE wave is incident, the reflected wave will consists of a TE and TM modes, similar for an incident TM mode.

The second configuration, that will be used in this paper, is the *Voigt* geometry where the magnetic field is parallel to the slabs. In this case there is no mode conversion upon reflection. Consider a slab parallel to the $x - z$ plane. In the Voigt geometry the external magnetic field points along the z axis. In this case, the components of the dielectric tensor are

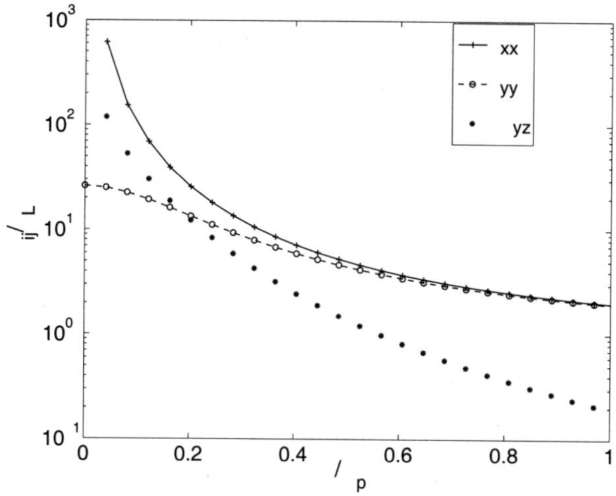

Fig. 1. Components of the dielectric function in the Voigt configuration (Eq. (4)) for $\Omega_c = 0.2$. The dielectric function is evaluated in the rotated frequency axis.

given by[10,11]

$$\epsilon_{xx} = \epsilon_L \left[1 - \frac{\omega_p^2}{\omega^2}\right],$$
$$\epsilon_{yy} = \epsilon_L \left[1 - \frac{\omega_p^2}{\omega^2 - \omega_c^2}\right],$$
$$\epsilon_{yz} = \epsilon_L \left[\frac{i\omega_c\omega_p^2}{\omega(\omega^2 - \omega_c^2)}\right], \tag{4}$$

and $\epsilon_{zz} = \epsilon_{yy}$ and $\epsilon_{zy} = -\epsilon_{yz}$. The other components are equal to zero. In these equations ϵ_L is the background dielectric function, ω_p the plasma frequency, In the absence of the magnetic field, $\omega_c = 0$ and the plates become isotropic. For simplicity we have not included the Drude damping parameter. In the rest of the paper we will use the dimensionless variable $\Omega_c = \omega_c/\omega_p$, that gives the relative importance of the external magnetic field. In Figure (1) we have plotted the dielectric function components as given by Eq. (4) after rotation of the frequency to the complex plane $\omega \to i\zeta$ for a value of $\Omega_c = 0.2$, showing the strong anisotropy of the system.

In the material slab, the dispersion relation is obtained from the wave equation

$$\nabla \times \nabla \times \mathbf{E} - q_0^2\tilde{\epsilon} \cdot \mathbf{E} = 0, \tag{5}$$

where $\tilde{\epsilon}$ is the dielectric tensor and $q_0^2 = \omega^2/c^2 = q_x^2 + q_y^2 + q_z^2$. Upon replacement of its components (Eq.(4)) into Eq.(5), the nontrivial solution of the resulting equation is obtained if

$$-q_y^2 = \beta^2 = q_z^2 - q_0^2(\epsilon_{zz} + \frac{\epsilon_{yz}^2}{\epsilon_{ss}}). \tag{6}$$

Outside the slab we assume there is vacuum and we have the usual dispersion

$$-q_y^2 = \alpha^2 = q_z^2 - q_0^2. \tag{7}$$

The reflection coefficient can now be calculated taking into account that outside the slab, for a TM polarized wave the field is of the form

$$B_x(r,t) = A_\pm e^{\pm \alpha y} e^{i(q_z z - \omega t)}, \tag{8}$$

and within the slab

$$B_x(y) = C_\pm e^{\pm \beta y} e^{i(q_z z - \omega t)}, \tag{9}$$

where β and α are given by Eqs.(6) and (7) respectively. From the corresponding electric fields and by applying the boundary conditions, the reflection coefficients can be found. The detailed procedure for the Voigt configuration can be found in Ref.([10]).

3. Reduction of the Casimir force with an external magnetic field

To study the effect of the external magnetic field on the Casimir force we use Lifhitz formula

$$F = \frac{k_B T}{8\pi L^2} \sum_{n=0}^{\infty} \int_{\zeta_n}^{\infty} q_y dq_y \frac{1}{r_s^{-2} e^{2q_y L} - 1} + (r_s \rightarrow r_p), \tag{10}$$

where $\zeta_n = 2\pi k_B T n/\hbar$ is the Matsubara frequencies and r_{sp} the reflectivities for p or s polarized modes (TM and TE, respectively). This expression for the Casimir force can be used only for the Voigt configuration since there is no mode conversion. The Faraday configuration, that will be presented elsewhere, requires a more general expression for the Casimir force, as the one discussed by Bruno.[12] The reflectivities r_p and r_s are replaced in Eq. (10) by a reflectivity matrix whose components are r_{pp}, r_{ps}, r_{ss}, r_{sp}, where the first subindex represents the polarization of the incident wave and the second sub-index the polarization of the reflected wave.

In Figure (2) we plot the Casimir force normalized to the ideal case $F_0 = -\hbar c\pi^2/240L^4$, for several values of the reduced frequency Ω_c. All

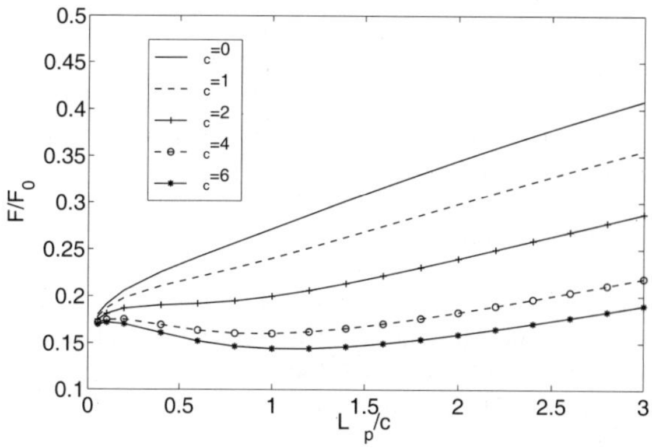

Fig. 2. The Casimir force normalized to the ideal case as a function of separation. The different curves correspond to different values of Ω_c. As the external magnetic field increases, Ω_c also increases and the Casimir force decreases. The separation between the plates is in terms of c/ω_p.

frequencies are normalized to the plasma frequency and the distances are normalized to the plasma wavelength c/ω_p. The value for the background dielectric function $\epsilon_L = 15.4$ is for $InSb$ as reported by Palik.[13] In general, $III - V$ semiconductors (e.g. $GaAs, GaN, InAs$) can be used, since they exhibit a strong magnetoplasmon response. The important feature of Fig. (2), is that as the magnetic field increases the Casimir force decreases. For high magnetic fields, there is a drop in F/F_0 as a function of separation, this drop is more significant with increasing magnetic field.

4. Conclusions

In this paper we have reviewed briefly the principles of magnetoplasmons in semiconductors and its effect on the calculations of the Casimir force between parallel slabs. In particular we consider the Voigt configuration, where the magnetic field is parallel to the surface of the slabs. This external magnetic field induces a strong optical anisotropy that reduces the reflectivity. This has the effect of reducing the Casimir force as the external magnetic field increases. In a future work, the effect of the external magnetic field on the Casimir torque will be considered.

Acknowledgments

The authors acknowledge partial support from DGAPA-UNAM grant no.project No. IN-113208 and CONACyT project No. 82474, VIEP-BUAP and SEP-BUAP-CA 191.

References

1. E. M. Lifshitz, *Sov. Phys. JETP* **2**, 73 (1956).
2. B. Sernelius, *Surface Modes in Physics*, (Wiley-VCH, Berlin, 2001).
3. C. Noguez, C.E. Romám-Velázquez, R. Esquivel-Sirvent and C. Villarreal, *Europhys. Lett.* **67**, 191 (2004).
4. C. Noguez, C.E. Romám-Velázquez, C. Villarreal and R. Esquivel-Sirvent, *Phys. Rev. A* **69**, 042109 (2004).
5. F. Intravaia and A. Lambrecht, *Phys. Rev. Lett.* **94**, 110404 (2005)
6. F. Intravaia, C. Henkel, and A. Lambrecht *Phys. Rev. A* **76**, 033820 (2007).
7. R. Garcia-Serrano, G. Martinez, P. Hernández, M. Palomino-Ovado, and G. H. Cocoletzi, *Phys. Status Solidi B* **246** , 1357 (2009).
8. R. Esquivel-Sirvent, M. Palomino-Ovado, and G. H. Cocoletz, *Appl. Phys. Lett.* **95** 051909 (2009).
9. E.D.Palik and J.K. Furdyna, *Rep. Prog. Phys.* **33**, 1193 (1970).
10. M. S. Kushwaha, *Surf. Sci. Rep.* **41**, 1 (2001).
11. J.J. Brion, R. F. Wallis, A. Hartstein, E. Burstein, *Phys. Rev. Lett* **28**, 1455 (1972).
12. P. Bruno, *Phys. Rev. Lett.* **88** , 240401 (2002).
13. E.D. Palik, R. Kaplan, R.W. Gammon, H. Kaplan, R. F. Wallis and J.J. Quinn, *Phys. Rev. B* **13**, 2497 (1976).

SUSPENDED GRAPHENE FILMS AND THEIR CASIMIR INTERACTION WITH IDEAL CONDUCTOR

I. V. FIALKOVSKY

Instituto de Física, Universidade de São Paulo, São Paulo, Brazil
Department of Theoretical Physics, Saint-Petersburg State University, Russia
ifialk@gmail.com

We adopt the Dirac model for graphene and calculate the Casimir interaction energy between a plane suspended graphene sample and a parallel plane ideal conductor. We employ both the Quantum Field Theory (QFT) approach, and the Lifshitz formula generalizations. The first approach turns out to be the leading order in the coupling constant of the second one. The Casimir interaction for this system appears to be rather weak but experimentally measurable. It exhibits a strong dependence on the mass of the quasi-particles in graphene.

Keywords: Casimir energy, graphene, QFT, Lifshitz formula.

1. Introduction

Graphene is a (quasi) two dimensional hexagonal lattice of carbon atoms. At present, it belongs to the most interesting materials in solid state physics due to its exceptional properties and importance for nano technology.[3,4] In the present article based on joint works[1,2] we consider the Casimir interaction between suspended graphene plane and a parallel ideal conductor. This setup was earlier considered in[5–7] using a hydrodynamical model for the electrons in graphene following.[8,9] Later it became clear that this model does not describe the electronic properties specific to this novel material.

Here we use a realistic and well-tested model where the quasi-particles in graphene are considered to be two-component Dirac fermions. This model incorporates the most essential and well-established properties of their dynamics: the symmetries of the hexagonal lattice, the linearity of the spectrum, a very small mass gap (if any), and a characteristic propagation velocity which is 1/300 of the speed of light.[3,10] By construction, this model should work below the energy scale of about 1eV, but even above this limit the optical properties of graphene are reproduced with a high precision.[15]

For each of the fermion species the Dirac action of the model reads

$$S_{\mathrm{D}} = \int d^3 x \bar{\psi}(\tilde{\gamma}^l (i\partial_l - eA_l) - m)\psi, \quad l = 0,1,2 \tag{1}$$

where $\tilde{\gamma}^l$ are just rescaled $2^{\times}2$ gamma matrices, $\tilde{\gamma}^0 \equiv \gamma^0$, $\tilde{\gamma}^{1,2} \equiv v_F \gamma^{1,2}$, $\gamma_0^2 = -(\gamma^i)^2 = 1$. In our units, $\hbar = c = 1$, and Fermi velocity $v_F \simeq (300)^{-1}$. The value of the mass gap parameter m and mechanisms of its generation are under discussion.[11–14] The upper limit on m is about 0.1eV at most.

The propagation of photons in the ambient $3+1$ dimensional space is described by the Maxwell action

$$S_{\mathrm{M}} = -\frac{1}{4}\int d^4 x F_{\mu\nu}F^{\mu\nu}, \quad \mu,\nu = 0,1,2,3. \tag{2}$$

In the following we shall suppose that the graphene sample occupies the plane $x^3 = a > 0$, and the conductor corresponds to $x^3 = 0$.

2. QFT approach

In the framework of QFT one evaluates the effective action in a theory described by the classical action $S_{\mathrm{D}} + S_{\mathrm{M}}$. Then the Casimir energy density per unit area of the surfaces at the leading order in the fine structure constant is given by

$$\mathcal{E}_1 = -\frac{1}{TS}$$ 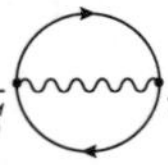 , (3)

where T is time interval, and S is the area of the surface. The solid line denotes the fermion propagator in 2+1 dimensions (i.e., inside the graphene sample), and the wavy line is the photon propagator in the ambient $3+1$ dimensional space subject to the perfect conductor boundary conditions at $x^3 = 0$: $A_0|_{x^3=0} = A_1|_{x^3=0} = A_2|_{x^3=0} = \partial_3 A_3|_{x^3=0} = 0$.

The fermion loop in $2+1$ dimensions has already been calculated in a number of papers.[11,13,14] It gives the quadratic order in A of the effective action for electromagnetic field

$S_{\mathrm{eff}}(A) = A$ 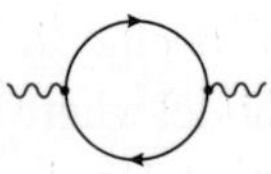 $A = \frac{1}{2}\int \frac{d^3p}{(2\pi)^3} A_j(p)\Pi^{jl}(p)A_l(p),$ (4)

where

$$\Pi^{mn} = \frac{\alpha\Phi(\tilde{p})}{v_F^2}\eta_j^m\left(g^{jl} - \frac{\tilde{p}^j\tilde{p}^l}{\tilde{p}^2}\right)\eta_l^n\,, \tag{5}$$

is the polarization tensor in the lowest, one loop, order. Here $\eta_j^m = \mathrm{diag}(1, v_F, v_F)$, $\tilde{p}$ denotes the rescaled momenta $\tilde{p}_j = \eta_j^k p_k$, The function $\Phi(p)$ is model dependent, and for graphene with four fermion species each described by (1) it reads $\Phi(p) = 2\left(2m\tilde{p} - (\tilde{p}^2 + 4m^2)\text{arctanh}(\tilde{p}/2m)\right)/\tilde{p}$. Note that here we assumed that all possible parity-odd parts of Π are canceled. Some observable effects invoked by their presence are considered in.[1]

To calculate the diagram (3) we only need to couple the kernel (5) to the photon propagator subject to conducting boundary conditions. In Fourier representation and for the Euclidean 3-momenta, i.e., after the Wick rotation $p \to p_E = (ip_0, p_1, p_2)$, the a-dependent part of the energy reads

$$\mathcal{E}_1 = -\frac{1}{4}\int \frac{d^3p_E}{(2\pi)^3}\frac{\Pi^j_j(p_E)}{p_\parallel}e^{-2ap_\parallel} = -\int \frac{d^3p_E}{(2\pi)^3}\frac{\alpha(p_\parallel^2 + \tilde{p}_\parallel^2)\Phi(p_E)}{4p_\parallel \tilde{p}_\parallel^2}e^{-2ap_\parallel}\,. \tag{6}$$

where we expanded $\Pi^j_j(p_E)$ explicitly with help of (5), and $p_\parallel \equiv |p_E|$.

3. Lifshitz formula approach

One can also consider the system as described by effective theory of the electromagnetic field with the action $S_{\text{M}} + S_{\text{eff}}$ subject to the conducting boundary conditions at $x^3 = 0$. Then at the surface of graphene, the Maxwell equations receive a singular contribution

$$\partial_\mu F^{\mu\nu} + \delta(x^3 - a)\Pi^{\nu\rho}A_\rho = 0 \tag{7}$$

following from S_{eff}. Here we set $\Pi^{3\mu} = \Pi^{\mu 3} = 0$. This contribution is equivalent to imposing the matching conditions

$$(\partial_3 A_\mu)|_{x^3=a+0} - (\partial_3 A_\mu)|_{x^3=a-0} = \Pi_\mu^{\ \nu}A_\nu|_{x^3=a}\,. \tag{8}$$

assuming that A_μ is continuous at $x^3 = a$. Now, one can forget the origin of $\Pi_\mu^{\ \nu}$ and quantize the electromagnetic field subject to these conditions or consider the plane waves propagation leading to the Lifshitz formula.

The original Lifshitz description[17] of interaction of finite-width slabs was later generalized[18,19] for the case of two plane parallel interfaces separated by the distance a and possessing arbitrary reflection coefficients $r^{(1,2)}_{\text{TE,TM}}$ of the TE and TM electromagnetic modes on each of the surfaces

$$\mathcal{E}_L = \int \frac{d^3p_E}{16\pi^3}\ln[(1 - e^{-2p_\parallel a}r^{(1)}_{\text{TE}}r^{(2)}_{\text{TE}})(1 - e^{-2p_\parallel a}r^{(1)}_{\text{TM}}r^{(2)}_{\text{TM}})]. \tag{9}$$

For graphene with help of matching conditions (8) we can obtain at the Euclidean momenta

$$r^{(1)}_{\text{TE}} = \frac{-\alpha\Phi}{2p_\parallel + \alpha\Phi}, \quad r^{(1)}_{\text{TM}} = \frac{\alpha p_\parallel \Phi}{2\tilde{p}_\parallel^2 + \alpha p_\parallel \Phi}, \tag{10}$$

while for the perfect conductor one has $r^{(2)}_{\text{TE}} = -1$, $r^{(2)}_{\text{TM}} = 1$. It is clear, that Φ must be rotated to Euclidean momenta as well. We also note that the perfect conductor case is recovered from (10) in the formal limit $\Phi \to \infty$.

One can show by a direct computation that the energy $\mathcal{E}_1$, Eq. (6), coincides with the leading α^1 order in a perturbative expansion of the Lifshitz formula (9)-(10), so that the two approaches are consistent.

4. Results and discussion

The formulae (6) and (9)-(10) are suitable for the numerical evaluation. The asymptotic expansion for short and long distances are readily obtained through uniform expansion of the integrands of (6), (9)

$$\mathcal{E}_L \underset{a\to\infty}{\sim} -\frac{\alpha}{24\pi^2}\frac{2+v_f^2}{ma^4}, \qquad \mathcal{E}_L \underset{a\to 0}{\sim} \frac{1}{16\pi a^3}\, h(\alpha, v_F) \tag{11}$$

Note that the asymptotics at large separations is of the first order in α while for small separations, it contains all powers of α through $h(\alpha, v_F)$, for the real values of parameters in graphene $h(1/137, 1/300) \approx 0.024$.

Therefore we see that at large separations Casimir energy does not turn into the ideal conductor case, while at small separation this case is indeed recovered up to rather small numerical factor. This is counter-intuitive since the main contribution at short separations shall come from the high momenta for which one would expect the graphene film to become transparent. Still the Dirac model used here is well established by optical[15] and electronical[16] experiments in graphene. We must also stress that this asymptotical behavior is drastically different from that in the hydrodynamic model.[5–7]

For numerical evaluation we normalize the results to the Casimir energy $\mathcal{E}_C = -\frac{\pi^2}{720\, a^3}$ for two plane ideal conductors separated by the same distance a. The results of calculations are depicted at Fig. 1. The scale is defined by the mass parameter m. For m of the order of 0.1eV, $ma = 1$ corresponds to $a = 1.97$ micrometer.

Thus, we can see that the magnitude of the considered Casimir interaction of graphene with a perfect conductor is rather small. Actual measurement of such weak forces is a challenging, but by no means hopeless, experimental problem. Strong dependence on the mass parameter m at large separation is also a characteristic feature of the Casimir force. Getting an independent measurement of m may be very important for our understanding of the electronic properties of graphene. The mass of quasi-particles in graphene is, probably, very tiny, which improves the detectability of the Casimir interaction since the energy increases with decreasing m.

Acknowledgments

The author gladly acknowledge the financial support of FAPESP, as well as of grants RNP 2.1.1/1575 and RFBR 07–01–00692.

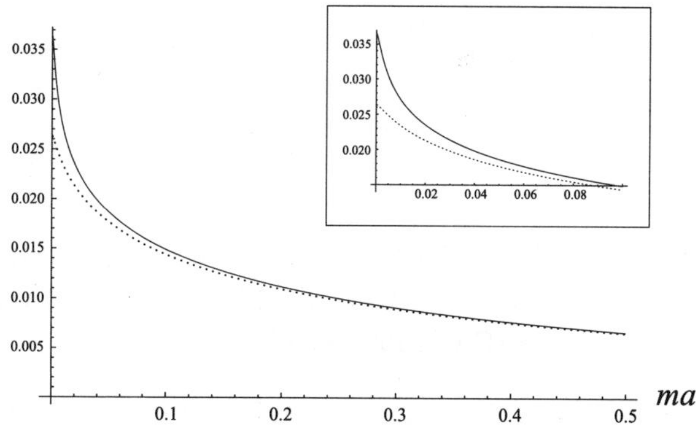

Fig. 1. The relative Casimir energy densities $\mathcal{E}_1/\mathcal{E}_C$ (solid line) and $\mathcal{E}_L/\mathcal{E}_C$ (dashed line) as functions of ma. Insert shows a zoom of the small-distances region.

References

1. I. V. Fialkovsky and D. V. Vassilevich, J. Phys. A: Math. Theor. **42** (2009) 442001, arXiv:0902.2570 [hep-th].
2. M. Bordag, I. V. Fialkovsky, D. M. Gitman, D. V. Vassilevich, Phys. Rev. B **80** (2009) 245406, rXiv: 0907.3242 [hep-th]
3. A. K. Geim and K. S. Novoselov, Nature Mater. **6**, 183 (2007); M. I. Katsnelson, Mater. Today **10**, 20 (2007); A. K. Geim, arXiv:0906.3799.
4. A. H. Castro Neto, et al., Rev. Mod. Phys. **81**, 109 (2009).
5. G. Barton. J. Phys., A **38** (2005) 2997.
6. M. Bordag, J. Phys. A **39** (2006) 6173, [arXiv:hep-th/0511269].
7. M. Bordag, et al. Phys. Rev. B **74**, 205431 (2006).
8. A. L. Fetter. Annals of Physics, **81**, 367, (1973).
9. G. Barton. J. Phys. A: Math. Gen., **37**, 1011, (2004).
10. G. W. Semenoff, Phys. Rev. Lett. **53**, 2449 (1984); D. P. DiVincenzo and E. J. Mele, Phys. Rev. B **29**, 1685 (1984);
11. T. W. Appelquist, et al. Phys. Rev. D **33**, 3704 (1986).
12. D. V. Khveshchenko, Phys. Rev. Lett. **87**, 206401 (2001).
13. E. V. Gorbar, et al., Phys. Rev. B **66**, 045108 (2002), V. P. Gusynin and S. G. Sharapov, Phys. Rev. B **73**, 245411 (2006)
14. P. K. Pyatkovskiy, J. Phys.: Condens. Matter **21**, 025506 (2009).
15. R. Nair, et al. Science **320**, 1308 (2008).
16. K.S. Novoselov et al., Nature **438** (2005) 197, arXiv:cond-mat/0509330v1 [cond-mat.mes-hall].
17. E. M. Lifshitz, Zh. Eksp. Teor. Fiz. **29**, 94 (1956) [Sov. Phys. JETP **2**, 73, (1956)]; E. M. Lifshitz and L. P. Pitaevskii, *Statistical Physics* (Pergamon Press, Oxford, 1980).
18. M. T. Jaekel and S. Reynaud. J. De Physique I, 1(10), 1395–1409, (1991).
19. M. Bordag. J. Phys., A **28**: 755–766, 1995.

CASIMIR ENERGY OF FINITE WIDTH MIRRORS: RENORMALIZATION, SELF-INTERACTION LIMIT AND LIFSHITZ FORMULA

I. V. FIALKOVSKY[†‡], V. N. MARKOV[¶] and YU. M. PIS'MAK[‡]

[†]*Instituto de Física, Universidade de São Paulo, São Paulo, Brazil*
[‡]*Department of Theoretical Physics, Saint-Petersburg State University, Russia*
[¶]*St. Petersburg Nuclear Physics Institute, Russia*
[†]*ifialk@gmail.com*

We study the field theoretical model of a scalar field in presence of spacial inhomogeneities in form of one and two finite width mirrors (material slabs). The interaction of the scalar field with the defect is described with position-dependent mass term. Within this model we derive the interaction of two finite width mirrors, establish the correspondence of the model to the Lifshitz formula and construct limiting procedure to obtain finite self-energy of a single mirror without any normalization condition.

Keywords: Casimir energy, QFT, finite width mirrors.

1. Introduction

The Casimir effect[1] was original considered for an extremely idealized configuration of electromagnetic field subject to ideal conducting boundary conditions on two parallel plates. Development of both theoretical and experimental techniques required consideration of more realistic systems which would describe real materials, both in their shape and properties. During last decades a number of such approaches were successfully developed, see Ref. 2 for a review.

One of the methods for investigation of (more) realistic situations was initially proposed by Symanzik[3] and consists of modeling and generalizing the rigid boundary conditions by introducing additional singular-potential terms into the action of the model. This approach was widely explored for the case of delta-type potentials. The essence of this paper is to present the results of elaboration of similar field-theoretical approach in the case of step-potentials. The detailed calculations and discussions are presented in Refs. 4, 5.

2. Single finite width mirror

Let us consider a scalar field interacting with a space defect with nonzero volume. Using the Symanzik's approach,[3] we add to the action an additional mass term being non-zero only inside the defect:

$$\begin{gathered} S = S_0 + S_{def} \\ S_0 = \frac{1}{2}\int d^4x\phi(x)(-\partial_x^2 + m^2)\phi(x), \quad S_{def} = \frac{\lambda}{2}\int d^4x\theta(\ell, x_3)\phi^2(x) \end{gathered} \tag{1}$$

where $\partial_x^2 = \partial^2/\partial x_0^2 + \ldots + \partial^2/\partial x_3^2$. In the simplest case the defect could be considered as homogenous and isotropic infinite plane layer of the thickness ℓ placed in the x_1x_2 plane (so called 'piecewise constant potential')

$$\theta(\ell, x_3) \equiv [\theta(x_3 + \ell/2) - \theta(x_3 - \ell/2)]/\ell. \tag{2}$$

To describe all physical properties of the systems it is sufficient to calculate the generating functional for the Green's functions

$$G[J] = N\int D\phi\, \exp\{-S[\phi] + J\phi\} \tag{3}$$

where J is an external source. The explicit calculation of $G[J]$ is performed with help of integral operators method developed by the authors, see Refs. 4, 5.

2.1. *Casimir self-energy and its renormalization*

The Casimir energy density per unit area of the defect S can be presented with the relation $\mathcal{E} = -\frac{1}{TS}\ln G[0]$, here T is the (infinite) time interval and S — the surface area of the defect.

For the case of a single finite–width mirror, as in (1), the Casimir energy is given by a sum of a UV finite, $\mathcal{E}_{fin}$, and UV divergent, $\mathcal{E}_{div}$, parts

$$\mathcal{E} = \mathcal{E}_{fin} + \mathcal{E}_{div}, \tag{4}$$

$$\mathcal{E}_{fin} = \frac{1}{4\pi^2}\int_0^\infty \left(\ln\frac{e^{-\ell(E+\sqrt{\rho})}}{4E\sqrt{\rho}\xi} - \frac{\lambda}{2E}\left(1 - \frac{\lambda}{4\ell E^2}\right)\right)p^2dp, \tag{5}$$

$$\mathcal{E}_{div} = \frac{\lambda\mu^{4-d}}{2(2\pi)^{d-1}}\int\frac{d^{d-1}p}{2E}\left(1 - \frac{\lambda}{4\ell E^2}\right),$$

where we used dimensional regularization, and put $\rho = E^2 + \kappa$, $\kappa = \lambda/l$, $\xi^{-1} = e^{2\ell\sqrt{\rho}}(E + \sqrt{\rho})^2 - (E - \sqrt{\rho})^2$. With appropriate redefinitions of the parameters this expression coincides with known results in the literature.[7]

For renormalization of the model at the one-loop level considered here we must add to the action the following field-independent counter-term δS

$$\delta S = f\lambda + g\lambda^2\ell^{-1}, \tag{6}$$

with bare parameters f and g. Within such 'minimal addition' renormalization scheme we obtain for the renormalized Casimir energy

$$\mathcal{E}_r = \mathcal{E}_{fin} + \lambda f_r + g_r \lambda^2 \ell^{-1} \tag{7}$$

where finite parameters f_r, g_r must be determined with appropriate experiments, or fixed with normalization conditions. The number of required conditions is dictated by the (in)dependence of the coupling constant λ on the slab thickness ℓ.

2.2. *Dirichlet limit and normalization condition*

Under the change of variables $\lambda \to \ell\kappa$ two counter-terms in (7) can be effectively combined into a single one $\tilde{g}_r$ of mass dimension one

$$\mathcal{E}_r = \mathcal{E}_{fin} + \kappa \ell \tilde{g}_r \tag{8}$$

One notes that putting $\kappa = -m^2$ and taking the $m \to \infty$ limit effectively converts the system under consideration into a massless scalar field confined between two plates at $x_3 = \pm\ell/2$ subject to Dirichlet boundary conditions at the boundaries. We can use this correspondance for fixing $\tilde{g}_r$.

The finite part of the Casimir energy (4) in the limit $m \to \infty$ yields

$$\mathcal{E}_{fin} = -\frac{m^4 \ell}{128\pi^2} + \left(\frac{\pi}{6} - \frac{4}{9}\right)\frac{m^3}{4\pi^2} - \frac{\pi^2}{1440\ell^3} + O(m^{-1}) \tag{9}$$

Now we require that the renormalized energy (8) in this limit lead to the same Casimir pressure as in the case of massless scalar field subject to Dirichlet boundary conditions $\mathcal{P}_{Dir} = -\frac{\pi^2}{480\ell^4}$. Then this condition fixes the renormalization parameter $\tilde{g}_r$ of (8)

$$\tilde{g}_r = -\frac{m^2}{128\pi^2} \tag{10}$$

Thus, using this Dirichlet limit procedure we are able to collate a particular limit of our results with a well known (unambiguous) physical situation.

3. Interaction of two finite width mirrors

Let us consider two plane slabs of thickness $\ell_{1,2}$ interacting over the distance r. The action can be written as

$$\begin{aligned} S = S_0 + S_{def}, \quad S_0 = \tfrac{1}{2}\int d^4x \phi(x)(-\partial_x^2 + m^2)\phi(x) \\ S_{def} = \int d^3x \big(\kappa_2 \int_{-a_2-l_2}^{-a_2} dx_3 \phi^2(x) + \kappa_1 \int_{a_1}^{a_1+l_1} dx_3 \phi^2(x)\big) \end{aligned} \tag{11}$$

here $a_1 + a_2 \equiv r$.

Proceeding along the lines of Refs. 4, 5 we get the final expression for the energy

$$\mathcal{E}_{2L} = \mathcal{E}_1 + \mathcal{E}_2 + \int \frac{d^3\vec{p}}{2(2\pi)^3} \log\left[1 - e^{-2Er} \prod_{i=1,2} \kappa_i \xi_i (1 - e^{2\ell_i \sqrt{\rho_i}})\right]. \tag{12}$$

Here $\mathcal{E}_{1,2}$ give the self-energy (4) of solitary layers 1, 2 correspondingly. All the notation here follows ones of Sect. 2 with subscript index corresponding to the layer number.

The third term in (12) represents the interaction of two layers and vanishes in the limit $r \to \infty$. We note that the interaction term is UV finite, and the removal of regularization made in (12) is indeed justified. This is in perfect accordance with general considerations[8] of the finiteness of Casimir interaction between disjoint bodies.

Basing on the general expression (12) we can calculate the vacuum energy in different limits such as self-pressure of the slab in presence of delta-spike and interaction between two of them. The two delta-spikes limit known in previous literature is also reproduced.

3.1. *Connection to the Lifshitz formula*

Now we consider in (12) the limit of slabs of infinite width separated by finite distance r. For the force in such a limit we obtain

$$\mathcal{F}_{Lif} \equiv -\frac{\partial \mathcal{E}_{Lif}}{\partial r} = -\int \frac{d^3\vec{p}}{(2\pi)^3} \frac{E}{e^{2Er}\frac{(E+\sqrt{\rho_1})(E+\sqrt{\rho_2})}{(E-\sqrt{\rho_1})(E-\sqrt{\rho_2})} - 1} \tag{13}$$

It is straightforward to see that the correspondence with the Lifshitz formula[6] achieved if we introduce particular dispersion into the interaction of quantum fields with the material defect,

$$\kappa_{1,2}^{TE}(p) = (\epsilon_{1,2} - 1)p_0^2, \ \ \kappa_{1,2}^{TM}(p) = \left(\frac{1}{\epsilon_{1,2}^2} - 1\right)\vec{p}^2 + \left(\frac{1}{\epsilon_{1,2}} - 1\right)p_0^2. \tag{14}$$

The parameter ϵ which enters these dispersion relations is to be identified with dielectric permittivity. Summing TE and TM contributions we immediately recover the Lifshitz formula[6]

$$\mathcal{F}_{Lif} = -\frac{1}{4\pi^2}\int_0^\infty dp_0 \int_0^\infty dp^2 E(d_{TE}^{-1} + d_{TM}^{-1}) \tag{15}$$

$$d_{(TE,TM)} = e^{2Er}\frac{(E + \sqrt{\rho_1^{(TE,TM)}})(E + \sqrt{\rho_2^{(TE,TM)}})}{(E - \sqrt{\rho_1^{(TE,TM)}})(E - \sqrt{\rho_2^{(TE,TM)}})} - 1 \tag{16}$$

with $\rho_{1,2}^{(TE,TM)} = E^2 + \kappa_{1,2}^{(TE,TM)}(p)$.

3.2. *Self-interaction limit*

For $\kappa_2 = \kappa_1$ the limit $\ell \to \infty$ taken in the two-slab action (11) reproduces the single-slab action (1) subject to the substitution $\ell \to r$, $m^2 \to m^2 + \kappa_1$,

$\lambda \to -\kappa_1 r$. One would expect that the Casimir energy of two slabs in this limit also reproduces the result for a single finite-width mirror.

However, one finds that $\mathcal{F}_{Lif}$ (13), being UV finite in this limit, differs from UV divergent r-derivative of (4) $\partial\mathcal{E}/\partial r$ by

$$\Delta = -\frac{\mu^{4-d}}{2(2\pi)^{d-1}} \int d^{d-1}\vec{p}\left(E - \sqrt{E^2+\kappa}\right) \tag{17}$$

Yet, this discrepancy does not signal any inconsistency. In Sec. 2 we presented a derivation of the Casimir *self*-energy of a single slab. It is known that in general the self-interactions of Casimir type do possess divergencies depending on the geometrical properties of the system. This fact is reflected in the presence of the counter terms (6).

On the other hand, it is well known,[8] that Casimir interaction of two *disjoint* bodies is always free of divergencies depending on the distance between them. This very system was considered in this section, and the force (13) between two distinct slabs was found to be finite and unambiguous. Consequently, the limit of infinite width of the slabs which recover the self-pressure of a single body, does not bring any divergencies or ambiguities.

Requiring that (8) coincides with (13) we can fix the counter terms $\tilde{g}_r$ and collate the two approaches. One finds that this condition fixes the same value for $\tilde{g}_r$ as one elaborated in Sec. 4 and given by (10).

Thus, we establish a new divergence free approach to calculate the self-pressure of a single finite width slab which lead to the same result as one obtained imposing a physically motivated normalization condition.

Acknowledgements

The authors gladly acknowledge the support of grants RNP 2.1.1/1575, RFBR 07–01–00692. IVF is especially grateful to support of University of Oklahoma and FAPESP.

References

1. H. B. G. Casimir, Proc. Kon. Nederl. Akad. Wet. 51 (1948) 793.
2. M. Bordag, G.L. Klimchitskaya, U. Mohideen, and V.M. Mostepanenko. *Advances in the Casimir Effect*, Oxford University Press, 2009
3. K. Symanzik, Nucl. Phys. B 190, 1 (1981).
4. I. Fialkovskiy, Y. Pis'mak, V. Markov, arXiv:0907.1058 [hep-th].
5. I. Fialkovskiy, Y. Pis'mak, V. Markov, Phys. Rev. **D79** (2009) 028701, arXiv:0812.2430v1 [hep-th].
6. E. M. Lifshitz, Zh. Eksp. Teor. Fiz. **29**, 94 (1956),
7. M. Bordag, J. Phys. A **28**, 755 (1995). R. A. Konoplya, D. V. Vassilevich, JHEP 01 (2008) 068, arXiv:0712.0360
8. T. Emig, et al., Phys. Rev. Lett. **99** (2007) 170403, arXiv:0707.1862.

ROLE OF SURFACE STATES IN THE CASIMIR FORCE BETWEEN SEMICONDUCTING FILMS

M. GOVONI* and C. CALANDRA

Dipartimento di Fisica, Università di Modena e Reggio Emilia,
Via Campi 213/A, I–41125 Modena, Italy
**E-mail: marco.govoni@unimore.it*

A. BENASSI

CNR/INFM–Natl. Research Center on nanoStructures and bioSystems at Surfaces S3,
Via Campi 213/A, I–41125 Modena, Italy
and
CNR/INFM–Natl. Simulation Center Democritos,
Via Beirut 2–4, I–34151 Trieste, Italy

We present results of first principle calculations of the Casimir force between Si films of nanometric size, which show that it depends significantly upon the configuration of the surface atoms, and give evidence of the importance of surface states.

Keywords: Casimir force; thin films; quantum confinement; surface effects; anisotropic media.

1. Introduction

Recently theoretical results have been reported which show the influence of the thickness on the Casimir force between thin slabs.[1–3] These studies are motivated by the role that thin films play in micro- and nano-devices. They rely on two approximations: (i) a local description of the dielectric properties of the material, (ii) the use of a plasma or Drude–Lorentz model to represent the bulk dielectric function. In Ref. 1,2 this same dielectric function has been adopted to represent the film dielectric properties.

In this paper we derive the dielectric tensor of a Si film starting from the microscopic description based on Density Functional Theory (DFT) in a pseudo-potential approach, thus automatically including size quantization and surface state effects. Here we present preliminary results on the forces between Si films of nanometric size.

2. Simulation of the Silicon film

For intrinsic Si the optical excitations at low energy involve the valence electrons. We are interested in determining how they are modified in a film of small thickness compared to the bulk due to: (i) the confinement of the electron gas (Quantum Size Effects, QSE), (ii) the presence of surface states (Surface Effects, SE).

The system we investigate is a film obtained by stacking 12 atomic planes of Si along the [111] direction. We take the surface normal as the z-axis. The film has 2D periodicity along the planes. If we terminate the film without allowing any rearrangement of the surface atoms the film turns out to be 1.9 nm thick and the surfaces show an half filled dangling bond per unit cell. On the other hand if we allow structural rearrangements at the surface, 2×1 surfaces may be generated, with two dangling bonds per unit cell. Indeed the Si(111) surface is the cleavage surface and, after cleavage, it shows a 2×1 reconstruction,[4,5] which produces zig-zag chains of dimers along a planar direction, that we take as the y-axis. In this case the film has a different two-dimensional periodicity and a smaller thickness (1.7 nm). One way to consider QSE only is to saturate the dangling bonds by

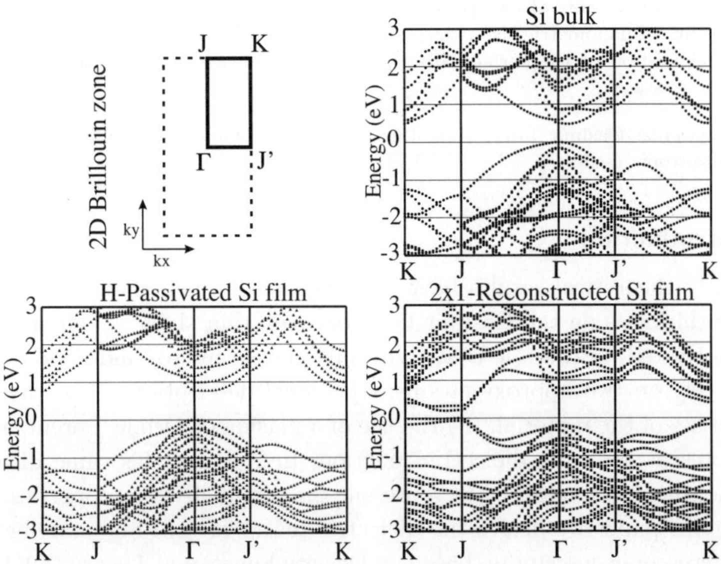

Fig. 1. Energy band structure of the H–passivated Si(111) film and the one with 2×1 reconstructed surfaces along the high symmetry directions of the 2D Brillouin zone. A sketch of the projected bulk band structure is given for comparison.

atomic H adsorption, which removes the surface reconstruction and leaves the band gap free of surface states. Therefore one can separate the SE from QSE just by comparing the electronic structure of the H terminated film with the one of the film with the 2×1 surfaces.

To calculate the film electronic structure we use the so-called repeated slab approach[6] and the DFT–KS implementation of the PWscf code[7] in the Local Density Approximation (LDA).

Figure 1 displays the 2D band structure along the high symmetry directions of the 2D Brillouin zone of the film with 2×1 surfaces. We give also the band structure of the film with the H–passivated surface and a sketch of the bulk band structure. One can see that while the passivated surface shows a behaviour in the band gap region rather similar the bulk, the band structure for the film with 2×1 surfaces has four bands in the band gap. Thė bands arise from surface states at the two sides of the film: one band of each couple is empty and the other is filled. These bands are responsible of the optical transitions that have been detected several years ago by surface optical absorption.[5]

3. The film dielectric tensor

Having at disposal the electron energies and wavefunctions we have calculated the microscopic dielectric matrix[8] and from it we have determined the macroscopic dielectric tensor of the film following the procedure indicated in Ref 9. In the case of bulk Si the dielectric tensor is isotropic and the diagonal comp1onents are given by the Si dielectric function. For the H–passivated film the tensor character arises essentially from size effects:

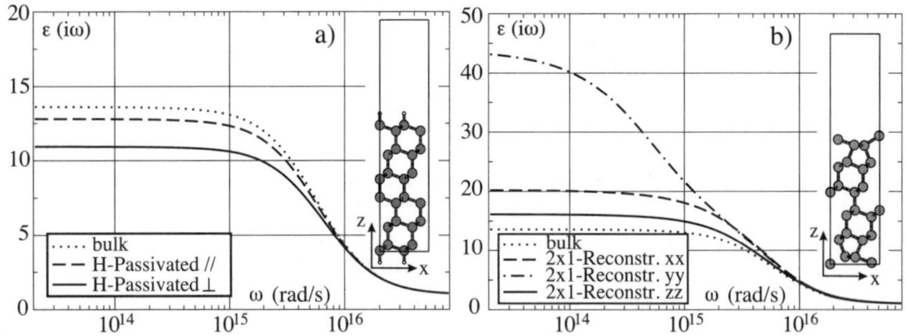

Fig. 2. London transform of the components of the calculated macroscopic dielectric tensor. 2a) case of H–terminated film; 2b) film with 2×1 surfaces. The insets show the film geometry.

the modifications induced by the H–Si bond in the electronic structure do not affect significantly the dielectric behaviour in the low energy range compared to the bulk. As a consequence of the symmetry the lateral components are equal $\epsilon_{xx} = \epsilon_{yy} = \epsilon_{\|}$ and take values close to the bulk dielectric function, while $\epsilon_{zz} = \epsilon_{\perp}$ is different. The London Transform of the dielectric tensor components are displayed in Fig. 2a. It is seen that the values are slightly lower that the bulk dielectric function.

The situation changes drastically for the 2×1 reconstructed film. As shown in Fig. 2b the components of the dielectric tensor are higher than the bulk, the discrepancy being much larger for ϵ_{yy}. The deviations with respect to the bulk are due to the presence of the surface bands, which reduce the band gap value leading to an increase of the static limit.

4. Calculation of the Casimir force

The calculations of the Casimir force can be done in the framework of the Lifshitz theory. For the H–passivated films this requires the extension of the formalism of Ref. 10 to films of uniaxial dielectric media. For the reconstructed surface one has to resort to a more complex theory.[11] A detailed discussion of the extension of Lifshitz theory to this case will be given elsewhere.[12]

To illustrate the importance of QSE and SE we give in Fig. 3 the calculated Casimir force per unit area between H–passivated and reconstructed Si films as a function of the distance. We plot the ratio between the force

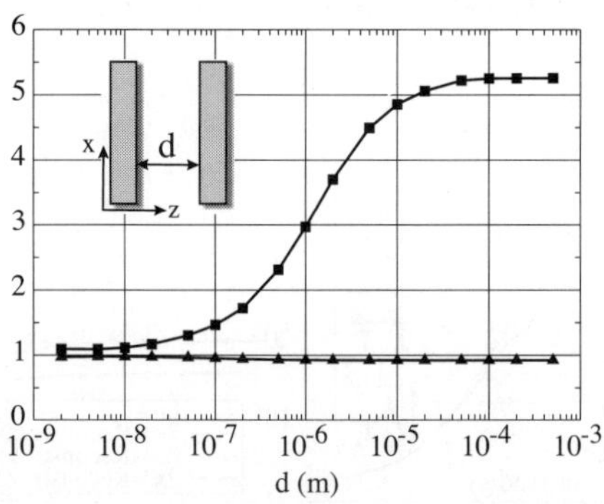

Fig. 3. Plot of the ratios of the calculated Casimir forces between Si films as a function of the film distance. The curve with ▲ (■) gives the ratio between the force calculated for the H–passivated (2×1 reconstructed) film and the one obtained with the bulk dielectric function.

determined using the appropriate dielectric tensor and the one obtained with the bulk dielectric function. The comparison shows that the effect of surface states is quite significant for distances larger than 10 nm and becomes very important at large distances, where the force is essentially determined by the value of the London transform at low frequencies. We believe that these results give evidence of the necessity of accounting for the modifications of the electronic structure in the determination of the force between nanometric objects.

Acknowledgments

M. Govoni acknowledges financial support by the European Science Foundation (ESF) within the activity 'New Trends and Applications of the Casimir Effect' (www.casimir-network.com).

References

1. I. Pirozhenko and A. Lambrecht, *Phys. Rev. A* **77**, 013811 (2008).
2. Z. Lenac and M. S. Tomaš, *Phys. Rev. A* **78**, 023834 (2008).
3. A. Benassi and C. Calandra, *J. Phys. Conf. Ser.* **161**, 012007 (2009).
4. K. C. Pandey, *Phys. Rev. Lett.* **47**, 1913 (1981).
5. K. C. Pandey, *Phys. Rev. Lett.* **49**, 223 (1982).
6. M. Schlüter and M. L. Cohen, *Phys. Rev. B* **17**, 716 (1978).
7. P. Giannozzi, S. Baroni, N. Bonini, M. Calandra, R. Car, C. Cavazzoni, D. Ceresoli, G. L. Chiarotti, M. Cococcioni, I. Dabo, A. Dal Corso, S. de Gironcoli, S. Fabris, G. Fratesi, R. Gebauer, U. Gerstmann, C. Gougoussis, A. Kokalj, M. Lazzeri, L. Martin-Samos, N. Marzari, F. Mauri, R. Mazzarello, S. Paolini, A. Pasquarello, L. Paulatto, C. Sbraccia, S. Scandolo, G. Sclauzero, A. P. Seitsonen, A. Smogunov, P. Umari and R. M. Wentzcovitch, *J. Phys. Condens. Mat.* **21**, 395502 (2009).
8. L. Martin-Samos and G. Bussi, *Comput. Phys. Commun.* **180**, 1416 (2009).
9. V. I. Gavrilenko and F. Bechstedt, *Phys. Rev. B* **54**, 13416 (1996).
10. F. Zhou and L. Spruch, *Phys. Rev. A* **52**, 297 (1995).
11. P. Yeh, *Optical waves in layered media* 1988.
12. M. Govoni, A. Benassi and C. Calandra, To be published, (2010).

ON THE CASIMIR ENTROPY BETWEEN "PERFECT CRYSTALS"

C. HENKEL* and F. INTRAVAIA†

Institut für Physik und Astronomie, Universität Potsdam,
Karl-Liebknecht-Str. 24/25, 14476 Potsdam, Germany
**E-mail: henkel@uni-potsdam.de*

We give a re-interpretation of an 'entropy defect' in the electromagnetic Casimir effect. The electron gas in a perfect crystal is an electromagnetically disordered system whose entropy contains a finite Casimir-like contribution. The Nernst theorem (third law of thermodynamics) is not applicable.

Keywords: Temperature; entropy; dissipation; overdamped mode.

1. Introduction

It is well known that fluctuation interactions at nonzero temperature are entropic in character, a prominent example being the critical Casimir effect in liquid mixtures close to a continuous phase transition (see Ref.1 for an overview). The electromagnetic Casimir interaction is also associated with an entropy that determines its limiting behaviour at high temperatures and/or large distances.[2–4] The Casimir entropy for two material plates has recently attracted much interest also for low temperatures, as for certain situations a violation of the third law of thermodynamics (the Nernst heat theorem) has been claimed.[5–7] This has been used to argue in favor of a description where the DC conductivity of the metallic plates is ignored. Although the result of this theoretical prescription provides a better fit to recent experiments,[8] the situation is, however, not satisfactory from the physical point of view. In addition, a similar analysis for an experiment with laser-irradiated semiconductors[9] leaves open the meaning of the threshold value above which the DC conductivity should be included in the theory.

†Present address: Theoretical Division, MS B213, Los Alamos National Laboratory, Los Alamos NM 87545, U. S. A.

Much has been said about spatially dispersive mirrors where the third law is verified, due to the anomalous skin effect,[10] and where a continuous cross-over from a dielectric to a perfectly conducting response has been found.[11] We focus in these proceedings on a strictly local framework, mainly for simplicity, but also to show that this case is thermodynamically consistent as well. We shall see that, indeed, spatial dispersion plays only a small role in the range of wave vectors that are relevant for current Casimir experiments.

We take up the interpretation of Ref.12 where the nonzero Casimir entropy found as $T \to 0$ was associated to two oscillators coupled via a third one. Following this idea, we consider two half-spaces filled by an ideal electron gas, separated by a distance L, and provide a direct calculation of the entropy per area $S(L) = \lim_{T\to 0} S(L,T)$ in one of the two field polarizations. This calculation highlights the following point:

The Casimir entropy $S(L) < 0$ results from the coupling between two systems that are not in equilibrium as $T \to 0$. They are filled with a *frozen* magnetization and, in the local limit, have separately divergent (bulk and surface) entropies that characterize the disorder of the electromagnetic configuration. The Casimir entropy is the correction to additivity when the two bodies are close enough for their frozen currents to be mutually coupled by the quasi-static magnetic fields that 'leak' through their surfaces. The Nernst theorem is clearly not applicable for this disordered system. The situation is quite similar to the 'ideal conductor' (in distinction to a superconductor, see Ref.13) that does not reach thermodynamical equilibrium as it is cooled, because its random bulk currents freeze. (See Ref.14 where it is argued how special this ideal conductor case is.)

2. Casimir Entropy from Frozen Medium Currents

2.1. *Motivation*

We have analyzed in a recent paper[15] the overdamped field modes to which the unusually large thermal corrections to the Casimir force between metals can be attributed. Substantiating previous observations,[16,17] we have interpreted the characteristic frequency, $\xi_L = D/L^2 = (\mu_0 \sigma L^2)^{-1}$, in terms of a diffusion equation (diffusion coefficient D) satisfied by the magnetic field and electric currents in a medium with DC conductivity σ.

There is no contradiction between thermodynamics and fluctuation electrodynamics in this case. Fields and currents induced in the metal are clearly damped and lose energy into the phonon bath, say. In equilibrium, however, this is compensated by field and current fluctuations that are created by the bath. This concept can be traced back to the Einstein–Langevin theory of Brownian motion[18] and is also the very essence of the fluctuation-dissipation theorem.[19,20] Moreover, for the quantum field theory, the quantum (or zero-point) fluctuations of the bath variables are an essential tool to establish at all times the commutation relations for the field operators.[21–25] In the field theory considered in Ref.15, we dealt with overdamped modes: if the wave equation were homogeneous, its eigenfrequencies would be purely imaginary, similar to free Brownian particles. The quantum theory of Brownian motion[26] provides a consistent scheme for the quantum thermodynamics of this damped system. In this setting, nonzero entropies and even negative heat capacities find a quite natural explanation (see, e.g., Refs.27,28).

In the particular case of an ideal electron gas (or 'perfect crystal'), the diffusion constant $D = D(T) = \mathcal{O}(T^2)$. As the temperature drops to zero, the diffusion-dominated modes of the electromagnetic field do not reach a unique ground state, but remain in the classical regime $\hbar\xi_L \ll T$. Indeed, we demonstrate here that the exact Casimir 'entropy defect' between perfect crystals is of classical origin.

The basic ingredient are transverse modes that extend throughout the bulk of the gas: static current waves interlocked with a magnetic field. The magnetic fields associated to the bulk currents leave one medium, by continuity, and cross the vacuum gap to the other medium in the form of (transverse) evanescent waves. This coupling between the two media changes slightly the wave vector of each current mode. Summing over all modes, we get a non-zero change in entropy that depends, quite naturally, on the separation L. It represents the distance-dependent change per unit area of the (much larger) entropy of the two frozen bulk systems.

2.2. *Lagrangian and conservation laws*

We start with the Lagrangian density

$$\mathcal{L} = \frac{nm}{2}\dot{\boldsymbol{\xi}}^2 + en\dot{\boldsymbol{\xi}} \cdot \mathbf{A} - \frac{1}{2\mu_0}(\nabla \times \mathbf{A})^2 \tag{1}$$

where the field $\boldsymbol{\xi}$ describes the displacement of a charged fluid element, $\mathbf{A}$ is the vector potential, n a constant background charge density and e a coupling constant with units of charge. The current density is $\mathbf{j} = en\dot{\boldsymbol{\xi}}$ [see Eq.(2) below], so that $\dot{\boldsymbol{\xi}}$ represents a velocity field. The first term in Eq.(1) is thus the kinetic energy (density), the second one a bilinear coupling, and the third one the magnetic energy. Note that we neglect electric fields here. This is consistent if we make the assumption that $\nabla \cdot \boldsymbol{\xi} = \nabla \cdot \mathbf{A} = 0$. The first equality ensures that the medium displacement does not produce any charge density, the second one is the Coulomb gauge. The variation of the Lagrangian (1) with respect to $\mathbf{A}$ gives the Faraday equation

$$en\dot{\boldsymbol{\xi}} - \frac{1}{\mu_0}\nabla \times (\nabla \times \mathbf{A}) = 0 \tag{2}$$

In addition, the displacement field $\boldsymbol{\xi}$ is a cyclic variable, hence we get a conserved momentum field

$$\frac{\partial \boldsymbol{\pi}}{\partial t} = 0, \qquad \boldsymbol{\pi} = nm\dot{\boldsymbol{\xi}} + en\mathbf{A} \tag{3}$$

There are two ways to implement this conservation law physically.[29]

(i) In a *London superconductor*, the current density is tied at all times to the vector potential, with the momentum $\boldsymbol{\pi}$ being zero:

$$\text{superconductor:} \quad \mathbf{j} = en\dot{\boldsymbol{\xi}} = -\frac{ne^2}{m}\mathbf{A} \tag{4}$$

The Maxwell–Faraday equation (2) becomes

$$\left(\lambda^{-2} - \nabla^2\right)\mathbf{A} = 0 \tag{5}$$

where the Meißner–London penetration depth λ is given by the familiar expression $\lambda^{-2} = \mu_0 ne^2/m = \Omega^2/c^2$ (Ω is the plasma frequency). Eq.(5) does only allow for solutions that start at the surface and exponentially decay into the bulk on a length scale λ (or shorter). Except for a surface layer of thickness $\sim \lambda$, the interior of the medium remains free of magnetic field: the Meißner–Ochsenfeld effect.

From the London equation (4), we can also conclude that the Meißner effect is maintained in time-dependent fields (at least with sufficiently slow variations; a detailed analysis clearly goes beyond the simple model considered here). For a given frequency component ω, the 'dielectric function' of the London superconductor can be read off from the polarization field associated to $\boldsymbol{\xi}$:

$$\mathbf{P} = en\boldsymbol{\xi} = \frac{\mathbf{j}}{-\mathrm{i}\omega} = -\frac{\varepsilon_0\Omega^2}{\omega^2}\mathbf{E} \tag{6}$$

leading to the so-called plasma model $\varepsilon(\omega) = 1 - \Omega^2/\omega^2$. There is no violation of causality here, if we read Eq.(4) as a retarded response function between the current density and the time integral of the electric field (i.e., the vector potential).

The option (ii) that complies with the conservation law (3) corresponds to an *ideal conductor*:

$$\text{ideal conductor:} \quad \frac{\partial \mathbf{j}}{\partial t} = 0 \quad \text{and} \quad \frac{\partial \mathbf{A}}{\partial t} = 0 \tag{7}$$

which means that currents, once created, are not damped and that the magnetic field is static. The value of the conserved momentum $\boldsymbol{\pi}$ is not restricted otherwise. The Faraday equation (2) then yields the field $\mathbf{A}$ in terms of its source $\mathbf{j}$. Note that the magnetic field is in this case tied to the current density, similar to the scalar potential and the charge density in Coulomb-gauge electrodynamics.[30] Let us switch to reciprocal space with wavevector $\mathbf{q}$: the vector potential $\mathbf{A_q}$ created by the current is

$$\mathbf{A_q} = \mu_0 \frac{\mathbf{j_q}}{\mathbf{q}^2} \tag{8}$$

so that the Lagrangian (1) becomes

$$L = \frac{\mu_0 V}{2} \sum_{\mathbf{q}} \left(\lambda^2 + \mathbf{q}^{-2}\right) |\mathbf{j_q}|^2 \tag{9}$$

where V is the quantization volume. The conjugate momentum becomes

$$\boldsymbol{\pi}_\mathbf{q} = \frac{m}{e} \left(1 + \frac{1}{\lambda^2 \mathbf{q}^2}\right) \mathbf{j_q} \tag{10}$$

2.3. *Normal modes and entropy*

The normal modes of the effective Lagrangian (9) can clearly be chosen as plane waves, labelled by wave vector $\mathbf{q}$ and polarization index μ. The associated Hamiltonian, expressed in terms of the canonical momentum field, is then

$$H = \frac{V}{2nm} \sum_{\mathbf{q},\mu} \left(1 + \frac{1}{\lambda^2 \mathbf{q}^2}\right)^{-1} |\pi_{\mathbf{q}\mu}|^2 \tag{11}$$

The (classical) thermodynamics of this system is determined by summing the free energies of the normal modes over the quantum numbers $\mathbf{q}$, μ. For one mode, we find by calculating the classical partition function ($\beta = 1/T$)

$$F_{\mathbf{q}\mu} = -T \log \int \mathrm{d}\pi_{\mathbf{q}\mu} \exp\left(-\frac{\beta}{2} \epsilon_q |\pi_{\mathbf{q}\mu}|^2\right) = \frac{T}{2} \log \frac{\beta \epsilon_q}{2\pi} \tag{12}$$

where $\epsilon_q = (V/nm)(1 + \lambda^{-2}\mathbf{q}^{-2})^{-1}$ determines the mode's energy. The entropy of this polarization mode is

$$S_{\mathbf{q}\mu} = -\frac{\partial F_{\mathbf{q}\mu}}{\partial T} = -\frac{1}{2} \log \frac{\beta \epsilon_q}{2\pi} + \frac{1}{2} \tag{13}$$

where the equipartition term $+1/2$ comes from the β in the logarithm.

When we sum this over all modes, the entropy becomes to leading order extensive in the volume V of the medium. The part that depends on the surface is calculated in the usual way. Consider two media (total volume V) with parallel surfaces of area $\mathcal{A}$ facing each other at a distance L and write the entropy of the total system in the form

$$S = Vs + 2\mathcal{A}S_{\rm surf} + \mathcal{A}S(L) \tag{14}$$

where s is the (intensive) bulk entropy density, $S_{\rm surf}$ is the entropy per area of one (isolated) surface and $S(L)$ the Casimir entropy per area. We can read the latter as the deviation from additivity in the system of two media: it thus describes how the disorder (or information content) of the two plates is changed by the coupling across the vacuum gap of thickness L.

The physical mechanism for this coupling is the penetration of magnetic fields through the medium surface, as allowed for by the electromagnetic boundary conditions. In the vacuum between the media, the fields satisfy the Laplace equation $\nabla^2 \mathbf{A} = 0$: for a given wave vector $\mathbf{k}$ parallel to the surface, they 'propagate' perpendicular to the surface (along the z-axis, say) as evanescent waves $\sim \exp(\pm kz)$ where $k = |\mathbf{k}|$. For a single surface, only solutions that decay into the vacuum are permitted. In the gap $0 \le z \le L$ between two surfaces, even and odd solutions $\cosh k(z - L/2)$ and $\sinh k(z - L/2)$ can be constructed. This is illustrated schematically in Fig. 1.

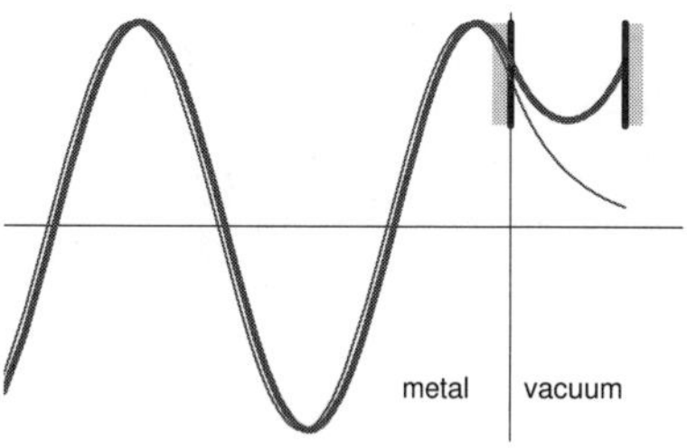

Fig. 1. Illustration of standing waves at the surface of an ideally conducting medium. We plot the component of the vector potential tangential to the surface. Thin line: isolated surface, thick line: mode between two surfaces with even parity.

Both the surface entropy and the Casimir entropy can be calculated from the phase shifts of standing wave modes (see Refs.31,32 for details). For the surface entropy,

$$S_{\rm surf} = \sum_{\mu} \int \frac{{\rm d}^2 k}{(2\pi)^2} \int_0^{\infty} {\rm d}k_z S_{\mathbf{q}\mu} \left(-\frac{1}{\pi} \frac{\partial \theta_\mu}{\partial k_z} \right) \tag{15}$$

where $\mathbf{q} = (\mathbf{k}, k_z)$, and the mode functions in the medium ($z \le 0$) are proportional to ${\rm e}^{{\rm i}\mathbf{k}\cdot\mathbf{r}_\parallel} \sin(k_z z + \theta_\mu)$ with $\mathbf{r}_\parallel = (x, y)$ the coordinates parallel to the surface. From this form, we can also read off a 'reflection coefficient' for (time-independent) waves from within the medium, $r_\mu = -\mathrm{e}^{-2\mathrm{i}\theta_\mu}$. From a physical point of view, we can interpret the phase derivative in Eq.(15) as a density of modes in $\mathbf{q}$-space, more precisely, its change due to the surface.

For an isolated interface, the usual matching of the component of the vector potential tangential to the surface and its derivative at the interface yields in TE-polarization (current perpendicular to the plane of incidence spanned by $\mathbf{k}$ and the surface normal)

$$r_{\mathrm{TE}} = \frac{k_z - \mathrm{i}k}{k_z + \mathrm{i}k}, \qquad \tan\theta_{\mathrm{TE}} = -\frac{k_z}{k} \tag{16}$$

Here, $k = |\mathbf{k}|$ gives the decay constant of the evanescent wave on the vacuum side. In the TM-polarization (current in the plane of incidence), the current has to satisfy the boundary condition $\lim\limits_{z\to 0} j_z(z) = 0$ to avoid the build-up of a surface charge sheet. (That case would require electrical field energy in the Lagrangian (1) and is best described within a spatially dispersive model.[31]) This boundary condition immediately leads to $r_{\mathrm{TM}} = 1$, and there is no phase shift. The surface entropy in TM-polarization hence vanishes, while it is logarithmically divergent at large $\mathbf{q}$ in the TE-polarization: $S_{\mathrm{surf}} \approx -(8\pi\lambda^2)^{-1}\log(q_c\lambda)$ with a short-range cutoff q_c. One needs a non-local description of the material response (spatial dispersion) to get a finite result, see, e.g., Ref.33 for the surface self-energy.

For the Casimir entropy, a local calculation is sufficient, as we shall see now: the reflection phases for even and odd modes in the vacuum gap are found as (we henceforward suppress the TE-polarization label)

$$\tan\theta_{\mathrm{even}}(L) = -\frac{k_z}{k}\coth\frac{kL}{2}, \qquad \tan\theta_{\mathrm{odd}}(L) = -\frac{k_z}{k}\tanh\frac{kL}{2} \tag{17}$$

The entropy per area for the two-surface system, $2S_{\mathrm{surf}} + S(L)$, is then given by Eq.(15) with θ replaced by $\theta_{\mathrm{even}}(L) + \theta_{\mathrm{odd}}(L)$. Subtract twice the single-interface phase shift and calculate the quantity

$$\exp 2\mathrm{i}[\theta_{\mathrm{even}}(L) + \theta_{\mathrm{odd}}(L) - 2\theta] = \frac{1 - r^2\,\mathrm{e}^{-2kL}}{1 - (r^*)^2\,\mathrm{e}^{-2kL}} \tag{18}$$

as can be checked with straightforward algebra. The Casimir entropy from TE-polarized bulk currents becomes [combining Eqs.(13, 15, 18)]

$$S(L) = \int\limits_0^\infty \frac{k\mathrm{d}k}{2\pi} \int\limits_0^\infty \frac{\mathrm{d}k_z}{2\pi} \left(\log\frac{\beta V\lambda^2\mathbf{q}^2}{2\pi n m(1+\lambda^2\mathbf{q}^2)} - 1\right) \frac{\partial}{\partial k_z}\mathrm{Im}\,\log\left(1 - r^2\,\mathrm{e}^{-2kL}\right) \tag{19}$$

The terms independent of k_z in the entropy per mode are irrelevant: after a partial integration, the integrated terms vanish because for $k_z \to 0, \infty$, the reflection coefficient r becomes real. Manifestly, short-wavelength modes with $2kL \gg 1$ are suppressed, and a local theory is sufficient unless L becomes comparable to the length scales typical for spatial dispersion (mean free path, Debye-Hückel screening length, Fermi wavelength).

2.4. *Calculation of the entropy*

Integrating Eq.(19) by parts, we have to evaluate the integral:

$$I_k = -\int_{-\infty}^{\infty} \frac{dk_z}{2\pi i\, \lambda^2} \frac{k_z \log\left(1 - r^2\, e^{-2kL}\right)}{(k_z^2 + k^2)(k_z^2 + k^2 + \lambda^{-2})} \tag{20}$$

where we recall that r is given by Eq.(16) above. We have extended the integration domain to $-\infty < k_z < +\infty$, using the property $r(-k_z) = [r(k_z)]^*$. Observe that $r(k_z)$, as a function of complex k_z, satisfies $|r(k_z)| \le 1$ in the upper half-plane, that the integrand vanishes at infinity, and evaluate the integral by closing the contour. There are simple poles at $k_z = ik$ and $k_z = i(k^2 + \lambda^{-2})^{1/2}$. At the first pole, the reflection coefficient (16) vanishes, and we get from the second one:

$$I_k = \frac{1}{2} \log\left[1 - r_{\rm pl}^2(k, 0) e^{-2kL}\right], \quad r_{\rm pl}(k, 0) = \frac{(k^2 + \lambda^{-2})^{1/2} - k}{(k^2 + \lambda^{-2})^{1/2} + k} \tag{21}$$

As it happens, the reflection coefficient $r_{\rm pl}(k, \omega)$ for electromagnetic waves from a plasma half-space [dielectric function after Eq.(6)] appears here, evaluated at zero frequency and in the TE-polarization. If Eq.(21) is integrated over k, we get the 'entropy defect' calculated in the Lifshitz theory of the Casimir effect using the local dielectric function of a 'perfect crystal' [see, e.g., Eq.(20) of Ref.6]:

$$S(L) = \int_0^{\infty} \frac{k dk}{2\pi} I_k = \int_0^{\infty} \frac{k dk}{4\pi} \log\left[1 - r_{\rm pl}^2(k, 0)\, e^{-2kL}\right] \tag{22}$$

A switch to the integration variable to $y = 2kL$ shows that

$$S(L) = -\frac{\zeta(3)}{16\pi} \frac{f(L/\lambda)}{L^2} \tag{23}$$

where the scaling function $f(L/\lambda)$, plotted in Fig. 2, is dimensionless and normalized to unity in the limit $L \gg \lambda$. Indeed, in this regime, one may expand $r_{\rm pl}(k, 0)$ in powers of k to get the asymptotic series [for higher terms, see Eq.(20) of Ref.6]:

$$L \gg \lambda : \quad f(L/\lambda) \approx 1 - 4\frac{\lambda}{L} + 12\frac{\lambda^2}{L^2} + \mathcal{O}(\lambda/L)^3 \qquad (24)$$

In the limit $\lambda \to 0$, we reproduce the entropy defect of the 'modified ideal metal' (MIM),[12] although from a quite different approach. This limit can be understood from an analysis of the Matsubara series:[7] indeed, for all terms in the series, the reflection coefficients for the perfect crystal become those of the MIM when the plasma frequency $\Omega = c/\lambda$ is taken to infinity. A direct adaptation of the present model to the MIM appears somewhat subtle, however, and will be presented elsewhere.

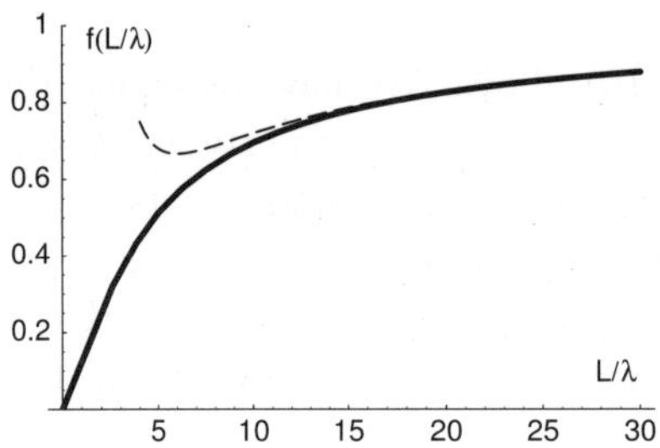

Fig. 2. Scaling function $f(L/\lambda)$ for the Casimir entropy of for the ideal electron gas, defined by Eqs.(22, 23). Dashed: Eq.(24). $\lambda = c/\Omega$ is the London-Meißner penetration depth (plasma wavelength). The same $f(L/\lambda)$ governs the 'entropy defect' for the electromagnetic Casimir effect between perfect crystals (i.e., Lifshitz theory with the Drude dielectric function and scattering rate $\gamma(T) = \mathcal{O}(T^2)$), see Eq.(20) in Ref.6.

3. Conclusions

We have analyzed the Casimir entropy for the ideal electron gas, in particular the contribution of electric currents frozen inside the bulk. This system shows (electromagnetic) disorder, and the third law of thermodynamic does not apply in its orthodox formulation. We have recovered the 'entropy defect' (negative Casimir entropy at zero temperature) reported in several places in the literature. Its thermodynamically consistent interpretation is the measure of the change in the disorder of the frozen currents due to their interaction through quasi-static magnetic fields.

Acknowledgments

We acknowledge discussions with H. Haakh and support from the European Science Foundation (programme 'New Trends and Applications of the Casimir Effect', www.casimir-network.com). F.I. thanks the Alexander von Humboldt Foundation for financial support.

References

1. A. Gambassi, C. Hertlein, L. Helden, S. Dietrich and C. Bechinger, *Europhys. News* **40**, 18 (Jan 2009).
2. R. Balian and B. Duplantier, *Ann. Phys. (N. Y.)* **112**, 165 (1978).
3. J. Feinberg, A. Mann and M. Revzen, *Ann. Phys. (N.Y.)* **288**, 103 (2001).
4. B. Jancovici and L. Šamaj, *Europhys. Lett.* **72** (2005) 35. P. R. Buenzli and Ph. A. Martin, *Europhys. Lett.* **72** (2005) 42.
5. G. L. Klimchitskaya and V. M. Mostepanenko, *Phys. Rev. A* **63**, 062108 (2001).
6. V. B. Bezerra, G. L. Klimchitskaya, V. M. Mostepanenko and C. Romero, *Phys. Rev. A* **69**, 022119 (2004).
7. F. Intravaia and C. Henkel. *J. Phys. A: Math. Gen.* **41**, 164018 (2008).
8. R. Decca, D. López, E. Fischbach, G. Klimchitskaya, D. Krause and V. Mostepanenko, *Eur. Phys. J. C* **51**, 963 (2007).
9. F. Chen, G. L. Klimchitskaya, V. M. Mostepanenko and U. Mohideen, *Phys. Rev. B* **76**, 035338 (2007).
10. V. B. Svetovoy, *Phys. Rev. Lett.* **101**, 163603 (2008).
11. L. P. Pitaevskii, *Phys. Rev. Lett.* **101**, 163202 (2008).
12. J. S. Hoye, I. Brevik, J. B. Aarseth and K. A. Milton, *Phys. Rev. E* **67**, 056116 (2003).
13. L. D. Landau, E. M. Lifshitz and L. P. Pitaevskii, *Electrodynamics of continuous media*, 2nd edn. (Pergamon, Oxford, 1984).
14. W. A. B. Evans and G. Rickayzen, *Ann. Phys. (N.Y.)* **33**, 275 (1965).
15. F. Intravaia and C. Henkel, *Phys. Rev. Lett.* **103**, 130405 (2009).
16. J. R. Torgerson and S. K. Lamoreaux, *Phys. Rev. E* **70**, 047102 (2004).
17. V. B. Svetovoy, *Phys. Rev. A* **76**, 062102 (2007).
18. A. Einstein, *Ann. Physik (Leipzig), Vierte Folge* **17**, 549 (1905).
19. H. B. Callen and T. A. Welton, *Phys. Rev.* **83**, 34 (1951).
20. X. L. Li, G. W. Ford and R. F. O'Connell, *Phys. Rev. E* **48**, 1547 (1993).
21. B. Huttner and S. M. Barnett, *Europhys. Lett.* **18**, 487 (1992).
22. T. Gruner and D.-G. Welsch, *Phys. Rev. A* **51**, 3246 (1995).
23. A. Tip, *Phys. Rev. A* **56**, 5022 (1997).
24. L. G. Suttorp, *J. Phys. A: Math. Gen.* **40**, 3697 (2007).
25. S. Scheel and S. Y. Buhmann, *acta phys. slov.* **58**, 675 (2008).
26. U. Weiss, *Quantum Dissipative Systems*, Series in Modern Condensed Matter Physics, Vol. 10, third edn. (World Scientific, Singapore, 2007).
27. G.-L. Ingold, P. Hänggi and P. Talkner, *Phys. Rev. E* **79**, 061105 (2009).
28. G.-L. Ingold, A. Lambrecht and S. Reynaud, *Phys. Rev. E* **80**, 041113 (2009).
29. F. London and H. London, *Proc. Roy. Soc. (London) A* **149**, 71 (1935).
30. C. Cohen-Tannoudji, J. Dupont-Roc and G. Grynberg, *Photons and Atoms — Introduction to Quantum Electrodynamics* (Wiley, New York 1989).
31. G. Barton, *Rep. Prog. Phys.* **42**, 963 (1979).
32. M. Bordag, U. Mohideen and V. M. Mostepanenko, *Phys. Rep.* **353**, 1 (2001).
33. N. J. Morgenstern Horing, E. Kamen and G. Gumbs, *Phys. Rev. B* **31**, 8269 (1985).

THEORY OF THE CASIMIR EFFECT FOR GRATINGS

VALERY N. MARACHEVSKY

V. A. Fock Institute of Physics, Saint-Petersburg State University,
198504 St.Petersburg, Russia
E-mail: maraval@mail.ru

The theory of the Casimir effect for two parallel gratings with a coinciding period is developed.

Keywords: Casimir effect; Rayleigh expansion; gratings; lateral force.

1. Introduction

The aim of this paper is a construction of the Casimir effect theory for grating geometries which can be applied to obtain exact results beyond the Lifshitz theory[1] and proximity force approximation.

The main idea of the present approach is the use of the equation for normal modes of two dielectrics (or metals) separated by a vacuum gap. This equation combined with the argument principle yields the Casimir energy of the system. The equation for normal modes can be immediately written if one knows the reflection matrices of the electromagnetic field for each of the bodies in the system. Thus the determination of the reflection matrices constitutes an important part of every Casimir problem for geometries of the complicated form. This is the basic idea of the theory which is usually called the scattering approach in the Casimir effect.

The scattering approach in the theory of the Casimir effect started from Ref. 2 where it was applied to obtain the Lifshitz formula.[1] Multipole scattering technique was developed in Ref. 3. This technique works well for large separations between the objects.[4] At small separations the leading asymptotic expansion of the Casimir energy at zero temperature within the scattering approach was developed in Ref. 5. Other techniques within the scattering approach to the Casimir effect were developed in Refs. 6–8. The technique developed in Refs. 9,10 uses the Rayleigh basis which is a generalization of the plane wave basis. The Rayleigh basis is a natural choice for grating geometries. In this paper we develop a theory of the Casimir effect for two parallel gratings with a coinciding period.

2. Theory of the Casimir effect for gratings geometries

Consider two parallel three-dimensional longitudinal (along z-axis) periodic (along x-axis) dielectric (or metallic) gratings of arbitrary form separated by a vacuum slit so that they form a waveguide and one grating is located above another one (we suppose that the edge of the corrugation region of each grating is perpendicular to y axis). The periods of both gratings d coincide. In general, the upper and lower gratings can have different shapes and different dielectric (or metallic) properties. In Fig.1 the height of the corrugation region is equal a for lower grating. The meaning of a lateral displacement s becomes obvious from the comparison of Fig.1 with Fig.2 and Fig.3. We suppose that the space between the two gratings is a vacuum with $\epsilon = \mu = 1$ and adopt the condition $\mu = 1$ inside the gratings.

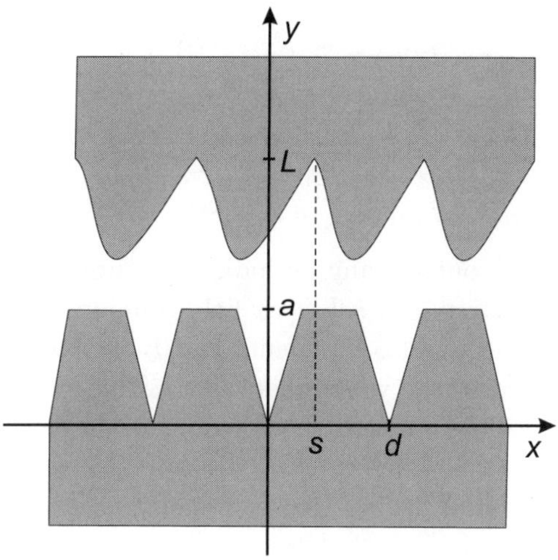

Fig. 1. Two gratings with a coinciding period d.

The physical problem is time and z invariant, so electric and magnetic fields can be written in the form:

$$E_i(x, y, z, t) = E_i(x, y) \exp(ik_z z - i\omega t), \tag{1}$$

$$H_i(x, y, z, t) = H_i(x, y) \exp(ik_z z - i\omega t). \tag{2}$$

The solutions of Maxwell equations should satisfy the quasi-periodicity conditions:

$$E_i(x+d,y) = e^{ik_x d} E_i(x,y) \tag{3}$$

$$H_i(x+d,y) = e^{ik_x d} H_i(x,y). \tag{4}$$

In every Casimir problem one needs to determine the complete basis of solutions. Let us suppose that the upper grating is absent. We consider a generalized conical diffraction problem ($k_z \neq 0$) on the lower grating. From now on we assume that k_x belongs to the interval $0 \leq k_x \leq 2\pi/d$ and define the x-component of the wave vector of the incident wave as $\alpha_p = k_x + 2\pi p/d$, p is an integer number. The solution of Maxwell equations with the given α_p for the longitudinal components of the electromagnetic field outside the corrugated region ($y \geq a$) may be written by making use of a Rayleigh expansion[11] for an incident monochromatic wave :

$$E_z(x,y) = I_p^{(e)} \exp(i\alpha_p x - i\beta_p^{(1)} y) + \sum_{n=-\infty}^{+\infty} R_{np}^{(e)} \exp(i\alpha_n x + i\beta_n^{(1)} y), \tag{5}$$

$$H_z(x,y) = I_p^{(h)} \exp(i\alpha_p x - i\beta_p^{(1)} y) + \sum_{n=-\infty}^{+\infty} R_{np}^{(h)} \exp(i\alpha_n x + i\beta_n^{(1)} y), \tag{6}$$

$$\alpha_n = k_x + 2\pi n/d, \quad \beta_n^{(1)2} = \omega^2 - k_z^2 - \alpha_n^2. \tag{7}$$

This solution is valid outside any periodic structure in x direction, in our notations it is valid for $y \geq a$. All other field components can be expressed in terms of the longitudinal components E_z, H_z by the standard formulas which are being used in the waveguide theory. This can be done since the factor $\exp(ik_z z)$ is conserved after the reflection of the electromagnetic wave from a grating.

At $y = 0$ the solution has to satisfy the expansions for a transmission matrix:

$$E_z(x,y) = \sum_{n=-\infty}^{+\infty} T_{np}^{(e)} \exp(i\alpha_n x - i\beta_n^{(2)} y), \tag{8}$$

$$H_z(x,y) = \sum_{n=-\infty}^{+\infty} T_{np}^{(h)} \exp(i\alpha_n x - i\beta_n^{(2)} y), \tag{9}$$

$$\beta_n^{(2)2} = \varepsilon\omega^2 - k_z^2 - \alpha_n^2, \tag{10}$$

which are valid for $y \leq 0$.

The E_z component of the electromagnetic field in the region $0 \le y \le a$ is defined as follows:

$$E_z(x, y) = \sum_{n=-\infty}^{+\infty} E_z^n(y) \exp(i\alpha_n x), \tag{11}$$

other components of the electromagnetic field in the region $0 \le y \le a$ are defined in analogy.

It is convenient to solve Maxwell equations with a fixed finite number of terms in the expansions (5), (6), (8), (9), (11), n is being changed from $-N$ to N. The solution can be obtained with needed accuracy for large enough N. In the limit $N \to +\infty$ the solution is exact.

Inside the corrugation region $0 \le y \le a$ it is convenient to rewrite Maxwell equations in the form of the first order differential equations, $\frac{\partial A}{\partial y} = M(y)A$, where $M(y)$ is a square matrix of a dimension $8N + 4$, $A^T = (E_z^N \dots E_z^{-N}, E_x^N \dots E_x^{-N}, H_z^N \dots H_z^{-N}, H_x^N \dots H_x^{-N})$. For a rectangular grating the matrix M is a constant one (independent on y). From Maxwell equations

$$ik_z E_x - \frac{\partial E_z}{\partial x} = i\omega H_y \tag{12}$$

$$ik_z H_x - \frac{\partial H_z}{\partial x} = -i\omega\varepsilon E_y \tag{13}$$

we get

$$E_y = (ik_z H_x - \partial H_z/\partial x)/(-i\omega\varepsilon) \tag{14}$$

$$H_y = (ik_z E_x - \partial E_z/\partial x)/(i\omega) \tag{15}$$

Then

$$E_y^n = \sum_m \left(\frac{1}{-i\omega\varepsilon}\right)_{n-m} (ik_z H_x^m - i\alpha_m H_z^m) \tag{16}$$

$$H_y^n = \frac{1}{i\omega}(ik_z E_x^n - i\alpha_n E_z^n), \tag{17}$$

$(\frac{1}{\varepsilon})_{n-m}$ is a notation for the Toeplitz matrix of the periodic in x function $1/\varepsilon(x, y)$.

The following equations follow from 4 remaining Maxwell equations:

$$dE_z^n/dy = ik_z E_y^n + i\omega H_x^n \tag{18}$$

$$dE_x^n/dy = i\alpha_n E_y^n - i\omega H_z^n \tag{19}$$

$$dH_z^n/dy = ik_z H_y^n - \sum_m (i\omega\varepsilon)_{n-m} E_x^m \tag{20}$$

$$dH_x^n/dy = i\alpha_n H_y^n + \sum_m (i\omega\varepsilon)_{n-m} E_z^m. \tag{21}$$

One can substitute equations (16), (17) into the equations (18)-(21) and obtain the system of the first order differential equations for Fourier components of the electromagnetic field $E_z^n, E_x^n, H_z^n, H_x^n$ in the region $0 \le y \le a$. For every $0 < y < a$ the functions $\varepsilon(x, y)$ or $1/\varepsilon(x, y)$ are periodic functions of x, Toeplitz matrices of these periodic functions can be evaluated analytically for any $0 < y < a$.

Now we have to determine the coefficients $R_{np}^{(e)}, R_{np}^{(h)}$ for a specific periodic geometry profile. One can determine the unknown Rayleigh coefficients by matching the solution of equations inside the corrugation region $0 \le y \le a$ with Rayleigh expansions (5),(6) at $y = a$ and expansions (8),(9) at $y = 0$. This can be done by imposing the continuity conditions on each Fourier component of the fields E_z, E_x, H_z, H_x at $y = 0$ and $y = a$.

There is no separation on TE and TM modes in general case. This is why the reflection matrix R_{1down} for a reflection from a lower grating can be defined as follows:

$$R_{1down}(k_x, k_z, \omega) = \\ = \begin{pmatrix} R_{n_1q_1}^{(e)}(I_p^{(e)} = \delta_{pq_1}, I_p^{(h)} = 0) & R_{n_2q_2}^{(e)}(I_p^{(e)} = 0, I_p^{(h)} = \delta_{pq_2}) \\ R_{n_3q_3}^{(h)}(I_p^{(e)} = \delta_{pq_3}, I_p^{(h)} = 0) & R_{n_4q_4}^{(h)}(I_p^{(e)} = 0, I_p^{(h)} = \delta_{pq_4}) \end{pmatrix}. \tag{22}$$

To obtain the Casimir energy we need to determine the eigenfrequencies of all normal modes of the electromagnetic field between two periodic gratings. These eigenfrequencies can be summed up by making use of an argument principle, which states:

$$\frac{1}{2\pi i}\oint \phi(\omega)\frac{d}{d\omega}\ln f(\omega)d\omega = \sum \phi(\omega_0) - \sum \phi(\omega_\infty), \tag{23}$$

where ω_0 are zeroes and ω_∞ are poles of the function $f(\omega)$ inside the contour of integration, degenerate eigenvalues are summed over according to their multiplicities. For the Casimir energy we have $\phi(\omega) = \hbar\omega/2$. The equation for eigenfrequencies of the corresponding problem of classical electrodynamics is $f(\omega) = 0$.

Consider first the plane-plane geometry when two dielectric parallel slabs (slab 1: $y < 0$, slab 2: $y > L$) are separated by a vacuum slit ($0 < y < L$). In this case TE and TM modes are not coupled. The equation for TE eigenfrequencies is:

$$f(\omega) = 1 - r_{1TEdown}(k_x, k_z, \omega) r_{2TEup}(k_x, k_z, \omega, L) = 0. \tag{24}$$

Here $r_{1TEdown}(k_x, k_z, \omega)$ is the reflection coefficient of a downward plane wave which reflects on a dielectric surface of a slab 1 at $y = 0$, while $r_{2TEup}(k_x, k_z, \omega, L)$ is the reflection coefficient of an upward plane wave which reflects on a dielectric surface of a slab 2 at $y = L$. One can deduce from Maxwell equations that $r_{2TEup}(k_x, k_z, \omega, L) = r_{2TEdown}(k_x, k_z, \omega)\exp(2ik_yL)$ ($r_{2TEdown}(k_x, k_z, \omega)$ is a reflection coefficient of a downward TE plane wave which reflects from a dielectric slab 2 temporarily located at the position of the slab 1, i.e. at $y < 0$). From (24) and the analogous equation for TM modes one immediately obtains the Lifshitz formula by making use of the argument principle (23).

For two periodic dielectrics or gratings separated by a vacuum slit one has to consider a reflection of downward and upward waves from a unit cell $0 < k_x < 2\pi/d$. Imagine we remove the upper grating from the system. The reflection matrix of the downward wave is defined as R_{1down} then. Imagine now that we remove the lower grating from the system. We denote the reflection matrix of the upward wave as R_{2up} then. The reflection matrices R_{1down}, R_{2up} depend on the wave vectors of the incident waves, parameters of the gratings and the mutual location of the gratings. The equation for normal modes states:

$$R_{1down}(k_x, k_z, \omega_i) R_{2up}(k_x, k_z, \omega_i, L, s)\psi_i = \psi_i, \tag{25}$$

where ψ_i is an eigenvector describing the normal mode with a frequency ω_i. Instead of equation (24) one obtains:

$$\det(I - R_{1down}(k_x, k_z, \omega) R_{2up}(k_x, k_z, \omega, L, s)) = 0. \tag{26}$$

For every k_x, k_z the solution of (26) yields possible eigenfrequencies ω_i of the solutions of Maxwell equations that should be substituted into the definition of the Casimir energy $E = \sum_i \hbar\omega_i/2$.

Suppose that the reflection matrix R_{2down} for a reflection from the fictitious (imaginary) grating located as in Fig.2 is known in the coordinates (x, y). Performing a change of coordinates $y = -y_1 + L$, $x = x_1 - s \quad (s < d)$ in (5), (6), it is possible to obtain a matrix R_{2up} for a reflection of upward waves from a grating with the same profile turned upside-down, displaced

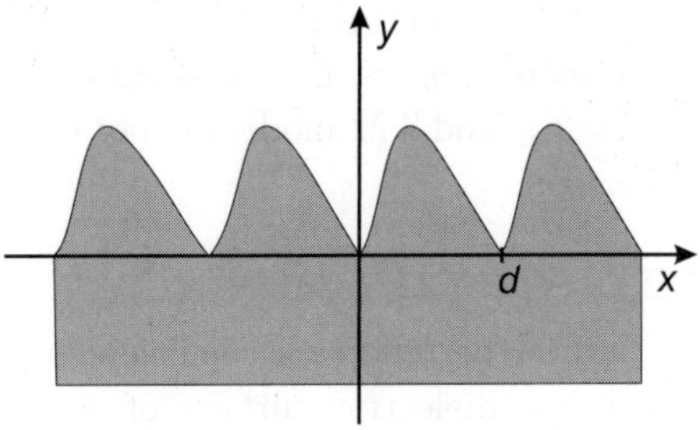

Fig. 2. A fictitious grating for which one evaluates R_{2down}.

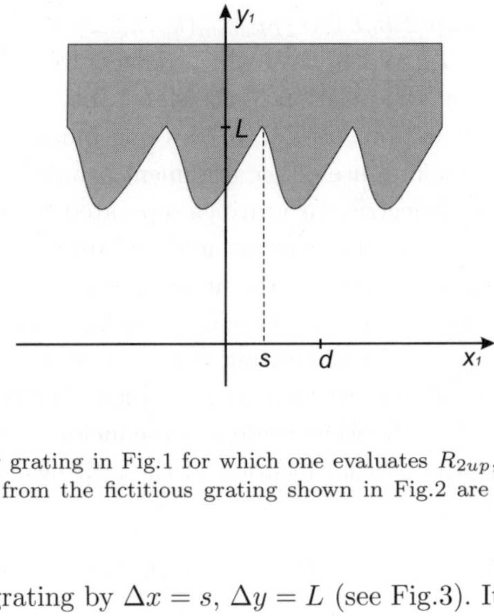

Fig. 3. The upper grating in Fig.1 for which one evaluates R_{2up}, the normal and lateral displacements from the fictitious grating shown in Fig.2 are denoted by L and s respectively.

from the lower grating by $\Delta x = s$, $\Delta y = L$ (see Fig.3). It follows that

$$R_{2up}(k_x, k_z, i\omega, L, s) = \\ = Q^*(s)K(k_x, k_z, i\omega, L)R_{2down}(k_x, k_z, i\omega)K(k_x, k_z, i\omega, L)Q(s), \quad (27)$$

where $R_{2down}(k_x, k_z, i\omega)$ is a reflection matrix of downward waves from the grating in the system of coordinates (x, y) depicted on Fig.2. Here $K(k_x, k_z, i\omega, L)$ is a diagonal $2(2N+1)$ matrix of the form:

$$K(k_x, k_z, i\omega, L) = \begin{pmatrix} G_1 & 0 \\ 0 & G_1 \end{pmatrix}, \quad (28)$$

with matrix elements $e^{-L\sqrt{\omega^2+k_z^2+(k_x+\frac{2\pi m}{d})^2}}$ $(m = -N \dots N)$ on a main diagonal of a matrix G_1. The lateral translation $2(2N+1)$ diagonal matrix

$Q(s)$ is defined as follows:

$$Q(s) = \begin{pmatrix} G_2 & 0 \\ 0 & G_2 \end{pmatrix}, \tag{29}$$

with matrix elements $e^{2\pi ims/d}$ $(m = -N \dots N)$ on a main diagonal of the matrix G_2.

The summation over the eigenfrequencies is performed by making use of the argument principle (23), which yields the Casimir energy of two parallel gratings on a unit cell of period d and unit length in z direction:

$$E = \frac{\hbar c\, d}{(2\pi)^3} \int_0^{+\infty} d\omega \int_{-\infty}^{+\infty} dk_z \int_0^{\frac{2\pi}{d}} dk_x \times \ln\det\Big(I - R_{1down}(k_x, k_z, i\omega) R_{2up}(k_x, k_z, i\omega, L, \varphi)\Big), \tag{30}$$

here $\varphi = 2\pi s/d$, s is a lateral displacement of two gratings, c is the speed of light in vacuum. This is an exact expression valid at zero temperature for two arbitrary parallel gratings with coinciding periods d separated by a vacuum slit.

The free energy on a unit surface $\mathcal{F}$ in the system of two gratings can be written as follows:

$$\mathcal{F}(L,\varphi) = \frac{k_B T}{\pi^2} \sum_{n=0}^{+\infty\,\prime} \int_0^{+\infty} dk_z \int_0^{\pi/d} dk_x \times \ln\det\Big(I - R_{1down}(k_x, k_z, i\omega_n) R_{2up}(k_x, k_z, i\omega_n, L, \varphi)\Big), \tag{31}$$

here $\omega_n = \frac{2\pi n T k_B}{\hbar}$ is a Matsubara frequency, k_B is a Boltzmann constant. The $n = 0$ term is multiplied by $1/2$. The formula (31) is valid for an arbitrary profile and an arbitrary dielectric permittivity of each grating.

When one grating has a curvature of the sphere of a radius R, $R \gg d$, the lateral force in this system F_{lat} can be obtained by combining the use of a proximity force approximation for sphere-plate and the exact formula for the free energy of two gratings (31):

$$F_{lat} = 2\pi R \, \frac{2\pi}{d} \int_{L_2}^{+\infty} dy' \, \frac{\partial \mathcal{F}(y',\varphi)}{\partial \varphi}. \tag{32}$$

Here L_2 is equal to the sum of the minimum distance between the sphere and the grating and the heights of both corrugations.

Thus the formula for the lateral force between the corrugated sphere and the grating is equal

$$F_{lat} = \frac{4k_B TR}{d} \int_{L_2}^{\infty} dy' \sum_{l=0}^{\infty}{}' \frac{\partial}{\partial\varphi} \int_0^{\infty} dk_z \int_0^{\pi/d} dk_x \times \ln\det[I - R_{1down}(k_x, k_z, i\omega_l) R_{2up}(k_x, k_z, i\omega_l, y', \varphi)]. \quad (33)$$

This formula was recently used for the comparison of the exact theory with the measurements of the lateral Casimir force between a golden sphere with sinusoidal corrugations on it and a golden grating with sinusoidal corrugations.[12]

Acknowledgments

The author is grateful to Kim Milton for hospitality in Norman during QFEXT-09. The author expresses gratitude to the organizing committee of QFEXT-09 for financial support. The research was supported by the Grants RNP 2.1.1/1575 and RFBR 07-01-00692-a.

References

1. E.M.Lifshitz, *Zh.Eksp.Teor.Fiz.* **29**, 94 (1955); *Soviet Phys.JETP* **2**, 73 (1956).
2. M.J.Renne, *Physica* **56**, 125 (1971).
3. T.Emig, R.L.Jaffe, M.Kardar and A.Scardicchio, *Phys.Rev.Lett.* **96**, 080403 (2006); T.Emig, N.Graham, R.L.Jaffe and M.Kardar *Phys.Rev.Lett.* **99**, 170403 (2007).
4. S.J.Rahi, T.Emig, N.Graham, R.L.Jaffe and M.Kardar, *Phys.Rev.D* **80**, 085021 (2009).
5. M.Bordag, *Phys.Rev.D* **73**, 125018 (2006).
6. O.Kenneth and I.Klich, *Phys.Rev.Lett.* **97**, 160401 (2006); *Phys.Rev.B* **78**, 014103 (2008).
7. A.Bulgac, P.Magierski and A.Wirzba, *Phys.Rev.D* **73**, 025007 (2006).
8. K.A.Milton and J.Wagner, *J.Phys.A: Math.Theor.* **41**, 155402 (2008).
9. A.Lambrecht and V.N.Marachevsky, *Phys.Rev.Lett.* **101**, 160403 (2008).
10. A.Lambrecht and V.N.Marachevsky, *Int.J.Mod.Phys.A* **24**, 1789 (2009).
11. O.M.Rayleigh, *Proc.Roy.Soc.A* **79**, 399 (1907).
12. H-C. Chiu, G.L.Klimchitskaya, V.N.Marachevsky, V.M.Mostepanenko and U.Mohideen, *Phys.Rev.B* **80**, 121402(R) (2009).

THE ROLE OF THE SURFACE MODES IN THE CASIMIR REPULSION THROUGH LIQUIDS

I. G. PIROZHENKO

Bogoliubov Labotarory of Theoretical Physics, JINR,
and Dubna International University,
Dubna, 141980, Russia
E-mail: pirozhen@theor.jinr.ru

A. LAMBRECHT

Laboratoire Kastler Brossel, CNRS, ENS, UPMC,
Campus Jussieu case 74, 75252 Paris, France
E-mail: astrid.lambrecht@lkb.ens.fr

In the present paper we consider surface modes in three layered systems modeled by dielectric functions guaranteeing repulsion. It is shown that surface modes play a decisive role in this phenomenon at short separations. For a toy plasma model we find the contribution of the surface modes at all distances.

Keywords: Casimir force, surface plasmons.

1. Introduction

The existence of repulsive dispersion forces follows straightforwardly from the Lifshitz theory.[1] The repulsion is achieved if the space between the bodies 1 and 3 with dielectric permittivities ε_1 and ε_3 is filled by a medium 2 (liquid) with a precisely chosen dielectric permittivity ε_2. The respective dielectric functions have to satisfy the relation

$$\varepsilon_1(i\omega) < \varepsilon_2(i\omega) < \varepsilon_3(i\omega) \tag{1}$$

in a wide frequency range relevant for the force measurement. There are several experiments testing the van der Waals repulsion at short separations (<20 nm).[2–4] The repulsive Casimir forces at larger distances have not been experimentally studied until very recently.[5,6]

The boundary-value problem for the electromagnetic fluctuations in this system has two types of solutions: the propagative waves (photonic modes) and the waves living on the interfaces and exponentially decaying outwards.[7–9] These surface modes exist only in the TM polarization. We call them surface plasmons for the materials described by the plasma model.

The Lifshitz formula comprises the contributions from both propagative and surface modes. At zero temperature the force in the three-layered system is given by[1,10]

$$F(L) = -\frac{\hbar}{2\pi^2}\sum_{\rho}\int_0^{\infty} dkk \int_0^{\infty} d\omega\kappa_2 \frac{r_{12}^{\rho}\, r_{32}^{\rho}}{\exp(2\kappa_2 L) - r_{12}^{\rho}\, r_{23}^{\rho}}. \tag{2}$$

Here L is the width of the gap between the plates; $r_{i2}^{TM}(i\omega) = (\varepsilon_2\kappa_i - \varepsilon_i\kappa_2)/(\varepsilon_2\kappa_i + \varepsilon_i\kappa_2)$, $r_{i2}^{TE}(i\omega) = -(\kappa_i - \kappa_2)/(\kappa_i + \kappa_2)$, are the reflection coefficients at imaginary frequencies for the surfaces facing the medium 2, with $\kappa_j(i\omega) = \sqrt{k^2 + \varepsilon_j(i\omega)\omega^2/c^2}$, $j = 1, 2, 3$. The dielectric functions respect the condition (1). The present paper pursues the roles of the propagative and surface modes in the system.

2. The interaction of the surface plasmons

Let us first formulate the interaction of the surface modes to understand the nature of Casimir repulsion. For simplicity we consider materials described by the plasma model, $\varepsilon_i = 1 - \omega_{pi}^2/\omega^2$, $i = 1..3$, where ω_{pi} is the material's plasma frequency. The frequency of the single surface plasmon living on the interface of medium i, $i = 1, 3$, with medium 2 is given by[11]

$$\omega_{i2}^{sp} = \frac{1}{\sqrt{2}}\left[2k^2c^2 + \omega_{p2}^2 + \omega_{pi}^2 - \sqrt{4k^4c^4 + (\omega_{p2}^2 - \omega_{pi}^2)^2}\right]^{\frac{1}{2}} \tag{3}$$

When $k \to 0$ the single surface plasmon frequency tends to ω_{p2}, provided $\omega_{pi}/\omega_{p2} < 1$, otherwise it approaches ω_{pi}; $\omega_{i2}^{sp}|_{k\to\infty} \to \sqrt{\omega_{pi}^2 + \omega_{p2}^2}/\sqrt{2}$. The surface modes evanescent in the direction of the gap between the plates are coupled through the cavity. Their frequencies satisfy the equation[12]

$$\prod_{i=1,3} \frac{\varepsilon_2 q_i + \varepsilon_i q_2}{\varepsilon_2 q_i - \varepsilon_i q_2} = e^{-2\,q_2\,L}, \quad q_i = \sqrt{k^2 - \varepsilon_i\omega^2/c^2}. \tag{4}$$

Figure 1(a) shows the plasmon modes inside the "sandwich" $\omega_{p1} < \omega_{p2} < \omega_{p3}$ (repulsive Casimir force). The coupled plasmon modes are plotted as solid lines. The dashed lines correspond to the plasmons living on the single interfaces. We denote the solutions of (4) by $\omega_\pm$, where ω_+ is usually referred to as the antisymmetric plasmon and ω_- as the symmetric plasmon. The dotted lines starting from $\omega/\omega_{p1} = 1$ and $\alpha = \omega_{p2}/\omega_{p1}$ separate the propagative and evanescent sectors respectively in the gap and in the slab

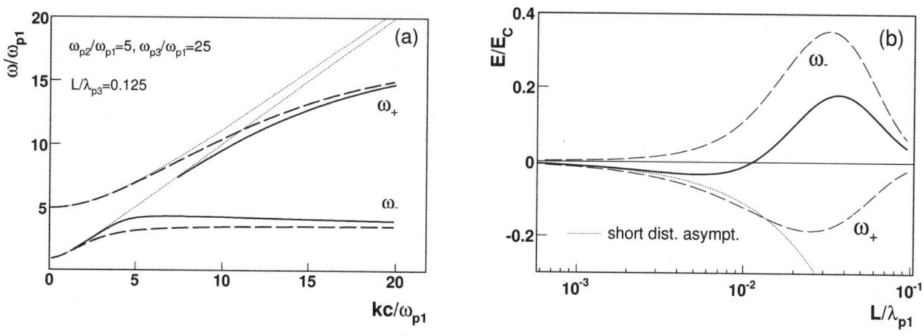

Fig. 1. (a)The plasmon modes in the set-up when $\omega_{p1} < \omega_{p2} < \omega_{p3}$ (repulsive Casimir force). (b) The normalized vacuum energy of the symmetric plasmon, ω_- (dashed line), antisymmetric plasmon, ω_+ (dash-dotted line), and the sum of both (solid line).

with ω_{p1}. In the area between these curves the equation (4) has no real solutions as $q_2^2, q_3^2 > 0$, but $q_1^2 < 0$. Solving (4) with $k^2 - \omega^2/c^2 + \omega_{p2}^2/c^2 \to 0$, one finds that the mode ω_+ exists for $k > k_+$,

$$k_+ = \frac{\omega_{p1}}{c}\left[\frac{\alpha^2 + \beta^2 f}{1+f} - 1\right]^{\frac{1}{2}}, \quad f = \frac{\sqrt{\alpha^2-1}}{\sqrt{\beta^2-1}} \tanh\left(\Lambda\sqrt{\alpha^2-1}\right). \tag{5}$$

The coupled plasmons $\omega_\pm$ lie inside the area enveloped by the single surface plasmon solutions ω_{32}^{sp} and ω_{12}^{sp} (3), and $\omega_-|_{k\to 0} = \omega_{p1}$, $\omega_+|_{k\to 0} = \omega_{p2}$.

The renormalized vacuum energy of the interacting surface plasmons living on the plane mirrors is formally defined by

$$E^{sp} = \frac{\hbar}{2}\int_0^\infty \frac{dk\,k}{2\pi}\left[\omega_- - \omega_{32}^{sp}\right] + \frac{\hbar}{2}\int_{k_+}^\infty \frac{dk\,k}{2\pi}\left[\omega_+ - \omega_{12}^{sp}\right], \tag{6}$$

where $\omega_\pm$ are given by Eq. (4), and k_+ is defined by (5). The numerical results for the plasmon energy are presented in Fig.1(b). For the details of calculation see Ref.[11]. The distance is normalized by the largest plasma wavelength in the system, $\lambda_{p1} = 2\pi c/\omega_{p1}$. The plot shows the energy reduction factor E^{sp}/E_C, $E_C = -\hbar c\pi^2/720L^3$. The reduction factor of the antisymmetric plasmon ω_+ is negative corresponding to repulsion, the reduction factor of the symmetric plasmon ω_- is positive. The antisymmetric plasmon dominates at short distances, while the symmetric plasmon makes decisive contribution at larger separations. The total plasmon interaction is repulsive at short separations and attractive at medium and long distances. Here we estimate the length scale with respect to λ_{p1}.

Figure 2(a) shows the reduction factor for the total Casimir energy when condition (1) is met. It comes out as the sum of the negative pho-

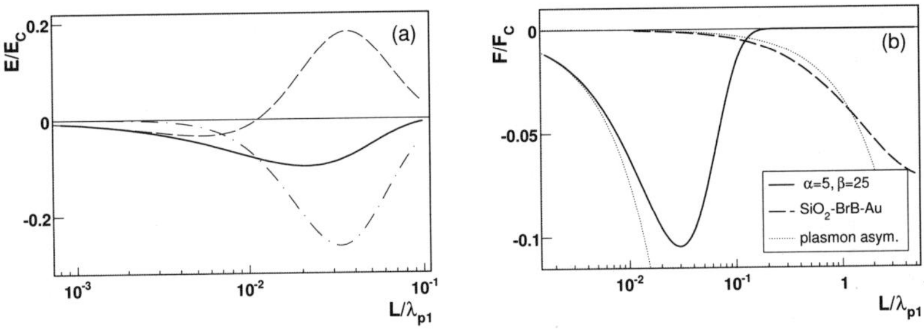

Fig. 2. (a) The normalized plasmon energy (dashed line), photon energy (dash-dotted line), total Casimir energy (solid line). $\alpha = \omega_{p2}/\omega_{p1} = 5, \beta = \omega_{p3}/\omega_{p1} = 25$. (b) Reduction factor $\eta_F = F/F_C$ as a function of dimensionless distance $\Lambda = L/\lambda_{p1}$. The dotted lines are the respective short distance asymptotes of the surface plasmon interaction. For the system SiO_2-C_6H_5Br-Au $\lambda_{p1} = 2\pi c/\omega^{eff}_{p,SiO_2} = 88.44$nm.

ton contribution and the plasmon contribution, plotted in Fig.2(a). The plasmon contribution dominates at short separations. At medium and long distances the photon contribution prevails assuring repulsion. At distances large with respect to λ_{p1} the energy decreases as the exponential integral Ei, $E \sim \text{Ei}(1, 2\alpha\Lambda)$, $\Lambda = 2\pi L/\lambda_{p1}$.

Let us now compare the repulsive force due to the interaction of the surface modes with the corresponding total Casimir force calculated according to (2). In Fig.2(b) we plot the force (2) normalized by the Casimir force between perfect conductors as a function of the dimensionless distance L/λ_{p1}, $\lambda_{p1} = 2\pi c/\omega_{p1}$. The calculation is done for $\varepsilon_i = 1 + \omega_{pi}^2/\omega^2$, $\alpha = \omega_{p2}/\omega_{p1} = 5, \beta = \omega_{p3}/\omega_{p1} = 25$ (solid line). At short distances the force coincides with the short distance asymptote of the interacting surface plasmons, $F^{sp}_{as}/F_C = -7.38L/\lambda_{p1}$ (dotted line). At long distances $(L >> \lambda_{p2})$ the force is repulsive and decays as $F \sim \exp(-2\omega_{p2}L/c)/L$.

Finally, as an illustration we study the system Silica-Bromobenzene-Gold. For gold we model the dielectric function by the Drude model.[10] For the permittivities of bromobenzene and silica we confine ourselves to a two-oscillator model with parameters taken from Ref.[2,5] With these values of the parameters the condition (1) is satisfied for $\omega < 9 \cdot 10^{15}$rad/s and for $\omega > 3 \cdot 10^{16}$rad/s. At close separations realized by a thin layer of bromobenzene, we may use the plasma model formulae with with $\omega^{eff}_{SiO_2} = 2.131 \cdot 10^{16}$rad/c, $\omega^{eff}_{C_6H_5Br} = 1.488 \cdot 10^{16}$rad/c. At medium and low frequencies the calculation of the vacuum energy corresponding to the surface modes is not so straightforward. The total reduction factor for the force is shown as the dashed line in Fig.2(b). The repulsion sets in for distances of the order of or larger than the effective plasma frequency. At large distances the reduction factor saturates at $\eta \approx -0.057$. For this system the plasmon approximation (dotted line) is valid for distances up to twice the effective plasma wavelength, that is about 170 nm.

3. Conclusion

In the present paper we have studied the Casimir force in the systems which meet the condition (1). We have shown that within the plasma model the surface modes repel at short distances. Moreover the Casimir repulsion is then completely due to the repulsion of the surface modes. At medium and long distances the interaction of the surface modes becomes attractive, but the dominating repulsive contribution of the propagative modes leads to a total repulsive force.

Acknowledgments

I.P. thanks the organizers of QFEXT-09 for kind hospitality in Oklahoma and financial support. The RFBR grant 09-02-08375 is acknowledged. A.L. thanks the ESF Research Network CASIMIR for providing excellent opportunities for discussion on the Casimir effect and acknowledges financial support from the French Research contract no. ANR-06-Nano-062.

References

1. I. E. Dzyaloshinski, E. M. Lifshitz and L. Pitaevski, *Usp. Fiz. Nauk* **73**, p. 381 (1961).
2. A. Milling, P. Mulvaney and I. Larson, *J. Colloid Interface Sci.* **180**, 460 (1996).
3. S.-W. Lee and W. M. Sigmund, *J. Colloid Interface Sci.* **243** (2001).
4. S.-W. Lee and W. M. Sigmund, *Colloids and Surfaces* **A 204** (2002).
5. J. N. Munday, F. Capasso and A. Parsegian, *Nature* **457** (2009).
6. P. J. van Zwol, G. Palasantzas and J. T. M. De Hosson, *Phys. Rev.* **E 79**, p. 041605 (2009).
7. C. Henkel, K. Joulain, J.-P. Mulet and J.-J. Greffet, *Phys. Rev.* **A 69**, p. 023808 (2004).
8. M. Bordag, *Journal Physics* **A 39**, 6173 (2005).
9. F. Intravaia and A. Lambrecht, *Phys. Rev. Lett.* **94**, p. 110404 (2005).
10. A. Lambrecht and S. Reynaud, *Eur. Phys. J.* **D 8**, p. 309 (2000).
11. I. G. Pirozhenko and A. Lambrecht, *Phys. Rev.* **A 80**, p. 042510 (2009).
12. Z. Lenac and M. S. Tomas, *Phys. Rev.* **A 78**, p. 023834 (2008).

CASIMIR STRESS IN AND FORCE ON A METAL SLAB IN A PLANAR CAVITY

M. S. TOMAŠ
Rudjer Bošković Institute,
P. O. Box 180, 10002 Zagreb, Croatia

Z. LENAC
Department of Physics, University of Rijeka,
51000 Rijeka, Croatia

Emphasizing first the utility of the generalized Fresnel coefficients in the theory of the Casimir effect in planar cavities, we complement our previous discussion of the ordinary Casimir force on and the Casimir stress in a metal (plasma) slab in a planar cavity. We demonstrate strong dependence of the Casimir stress in a thin slab on properties of the bounding medium in the symmetric Lifshitz configuration. Contrary to this, the stress in a thick slab gradually becomes insensitive on external boundary conditions. We also consider the position dependence of the Casimir force on and stress in a thin metal slab in a planar cavity. Whereas the force per unit area on the slab strongly increases when it approaches a mirror the stress in the slab decreases and eventually changes the sign. Generally, the stress decreases with the cavity width and decreasing reflectivity of the mirrors.

Keywords: Casimir stress, metal slab, plasma model, planar cavity.

1. Introduction

In addition to the ordinary Casimir forces[1] acting between the layers of a multilayered system, vacuum fluctuations of the electromagnetic field cause a stress in each layer. This (often disregarded) effect is important when considering mechanical stability of thin layers and components[2] and is therefore, besides being of fundamental interest,[3] relevant in micro- and nano-technology. In our previous work[4] (referred to as I), we have considered the stress (referred in that work as pressure) and its modal structure in a metal (plasma) slab in the center of an ideal planar cavity and demonstrated their strong dependence on the cavity width. Upon emphasizing the utility of the concept of the generalized Fresnel coefficients[5] (in conjunction with their recurrence relations[6]) in the theory of the Casimir effect in planar cavities,[7] in this work we discuss the dependence of the vacuum-field stress in a metal (plasma) layer in the symmetric Lifshitz[8] configuration on the properties of the bounding medium. To emphasize the difference between the standard Casimir force per unit area and the Casimir stress, we also consider the position dependence of these quantities for a metal slab in a planar cavity.

2. Theory

Consider a dielectric slab inserted in a planar cavity, as depicted in Fig. 1. The vacuum-field forces (per unit area) acting on the slab consist of the

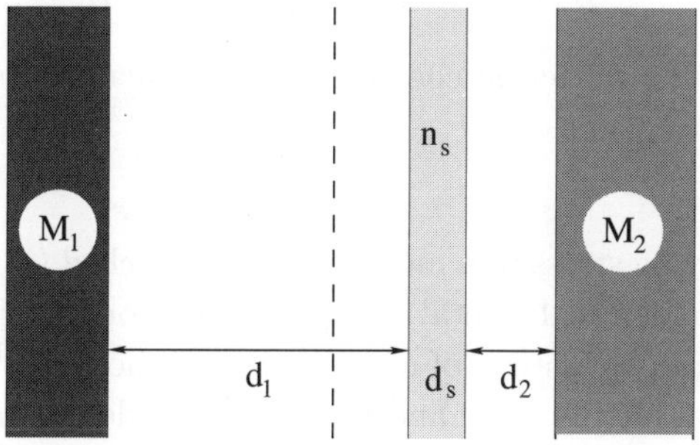

Fig. 1. A dielectric slab in na empty ($n_1 = n_2 = 1$) planar cavity schematically.

stress F_s in the slab and the net slab-mirror interaction force per unit area $F = F_2 - F_1$,[2] where according to the theory of the Casimir force in multilayers[7]

$$F_j = \mathrm{T}_{\mathrm{zz}}^{(\mathrm{j})} = \frac{\hbar}{2\pi^2} \int_0^\infty d\xi \int_0^\infty dk k \kappa_j \sum_{\mathrm{TM,TE}} \frac{r_{j-} r_{j+} e^{-2\kappa_j d_j}}{1 - r_{j-} r_{j+} e^{-2\kappa_j d_j}}. \tag{1}$$

Here $\kappa_j(i\xi, k) = \sqrt{\varepsilon_j(i\xi)\xi^2/c^2 + k^2}$ is the perpendicular wave vector at the imaginary frequency in the jth layer and $r_{j\pm}(i\xi, k)$ are the reflection coefficients of the right and left stack of layers bounding the layer. F can be conveniently expressed in terms of the Fresnel coefficients $r \equiv r_{1/2} = r_{2/1}$ and $t \equiv t_{1/2} = t_{2/1}$ of the *whole* slab using the recurrence relation[6,7]

$$r_{1+(2-)}(i\xi, k) = r + \frac{t^2 R_{2(1)} e^{-2\kappa d_{2(1)}}}{1 - r R_{2(1)} e^{-2\kappa d_{2(1)}}}, \tag{2}$$

where $\kappa(i\xi,k) \equiv \kappa_1 = \kappa_2 = \sqrt{\xi^2/c^2+k^2}$ and $R_{1(2)}(i\xi,k)$ are reflection coefficients of the mirrors, and noting that $r_{1-(2+)} = R_{1(2)}$. We find[7]

$$F = \frac{\hbar}{2\pi^2}\int_0^\infty d\xi \int_0^\infty dkk\kappa \sum_{\mathrm{TM,TE}} r\frac{R_2e^{-2\kappa d_2} - R_1e^{-2\kappa d_1}}{N},$$

$$N = 1 - r(R_1e^{-2\kappa d_1} + R_2e^{-2\kappa d_2}) + (t^2 - r^2)R_1R_2e^{-2\kappa(d_1+d_2)}, \quad (3)$$

which agrees with the result obtained through a conventional way.[9]

Following Benassi and Calandra,[2] we ignore the electostriction and magnetostrition forces in the slab. The stress in the slab is then determined solely by the Minkowski stress tensor[10] and is therefore given by Eq. (1), with the reflection coefficients for the waves reflected within the slab

$$r_{s-(+)}(i\xi,k) = \frac{-\rho + R_{1(2)}e^{-2\kappa d_{1(2)}}}{1-\rho R_{1(2)}e^{-2\kappa d_{1(2)}}}, \quad (4)$$

where $\rho(i\xi,k)$ is the reflection coefficient of the vacuum-slab interface.

3. Discussion

We first consider the stress in a metal layer sandwiched ($d_1 = d_2 = 0$ in Eq. (4)) between two identical (metal) mirrors corresponding to the symmetric Lifshitz configuration. Instead of a sophisticated model,[11,12] we adopt here the plasma model for the layer and the Drude model for mirrors

$$\varepsilon_s(i\xi) = 1 + \frac{\omega_P^2}{\xi^2}, \qquad \varepsilon_m(i\xi) = 1 + \frac{\Omega_P^2}{\xi^2+\Gamma^2}, \quad (5)$$

where ω_P and Ω_P are the corresponding plasma frequencies and Γ is the damping parameter of the mirrors (in this work we use $\Gamma = 10^{-3}\Omega_P$). The thickness dependence of the stress is presented on the left side of Fig. 2 for several values of the (contrast) ratio Ω_P/ω_P. The uppermost curve practically coincides with the result obtained assuming perfect ($\Omega_P = \infty$) mirrors. In that case the stress in the layer can be calculated exactly and is, in the thin layer ($k_Pd_s \ll 1$) limit, given by the Casimir force per unit area F_C,[4] which we used to scale the stress in the figure. As seen, with decreasing reflectivity of the mirrors the stress strongly drops and acquires in this range of the layer thicknesses its nonretarded (nr) value $\sim d_s^{-3}$. Ultimately, when $\Omega_P = 0$, we obtain the stress in a free-standing metal slab $F_s^{\mathrm{nr}} = 0.19k_Pd_sF_C$.[3,4] Plotted on the right side of Fig. 2 is the stress in absolute units in a gold ($\omega_P = 9$ eV[13]) slab for two extreme cases of the mirrors. Thus, the stress in an Au slab sandwiched between a couple of realistic mirrors lies in between these two curves.

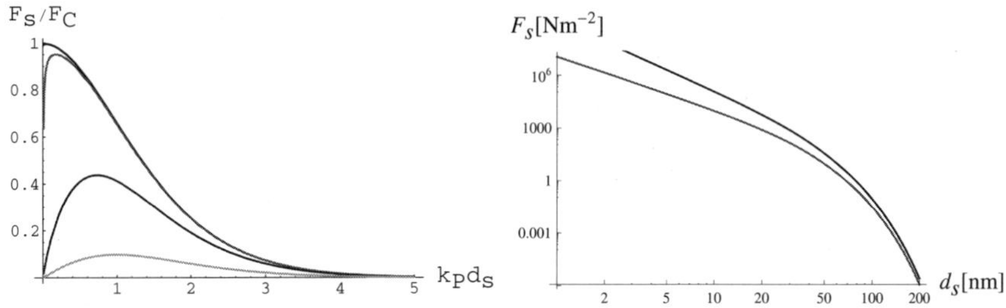

Fig. 2. Left: Thickness dependence of the stress in a metal (plasma) layer. From top to bottom, the curves correspond to $\Omega_P/\omega_P = 10^5, 10^3, 10$ and 1, respectively. $k_P = \omega_P/c$ and $F_C = \pi^2\hbar c/240 d_s^4$. Right: Stress in a gold layer between perfect ($\Omega_P = \infty$) mirrors (upper curve) and in a free-standing ($\Omega_P = 0$) gold layer (lower curve) in absolute units.

As noted already by Dzyaloshinski *et al.*,[14] the stress in a thick metal (plasma) layer exponentially decreases. Figure 2 reveals, however, that for $k_P d_s \gg 1$ it becomes gradually insensitive to the properties of the mirrors. For a thick enough layer, it is therefore given by the result obtained for perfect mirrors[4] $F_s = (\hbar c k_P^4/4\sqrt{(\pi k_P d_s)^3})\exp(-2k_p d_s)$. Since the large d_s behaviour of the stress is determined by small ξ values of $F_s(\xi, k)$, we note that the same conclusion applies to layers described by a dielectric function of the form $\tilde{\varepsilon}(i\xi) + \omega_P^2/\xi^2$, where $\tilde{\varepsilon}(i\xi)$ behaves regularly at the origin. This implies that addition of a salt into the liquid between the plates, as in recent experiments on screened Casimir force,[15] will cause exponential decay of the force at large liquid layer thicknesses since salt brings a plasma-like component to the dielectric function of the solution.

We end this discussion by briefly considering the stress in a metal slab in a planar cavity. According to Eq. (4), removing the mirrors from the slab ($d_1 = d_2 \neq 0$ in Eq. (4)) decreases its internal reflectivity. Accordingly, with increasing slab-mirror distance, the stress in the slab behaves similarly as in Fig. 2 with decreasing reflectivity of the mirrors (cf. with Fig. 4 of I). The position dependence of the stress in a thin slab is illustrated in Fig. 3. As seen on left side of this figure, the stress is largest in the center of the cavity (where the force vanishes) and decreases with the cavity width until it saturates to $F_s^{\rm nr}$ (this practically occurs already at $L = 10d_s$). Near a mirror the stress changes sign since $1 < \varepsilon_s(i\xi) < \varepsilon_m(i\xi)$ is fulfilled.[14] On the right side of Fig. 3, we compare the stress in and the force per unit area (scaled by a factor of 10^{-4}) on the slab in the $L = 3d_s$ cavity. As discussed by Benassi and Calandra,[2] these quantities become approximately of the same order when the slab-mirror distance is comparable with the slab thickness whereas at smaller slab-mirror distances the Casimir force dominates.

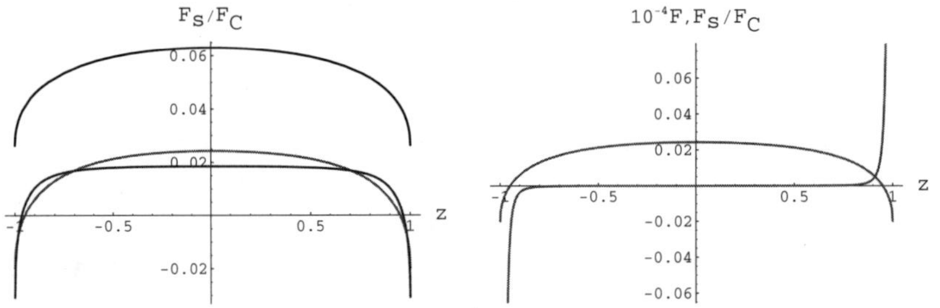

Fig. 3. Left: Position dependence of the stress in a $k_p d_s = 0.1$ thick metal slab in a cavity formed by mirrors with $\Omega_P = 10^3\omega_P$. From top to bottom, the curves correspond to the cavity width $L = 2d_s, 3d_s$ and $10d_s$, respectively. Right: Stress in vs net force per unit area on the slab in the $L = 3d_s$ cavity. Parameter z is defined with $d_{1(2)} = \frac{L-d_s}{2}(1 \pm z)$.

To summarize, in this work we have demonstrated strong thickness and medium dependence of the Casimir stress in a metal layer in the Lifshitz configuration as well as strong dependence of the Casimir stress in and force on a thin metal slab in a planar cavity on its position and cavity properties.

Acknowledgments

This work was supported by the Ministry of Science, Education and Sport of the Republic of Croatia under contract No. 098-1191458-2870.

References

1. H. B. G. Casimir, *Proc. Kon. Ned. Akad. Wet.* **51**, 793 (1948).
2. A. Benassi and C. Calandra, *J. Phys. A: Math. Theor.* **40**, 13453 (2007).
3. Y. Imry *Phys. Rev. Lett.* **95**, 080404 (2005).
4. M. S. Tomaš and Z. Lenac, *J. Phys: Conf. Ser.* **161**, 012017 (2009).
5. O. H. Crawford, *J. Chem. Phys.* **89**, 6017 (1988).
6. M. S. Tomaš, *Phys. Rev. A* **51**, 2545 (1995).
7. M. S. Tomaš, *Phys. Rev. A* **66**, 052103 (2002).
8. E. M. Lifshitz, *Sov. Phys. JETP* **2**, 73 (1956).
9. S. A. Ellingsen and I. Brevik, *J. Phys. A: Math. Theor.* **40**, 3643 (2007).
10. L. D. Landau, E. M. Lifshitz and L. P. Pitaevskii, *Electrodynamics of Continuous Media* (Pergamon, Oxford, 1984).
11. A. Benassi and C. Calandra, *Europhys. Lett.* **82**, 61002 (2008).
12. R. Esquivel-Sirvent, *Phys. Rev. A* **77**, 042107 (2008).
13. M. Bordag, U. Mohiden and V. M. Mostepanenko, *Phys. Rep.* **353**, 1 (2001).
14. I. E. Dzyaloshinskii, E. M. Lifshitz and L. P. Pitaevskii, *Adv. Phys.* **10** 165 (1961).
15. J. N. Munday, F. Capasso, V. A. Parsegian and S. M. Bezrukov, *Phys. Rev.* A **78**, 032109 (2008).

BOSE-EINSTEIN CONDENSATION IN QUASI-2D SYSTEMS: APPLICATIONS TO HIGH-T_c SUPERCONDUCTIVITY

C. VILLARREAL* and M. DE LLANO†

Instituto de Física, Universidad Nacional Autónoma de México,
Mexico City, DF 04510, Mexico
**E-mail: carlos@fisica.unam.mx*

Physics Department, University of Connecticut, Storrs, CT 06269 USA
and
Instituto de Investigaciones en Materiales,
Universidad Nacional Autónoma de México,
Mexico City, DF 04510, Mexico
†*E-mail: dellano@servidor.unam.mx*

We calculate the number and energy densities of a quasi-2D Bose-Einstein gas constrained within a thin region of infinite extent but of finite width δ. The BEC critical transition temperature then becomes an explicit function of δ. We use this result to construct a model of high-T_c superconductivity in cuprates with a periodic layered atomic structure. The predicted behavior of the BEC T_c agrees with recent experimental findings in severely underdoped cuprates.

Keywords: High-T_c superconductivity; BEC condensates; BCS theory.

1. Introduction

Since the discovery of high-T_c superconductivity (HTSC) in cuprates by Bednorz and Müller in 1986 many studies to explain the phenomenon have been reported. Its physical origin is not yet clear. Recent measurements[1] of photoelectron emission spectral intensities from HTSCs suggest that bound electron Cooper pairs (CPs) form already at temperatures higher than the critical T_c. This finding is consistent with several theoretical efforts[3,4] proposing that HTSC originates from a 2D Bose-Einstein condensate (BEC) of CPs pre-existing above T_c and formed through a BCS-like phonon mechanism. The 2D character of the phase transition is associated with the layered structure of cuprates, which in the case of $YBa_2Cu_3O_{7-y}$ (YBCO) consists of a succession of parallel layers perpendicular to the vertical c-axis with a unit cell of height $\sim 12\mathring{A}$, and the chemical composition $CuO - BaO - CuO_2 - Y - CuO_2 - BaO - CuO$. It is generally agreed that the CuO_2 planes, which in the case of YBCO are equidistant from the central Y atom with separation $\simeq 1.5\mathring{A}$, are mainly responsible for the superconductivity in cuprates. Contour plots of the charge distribution derived from energy-band-structure calculations for YBCO reveal[5] that the SC charge carriers are mainly concentrated within a slab of width $\delta \simeq 2.15\mathring{A}$ about the CuO_2 plane.

BCS-like theories[3,4] contemplate a Hamiltonian H_{BCS} containing the kinetic energies of electrons and holes, and a pairing interaction arising from phonon-exchange attractions that overwhelm and Coulombic repulsion. As a consequence, bound CPs of electrons or holes with antiparallel spins and charge $\pm 2e$ arise with an energy-momentum relation linear at leading order, rather than quadratic, namely $\mathcal{E}_K \simeq \mathcal{E}_0 + c_1 \hbar K$ with $c_1 = 2v_F/\pi$ in 2D with v_F the Fermi speed, while K is the CP center-of-mass momentum wavenumber, and $\mathcal{E}_0$ the familiar weak-coupling energy $\mathcal{E}_0 = -2\hbar\omega_D \exp[-2/v_0 N_0]$ for $K = 0$ CPs, where ω_D is the Debye frequency, v_0 the positive BCS electron-phonon coupling constant, and N_0 the electron density of states for one spin at the Fermi level. The linear dispersion relation is induced[2] by the Fermi sea medium so that CPs propagate like free massless composite particles in the Fermi sea (whereas *in vacuo* they would do so quadratically as $\hbar^2 K^2/4m^*$ if m^* is the effective electron mass). Their Bose statistical nature allows them to undergo BEC.

2. Formalism

We study a model of HTSC in cuprate materials comprising a quasi-2D BEC of excited CPs of energy $\mathcal{E}_K \simeq \mathcal{E}_0 + c_1 \hbar K$ constrained to propagate within quasi-2D layers of infinite extent in the a_1, a_2 directions, but of finite width $a_3 = \delta$ in the perpendicular direction. We assume that the CP field satisfies periodic boundary conditions (BC) along the a_1, a_2 and a_3 directions, although the more restrictive Dirichlet or Neumann BCs may be straightforwardly implemented. The average number density and the energy per unit volume of the CP field are given by

$$n(T) = \frac{1}{V}\int dK \frac{g(K)}{e^{\beta(\varepsilon_K - \mu)} - 1} \quad \text{and} \quad u(T) = \frac{1}{V}\int dK \frac{g(K)\varepsilon_K}{e^{\beta(\varepsilon_K - \mu)} - 1} \tag{1}$$

where $g(K)$ is the *exact* eigenmode distribution of the field defined by $g(K) = \sum_{\{\mathbf{n}\}} \delta(K - K_\mathbf{n})$ with $K_\mathbf{n}^2 = (2n_1\pi/a_1)^2 + (2n_2\pi/a_2)^2 + (2n_3\pi/a_3)^2$. By introducing Poisson's summation formula it may be shown that $g(K)$

can be expressed in the more tractable way

$$g(K) = \frac{V}{2\pi^2} K \sum_{m_1,m_2,m_3} \frac{\sin\left[K\left((a_1m_1)^2 + (a_2m_2)^2 + (a_3m_3)^2\right)^{1/2}\right]}{[(a_1m_1)^2 + (a_2m_2)^2 + (a_3m_3)^2]^{1/2}}. \quad (2)$$

An alternative derivation based on properties of Bessel functions has been formerly employed to study the Casimir energy-momentum tensor in rectangular cavities, both at zero and finite temperatures.[6] We then evaluate (1) by introducing the CP mode distribution (2) and the CP excitation energy $\varepsilon_K \equiv \mathcal{E}_K - \mathcal{E}_0 \simeq c_1\hbar K$. The Bose-Einstein denominators in the integrals render a rapid convergence so that they may safely be extended to infinity. The integrals can be computed by expanding the integrand in powers of ze^{-x} with the result

$$n(T) = n_0(T) + \frac{(k_BT)^3}{\pi^2\hbar^2c^3} \sum_{m_1,m_2,m_3} \sum_{m=1}^{\infty} \frac{m\ z^m}{\left(m^2 + \alpha^2_{m_1,m_2,m_3}\right)^2} \quad (3)$$

with $\alpha^2_{m1,m2,m3} = (k_BT/\hbar c_1)^2[(m_1a_1)^2 + (m_2a_2)^2 + (m_3a_3]^2$ while the energy density is

$$u(T) = \frac{(k_BT)^4}{\pi^2\hbar^2c_1^3} \sum_{m_1,m_2,m_3} \sum_{m=1}^{\infty} \frac{\left(3m^2 - \alpha^2_{m_1,m_2,m_3}\right)z^m}{\left(m^2 + \alpha^2_{m_1,m_2,m_3}\right)^3}. \quad (4)$$

It is easily checked that the usual thermodynamic limit is attained by considering the terms with $m_1 = m_2 = m_3 = 0$ in the former expressions. On the other hand, for CPs constrained to move within a layer of finite width $\delta \ll a_1, a_2$ but unconstrained along the infinite a_1, a_2 directions we must set $m_1 = m_2 = 0$ in(3) and (4). In that case, we introduce the dimensionless thickness variable $\eta \equiv k_BT\delta/\hbar c_1$ and the remaining summations over m_3 may be performed analytically to give

$$n(T) = n_0(T) + \frac{(k_BT)^3}{\pi^2\hbar^3c^3}\Psi_3(z,\eta) \quad \text{and} \quad u(T) = 3\frac{(k_BT)^4}{\pi^2\hbar^3c^3}\Phi_4(z,\eta) \quad (5)$$

where $\Psi_s(z,\eta) \equiv \sum_{m=1}^{\infty} \frac{z^m}{m^s} f_m(\eta)$ and $\Phi_s(z,\eta) \equiv \sum_{m=1}^{\infty} \frac{z^m}{m^s} g_m(\eta)$ with $f_m(\eta) = \frac{1}{2}\left[h_m(\eta) + \frac{m\pi}{\eta}\coth\left(\frac{m\pi}{\eta}\right)\right]$, $h_m(\eta) = \left(\frac{m\pi}{\eta}\right)^2 \sinh^{-2}\left(\frac{m\pi}{\eta}\right)$, and $g_m(\eta) = \frac{1}{3}\left[\left(h_m(\eta) + \left(\frac{m\pi}{\eta}\right)\coth\left(\frac{m\pi}{\eta}\right)(1 + h_m(\eta)\right)\right]$.

3. Thick- and thin-layer limits

The thick-layer limit given by $\eta \gg 1$ represents unconstrained propagation of CPs throughout the entire volume of the material as in conventional 3D superconductivity. In this limit $h_m(\eta) \to 1$, $f_m(\eta) \to 1$, $g_m(\eta) \to 1$ and we recover well-known expressions describing 3D BECs:[3,4]

$$n(T) = n_0(T) + \frac{(k_BT)^3}{\pi^2\hbar^3c^3}\sum_{m=1}^{\infty}\frac{z^3}{m^3} \quad \text{and} \quad u(T) = 3\frac{(k_BT)^4}{\pi^2\hbar^3c^3}\sum_{m=1}^{\infty}\frac{z^4}{m^4} \tag{6}$$

The critical temperature T_c follows from the conditions $n_0(T_c) \to 0$ and $z(T_c) \to 1$ leading to $k_BT_c^{3D} = [\pi^2\hbar^3c_1^3n^{3D}/\zeta(3)]^{1/3}$ where $\zeta(n)$ is Riemann's ζ-function. The molar heat capacity $C(T) = R(nk_B)^{-1}\partial u(T)/\partial T$ (with R the gas constant) is straightforwardly obtained from (6) as $C(T) = \left(\frac{12R\zeta(4)}{\zeta(3)}\right)\left(\frac{T}{T_c}\right)^3$ for $T < T_c$ and $C(T) = \left(\frac{12R\zeta(4)}{\zeta(3)}\right)\left(\frac{T}{T_c}\right)^3 - \left(\frac{12R\zeta(3)}{\zeta(2)}\right)$ for $T > T_c$ which is consistent with measurements in conventional 3D superconductors[8] since at $T = T_c$ the heat capacity shows a discontinuous drop $\Delta C = 6.57R$ indicative of a second-order phase transition.

In the thin-layer limit $\eta \ll 1$ associated with HTSC we get $h_m(\eta) \to 0$, $f_m(\eta) \simeq m\pi/2\eta$, and $g_m(\eta) \simeq m\pi/3\eta$. Simple algebra leads to

$$n^{2D}(T) = n_0^{2D}(T) + \frac{(k_BT)^2}{2\pi\hbar^2c_1^2}\sum_{m=1}^{\infty}\frac{z^m}{m^2} \quad \text{and} \quad u^{2D}(T) = \frac{(k_BT)^3}{\pi\hbar^2c_1^2}\sum_{m=1}^{\infty}\frac{z^m}{m^3}, \tag{7}$$

where $n^{2D} \equiv n\delta$ and $u^{2D} \equiv u\delta$. The critical BEC temperature is now given by $k_BT_c^{2D} = [2\pi\hbar^2c_1^2n^{2D}/\zeta(2)]^{1/2}$ where $\zeta(2) = \pi^2/6$. In this case the molar heat capacity is $C(T) = [6R\zeta(3)/\zeta(2)]\,(T/T_c)^2$ for $T < T_c$ but it must be evaluated numerically for $T > T_c$. It turns out that the $C(T)$ is continuous at $T = T_c$ although its derivative $\partial C/\partial T$ is discontinuous. The linear behavior $C(T)/T \propto T$ for $T \leq T_c$ is characteristic of cuprate materials.[9]

Of crucial importance in evaluating T_c is to reliably estimate the fraction of charge carriers that actually contribute to the supercurrent. The charge carrier density is usually determined from measurements of London penetration depth λ_{ab} along the CuO_2 planes. It gives an estimate of the supercurrent that causes partial rejection of an applied external magnetic field in the superconductor. Within the framework of the present model the supercurrent is due to massless-like CPs of charge $2e$ moving with the CP speed c_1, so that the surface supercurrent $\mathbf{J}_s = n^{2D}(2e)c_1\hat{\mathbf{k}}$ with $\hat{\mathbf{k}} \equiv \mathbf{k}/k$.[4] A straightforward calculation[7] shows that $n^{2D} = (e^2/32\pi c_1^2c^2)\delta\Delta_0^2/\hbar\omega_D\lambda_{ab}^2$ where c is the speed of light and Δ_0 is the zero-temperature energy gap. The final expression of the BEC critical temperature is

$$T_c = \frac{\hbar c}{2\pi k_Be}\left(\frac{3\delta}{2\hbar\omega_D}\right)^{1/2}\frac{\Delta_0}{\lambda_{ab}}. \tag{8}$$

4. Conclusions

Introducing in (8) the YBCO parameters tabulated in Ref.[8] $\Theta_D = 410$ K , $\Delta_0 = 14.5$ meV, and $\delta = 2.15$ Å[5] we get the relation $T_c = 16.79/\lambda_{ab}$ (μm-K) which accurately reproduces experimental data reported by Zuev *et al.*[10] in measurements performed in YBCO films with T_cs ranging from 6 to $50K$. They conclude that, within some noise their data fall on the same curve $\lambda_{ab}^{-2} \propto T_c^{2.3\pm0.4}$ regardless of annealing procedure, oxygen content, etc. In an independent study, Broun *et al.*[11] found that their samples of high-purity single-crystal YBCO followed also the rule $T_c \propto \lambda_{ab}^{-1}$. A forthcoming paper[7] gives a more detailed discussion of the model presented here applied also to several other cuprates including the l-wave extension of the present formalism valid only for $l = 0$.

Acknowledgments

We thank D.M. Eagles, M. Fortes, S. Fujita and M.A. Solís for fruitful discussions. MdeLl thanks UNAM-DGAPA-PAPIIT (Mexico) IN106908 for partial support and is grateful to W.C. Stwalley for discussions and the University of Connecticut for its hospitality while on sabbatical leave.

References

1. Yang H B, Rameau J D, Johnson P D, Valla T, and Gu G D 2008 *Nature* **456,** 77.
2. de Llano M and Annett J F 2007 Int. J. Mod. Phys. B **21** 3657
3. Casas M, Rigo A, de Llano M, Rojo O and Solís M A 1998 *Phys. Lett.* A **245**, 55
4. Fujita S, Ito K and Godoy S 2009 *Quantum Theory of Conducting Matter: Superonductivity* (Springer-Verlag, Heidelberg)
5. Krakauer H, Pickett W E and Cohen R E 1998 *J. Supercond.* **1**, 111
6. Hacyan S, Jáuregui R and Villarreal C 1993 *Phys. Rev.* A **47** 4204; Jáuregui R, Villarreal C and Hacyan S 2006 *Ann. Phys.* **321**, 2156
7. Villarreal C and de Llano M, in preparation
8. Poole C P, Farach H A and Creswick R J 1995 *Superconductivity* (Academic Press, London)
9. Fisher R A, Gordon J E, and Phillips N E 1988 *J. Supercond.* **1** 231
10. Zuev Y, Kim M S and Lemberger T R 2005 *Phys. Rev. Lett.* **95** 137002-1
11. Broun D M, Huttema W A, Turner P J, Öscan S, Morgan B, Liang R, Hardy W N and Bonn D A 2007 *Phys. Rev. Lett.* **99** 237003-1

NOBLE GAS, ALKALI AND ALKALINE ATOMS INTERACTING WITH A GOLD SURFACE

GRZEGORZ ŁACH and MAARTEN DEKIEVIET

Physikalisches Institut der Universität Heidelberg, Albert-Ueberle-Strasse 3-5, 69120 Heidelberg, Germany

ULRICH D. JENTSCHURA

Department of Physics, Missouri University of Science and Technology, Rolla, Missouri 65409-0640, USA

The attractive branch of the interaction potentials with the surface of gold have been computed for a large variety of atomic systems: the hydrogen atom, noble gases (He, Ne, Ar, Kr, Xe), alkali atoms (Li, Na, K, Rb, Cs) and alkaline atoms (Be, Mg, Ca, Sr, Ba). The results include highly accurate dynamic polarizabilities for the helium atom calculated using a variational method and explicitly correlated wavefunctions. For other atoms considered we used the data available in the literature. The interaction potentials include both the effects of retardation of the electromagnetic interactions and a realistic representation of the optical response function of gold (beyond the approximation of a perfect conductor). An explicit comparison of our result to the interaction between an atom and a perfect conductor is given.

Keywords: Atom-surface interactions; retardation; Van der Waals forces; Casimir forces; quantum reflection; helium; gold; dynamic polarizability.

1. Motivation

In the last decade, there has been drastic experimental progress in the field of (ultra-) cold collisions in atomic systems. These may involve interactions among two atoms as well as those between an atom and a solid wall. Among the latter, one of the experimental methods for studying the long range ($z \gg a_0$), attractive part of the interaction potential makes use of the phenomenon of quantum reflection. It was demonstrated, that scattering cold atomic beams under grazing angles from well defined, single crystal solid surfaces is especially useful for probing Casimir-Polder forces. This is due to the fact that quantum reflectivity is particularly sensitive to the shape of the potential, at distances where it is heavily modified by retardation.[1] Other techniques involving the manipulation of clouds of cold atomic gases, or condensates in the presence of a solid wall[2] heavily depend on the details of the interaction potential. Despite the abundance of experimental results, the theoretical analysis of the interaction potentials was often based on simple model potentials with parameters chosen to fit the data. Here, we provide accurate atom-surface interaction potentials for the surface of gold based on ab-initio computed atomic dynamic polarizabilities and a compilation of all available optical data in literature.

2. Atom-Surface Interactions

For distances where the exchange effects become negligible, the atom-surface can be computed by considering a polarizable particle interacting with quantum electromagnetic field fluctuations. In this approach, the solid is treated as a continuous medium having a frequency dependent permittivity $\epsilon(\omega)$. The derivation of the dipolar term $V_1(z)$, dominant at the long-distance limit, has been first performed by Lifshitz,[3] and the result when given in atomic units ($m_e = e = \hbar = 1$) reads:

$$V_1(z) = -\frac{\alpha^3}{2\pi}\int_0^\infty \mathrm{d}\omega\, \omega^3\, \alpha_1(\mathrm{i}\omega) \int_1^\infty \mathrm{d}\xi\, \mathrm{e}^{-2\alpha\xi\omega z}\, \mathrm{H}(\xi, \epsilon(\mathrm{i}\omega))\ , \tag{1}$$

where α is the fine structure constant and $\alpha_1(\mathrm{i}\omega)$ is the dipole polarizability of the atom [for the fundamental physical constants we take the CODATA[4] recommended values, e.g., $\alpha = 137.035999679(94)$]. The function $\mathrm{H}(\xi, \epsilon)$ is given by:

$$\mathrm{H}(\xi, \epsilon) = (1 - 2\xi^2)\frac{\sqrt{\xi^2 + \epsilon - 1} - \epsilon\,\xi}{\sqrt{\xi^2 + \epsilon - 1} + \epsilon\,\xi} + \frac{\sqrt{\xi^2 + \epsilon - 1} - \xi}{\sqrt{\xi^2 + \epsilon - 1} + \xi}\,. \tag{2}$$

This expression simplifies considerably in the limit of a perfect conductor $[\epsilon(\omega) \to \infty]$, for which case the potential becomes:

$$V_1^{(\infty)}(z) = -\frac{1}{4\pi}\int_0^\infty \mathrm{d}\omega\, \alpha_1(\mathrm{i}\omega)\, \mathrm{e}^{-2\alpha\omega z}\, P_1^{(\infty)}(\alpha\omega z)\ , \tag{3}$$

where $P_1^{(\infty)}(z) = 1 + 2z + 2z^2$. Another simplification takes place in the short-distance and long-distance limits. For $z \to 0$, the interaction potential behaves as:

$$V_1(z) \overset{z\to 0}{\sim} -\frac{1}{4\pi z^3}\int_0^\infty \mathrm{d}\omega\, \alpha_1(\mathrm{i}\omega)\frac{\epsilon(\mathrm{i}\omega) - 1}{\epsilon(\mathrm{i}\omega) + 1} \equiv -\frac{C_3}{z^3}\,. \tag{4}$$

The long-distance limit the interaction potential for a generic $\epsilon(\omega)$ is:

$$V_1(z) \overset{z\to\infty}{\sim} -\frac{3\,\alpha_1(0)}{8\pi\alpha z^4}\frac{\epsilon(0) - 1}{\epsilon(0) + 1} \equiv -\frac{C_4}{z^4}\,. \tag{5}$$

but this result cannot be used for the case of conductors, where ϵ has a pole at ω=0. We have checked that when the $\epsilon(i\omega)$ (a real function) is bounded from below by a/ω, as it is in the case of conductors, the long distance behavior of the potential is equal to the one of perfect conductor, asymptotically parametrized by $C_4^{(\infty)}$:

$$V_1^{\text{cond}}(z) \overset{z\to\infty}{\sim} V_1^{(\infty)}(z) \overset{z\to\infty}{\sim} -\frac{3\,\alpha_1(0)}{8\pi\alpha z^4} \equiv -\frac{C_4^{(\infty)}}{z^4}\,. \tag{6}$$

3. Frequency Dependent Dielectric Permittivity of Gold

The *ab-initio* computation of the frequency dependent dielectric response function of metals is beyond the reach of present day electronic structure calculations. In this work, the complex frequency dependent permittivity of gold was reconstructed from the experimental optical data. The imaginary part of the dielectric constant was modeled using a function:

$$\text{Im}\,\epsilon(\omega) = \frac{\omega_{\text{p}}^2\,\omega_\tau}{\omega(\omega^2+\omega_\tau^2)} + \sum_n c_n\,\text{f}_{\text{TL}}(\omega_{0,n},\omega'_{0,n},\gamma_n;\omega)\,, \tag{7}$$

where the first term is the Drude model of the free-electron contribution, and the second is a sum of empirical Tauc-Lorentz[5] functions:

$$\text{f}_{\text{TL}}(\omega_0,\omega'_0,\gamma;\omega) = \frac{\omega_0\,\gamma\,(\omega-\omega'_0)^2\,\theta(\omega-\omega_0)}{\omega\,(\omega^2-\omega_0^2)^2+\gamma^2\,\omega^2}\,, \tag{8}$$

where $\theta(x)$ is the Heaviside step function [$\theta(x)=0$ for $x<0$, and $\theta(x)=0$ otherwise]. Once a satisfactory representation of the imaginary part of $\epsilon(\omega)$ is found, the real part of the permittivity can be calculated using the Kramers-Kroenig relation. The parameters of Eq. (7) have been fitted to the experimental optical data. The data used included the data sets collected in the handbook by Palik[6] which cover the visible, ultraviolet and X-ray regimes, and various compilations of optical data in the microwave and in the infrared regions of the electromagnetic spectrum.[7]

The results of a global fit of the model (7) to the experimental data for gold are listed (in atomic units) in Table 1 and depicted in Fig 1. Our fitted values of the plasma frequency (ω_{p}), and the damping frequency (ω_τ) can be compared to the values used by Lambrecht *et al.*[8] who used values of ω_{p}=0.330, ω_τ=0.00108 and ω_{p}=0.276, ω_τ=0.00478.

When using the model defined in Eq. (7), we observe a slight, but systematic deviation of the fitted function with respect to the experimental data for the imaginary part of the permittivity in the very low frequency region. In order to improve the fit, we decided to further refine our model

Table 1. Best fit parameters for the analytic model of $\operatorname{Im}\epsilon(\omega)$ for gold according to Eq. (7).

ω_p	0.357091			ω_τ	0.001636		
c_1	3.177115	ω_1	0.117730	ω_1'	0.061100	γ_1	0.114560
c_2	0.483874	ω_2	0.337153	ω_2'	$=\omega_1'$	γ_2	0.262558
c_3	0.106614	ω_3	0.777090	ω_3'	$=\omega_1'$	γ_3	0.140060
c_4	1.988346	ω_4	1.014223	ω_4'	$=\omega_1'$	γ_4	2.017446
c_5	2.220095	ω_5	5.242422	ω_5'	$=\omega_5$	γ_5	10.076456

for the dielectric function by including an additional Drude-model term, corresponding to a conductor with two types of carriers:

$$\operatorname{Im}\epsilon(\omega) = \frac{\omega_{p_1}^2\,\omega_{\tau_1}}{\omega(\omega^2+\omega_{\tau_1}^2)} + \frac{\omega_{p_2}^2\,\omega_{\tau_2}}{\omega(\omega^2+\omega_{\tau_2}^2)} + \sum_n c_n\,\mathrm{f}_{\mathrm{TL}}(\omega_n,\omega_n',\gamma_n;\omega)\,. \quad (9)$$

The parameter of the model function have been reoptimized using nonlinear least squares fit to the optical data, and their values are presented in Table 2. The double Drude model leads to a significant improvement of the

Table 2. Best fit parameters for the two plasma frequency model of $\operatorname{Im}\epsilon(\omega)$ for gold according to Eq. (9).

ω_{p_1}	0.327756	ω_{τ_1}	0.001127	ω_{p_2}	0.107482	ω_{τ_2}	0.019638
c_1	4.084274	ω_1	0.110273	ω_1'	0.066774	γ_1	0.108472
c_2	0.478826	ω_2	0.337918	ω_2'	$=\omega_1'$	γ_2	0.259896
c_3	0.108575	ω_3	0.776982	ω_3'	$=\omega_1'$	γ_3	0.140290
c_4	2.001595	ω_4	1.011645	ω_4'	$=\omega_1'$	γ_4	2.010662
c_5	2.193675	ω_5	5.168412	ω_5'	$=\omega_5$	γ_5	10.819556

low-frequency behavior of ϵ, but has a negligible influence on the calculated potential. Since there is no physical justification for Eq. (9), we use Eq. (7) for the results obtained in the rest of this paper.

4. Numerical Results

The atom-surface interaction potentials have been computed by numerically evaluating Eq. (1) using Eqs. (7) and (9) and *ab-initio* calculated values of dynamic dipole polarizabilities. For the case of helium the respected polarizabilities have been computed variationally using basis sets of explicitly correlated functions.[9] For the other atoms considered we used dynamic polarizabilities published by Derevianko, Porsev and Babb.[10]

We find it convenient to represent the potential by its short distance limit multiplied by a "damping function", accounting for the effects of retardation, which for the dipole case reads:

$$V_1(z) = -\frac{C_3}{z^3} f_3(z) \,, \tag{10}$$

For the damping function $f_3(z)$, we found the following functional form to lead to an accurate representation of all interaction potential for all atom-wall distances:

$$f_3(z) = \frac{1 + a_1\,\alpha z + a_2(\alpha z)^2}{1 + a_1\,\alpha z + b_2(\alpha z)^2 + b_3(\alpha z)^3} \,, \tag{11}$$

with α being the fine structure constant. The equality of the linear terms in both the numerator and in the denominator of this rational function is due to the requirement of the potential having the correct short distance limit: $f_3(z) = 1+\mathcal{O}[(\alpha z)^2]$. The best values of the parameters a_1, a_2, b_2, b_3 in (11) fitted to the calculated interaction potentials for different atomic species are presented in Tables 3-5, together with the corresponding values of C_3.

Table 3. Long range potentials for noble gas atoms interacting with a surface of gold. The second row contains the C_3 constants, and the following rows give the values of the $f_3(z)$ damping function, defined in Eq. (10) for different distances. In the last four rows, we present the best fit parameters for the rational function (11) which covers all distances.

	He	Ne	Ar	Kr	Xe
C_3	0.062(3)	0.127(1)	0.415(4)	0.588(5)	0.870(6)
z	$f_3(z) = V(z)/(C_3/z^3)$				
1×10^1	0.99108	0.98804	0.99323	0.99348	0.99405
2×10^1	0.97388	0.96653	0.98014	0.98112	0.98275
5×10^1	0.90949	0.89155	0.92931	0.93352	0.93907
1×10^2	0.80490	0.77796	0.84192	0.85135	0.86320
2×10^2	0.64602	0.61531	0.69933	0.71522	0.73546
5×10^2	0.40337	0.37942	0.45908	0.47873	0.50522
1×10^3	0.25046	0.23481	0.29294	0.30931	0.33232
2×10^3	0.14492	0.13575	0.17185	0.18274	0.19849
5×10^3	0.06539	0.06124	0.07800	0.08322	0.09088
1×10^4	0.03441	0.03222	0.04109	0.04387	0.04796
2×10^4	0.01770	0.01657	0.02114	0.02257	0.02469
5×10^4	0.00721	0.00675	0.00861	0.00919	0.01005
1×10^5	0.00363	0.00339	0.00433	0.00462	0.00506
a_1	4.19366	5.91822	3.76881	3.95140	4.21331
a_2	0.29208	0.47236	0.21430	0.20804	0.20365
b_2	2.13922	3.33497	1.53023	1.47551	1.41033
b_3	0.10861	0.18758	0.06671	0.06066	0.05430

Table 4. Long range potentials for alkali atoms interacting with a surface of gold. The second row contains the C_3 constants, and the following rows give the values of the $f_3(z)$ damping function, defined in Eq. (10) for different distances. In the last four rows, we present the best fit parameters for the rational function (11) which covers all distances.

	Li	Na	K	Rb	Cs
C_3	1.210(5)	1.356(6)	2.058(9)	2.31(1)	2.79(1)
z			$f_3(z) = V(z)/(C_3/z^3)$		
1×10^1	0.99899	0.99845	0.99829	0.99801	0.99780
2×10^1	0.99740	0.99590	0.99542	0.99465	0.99408
5×10^1	0.98992	0.98533	0.98333	0.98086	0.97903
1×10^2	0.97415	0.96515	0.96065	0.95541	0.95169
2×10^2	0.94032	0.92509	0.91803	0.90850	0.90210
5×10^2	0.84866	0.82359	0.82012	0.80456	0.79521
1×10^3	0.73176	0.69946	0.70857	0.69126	0.68309
2×10^3	0.57369	0.53721	0.56295	0.54776	0.54485
5×10^3	0.34082	0.30970	0.34388	0.33492	0.33918
1×10^4	0.19776	0.17688	0.20301	0.19808	0.20323
2×10^4	0.10552	0.09370	0.10926	0.10672	0.11029
5×10^4	0.04353	0.03854	0.04525	0.04422	0.04585
1×10^5	0.02196	0.01943	0.02285	0.02233	0.02318
a_1	6.96178	19.81680	1.50957	1.90225	2.21418
a_2	0.10676	0.30362	2.97205	2.06782	1.55523
b_2	0.44672	1.43152	3.15260	2.30121	1.83158
b_3	0.00659	0.02118	0.15449	0.11044	0.07999

The potentials presented above can be compared to simplified ones used previously[1] in the analysis of the atomic beam experiment measuring the quantum reflectivity. There, retardation has been accounted for using:

$$V(z) = -\frac{C_4}{z^3(z+\lambda)}\,, \tag{12}$$

in which λ represents the reduced wavelength of the first electronic dipole transition in He, i.e. $\lambda = 178$ Bohr (93Å). This function has also been used in the preliminary evaluation of the experimental results for ^{3}He atoms quantum reflecting from a single crystal gold surface. Here however, significant deviations have been found indicating, that the coarse-grained model (12) is not sufficiently complex to mimic the more involved rational structure given in Eqs. (10) and (11).

An independent check of our potentials has been performed by using our model for the dielectic constant $\epsilon(\mathrm{i}\omega)$ of gold to calculate the interaction energy between two surfaces of gold, for both semi-infinite solids and

Table 5. Long range potentials for alkaline atoms interacting with a surface of gold. The second row contains the C_3 constants, and the following rows give the values of the $f_3(z)$ damping function, defined in Eq. (10) for different distances. In the last four rows, we present the best fit parameters for the rational function (11) which covers all distances.

	Be	Mg	Ca	Sr	Ba
C_3	0.650(5)	1.067(6)	1.82(1)	2.19(1)	2.75(1)
z			$f_3(z) = V(z)/(C_3/z^3)$		
1×10^1	0.99797	0.99790	0.99805	0.99788	0.99778
2×10^1	0.99388	0.99394	0.99442	0.99404	0.99382
5×10^1	0.97571	0.97722	0.97909	0.97810	0.97764
1×10^2	0.93887	0.94441	0.94957	0.94792	0.94755
2×10^2	0.86495	0.87903	0.89248	0.89054	0.89129
5×10^2	0.69195	0.72244	0.75855	0.75885	0.76468
1×10^3	0.51642	0.55531	0.61130	0.61588	0.62880
2×10^3	0.34021	0.37699	0.44043	0.44893	0.46827
5×10^3	0.16580	0.18850	0.23484	0.24315	0.26173
1×10^4	0.08878	0.10168	0.12963	0.13511	0.14765
2×10^4	0.04590	0.05270	0.06773	0.07078	0.07784
5×10^4	0.01872	0.02151	0.02772	0.02899	0.03196
1×10^5	0.00942	0.01082	0.01395	0.01460	0.01610
a_1	3.00255	3.88303	12.61180	3.09874	2.83515
a_2	0.12664	0.12802	0.19042	4.88813	2.70340
b_2	0.51498	0.55406	1.27280	5.20283	3.03090
b_3	0.01818	0.01601	0.01848	0.41434	0.20650

for thin foils. Our computations are within 1% of the revisited results of Lambrecht *et al.*[11,12] This gives us confidence, that the ready-to-use interaction potentials for the various atomic species presented in this paper are accurate. In addition, it demonstrates that our simple analytic model of the permittivity of gold may be very useful for future computation of Casimir potential involving gold, for both microscopic and macroscopic bodies.

Acknowledgments

This project was supported by the National Science Foundation (Grant PHY–8555454) and by a precision measurement grant from the National Institute of Standards and Technology. M.D. acknowledges financial support by the European Science Foundation (ESF) within the activity 'New Trends and Applications of the Casimir Effect' (www.casimir-network.com). G.L. acknowledges support by the Deutsche Forschungsgemeinschaft (DFG, contract Je285/5–1).

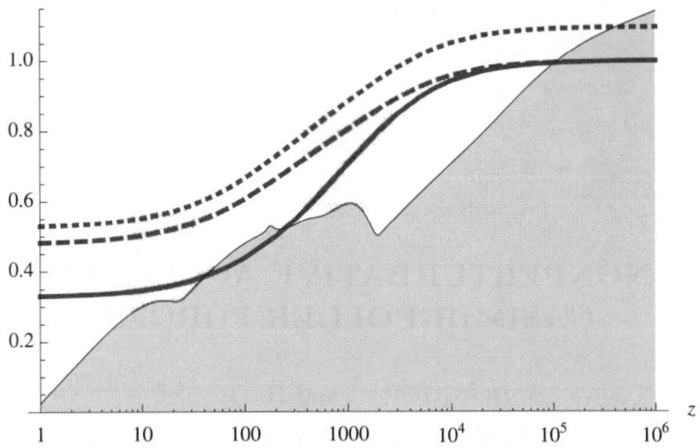

Fig. 1. Ratio between the full He-Au potential $V_1(z)$ calculated using the optical data in (7) and $V_1^{(\infty)}(z)$ (solid curve), between $V_1(z)$ and the approximate model (12) using $C_4 = C_4^{(\infty)}$ (dashed curve) or using the best fit value of $C_4 = 44$ eVÅ^4 (dotted curve). The shaded spectrum in the background depicts $\log \operatorname{Im} \epsilon(\omega)$ of the fit obtained in Table 1, for $\omega = c/z$.

References

1. V. Druzhinina and M. DeKieviet, Phys. Rev. Lett. **91**, 193202 (2003).
2. D. M. Harber, J. M. Obrecht, J. M. McGuirk, and E. A. Cornell, Phys. Rev. A **72**, 033610 (2005).
3. E. M. Lifshitz, Zh. Éksp. Teor. Fiz. **29**, 94 (1955), [JETP **2**, 73 (1956)].
4. P. J. Mohr, B. N. Taylor, and D. B. Newell, Rev. Mod. Phys. **80**, 633 (2008).
5. G. E. Jellison, Jr. and F. A. Modine, Appl. Phys. Lett. **69**, 371 (1996).
6. E. D. Palik, *Handbook of Optical Constants of Solids* (Academic Press, San Diego, 1985).
7. H. E. Bennett and J. M. Bennett, *Optical Properties and Electronic Structure of Metals and Alloys edited by F. Abeles* (North-Holland, Amsterdam, 1966); L. G. Schulz, J. Opt. Soc. Am. **44**, 357 and 362 (1954); G. P. Motulevich and A. A. Shubin, Soviet Phys. JETP **20**, 560 (1965); V. G. Padalka and N. Shklyarevskii, Opt. Spectr. USSR **11**, 285 (1961); G. A. Bolotin *et. al.*, Phys. Met. and Mt. **13**, 823 (1962); B. Brändli and A. J. Sievers, Phys. Rev. B **5**, 3550 (1972); J. H. Weaver *et. al.*, *Physics Data, Optical Properties of Metals*, (Fach-Information Zentrum, Kalsruhe, 1981).
8. A. Lambrecht, P. A. Maia Neto and S. Reynaud, New J. Phys. **8**, 243 (2006).
9. G. Łach, B. Jeziorski and K. Szalewicz, Phys. Rev. Lett. **92**, 233001 (2004).
10. A. Derevianko, S. G. Porsev and J. F. Babb, preprint: `arXiv:0902.3929`
11. A. Lambrecht, I. Pirozhenko, L. Duraffourg and Ph. Andreucci. Eur. Phys. Lett. **77**, 44006 (2007).
12. A. Lambrecht, I. Pirozhenko, L. Duraffourg and Ph. Andreucci. Eur. Phys. Lett. **81**, 19901 (2008).

NONPERTURBATIVE ACCESS TO CASIMIR-POLDER FORCES

B. DÖBRICH[1,*] and H. GIES[1,2]

[1] *Theoretisch-Physikalisches Institut, Friedrich-Schiller-Universität Jena,*
[2] *Helmholtz Institute Jena, Max-Wien-Platz 1,D-07743 Jena, Germany*
[*] *E-mail: babette.doebrich@uni-jena.de*

M. DEKIEVIET

Physikalisches Institut, Universität Heidelberg,
Philosophenweg 12, D-69120 Heidelberg, Germany

We discuss the scalar analogue of the Casimir-Polder force between a sphere and a uniaxially corrugated surface with Dirichlet boundary conditions. Presenting a formulation that is nonperturbative in the height profile of the surface, we give explicit numerical results for a sinuosoidal corrugation profile.

Keywords: Casimir-Polder, corrugated surfaces, nonperturbative techniques.

1. Introduction

Past years have witnessed great progress in the study of Casimir-Polder[1] forces between an atom and a surface. On the theoretical side, the dependence of the Casimir-Polder force on the surface geometry is an important problem.[2–5] Often, its deviation from the standard planar situation is accounted for in a perturbative manner: the (mean) amplitude A of the surface corrugation is assumed to be the smallest length scale of the system. However, in recent high resolution experiments using the atomic beam spin echo technique,[6] the atom-wall distance can become much smaller than the amplitude of the surface corrugation. Thus, there is an urgent need for nonperturbative calculations of the Casimir-Polder potential. Starting with the scalar Dirichlet situation, we present such a treatment in the following.

2. Nonperturbative access to scalar fields

The presence of bodies or surfaces imposes boundary conditions on fluctuating quantum fields. This gives rise to a shift in the energy of the ground state, the Casimir energy. A substraction of the Casimir self-energy of the bodies then yields the Casimir interaction energy between the surfaces which serves as a potential energy for the Casimir force. Using the constrained-functional integral approach,[7,8] the boundaries on the fluctuating field are implemented through a δ functional. Upon integration over the fields, the Casimir interaction energy between two surfaces S_1 and S_2, separated by a (mean) distance H can be written as

$$E(H) = -\frac{\hbar c}{T_{\mathrm{E}}}\frac{1}{2}\sum_{n=1}^{\infty}\frac{1}{n}\mathsf{Tr}\left(\mathcal{M}_{11}^{-1}\mathcal{M}_{12}\mathcal{M}_{22}^{-1}\mathcal{M}_{21}\right)^n , \tag{1}$$

where T_{E} denotes the length in Euclidean time direction. $\mathcal{M}_{\alpha\beta}$ is the propagator of the fluctuations, i.e.

$$\mathcal{M}_{\alpha\beta}(\zeta, \vec{x}-\vec{x}') = \frac{1}{4\pi|\vec{x}-\vec{x}'|}\exp\left(-|\vec{x}-\vec{x}'||\zeta|\right) \tag{2}$$

for the scalar Dirichlet case.

In Eq. (2), ζ denotes the imaginary frequency, while $\vec{x} = (x_1, x_2, x_3)$ and $\vec{x}'$ are three-vectors pointing onto the surfaces S_α and S_β, respectively. As the surfaces respond to the field by charge fluctuations, the inverse propagator $\mathcal{M}_{\alpha\beta}^{-1}$ can be interpreted as the propagator of charge fluctuations within the surface. The trace in Eq.(1) has to be taken over the coordinates of the surfaces, demanding the inclusion of appropriate metric factors for the integration measures. Furthermore, the functional inverse of $\mathcal{M}_{\alpha\beta}$ is generally not analytically known for nontrivial surfaces.

In the following, we evaluate the Casimir energy between a surface S_1 which is uniaxially corrugated along the direction x_1 and a sphere S_2 with radius r, cf. left panel of Fig. 1. We are interested in the Casimir-Polder limit ($r \ll H$), where the analytical result for a flat surface S_1 is known to be $\mathcal{O}(\frac{r}{H^2})$ to leading order.[9]

We compute $\mathcal{M}_{22}^{-1}$ from $\mathcal{M}_{22}^{-1}\mathcal{M}_{22} = \mathbb{1}$, where $\mathcal{M}_{22}$ is given through (2). By expansion of the equation in terms of spherical harmonics Y_{lm}, $\mathcal{M}_{22}^{-1}$ can be calculated to arbitrary order in l. For the computation of the leading order Casimir energy, however, it suffices to consider the monopole contribution $l = 0 = m$, which reads $\mathcal{M}_{22}^{-1}(\zeta) = |\zeta|\exp(r|\zeta|)/\left[4\pi r^2\sinh(r|\zeta|)\right]$.

Next, we go over to dimensionless variables by a rescaling with the distance parameter H: $\vec{x} \to \tilde{\vec{x}}H$, $\zeta \to \tilde{\zeta}/H$. We find that to first order $\mathcal{O}(\frac{r}{H^2})$ in the Casimir energy, it is sufficient to consider the $n = 1$ term of the sum in Eq. (1). Furthermore, in this limit the propagators $\mathcal{M}_{12}$ and $\mathcal{M}_{21}$ become independent of the coordinates on the sphere S_2. As the monopole order of $\mathcal{M}_{22}^{-1}$ is also independent of these coordinates, the two integrations over the surface of the sphere contribute only a factor of $16\pi^2$ in Eq. (1). Only after this step, the translational invariance of the surface S_1 along the 2-component can be exploited by a Fourier transformation of Eq. (1) to momentum space.

In a final step, substituting $\tilde{q} = \sqrt{\tilde{\zeta}^2 + \tilde{p}_2^2}$, Eq. (1) reduces to:

$$E = -\frac{\hbar c r}{H^2} \int_0^\infty \mathrm{d}\tilde{q} \int_{-\infty}^{\infty} \mathrm{d}\tilde{x} \, \sqrt{g(\tilde{x})} \tilde{q} \Delta\tilde{\mathcal{M}}_{12}(\tilde{q};\tilde{x}) \tilde{\mathcal{M}}_{21}(\tilde{q};\tilde{x}) + \mathcal{O}\left(\frac{r^2}{H^3}\right), \quad (3)$$

where we have defined $\Delta\tilde{\mathcal{M}}_{12} = \tilde{\mathcal{M}}_{11}^{-1}\tilde{\mathcal{M}}_{12}$ and dropped the coordinate subscript "1": $\tilde{x}_1 \rightarrow \tilde{x}$. The metric factor is related to the height profile $h(\tilde{x})$ by

$$\sqrt{g(\tilde{x})} = \sqrt{1 + \left(\partial_{\tilde{x}} \tilde{h}(\tilde{x})\right)^2}, \quad \tilde{h}(\tilde{x}) = \frac{1}{H} h(\tilde{x} H) \; . \quad (4)$$

In principle, computing the energy in Eq. (3) is now very simple. The combined propagator $\Delta\mathcal{M}_{12}$ can be obtained by solving

$$\int_{\tilde{x}} \sqrt{g(\tilde{x})} \tilde{\mathcal{M}}_{11}(\tilde{q};\tilde{x}';\tilde{x}) \Delta\tilde{\mathcal{M}}_{12}(\tilde{q};\tilde{x}) = \tilde{\mathcal{M}}_{12}(\tilde{q};\tilde{x}') \quad (5)$$

numerically. However, the treatment of the above equation is nontrivial due to the singular structure of $\mathcal{M}_{11}$ at the origin, see Eq. (6). Thus, a suitable regularization scheme has been worked out.[10]

The dimensionless propagators $\tilde{\mathcal{M}}_{12} \equiv \tilde{\mathcal{M}}_{21}$ and $\tilde{\mathcal{M}}_{11}$ that enter Eqs. (5) and (3) are given in terms of Bessel functions:

$$\tilde{\mathcal{M}}_{11}(\tilde{q};\tilde{x}';\tilde{x}) = \frac{1}{2\pi} K_0 \left(\tilde{q} \sqrt{(\tilde{x}' - \tilde{x})^2 + \left(\tilde{h}(\tilde{x}') - \tilde{h}(\tilde{x})\right)^2} \right), \quad (6)$$

$$\tilde{\mathcal{M}}_{12}(\tilde{q};\tilde{x}') = \frac{1}{2\pi} K_0 \left(\tilde{q} \sqrt{(\tilde{x}')^2 + \left(\tilde{h}(\tilde{x}') - 1\right)^2} \right) \; . \quad (7)$$

3. Results for a sinusoidal surface corrugation

As a concrete example, we consider the case of a sinusoidal surface corrugation. To this end, we employ $h(x) = A \sin(\omega x + \phi)$ as height function in Eq. (4). We fix the center of the sphere at $x = 0$ and use the phase ϕ to effectively vary the sphere's position above the structure.

In the following, we present numerical results for the energy above a minimum of the structure, i.e. for $\phi = -\frac{\pi}{2}$. In order to highlight the geometry-induced effects, we normalize the results for the Casimir energy E_{sine} with respect to the energy of the planar-surface situation E_{planar}. For consistency, E_{planar} is also evaluated numerically. On the right panel of Fig. 1, we display $E_{\text{sine}}/E_{\text{planar}}$ as a function of the normalized distance H/A for three different corrugation frequencies $\omega A = 1, 2, 3$. In the limiting cases of $H/A \to \infty$ and $H/A \to 0$, we find that $E_{\text{sine}}/E_{\text{planar}} \to 1$, as can be expected: For $H/A \to \infty$, the corrugation cannot be resolved by the sphere as it is much smaller than the distance, whereas for $H/A \to 0$, the corrugation is much larger than the distance and is thus not seen locally. By contrast, for distances $H \sim A$, a distinct deviation from the planar-surface case is found. As the surface at the structure minimum curves towards the sphere, one finds $E_{\text{sine}}/E_{\text{planar}} > 1$. This effect becomes more pronounced as the structure wells become more narrow, i.e. for larger ωA.

It is useful to parameterize the deviation of the Casimir energy in the non-planar situation from the flat-surface setup in terms of an anomalous dimension η, by defining $E_{\text{corrugation}} \sim 1/H^{2+\eta}$, where $\eta = 0$ for a flat surface S_1. The increase of the normalized Casimir-Polder energy towards the peak is found to scale linearly, corresponding to $\eta = -1$, with ω-dependent linear coefficients (not shown in Fig. 1). In the drop-off region of the potential right beyond the peak, the anomalous dimension depends on ω: For $\omega A = 1, 2, 3$ we find $\eta \simeq 0.4, 1.0, 1.6$, respectively. Most interestingly, at larger distances $H/A \simeq 10$, all curves converge towards a universal curve characterized by an anomalous dimension of $\eta \simeq 0.2$, irrespectively of the frequency ωA.

Within the worldline picture of quantum field theory[11] this can be attributed to the fact that the quantum vacuum fluctuations average over the surface geometry as they are isotropic in space. With growing separation between sphere and plate, the effect of higher corrugation frequencies $H\omega \gg 1$ is not resolved anymore, as has been also confirmed by studies of a sawtooth-like corrugation.[10]

4. Conclusions

In this work we have investigated the scalar analogue of Casimir-Polder energies between a sphere and a uniaxially corrugated surface, using a sinusoidal surface profile as an example. In particular, our study was not based on a perturbative ordering of length scales and thus allows for arbitrary ratios of the objects' separation H and the deformation parameters ω and A in the limit of vanishing sphere radius r. In a numerical study we have parameterized the geometry-dependence of the Casimir energy by introducing an anomalous dimension η, which was shown to be non-integer

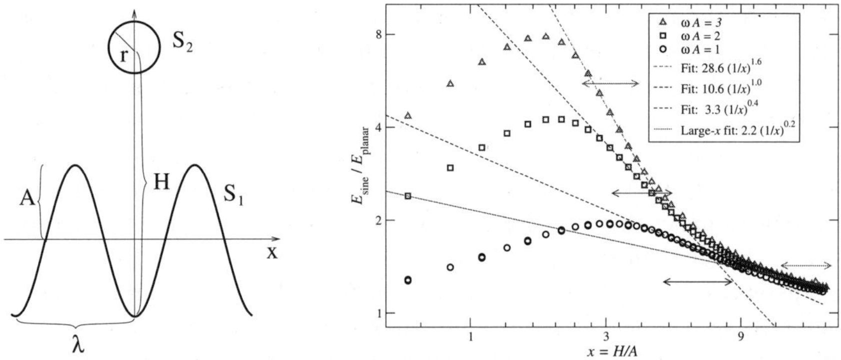

Fig. 1. Left: Involved length scales of the setup: Sphere S_2 of radius r at distance H above a corrugated surface S_1 with amplitude A and corrugation wavelength $\lambda = \frac{2\pi}{\omega}$. Right: $E_{\text{sine}}/E_{\text{planar}}$ as a function of separation H/A for corrugation frequencies $\omega A = 1, 2, 3$

valued in the regime of $H \sim A$. This result is not accessible through a perturbative calculation. Although our results for the Dirichlet scalar case should not be viewed as a quantitative estimate for the electromagnetic case, we expect analogous results for the anomalous dimensions also for the latter case. This is currently under investigation.

Acknowledgments

B.D. would like to thank the organizers of QFEXT 09 for the opportunity to present this work. Financial support through DFG/TR18, DFG/GRK1523 and DFG/Gi328/5-1 is greatfully acknowledged.

References

1. H.B.G. Casimir and D. Polder, Phys. Rev. **73**, 360 (1948).
2. V.B. Bezerra, G.L. Klimchitskaya, and C. Romero, Phys. RevȦ **61**, 022115 (2000).
3. R. Messina et. al. Phys. Rev. A **80**, 022119 (2009).
4. D. A. R. Dalvit et. al Phys. Rev. Lett. **100**, 040405 (2008).
5. S. Y. Buhmann and D. G. Welsch, Prog. Quant. Electron. **31**, 51 (2007).
6. M. DeKieviet et. al, Phys. Rev. Lett. 75, 1919 (1995); Surf. Sci. 377-379, 1112 (1997).
7. M. Bordag, D. Robaschik and E. Więczorek, Annals Phys. **165**, 192 (1985).
8. T. Emig, A. Hanke and M. Kardar, Phys. Rev. Lett. **87**, 260402 (2001).
9. A. Bulgac, P. Magierski and A. Wirzba, Phys. Rev. D **73**, 025007 (2006).
10. B. Dobrich, M. DeKieviet and H. Gies, Phys. Rev. D **78**, 125022 (2008).
11. H. Gies, K. Langfeld and L. Moyaerts, JHEP **0306**, 018 (2003).

GRAPHENE VAN DER WAALS INTERACTIONS WITH AN ADATOM AND WITHIN A DOUBLE LAYER

NORMAN J. M. HORING

Department of Physics and Engineering Physics, Stevens Institute of Technology, Hoboken, New Jersey 07030, USA

VASSILIOS FESSATIDIS

Department of Physics, Fordham University, Bronx, New York 10458, USA

We analyze the van der Waals interaction between a Graphene monolayer and (i) an adatom/molecule, and also (ii) a second monolayer in a double layer array, to second order in the Coulomb interaction between the internally correlated electrons of the two distinct subsystems. In both cases of van der Waals interaction, (i) and (ii), we obtain explicit analytic results for the plasmon contributions and for the undoped electron-hole contributions to the energy.

1. Introduction

In this paper we report on the van der Waals interactions of Graphene, a single-atom-thick planar layer of Carbon atoms in a hexagonal honeycomb lattice. This study includes (i) the interaction of a single two-dimensional sheet of Graphene with an adsorbed atom/molecule, and (ii) the van der Waals interaction between the two sheets of a Graphene double layer. Graphene's surface interactions have taken on special importance because of its exceptional device-friendly properties[1,2] in regard to charge transport and sensing. With mobility reaching $200,000$ cm^2/Vs, high electron density on the order of 10^{13}cm^{-2}, long mean free path $l \sim 400$nm, stability up to $3,000$K and a quantum hall effect at room temperature, it holds great promise for nanoelectronic applications.[2] Such applications include sensors,[3] field-effect transistors,[4] spin valves,[5] electromechanical resonators,[6] quantum interference devices and others. Furthermore, as a planar layer of carbon atoms, it is amenable to highly developed top-down CMOS compatible process flows, a substantial advantage over carbon nanotubes.

This calculation of van der Waals energy proceeds from the Gell-Mann/Low theorem applied to the Coulomb interaction (i) between electrons of a single sheet of Graphene and those of an adatom, and (ii) between the electrons of one sheet of Graphene double layer with those of the second sheet of the double layer. In each of the cases (i) and (ii) the two distinct sets of electrons in mutual interaction are labeled A and B. To second order in these Coulomb interactions between the electrons of A with those of B, the interaction energy is given by[7] ($v(\mathbf{r}_1 - \mathbf{r}_2)$ is the Coulomb potential; $\hbar \to 1$)

$$E_{int} = -i \int_{-\infty}^{0} dt \int d\vec{r_1} \int d\vec{r_2} \int d\vec{r_3} \int d\vec{r_4} v(\vec{r_1} - \vec{r_2}) v(\vec{r_3} - \vec{r_4}) \times G_2^A(\vec{r_1}0, \vec{r_3}t; \vec{r_1}0^+, \vec{r_3}t^+) G_2^B(\vec{r_2}0, \vec{r_4}t; \vec{r_2}0^+, \vec{r_4}t^+). \quad (1)$$

Here, we assume no wave function overlap and/or charge exchange, and $G_2^{A,B}$ are the two-electron equilibrium Green's functions of the disjoint subsystems A, B, respectively, and we have used overall charge neutrality to exactly eliminate a number of terms involving one-electron equilibrium Green's functions, $G_1^{A,B}$. An interpretation in terms of self-energy can be advanced by employing a Kubo-type linear response relation for the screening function $K(\vec{r_1}t_1, \vec{r_2}, t_2)$ that is the space-time matrix inverse of the direct dielectric function $\varepsilon(\vec{r_1}t_1, \vec{r_2}t_2)$, written for subsystem A as

$$\int d\vec{r_3} \int dt_3\, \varepsilon_A(\vec{r_1}t_1, \vec{r_3}t_3) K_A(\vec{r_3}t_3, \vec{r_2}t_2) = \delta^{(3)}(\vec{r_1} - \vec{r_2}) \delta(t_1 - t_2). \quad (2)$$

The linear response relation for A states that K_A relates to the density-density correlation function of A as ($v(1-3) = v(\vec{r_1} - \vec{r_3})\delta(t_1 - t_3)$; $1 = \vec{r_1}, t_1$,etc.)

$$K_A(1,2) = \delta^{(4)}(1-2) + \int \delta^{(4)}3\, v(1-3) \times [G_2^A(2,3;2^+,3^+) - G_1^A(2,2^+)G_1^A(3,3^+)]. \quad (3)$$

Explicitly recognizing its role in a screening capacity, it generates the effective potential due to Coulomb interaction, V_{eff}, and its image part, V_{image}, in accordance with

$$V_{eff}^A(3,1) = \int d^{(4)}2\, K(3,2) v(2-1) \quad (4)$$

and

$$V_{image}^A(3,1) \equiv V_{eff}^A(3,1) - v(3-1). \quad (5)$$

Employing this in Eq. (1), we have

$$E_{int} = -\int_{-\infty}^{0} dt \int d\vec{r_1} \int d\vec{r_3} G_2^B(\vec{r_1}0, \vec{r_3}t; \vec{r_1}0^+, \vec{r_3}t^+) V_{image}^A(\vec{r_3}t; \vec{r_1}0). \quad (6)$$

In all of these considerations $G_2^{A,B}$ carry the full correlations internal to their individual subsystems, A, B, respectively, with no relation to the other subsystem, B, A, respectively. As the operators internal to the statistical trace matrix elements of G_2^B in Eq. (6) involve the charge density operators of subsystem B, $\hat{\rho}_B(\vec{r_1}0), \hat{\rho}_B(\vec{r_3}, t)$, it is immediately apparent that $G_2^B V_{image}^A$ as it occurs in E_{int} in Eq. (6) represents a correlation self-energy of subsystem B electrons mediated by the potential V_{image}^A induced through the polarization of subsystem A (which occurs mutually with the polarization of subsystem B when A and B are in proximity). Of course, this image potential is dynamic, nonlocal and inhomogeneous.

As attractive as the foregoing interpretation is for case (i), it is preferable to use a more symmetric formulation for the problem of van der Waals interaction for case (ii). To this end, we reconsider Eq. (1) using Eq. (3), obtaining

$$\begin{aligned} E_{int} = -i \int_{-\infty}^{0} dt \int d\vec{r_2} \int d\vec{r_3} \{ & [K_A(\vec{r_2}, 0; \vec{r_3}, t) - \delta^{(3)}(\vec{r_2} - \vec{r_3})\delta(t)] \\ & \times [K_B(\vec{r_3}, t; \vec{r_2}, 0) - \delta^{(3)}(\vec{r_3} - \vec{r_2})\delta(t)]\}, \end{aligned} \quad (7)$$

where irrelevant terms involving one-electron Green's functions have again been eliminated.

2. Graphene-Atom van der Waals Interaction

It is clear from the foregoing discussion that the van der Waals interaction of a two-dimensional system devolves upon its screening function, K, which may be determined from the direct dielectric function, ε, using Eq. (2) jointly with (Fourier transform in the plane of the translationally invariant sheet of plasma $[x, y \to q_x, q_y \to \bar{q}]$ and in time $[t_1 - t_2 \to \omega]$)

$$\varepsilon(\bar{q}, z, z_2; \omega) = \delta(z_1 - z_2) + \delta(z_2)\alpha_{2D}(\bar{q}, \omega)e^{-q|z_1|}, \quad (8)$$

which describes a strictly two-dimensional dielectric confined to the plane $z = 0$ having polarizability $\alpha_{2D}(\bar{q}, \omega)$ *on the plane*. The resulting screening function is readily found to be[8]

$$K(\bar{q}, z, z_2; \omega) = \delta(z_1 - z_2) + \delta(z_2 - z_0)e^{-|\bar{q}||z_1 - z_0|}[\tilde{K}_{2D}(\bar{q}, \omega) - 1] \quad (9)$$

where

$$\tilde{K}_{2D} = (1 + \alpha_{2D})^{-1} = [\tilde{\varepsilon}_{2D}]^{-1}. \tag{10}$$

We take $\alpha_{2D}(\bar{q}, \omega)$ to be the polarizability of Graphene in the random phase approximation (RPA) in the degenerate limit of zero temperature[9−13]. The atomic electron G_2^a-function involved in E_{vdW} has the form

$$G_2^a \to \langle \Phi_0^a | \rho_a(\mathbf{x}'', t'') \rho_a(\mathbf{x}', t') | \Phi_0^a \rangle \tag{11}$$

where $\rho_a(\mathbf{x}, t)$ is the density operator for the atomic electrons, and we denote the E_n^a- energy eigenstates of the atomic electrons by $|\Phi_n^a\rangle$, with ground state $|\Phi_0^a\rangle$. These states of the atomic electrons bear the full complement of correlations due to electron-electron interactions of the electrons within the atom. Denoting the atomic matrix elements of $\rho_a(\mathbf{x}'', t'')$ by

$$\langle \Phi_0^a | \, \rho_a(\mathbf{x}'', t'') \, | \Phi_n^a \rangle = \langle \rho_a(\mathbf{x}'') \rangle_{0n} \tag{12}$$

with

$$\omega_{n0}^a = E_n^a - E_0^a, \tag{13}$$

and using the atom's time translation operator, we have

$$G_2^a(\mathbf{x}''t''; \mathbf{x}'t'; \mathbf{x}''t''^+; \mathbf{x}'t'^+) = \sum_n{}' \langle \rho_a(\mathbf{x}'') \rangle_{0n} \langle \rho_a(\mathbf{x}') \rangle_{n0} \, e^{i\omega_{no}^a(t'-t'')}. \tag{14}$$

(The prime on the $\Sigma \to \Sigma'$ indicates that the $n = 0$ term is excluded since it has no time dependence and yields a constant, static contribution, which is irrelevant to the van der Waals interaction.) Substitution of this into Eq. (6), yields E_{vdW} in terms of the Fourier time transform of $\mathbf{V}_{image}(t' - t'') \to \mathbf{V}_{image}(\omega)$ as

$$E_{vdW} = \int d^3\mathbf{x}' \int d^3\mathbf{x}'' \int_{-\infty}^{\infty} \frac{d\omega}{2\pi i} \mathbf{V}_{image}(\mathbf{x}', \mathbf{x}''; \omega) \times \sum_n{}' \left(\frac{\langle \rho_a(\mathbf{x}'') \rangle_{0n} \langle \rho_a(\mathbf{x}') \rangle_{n0}}{\omega - \omega_{n0}^a} \right), \tag{15}$$

where we have taken account of nonlocality in time in performing the time integration. Considering spatial translational invariance in the lateral plane$(x, y) \to \bar{\mathbf{q}}$ and using $\mathbf{V}_{image}(z', z''; \bar{q}, \omega)$ obtained from Eqns. (4), (5), (9) and (10), we obtain the result for E_{vdW} using a multipole expansion mandated by the spatially confined distribution of the atomic electrons. To dipole-dipole terms, it is given by ($\mathbf{D}_{0n}$ is the matrix element of the atom's dipole moment operator between its electronic eigenstates 0, n; we take the background dielectric constant as unity, $\varepsilon_0 \to 1$):

$$E_{vdW} = \frac{4}{3\hbar} \sum_n{}' \int_0^\infty \frac{du}{2\pi} \frac{\omega_{n0}^a \left|\mathbf{D}_{0n}\right|^2}{u^2 + (\omega_{n0}^a)^2} \int_0^\infty dq q^2 e^{-2q|Z|} \frac{\alpha_{2D}(q, iu)}{1 + \alpha_{2D}(q, iu)}, \tag{16}$$

where $|Z|$ is the distance of the atom from the $2D$ planar quantum well.

Eq. (16) is a useful point of departure to determine both local and nonlocal structure of the Atom-Graphene van der Waals interaction. $\alpha_{2D}(\overline{q}, \omega)$ was determined for Graphene in references 9-13 and we generally use the notation of Ref. 12. A principle plasmon contribution to E_{vdw} occurs in the local limit, $q \to 0$, in which

$$\alpha_{2D} \to -\frac{\omega_p^2}{\omega^2} = \frac{\omega_p^2}{u^2}, \tag{17}$$

where the 2D local plasma frequency is given by $\omega_p = \lambda\sqrt{q}$ with $\lambda = (\gamma e^2 \sqrt{\rho_{2D} \pi g_s g_v}/\epsilon_0 \hbar)^{1/2}$ and ρ_{2D} is the unperturbed equilibrium density. Here, γ is the Graphene Fermi velocity (independent of density) and $g_s = g_v = 2$ are spin and valley degeneracies. Accordingly, the local plasmon contribution to the q-integral of E_{vdw} in Eq. (16) is

$$\int_0^\infty dq ... = \lambda^2 \int_0^\infty dq q^3 \frac{e^{-2q|Z|}}{u^2 + \lambda^2 q} = -\frac{u^6}{\lambda^6} \frac{\partial^3}{\partial \mu^3} \int_0^\infty \frac{dx e^{-\mu x}}{1 + x}, \tag{18}$$

where $\mu = 2|Z|u^2/\lambda^2$, so that

$$\int_0^\infty dq ... = \frac{u^6}{\lambda^6} \frac{\partial^3}{\partial \mu^3} [e^\mu Ei(-\mu)], \tag{19}$$

with $Ei(x)$ as the Exponential-integral function.[14]

For the undoped case, $\alpha_{2D}(q, iu)$ is presented in Ref. 13 (Eq. (10)) as

$$\alpha_{2D}(q, iu) = \left(\frac{g_s g_v \pi e^2}{8\hbar}\right) \frac{q}{\sqrt{u^2 + \gamma^2 q^2}}, \tag{20}$$

which we employ to estimate higher wavenumber nonlocal contributions to E_{vdw}. It should be noted that an expansion of E_{vdw} in inverse powers of $|Z|$ is *not* available since it would involve expanding the q-integrand of Eq. (16) in powers of q/u, which would cause divergences in the final u-integration. However, for atomic frequencies (ω_{n0}^a), the polarizability is small, $\alpha_{2D}(q, iu) \ll 1$, and we have the q-integral of Eq. (16) approximately as

$$\int_0^\infty dq \ldots \cong \left(\frac{g_s g_v \pi e^2}{8\hbar}\right) \int_0^\infty dq e^{-2q|Z|} \frac{q^3}{\sqrt{u^2 + \gamma^2 q^2}}. \tag{21}$$

This integral may be rewritten using[15]

$$\int_0^\infty \frac{xe^{-\mu x}}{\sqrt{x^2+\beta^2}}dx = \frac{\beta\pi}{2}\left[\mathbf{H}_1(\beta\mu) - Y_1(\beta\mu)\right] - \beta, \tag{22}$$

where $\mathbf{H}_1(x)$ is the Struve function and $Y_1(x)$ is the Bessel function of the second kind, with the result

$$\int_0^\infty dq \ldots \cong \left(\frac{g_s g_v \pi e^2}{8\hbar}\right)\frac{u\pi}{8\gamma^2}\frac{\partial^2}{\partial|Z|^2}\left[\mathbf{H}_1\left(\frac{2|Z|}{\gamma}u\right) - Y_1\left(\frac{2|Z|}{\gamma}u\right)\right]. \tag{23}$$

3. Graphene-Graphene van der Waals Interaction

To analyze the van der Waals interaction between two Graphene sheets at $z = \pm z_0$ (separation $2z_0$) we employ Eq. (7) with the notation $\overline{\overline{K}} \equiv K - \delta$, obtaining

$$\frac{E_{\text{int}}}{\text{Area}} = \frac{1}{\pi}\int_0^\infty du \int \frac{d^2\bar{q}}{(2\pi)^2}\int dz_2 \int dz_3 \overline{\overline{K}}_A(z_2, z_3; \bar{q}, iu)\overline{\overline{K}}_B(z_3, z_2; \bar{q}, iu). \tag{24}$$

In this matter, we ignore tunneling between the two Graphene sheets. Because of the shifted z_0-origins of the sheets,

$$\overline{\overline{K}}_{A,B}(z_1, z_2; \bar{q}, \omega) = -\delta(z_2 \pm z_0)e^{-|\bar{q}||z_1 \pm z_0|}\frac{\alpha_{2D}(q, iu)}{1 + \alpha_{2D}(q, iu)}, \tag{25}$$

and considering plasmonic contributions in the local regime,

$$\frac{\alpha_{2D}(q, iu)}{1 + \alpha_{2D}(q, iu)} = \frac{\omega_p^2}{u^2 + \omega_p^2} \quad \text{with } \omega_p^2 = \lambda^2 q, \tag{26}$$

we have the second order van der Waals interaction energy as

$$\frac{E_{\text{int}}}{\text{Area}} = \frac{\lambda^4}{2\pi^2}\int_0^\infty du \int_0^\infty dq q^3 e^{-4q|z_0|}\frac{1}{(u^2 + \lambda^2 q)^2}. \tag{27}$$

Executing the elementary u-integration, we obtain

$$\frac{E_{\text{int}}}{\text{Area}} = \frac{\lambda}{8\pi}\int_0^\infty dq e^{-4q|z_0|}q^{3/2} = \frac{3}{128\pi^{1/2}}\frac{\hbar\lambda}{|z_0|^{5/2}}, \tag{28}$$

where we have restored $\hbar$ at the end of Eq. (28).

Finally, we estimate higher order nonlocal wavenumber contributions in the undoped case involving the excitation of electron-hole pairs by the mutual polarization of the Graphene sheets, to the exclusion of plasmons.

In this, we employ Eq. (20) and generate an order of magnitude estimate using Eq. (25), with the result

$$\frac{E_{\rm int}}{\rm Area} = \frac{1}{2\pi^2}\left(\frac{g_s g_v \pi e^2}{8\kappa\hbar}\right)^2 \int_0^\infty du \int_0^\infty dq \, \frac{q^3 e^{-4q|z_0|}}{\left[\sqrt{u^2+\gamma^2 q^2} + \left(\frac{g_s g_v \pi e^2}{8\kappa\hbar}\right) q\right]^2}, \tag{29}$$

and introducing dimensionless quantities $\nu = u/E_F$ and $x = q/q_F$ we have

$$\frac{E_{\rm int}}{\rm Area} = \frac{(g_s g_v)^2 q_F^3 e^4}{128\kappa^2\gamma\hbar} \int_0^\infty d\nu \int_0^\infty dx \frac{x^3 e^{-4x|z_0|q_F}}{\left[\sqrt{\nu^2+x^2} + \left(\frac{g_s g_v \pi e^2}{8\kappa\gamma\hbar}\right) x\right]^2}. \tag{30}$$

Making the substitutions $\nu = kx \Rightarrow d\nu = xdk$ and $\beta = \frac{g_s g_v \pi e^2}{8\kappa\gamma\hbar} = 1.432$, we obtain

$$\begin{aligned}\frac{E_{\rm int}}{\rm Area} &= \frac{(g_s g_v)^2 q_F^3 e^4}{128\kappa^2\gamma\hbar} \int_0^\infty dk \frac{1}{\left(\sqrt{k^2+1}+\beta\right)^2} \int_0^\infty dx\, x^2 e^{-4x|z_0|q_F} \\ &= \frac{(g_s g_v)^2 e^4}{4096\kappa^2\gamma\hbar\,|z_0|^3} \int_0^\infty dk \frac{1}{\left(\sqrt{k^2+1}+\beta\right)^2}.\end{aligned} \tag{31}$$

The last integral on the right of Eq. (31) is evaluated in the Appendix with the result

$$\frac{E_{\rm int}}{\rm Area} = \frac{(g_s g_v)^2 e^4}{4096\kappa^2\gamma\hbar\,|z_0|^3}\left[\frac{\beta}{(\beta^2-1)} - \frac{1}{(\beta^2-1)^{3/2}} \ln\left(\frac{\beta+1+\sqrt{\beta^2-1}}{\beta+1-\sqrt{\beta^2-1}}\right)\right], \tag{32}$$

and the value of the quantity in the square brackets is 0.528.

4. Conclusions: Summary

We have analyzed the van der Waals interaction energy between a two-dimensional sheet of Graphene and (i) an adatom, and (ii) a second sheet in a Graphene double layer. These calculations have been carried out to second order in the Coulomb interaction of the two-dimesional electrons of the Graphene and the electrons of the second system (i) an adatom, and (ii) a second Graphene sheet. For the case of the adatom (i), we obtain the plasmon contribution in the local regime as

$$E_{vdW} = \frac{4}{3\hbar\lambda^6} \sum_n{}' \int_0^\infty \frac{du}{2\pi} u^6 \frac{\omega_{n0}^a |\mathbf{D}_{0n}|^2}{u^2+(\omega_{n0}^a)^2} \frac{\partial^3}{\partial\mu^3}[e^\mu Ei(-\mu)], \tag{33}$$

where $Ei(x)$ is the Exponential-integral function, $\mu = 2|Z|u^2/\lambda^2$, $\lambda = (\gamma e^2\sqrt{\rho_{2D}\pi g_s g_v}/\varepsilon_0\hbar)^{1/2}$, γ is the Graphene Fermi velocity (independent of density), etc. For the undoped case, we estimate the higher wavenumber nonlocal electron-hole contribution to be

$$E_{vdW} = \frac{\pi e^2}{24\hbar^2\gamma^2}\sum_n{}' \int_0^\infty duu \frac{\omega_{n0}^a |\mathbf{D}_{0n}|^2}{u^2 + (\omega_{n0}^a)^2} \times \frac{\partial^2}{\partial |Z|^2}\left[\mathbf{H}_1\left(\frac{2|Z|}{\gamma}u\right) - Y_1\left(\frac{2|Z|}{\gamma}u\right)\right], \quad (34)$$

where $\mathbf{H}_1(x)$ is the Struve function and Y_1 is the Bessel function of the second kind.

For the case of van der Waals attraction of the two sheets of a Graphene double layer (ii) which are understood to be separated sufficiently so that tunneling between them is negligible, we obtain the local plasmon contribution to the interaction energy as

$$\frac{E_{\text{int}}}{\text{Area}} = \frac{3}{128\pi^{1/2}}\frac{\hbar\lambda}{|z_0|^{5/2}}, \quad (35)$$

whereas we estimate the higher wavenumber nonlocal undoped electron-hole contribution as

$$\frac{E_{\text{int}}}{\text{Area}} = \frac{(g_s g_v)^2 e^4}{4096\kappa^2\gamma\hbar|z_0|^3}\left[\frac{\beta}{(\beta^2-1)} - \frac{1}{(\beta^2-1)^{3/2}}\ln\left(\frac{\beta+1+\sqrt{\beta^2-1}}{\beta+1-\sqrt{\beta^2-1}}\right)\right]. \quad (36)$$

Figure 1 exhibits the behavior of the right hand sides of Eqs. (35) and (36) in eV/nm^2 as functions of z_0 for the range $50\,\text{nm} \le z_0 \le 500\,\text{nm}$. Using Graphene parameters: $\gamma = 9\times 10^5\,\text{m/s} \rightarrow \hbar\gamma = 0.5924$ eV·nm, $\kappa = 2.4$, $n = 10^{12}\ \text{cm}^{-2} = 10^{16}\text{m}^{-2} \Rightarrow q_F = 1.77\times 10^4\,\text{m}^{-1} = 1.77\times 10^{-5}\,\text{nm}^{-1}$, $e^2 = 1.44$ eV·nm, and $\hbar = 6.582\times 10^{-16}\,$eV·s, the value of λ is

$$\lambda = \sqrt{\frac{\gamma e^2}{\kappa\hbar}\sqrt{n\pi g_s g_v}} = 5.39\times 10^{14}\,\text{nm}^{1/2}/\text{s}, \quad (37)$$

and the plasmon contribution estimate for $z_0 = 100$ nm is $E_{\text{int}}/\text{Area} = 4.691\times 10^{-8}\,\text{eV/nm}^2$, whereas the electron-hole estimate of Eq. (36) is $E_{\text{int}}/\text{Area} = 1.254\times 10^{-9}\,\text{eV/nm}^2$. Thus, the plasmon contribution dominates by about a factor of 40. The ratio of the plasmon contribution to the electron-hole estimate is exhibited in Fig. 2 for the same range of z_0.

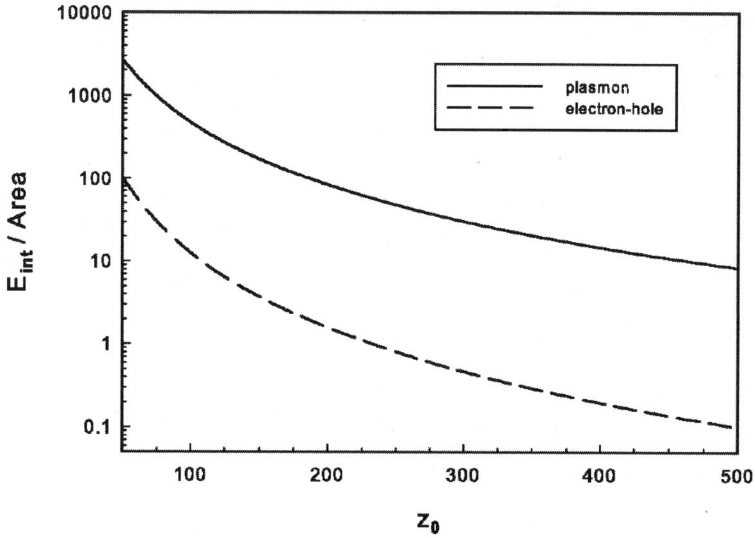

Fig. 1. Plot of E_{int}/Area (in eV/nm^2) versus distance z_0 (in nm). The values of the various parameters are given in the text. The solid line represents the plasmon contribution and the dashed line represents the electron-hole estimate.

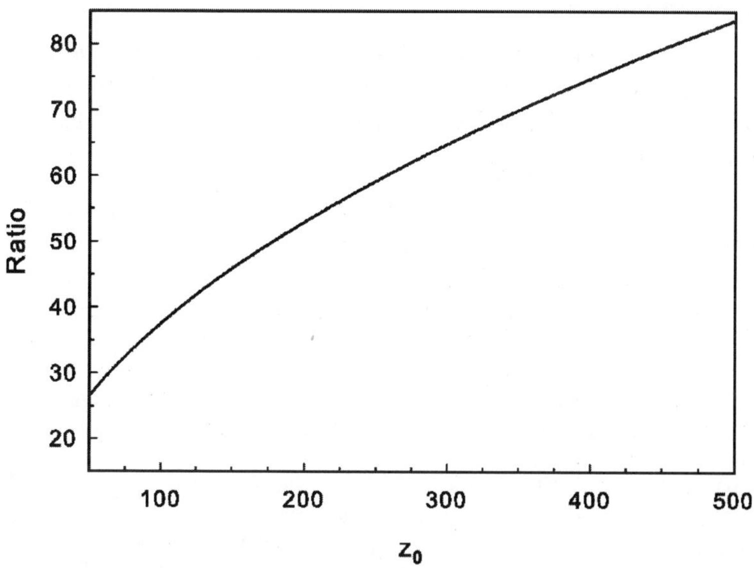

Fig. 2. Ratio of plasmon contribution to electron-hole estimate as a function of z_0.

Appendix A.

The final k-integral of Eq. (31)

$$I = \int_0^\infty dk \frac{1}{\left(\sqrt{k^2+1}+\beta\right)^2}, \tag{A.1}$$

may be rewritten setting $k = \sinh z$, $dk = \cosh z\, dz$, as

$$I = \int_0^\infty dz \frac{\cosh z}{(\cosh z + \beta)^2} = -\int_0^\infty dz \frac{\partial}{\partial \beta}\left(1 - \frac{\beta}{\cosh z + \beta}\right). \tag{A.2}$$

The final z-integral is evaluated using[16]

$$\int_0^\infty \frac{dz}{a + b\cosh z} = \frac{1}{\sqrt{a^2-b^2}} \ln\left(\frac{a+b+\sqrt{a^2-b^2}}{a+b-\sqrt{a^2-b^2}}\right) \quad \text{for } a^2 > b^2, \tag{A.3}$$

and differentiation with respect to β yields

$$\int_0^\infty \frac{\cosh z}{(a+b\cosh z)^2} dz = \frac{a}{b\,(a^2-b^2)} - \frac{b}{(a^2-b^2)^{3/2}} \ln\left(\frac{a+b+\sqrt{a^2-b^2}}{a+b-\sqrt{a^2-b^2}}\right). \tag{A.4}$$

Substitution of this into Eq. (31) yields Eq. (32).

Acknowledgments

This work was supported in part by DARPA grant #HR0011-09-1-0008. We are pleased to acknowledge our referees for their advice concerning related work,[17–21] which we were unaware of during our calculations and the preparation of this paper.

References

1. K.S.Novoselov, A.K. Geim, et al., Science **306**, 666 (2004); also see A.H. Castro Neto, *et al.*, arXiv:0709.1163v1[cond-mat. other] 7 Sep 2007.
2. S.V. Morozov, *et al* Phys. Rev. Lett. **100**, 016602 (2008).
3. F. Schedin & K.S. Novoselov, *et al.*, arXiv.cond-mat/0610809 (unpublished).
4. Georgia Tech Research News - `http://gtresearchnews.gatech.edu/newsrelease/graphene.html`; also, M.C.Lemme, IEEE Electron Device Letters **28**, 282 (2007).
5. E.W. Hill, *et al.*, IEEE Trans. Magn. **42**, 2694 (2006).
6. J. Scott Bunch, *et al.*[5] Science **315**, 490 (2007).
7. N.J.M. Horing & L.Y. Chen, Phys. Rev. A **66**, 042905 (2002).
8. N.J.M. Horing, H.C. Tso & G. Gumbs, Phys. Rev. B**36**, 1588 (1987, Appendix).
9. K. W.-K. Shung, Phys. Rev. B **34**, 979 (1986).

10. K. W.-K. Shung, Phys. Rev. B **34**, 1264 (1986).
11. T. Ando, J. Phys. Soc. Jpn. **75**, 074716 (2006).
12. E. H. Hwang and S. Das Sarma, Phys. Rev. B **75**, 205418 (2007).
13. B. Wunsch, T. Stauber, F. Sols and F. Guinea, New J. Phys. **8**, 318 (2006).
14. I.S. Gradshteyn and I.M. Ryzhik, *Table of Integrals, Series, and Products*, 7th edition (Academic Press, New York, 2007), p.341, Eq. 3.352.4 and p.883.
15. Ref. 12 page 345, Eq. 3.366.3.
16. Ref. 12 page 372, Eq. 3.513.2.
17. M. Bordag, I.V. Fialkovsky, D.M. Gitman & D.V. Vassilevich, arXiv:0907.3242v2 [hep-th] 15 Nov 2009.
18. M. Bordag, B. Geyer, G.L. Klimchitskaya & V.M. Mostepanenko, Phys. Rev. B**74**, 205431 (2006).
19. M. Bordag, Phys. Rev. D**76**, 065011 (2007).
20. Y.J. Dappe, M.A. Basanta, F. Flores & J. Ortega, Phys. Rev. B**74**, 205434 (2006).
21. T. Gould, E. Gray & J.F. Dobson, Phys. Rev. B**79**, 113402 (2009).

CASIMIR-POLDER FORCES BETWEEN TWO ACCELERATING ATOMS AND THE UNRUH EFFECT

JAMIR MARINO* and ROBERTO PASSANTE†

Dipartimento di Scienze Fisiche ed Astronomiche,
dell'Università degli Studi di Palermo and CNISM, Via Archirafi 36,
I-90123 Palermo, Italy
** E-mail: jamirmarino@gmail.com*
† E-mail: roberto.passante@fisica.unipa.it

The Casimir-Polder force between two atoms with equal uniform acceleration and separated by a constant distance R is considered. We show that, in the low-acceleration limit, while the near-zone R^{-6} behavior of the interatomic interaction energy is not changed by the acceleration of the atoms, the far-zone interaction energy decreases as R^{-5} instead of the well-known R^{-7} behavior for inertial atoms. Possibility of an indirect detection of the Unruh effect through measurements of the Casimir-Polder force between the two accelerating atoms is also suggested. We also consider a heuristic model for calculating the Casimir-Polder potential energy between the two atoms in the high-acceleration limit.

Keywords: Unruh effect; Casimir-Polder forces.

1. Introduction

Uniformly accelerated observers in Minkowski spacetime, i.e. linearly accelerated observers with constant proper acceleration, called *Rindler observers*, associate a thermal bath of Rindler particles to the scalar-field no-particle state of inertial observers, the Minkowski vacuum. This effect was discovered by W.G. Unruh[1] and has played an important role in our understanding that the particle content of a field theory is observer-dependent.[2] Detection of the Unruh effect however requires very high accelerations, $a \simeq 10^{20} m/s^2$, and many proposals have been made in order to get a direct measurement of the Unruh effect, analyzing for example electrons in particle accelerators[3] or atoms accelerating in microwave cavities[4] .

A change in the particle content of the vacuum state can in principle produce a change in any physical phenomena directly related to the vacuum fluctuations, such as the Lamb Shift of atomic levels or Casimir forces. Recently, the corrections to the Casimir-Polder force between an accelerating atom and a conducting plate due to the Unruh effect have been calculated, both in the scalar- and in the electromagnetic-field case,[5,6] using an appropriate extension of an approach which allows separation of vacuum fluctuations and radiation reaction contributions[7] . These works have shown that modifications to the atom-plate Casimir-Polder force are relevant only for accelerations of the order of $10^{20} m/s^2$, confirming the necessity of high accelerations for a direct detection of the Unruh effect. Also, it has been shown that in the high-acceleration regime, atoms are spontaneously excited, absorbing a quantum from the Unruh bath;[5] in this way the atom-wall force becomes reminiscent of that between a static excited atom and a plate.

In this paper, we consider the case of the Casimir-Polder force between two atoms with the same uniform acceleration and separated by a constant distance R. Both the low-acceleration and the high-acceleration limits ($a \ll \omega_0 c$ and $a \gg \omega_0 c$, respectively) are discussed, using an approach first developed by Goedecke and Wood[8] for the Casimir-Polder force between two atoms at finite temperature. We show that in the low-acceleration regime the Unruh effect changes the far-zone law of the interatomic potential energy from the R^{-7} dependence characteristic of two inertial atoms, to a R^{-5} dependence; in the near zone we find a correction to the potential energy due to the acceleration scaling as R^{-6}, as for inertial atoms. We also argue about the possibility of an indirect measurement of the Unruh effect, exploiting this change in the behavior of the interatomic potential. In the high-acceleration regime, we obtain an interatomic potential energy reminiscent of that between two excited-state atoms.

2. The Casimir-Polder force between two accelerating atoms

When a neutral polarizable atom uniformly accelerates in the electromagnetic vacuum, a radiative shift of its energy levels occurs due to the presence of the electromagnetic Unruh bath. These energy shifts are known in the literature,[9] and they provide the following expression for the average field energy of the Unruh bath in each mode divided by $\hbar\omega_k$,

$$\langle n(\omega)\rangle_a = \frac{1}{2}\left(1 + \frac{a^2}{c^2\omega^2}\right)\left(1 + \frac{2}{e^{2\pi c\omega/a} - 1}\right). \tag{1}$$

In (1) there is a contribution proportional to $a^2/c^2\omega^2$, that in the case of the electromagnetic field makes the response of the atom in the Unruh bath different from that in a thermal bath in Minkowski space at an equivalent temperature.[9]

We now consider a pair of neutral polarizable atoms, both accelerating with the same uniform acceleration a in a direction orthogonal to their separation, measured from the laboratory reference frame. Because the two atoms have the same acceleration, they see the same electromagnetic Unruh bath, and in order to estimate the Casimir-Polder force between them we use a technique developed by Goedecke and Wood[8] for the interatomic force at temperature T. Following this approach, and taking into account that the two atoms see the same Unruh bath, we consider the expression for the Casimir-Polder potential for two atoms in the vacuum

$$V(R) = -\frac{2\hbar c}{\pi R^2}\mathrm{Im}\int_0^\infty dk k^4\left(\frac{1}{2}\right)e^{i2kR}U(kR)\alpha^2(k) \tag{2}$$

($\alpha(k)$ is the dynamical atomic polarizability) with

$$U(x) = 1 + 5x^{-2} + 3x^{-4} + i\left(2x^{-1} + 6x^{-3}\right), \tag{3}$$

and we replace in (2) the average energy per mode in the vacuum state divided by $\hbar\omega_k$ (which is equal to $\frac{1}{2}$), by the analogous quantity for the Unruh bath of Rindler particles, given by (1). This gives the following expression for the Casimir-Polder potential between two accelerating atoms

$$V(R) = -\frac{2\hbar c}{\pi R^2}\mathrm{Im}\int_0^\infty dk k^4 \langle n(\omega)\rangle_a e^{i2kR}U(kR)\alpha^2(k), \tag{4}$$

in which we have assumed that the acceleration is so small that no spontaneous excitation occurs in the two atoms, that is $a \ll \omega_0 c$. We can develop the calculations in (4) similarly as in Ref.[8] by a rotation in the first quadrant of the complex k-plane and separating the contributions from the principal part of the integral from the contributions from the poles of the integrand. If we go straightforwardly with this method, we get two different expressions for the potential energy in two different zones, $\frac{aR}{c^2} \ll 1$ and $\frac{aR}{c^2} \gg 1$, in a similar way to the case of the thermal Casimir-Polder potential energy.[8]

For $\frac{aR}{c^2} \ll 1$, we find in the *near zone* ($R \ll c/\omega_0$) the following expression for the interatomic potential energy

$$\begin{aligned} V(R) = &-\frac{2}{3R^6}\sum_{r,s}\frac{\mu_{r0}^2\mu_{s0}^2}{E_{r0}+E_{s0}} \\ &-\frac{4}{3}a^2\frac{\hbar}{\pi}\frac{1}{c^3}\frac{1}{R^6}\sum_{r,s}E_{r0}E_{s0}\mu_{r0}^2\mu_{s0}^2 f(E_{r0},E_{s0}) \end{aligned} \tag{5}$$

where μ_{r0} are matrix elements of the atomic dipole moment and $f(E_{r0}, E_{s0})$ is a function of the atomic transition frequencies. This shows that the Unruh effect introduces in the *near zone* only a correction proportional to the square of the acceleration of the two atoms, which scales with the same power law as the inertial Casimir-Polder potential energy. In the *far zone* ($R \gg c/\omega_0$) we instead get a correction to the static Casimir-Polder force, which scales as R^{-5},

$$V(R) = -\frac{23\hbar c}{4}\frac{\alpha^2(0)}{R^7} - \frac{\hbar a^2}{4\pi c^3}\frac{\alpha^2(0)}{R^5}. \tag{6}$$

The new acceleration-dependent term introduced here by the Unruh effect changes the dependence of the potential on the interatomic distance R, and in principle it could be used to make an indirect measurement of the Unruh effect, for example by measuring some macroscopic quantity of an ensemble of atoms sensitive to the form of the potential. This new term gives an attractive contribution to the interatomic force, increasing its strength.

For $aR/c^2 \gg 1$ we find in both the near and far zones an expression of the potential energy that depends on the interatomic distance as R^{-6},

$$V(R) = -\frac{6\hbar a\alpha^2(0)}{\pi R^6 c}\left(\frac{1}{4} + \frac{\pi^2}{12}\right). \tag{7}$$

This result is similar to that obtained by Goedecke and Wood[8] for the Casimir-Polder interaction at finite temperature in the limit $2\pi k_B TR/\hbar c \gg 1$.

The results (6) and (7) are obtained in the hypothesis of small accelerations, such that the spontaneous excitation probability of the atom be negligible. For $a \gg \omega_0 c$, spontaneous excitation must be taken into account.[11] In such a case, we can obtain the interatomic interaction energy by substituting the average field energy of the Unruh bath in each mode, divided by $\hbar\omega_k$, (1) in the expression for the Casimir-Potential between two excited atoms in terms of vacuum field correlations as obtained by Power and Thirunamachandran.[10]

The most relevant contribution in this case is given by the Rindler particles resonant with the atomic transition frequency ck_A. For $a \gg \omega_0 c$ and $k \simeq k_A$, from (1) we get

$$\langle n(\omega_A)\rangle_a \simeq \frac{a^3}{2\pi c^6 k_A^3}. \tag{8}$$

Thus, with our method, the final expression for the interatomic potential in the high-acceleration regime is

$$\Delta E = -\frac{2}{3}\frac{\mu_A^2 \alpha_B(k_A) a^3 k_A}{\pi R^2 c^6}\left[1 + \frac{1}{k_A^2 R^2} + \frac{3}{k_A^4 R^4}\right], \tag{9}$$

where μ_A is the dipole moment of atom A. This expression shows a R^{-6} behaviour in the near zone, and a R^{-2} behavior in the far zone, reminiscent of the interatomic potential energy between two excited static atoms[10] and coherent with the results for the atom-plate Casimir-Polder interaction.[5]

Finally, a few words are necessary on the limits that our approximations imply. The non-relativistic approach, $v \ll c$, implies a constraint on the timescale in which our calculations are valid, that is $\tau \ll c/a$; on the other hand, this timescale must be greater than the characteristic time of atomic transitions ω_0^{-1}, in order to let Casimir-Polder forces between the two atoms to be established. The limits above are the same assumed also in previous works on the role of the Unruh effect in atomic systems.[12]

3. Conclusions

We have studied the Casimir-Polder force between two uniformly accelerating atoms both in the low- ad high-acceleration regime. In the low acceleration limit, $a \ll \omega_0 c$, we have obtained the Unruh correction to the interatomic Casimir-Polder force. Our results show that in the *near zone* the Unruh electromagnetic bath only gives a correction to the R^{-6} law of the interatomic interaction energy, while in the *far zone* the Unruh correction changes the R-dependence of the potential energy, adding an attractive acceleration-dependent R^{-5} contribution to the inertial R^{-7} term. In the high-acceleration limit, we find that the Casimir-Polder potential energy shows a R^{-2} dependence in the far zone, related to the possibility of spontaneous excitation of the atoms.

Acknowledgments

J.M. acknowledges financial support by the European Science Foundation (ESF) within the activity 'New Trends and Applications of the Casimir Effect' (www.casimir-network.com). Partial financial support from Ministero dell'Istruzione, dell'Università e della Ricerca and by Comitato Regionale di Ricerche Nucleari e di Struttura della Materia is also acknowledged.

References

1. Unruh, W.G., Phys. Rev. D **14**, 870-892 (1976).
2. Fulling, S.A., Phys. Rev. D **7**, 2859-2862 (1973).
3. Bell, J. S., and J. M. Leinaas, Nucl. Phys. B **284**, 488508 (1987).
4. Belyanin, A., V. V. Kocharovsky, F. Capasso, E. Fry, M. S. Zubairy, and M. O. Scully, Phys. Rev. A **74**, 023807 (2006).
5. Rizzuto, L., Phys. Rev. A **76**, 062114 (2007).
6. Rizzuto, L. and Spagnolo, S., Phys. Rev. A **79**, 062110 (2009).
7. Dalibard, J., Dupont-Roc, J., and Cohen-Tannoudji, C., J. Phys. (Paris) **43**, 1617 (1982).
   Dalibard, J., Dupont-Roc, J., and Cohen-Tannoudji, C., J. Phys. (Paris) **45**, 637 (1984).
8. Goedecke, G.H. and Wood, R. Phys. Rev. A **60**, 2577 (1999).
9. Passante, R., Phys. Rev. A **57**, 1590 (1998).
10. Power, E.A. and Thirunamachandran, T., Phys. Rev. A **48**, 4761 (1993).
11. Audretsch J. and Muller R., Phys. Rev. A **50**, 1755 (1994).
12. Audretsch, J. and Muller, R., Phys. Rev. A **52**, 629 (1995).

DYNAMICAL CASIMIR EFFECT WITH ROBIN BOUNDARY CONDITIONS IN A THREE DIMENSIONAL OPEN CAVITY

C. FARINA* and D. AZEVEDO

Instituto de Física, Universidade Federal do Rio de Janeiro, Rio de Janeiro, CEP 21941-972, Brasil
**E-mail: farina@if.ufrj.br*

F. PASCOAL

Universidade Federal do Rio de Janeiro, Campus de Macaé, Macaé, Rio de Janeiro, State ZIP/Zone, Brasil
E-mail: fabiopr@df.ufscar.br

We consider a massless scalar field in 1+1 dimensions inside a cavity composed by a fixed plate, which imposes on the field a Robin BC, and an oscillating one, which imposes on the field a Dirichlet BC. Assuming that the plate moves for a finite time interval, and considering parametric resonance, we compute the total number of created particles inside the cavity. We generalize our results to the case of two parallel plates in 3+1 dimensions.

Keywords: Dynamical Casimir effect; Robin boundary conditions.

1. Introduction

The dynamical Casimir effect (DCE) consists of two related phenomena: real particle creation due to moving boundaries and radiation reaction forces on moving boundaries. This effect already manifests itself for a unique moving plate and, for a non-relativistic motion, the frequencies of the created particles (photons in the case of the quantized electromagnetic field) are smaller or equal than the mechanical frequency of the moving plate. Since Moore's pioneering paper,[1] the DCE has been studied in many different situations by many authors (for a review on this subject see[2] and the special issue[3]). Particularly, many distinct boundary conditions (BC) have been considered, from the idealized Dirichlet and Neumann ones to more realistic ones. However, the so called Robin boundary conditions (RBC), which interpolate continuously Dirichlet and Neumann ones, have rarely been used explicitly in the context of the DCE (though they have been considered by many authors in the context of the static Casimir effect, see for instance[4]). As far as we know, RBC appeared in the DCE only for the situation of one moving plate in 1+1 dimensions.[5,6] Our purpose here is to consider RBC in one-dimensional cavities with one oscillating wall and in three-dimensional (open) cavities formed by two parallel plates with one of them oscillating in time. For a scalar field ϕ in 3+1 dimensions, RBC are defined by $\phi|_{bound.} = \beta\frac{\partial\phi}{\partial n}|_{bound.}$ where β is a constant parameter with dimension of length. They interpolate continuously Dirichlet ($\beta \to 0$) and Neumann ($\beta \to \infty$) BC. They appear in different areas of physics: from Mechanics, Electromagnetism and Quantum Mechanics to Quantum Field Theory, among others. These BC were used as a phenomenological model for penetrable surfaces.[7] In fact, for $\omega \ll \omega_P$, parameter β plays the role of the plasma wavelength. In Classical Mechanics, RBC may appear in a vibrating string coupled to a harmonic oscillator at one of its edges.[5,8] In the context of the static Casimir effect, RBC lead to eigenfrequencies for the cavity modes that are roots of a transcendental equation.

In the context of the DCE, Mintz *et al*[5] considered a massless scalar field ϕ in $1+1$ dimensions under the influence of one moving boundary in a prescribed and non-relativistic motion with small amplitudes, namely, $|\delta\dot{q}(t)| << c$ and $|\delta q(t)| << c/\omega_0$, where $\delta q(t)$ is the position of the moving boundary at instant t and ω_0 is the dominant mechanical frequency. Using the perturbative approach of Ford and Vilenkin,[9] one can show that the solution of the wave equation, $\partial^2\phi(t,x) = 0$, submitted to a RBC, leads to a susceptibility with both real and imaginary parts, so that, $\delta\mathcal{F}(\omega) = \chi(\omega)\delta Q(\omega)$, with $\chi(\omega) = \mathcal{R}e\chi(\omega) + i\mathcal{I}m\chi(\omega)$. Recall that, for the same situation, the use of a Dirichlet (or Neumann) BC would lead to a purely imaginary susceptibility ($\chi_{_D}(\omega) = \omega^3/(6\pi)$, with $c = \hbar = 1$). For a typical oscillatory motion, given by $\delta q(t) = \delta q_0\, e^{-|t|/T}\cos(\omega_0 t)$, with $\omega_0 T \gg 1$, Mintz *et al*[5] showed that the dissipative force on the moving boundary can be enormously suppressed for $\beta\omega_0 \approx 2$. In a subsequent paper,[6] these authors analyzed the particle creation phenomenon for the same situation and found that, for the above mentioned relation between β and ω_0, there is also an enormous suppression of particle creation.

2. One-dimensional cavities with Robin BC

For simplicity, we consider a one-dimensional cavity composed by a fixed plate at $x = 0$, which imposes on the massless scalar field a RBC, and a moving plate whose position at instant t is given by $q(t)$, which imposes on the field a Dirichlet BC. Then, we must solve the wave equation with $\hat{\phi}$ submitted to $\left(\hat{\phi}(x,t) - \gamma \frac{\partial}{\partial x}\hat{\phi}(x,t)\right)|_{x=0} = 0$ and $\hat{\phi}(x,t)\big|_{x=q(t)} = 0$. The scalar field satisfies the usual field commutation relations, namely, $\left[\hat{\phi}(x;t), \hat{\pi}(x';t)\right] = i\delta(x'-x)$, and $\left[\hat{\phi}(x;t), \hat{\phi}(x';t)\right] = [\hat{\pi}(x;t), \hat{\pi}(x';t)] = 0$,

Our anstaz for the field operators $\hat{\phi}$ and $\hat{\pi} = \dot{\hat{\phi}}$ in terms of the instantaneous basis introduced by Law,[11] in 1994, is

$$\hat{\phi}(x;t) = \sum_n \frac{1}{\sqrt{2k_n(t)}}\, u_n(x;t)\Big\{\hat{\mathrm{a}}_n(t) + \hat{\mathrm{a}}_n^\dagger(t)\Big\} \tag{1}$$

$$\hat{\pi}(x;t) = -i\sum_n \sqrt{\frac{k_n(t)}{2}}\, u_n(x;t)\Big\{\hat{\mathrm{a}}_n(t) - \hat{\mathrm{a}}_n^\dagger(t)\Big\}, \tag{2}$$

where the modes $\{u_n(x,t)\}$ of the instantaneous basis must satisfy Helmholtz equation, $\left(\frac{\partial^2}{\partial x^2} + k_n^2(t)\right) u_n(x;t) = 0$, the boundary conditions $\left(u_n(x,t) - \gamma\frac{\partial}{\partial x}u_n(x,t)\right)|_{x=0} = 0$ and $u_n(x,t)|_{x=q(t)} = 0$, and the orthonormality condition $\int_0^{q(t)} \mathrm{d}x\; u_n(x;t)\, u_m(x;t) = \delta_{nm}$. With these properties, it follows that $\left[\mathrm{a}_n(t), \mathrm{a}_m^\dagger(t)\right] = \delta_{nm}$ and $\left[\mathrm{a}_n^\dagger(t), \mathrm{a}_m^\dagger(t)\right] = [\mathrm{a}_n(t), \mathrm{a}_m(t)] = 0$. The instantaneous basis can be obtained, with modes $u_n(x,t)$ given by

$$u_n(x;t) = \frac{A_n(t)}{\sqrt{2q(t)}}\sin\Big[k_n(t)(x - q(t))\Big], \tag{3}$$

where $A_n(t) = 2\left[1 + \frac{\gamma/q(t)}{1+\gamma^2 k_n^2(t)}\right]^{-1/2}$ and $\{k_n(t)\}$ are the roots of the following transcendental equation $\sin[q(t)k_n(t)] + \gamma k_n(t)\cos\big[q(t)k_n(t)\big] = 0$.

Time evolution equations for $\hat{a}_n(t)$ and $\hat{a}_n^\dagger(t)$ can be found,

$$\dot{\hat{\mathrm{a}}}_n(t) = -ik_n(t)\hat{\mathrm{a}}_n(t) + \sum_j \Xi_{jn}(t)\hat{\mathrm{a}}_j(t) + \sum_j \Lambda_{jn}(t)\hat{\mathrm{a}}_j^\dagger(t)\,, \tag{4}$$

where

$$\Xi_{mn}(t) := -\frac{1}{2}G_{mn}(t)\left(\sqrt{\frac{k_n(t)}{k_m(t)}} + \sqrt{\frac{k_m(t)}{k_n(t)}}\right); \tag{5}$$

$$\Lambda_{mn}(t) := \frac{\dot{k}_n(t)}{2k_n(t)}\delta_{mn} - \frac{1}{2}G_{mn}(t)\left(\sqrt{\frac{k_n(t)}{k_m(t)}} - \sqrt{\frac{k_m(t)}{k_n(t)}}\right), \tag{6}$$

with $G_{nm}(t) := \int_0^{q(t)} \mathrm{d}x\; \dot{u}_n(x;t)u_m(x;t)$ (an analogous equation holds for $\hat{a}_n^\dagger(t)$)). Relating $\hat{\mathrm{a}}_n$ and $\hat{\mathrm{a}}_n^\dagger$ for different times, we write

$$\hat{\mathrm{a}}_n(t) = \sum_m \alpha_{nm}(t)\hat{\mathrm{a}}_m(t_0) + \sum_m \beta_{nm}(t)\hat{\mathrm{a}}_m^\dagger(t_0)\,, \tag{7}$$

where the Bogoliubov coefficients must satisfy $\alpha_{nm}(t_0) = \delta_{nm}$ and $\beta_{nm}(t_0) = 0$. The time evolution of these coefficients can be established,

$$\dot{\alpha}_{nm}(t) = -ik_n(t)\alpha_{nm}(t) + \sum_j \Xi_{jn}(t)\alpha_{jm}(t) + \sum_j \Lambda_{jn}(t)\beta^*_{jm}(t); \quad (8)$$

$$\dot{\beta}_{nm}(t) = -ik_n(t)\beta_{nm}(t) + \sum_j \Xi_{jn}(t)\beta_{jm}(t) + \sum_j \Lambda_{jn}(t)\alpha^*_{jm}(t). \quad (9)$$

Previous equations may be simplified with the aid of definitions:

$$\begin{aligned} &\alpha_{nm}(t) =: \mathrm{e}^{-iK_n(t)}\tilde{\alpha}_{nm}(t)\,; \quad \beta_{nm}(t) =: \mathrm{e}^{-iK_n(t)}\tilde{\beta}_{nm}(t)\,; \\ &K_n(t) := \int_{t_0}^{t} \mathrm{d}t'\ k_n(t'); \qquad (10) \\ &\Xi_{mn}(t) =: \tilde{\Xi}_{mn}(t)\mathrm{e}^{i[K_m(t)-K_n(t)]}\,; \ \ \Lambda_{mn}(t) =: \tilde{\Lambda}_{mn}(t)\mathrm{e}^{-i[K_m(t)+K_n(t)]}\,, \end{aligned}$$

Consequently, the time evolution for coefficients $\tilde{\alpha}_{nm}$ and $\tilde{\beta}_{nm}$ are

$$\dot{\tilde{\alpha}}_{nm}(t) = \sum_j \tilde{\Xi}_{jn}(t)\tilde{\alpha}_{jm}(t) + \sum_j \tilde{\Lambda}_{jn}(t)\tilde{\beta}^*_{jm}(t); \quad (11)$$

$$\dot{\tilde{\beta}}_{nm}(t) = \sum_j \tilde{\Xi}_{jn}(t)\tilde{\beta}_{jm}(t) + \sum_j \tilde{\Lambda}_{jn}(t)\tilde{\alpha}^*_{jm}(t)\,. \quad (12)$$

Up to this point, our calculations are exact. However, from now on, we shall consider only oscillating motions with small amplitudes, so we write $q(t) = q_0[1 + \epsilon\xi(t)]$, with $\epsilon \ll 1$ and $\xi(t)$ given, for a typical motion, by

$$\xi(t) = \begin{cases} \sin(\omega_0 t) & 0 < t < t_f \\ 0 & t \le 0 \text{ or } t \ge t_f\,. \end{cases}$$

Expansions in powers of ϵ (recall that all quantities get an implicit ϵ-dependence through $q(t)$) lead to

$$\tilde{\alpha}^{(\ell)}_{nm}(t) = \sum_{\ell'=1}^{\ell}\sum_j \int_{t_0}^{t} \mathrm{d}\tau \left[\tilde{\Xi}^{(\ell')}_{jn}(\tau)\tilde{\alpha}^{(\ell-\ell')}_{jm}(\tau) + \tilde{\Lambda}^{(\ell')}_{jn}(\tau)\tilde{\beta}^{*(\ell-\ell')}_{jm}(\tau)\right]; \quad (13)$$

$$\tilde{\beta}^{(\ell)}_{nm}(t) = \sum_{\ell'=1}^{\ell}\sum_j \int_{t_0}^{t} \mathrm{d}\tau \left[\tilde{\Xi}^{(\ell')}_{jn}(\tau)\tilde{\beta}^{(\ell-\ell')}_{jm}(\tau) + \tilde{\Lambda}^{(\ell')}_{jn}(\tau)\tilde{\alpha}^{*(\ell-\ell')}_{jm}(\tau)\right]. \quad (14)$$

where the superscritps mean the order of the derivative respect to ϵ of the quantity in question and conditons $\tilde{\alpha}^{(0)}_{nm}(t) = \delta_{nm}$ and $\tilde{\beta}^{(0)}_{nm}(t) = 0$ are satisfied. The number of particles created inside the cavity, with energy $\omega_n = k_n$, after the motion is finished is given by

$$\mathcal{N}_n(t_f) = \langle 0|\, \hat{a}^\dagger_n(t_f)\, \hat{a}_n(t_f)\, |0\rangle = \sum_j \left|\sum_\ell \epsilon^\ell \beta^{(\ell)}_{nj}(t_f)\right|^2. \quad (15)$$

The first correction to $\mathcal{N}_n(t_f)$ occurs at order ϵ^2,

$$\mathcal{N}_n(t_f) = \epsilon^2 \sum_m \left|\beta^{(1)}_{nm}(t_f)\right|^2 = \epsilon^2 \sum_m \left|\tilde{\beta}^{(1)}_{nm}(t_f)\right|^2 , \tag{16}$$

For the motion in consideration, we have

$$\mathcal{N}_n(t_f) = \sum_m \left|C_{nm}(\gamma) f_{nm}(\omega_0, t_f)\right|^2 (\epsilon\omega_0 t_f)^2 . \tag{17}$$

where

$$\begin{aligned} f_{nm}(\omega_0; t) &:= \frac{e^{i(\omega_0+\kappa_{nm})t} - 1}{(\omega_0 + \kappa_{nm})t} - \frac{e^{-i(\omega_0-\kappa_{nm})t} - 1}{(\omega_0 - \kappa_{nm})t} \\ C_{nm}(\gamma) &= \frac{1}{8} A_n(0) A_m(0) \frac{\sqrt{k_n(0) k_m(0)}}{k_n(0) + k_m(0)} , \\ \kappa_{nm} &= k_n(0) + k_m(0) . \end{aligned}$$

At this order, the total number of particles created inside the cavity is given by $\mathcal{N} = \sum_{n,m} \left|C_{nm}(\gamma) f_{nm}(\omega_0, t_f)\right|^2 (\epsilon\omega_0 t_f)^2$ while the total energy of the created particles is given by $\mathcal{E} = \sum_{n,m} k_n \left|C_{nm}(\gamma) f_{nm}(\omega_0, t_f)\right|^2 (\epsilon\omega_0 t_f)^2$. The behavior of $|f_{nm}(\omega_0; t_f)|^2$ is shown in Figure 1. For $\omega_0 t_f \gg 1$, it has a peak around $\omega_0 = \kappa_{mn}$ whose width δ is proportional to $1/(\kappa_{nm} t_f)$ (a simple estimative gives $\delta \approx 5.6/(\kappa_{nm} t_f)$). Hence, in a first approximation, $|f_{nm}(\omega_0; t_f)|^2$ behaves like a delta function, showing that whenever the oscillation frequency ω_0 equals the sum of two energy levels of the corresponding static cavity we have the best conditions for particle creation.

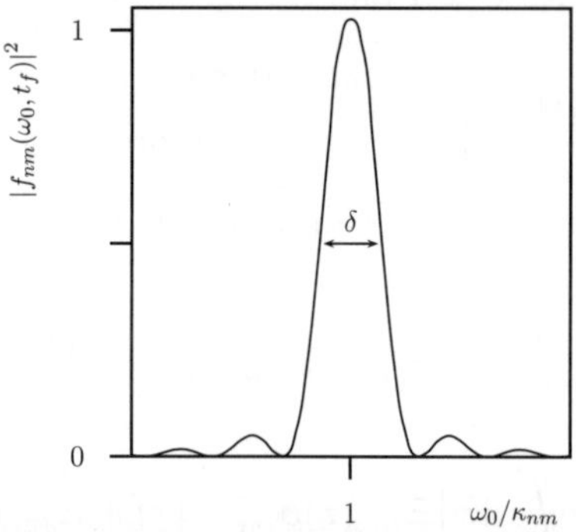

Fig. 1. $|f_{nm}(\omega_0; t_f)|^2$ as a function of ω_0/κ_{nm} for $\omega_0 t_f \gg 1$.

The fact that κ_{mn} is given by a sum of 2 terms means that particles are created in pairs. The set of values of κ_{mn} are called the resonances of the problem. Note that, for each value of the Robin parameter, γ, we have a different set of resonances. Figure 2 shows how the resonances vary with γ. Since γ varies from 0 (Dirichlet BC) to ∞ (Neumann BC), it is convenient to make the plot against $log_{10}(\gamma/q_0)$, instead of γ.

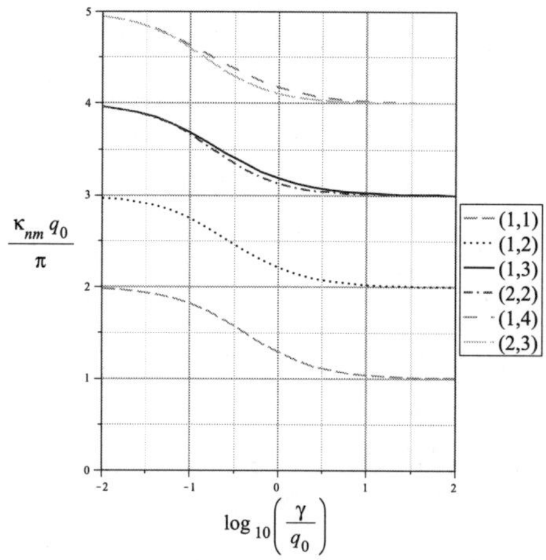

Fig. 2. Resonances (in units of π/q_0) as functions of $\log_{10}(\gamma/q_0)$.

For a given value of γ, the resonances are obtained by tracing a vertical line and looking at the intersections in Figure 2. The values obtained this way for $\log_{10}(\gamma/q_0) = -2$ (extreme left on the graph) are, approximately, the resonances for Dirichlet-Dirichlet BC since, for this case, $\gamma \ll q_0$. By the same token, the values obtained this way for $\log_{10}(\gamma/q_0) = 2$ (extreme right on the graph) are, approximately, the resonances for Neumann-Dirichlet BC since, for this case, $\gamma \gg q_0$. Adjacent resonances are equally spaced only for D-D and N-D cases. For these cases we have degeneracies, which are broken in the Robin-Dirichlet case. For instance, for this last case, $\kappa_{13} = k_1 + k_3 \neq k_2 + k_2 = \kappa_{22}$, as can be seen in Figure 2 near $\log_{10}(\gamma/q_0) = 0$. Note, also, the monotonic behavior of the curves with γ/q_0.

Figure 3 shows the number of created particles with energy k_1 for different resonant values of the mechanical frequency as a function of $\log_{10}(\gamma/q_0)$ (recall that particles are created in pairs, so that there are many ways of creating particles with energy k_1, namely, $\omega_0 = \kappa_{11}$, $\omega_0 = \kappa_{12}$, etc.). For the resonance $\omega_0 = \kappa_{1m}$, we have $N_{1m} = \mathcal{C}_{1m}^2(\gamma)\left(\epsilon\kappa_{1m}t_f\right)^2$.

It is worth saying a few words about how the curves in Figure 3 are traced. For each value of γ, we compute numerically the set of corresponding resonances. Then, we compute N_{11}, N_{12}, N_{13}, ..., for that value of γ. We, then, take another value of γ and compute the new values of the resonances. Taking ω_0 equal to the new values of resonances we compute again N_{11}, N_{12}, N_{13}, and so on. Hence, distinct points of a given curve, for instance N_1, are computed with distinct values of ω_0, but with ω_0 always equal to

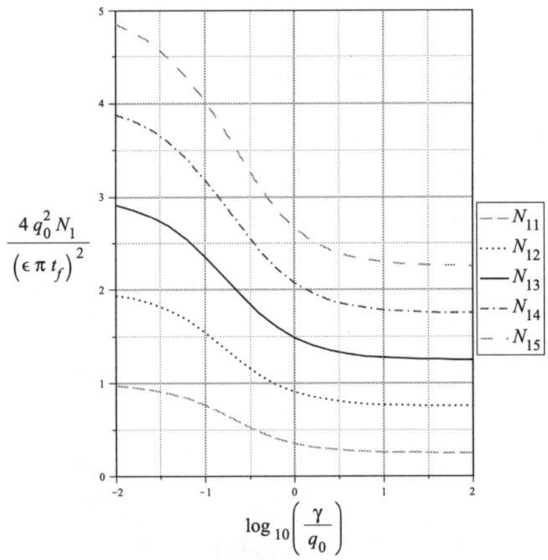

Fig. 3. Number of created particles with energy $k_1(0)$ (in units of $\left(\frac{\epsilon\pi t_f}{2q_0}\right)^2$ for the first five resonant values of ω_0p as a function of $\log_{10}(\gamma/q_0)$.

the first resonance (κ_{11}, which depends on γ). Note, also, the monotonic behavior of curves in Figure 3.

Let us check some particular cases. For $\omega_0 = \frac{2\pi}{q_0}$ and $\gamma = 0$, which corresponds to the D-D case with $\omega_0 = 2k_1$ (parametric resonance with the lowest level of the static cavity), we have $\mathcal{N}_1 \simeq \left(\frac{\epsilon\pi t_f}{2q_0}\right)^2$ and $\mathcal{E}_1 \simeq \frac{\pi}{q_0}\left(\frac{\epsilon\pi t_f}{2q_0}\right)^2$ in agreement with Dodonov and Klimov.[12] For $\omega = \frac{\pi}{q_0}$ and $\gamma \to \infty$, which corresponds to the N-D case with $\omega_0 = 2k_1$ (parametric resonance with the lowest level of the static cavity, which is $\frac{1}{2}$ the value for the D-D case), we have $\mathcal{N}_1 \simeq \left(\frac{\epsilon\pi t_f}{4q_0}\right)^2$ and $\mathcal{E}_1 \simeq \frac{\pi}{2q_0}\left(\frac{\epsilon\pi t_f}{4q_0}\right)^2$, in agreement with Alves *et al.*[13]

3. Parallel plates in 3+1 dimensions with Robin BC

Here, we shall generalize some of the previous results to 3+1 dimensions. We, then, consider a fixed plate at $z = 0$, which imposes on a massless scalar field a RBC and a moving plate, parallel to the first one, which imposes on the field a DBC. Let $q(t)$ be the position of the moving plate at instant t. Operators $\hat{\phi}(\mathbf{x};t)$ and $\hat{\pi}(\mathbf{x};t)$ are given, in terms of instantaneous basis, by

$$\hat{\phi}(\mathbf{x};t) = \sum_{n=1}^{\infty} \int \frac{d^2\mathbf{k}_{\parallel}}{\sqrt{2\omega_n(k_{\parallel},t)}} u_n(x;t) \left[\frac{e^{i\mathbf{k}_{\parallel}\cdot\mathbf{x}}}{(2\pi)^2}\hat{a}_n(\mathbf{k}_{\parallel},t) + \text{h.c.}\right], \quad (18)$$

$$\hat{\pi}(\mathbf{x};t) = -i\sum_{n=1}^{\infty}\int d^2\mathbf{k}_{\parallel}\sqrt{\frac{\omega_n(k_{\parallel},t)}{2}}u_n(x;t)\left[\frac{e^{i\mathbf{k}_{\parallel}\cdot\mathbf{x}}}{(2\pi)^2}\hat{a}_n(\mathbf{k}_{\parallel},t) - \text{h.c.}\right] \quad (19)$$

where $\omega_n^2\left(k_\parallel,t\right)=k_\parallel^2+k_n^2(t)$ and $u_n\left(x,t\right)=\sqrt{\frac{2}{q(t)}}A_n(t)\sin\left[k_n(t)\left(x-q(t)\right)\right]$, with $A_n(t)$ and $k_n(t)$ defined as in the 1+1 case. We shall consider the same motion as in the 1+1 case. The Bogoliubov coefficients are now defined by

$$\hat{a}_n\left(\mathbf{k}_\parallel,t\right)=\sum_{m=0}^{\infty}\left[\alpha_{nm}\left(k_\parallel,t\right)\hat{a}_m\left(\mathbf{k}_\parallel,0\right)+\beta_{nm}\left(k_\parallel,t\right)\hat{a}_m^\dagger\left(-\mathbf{k}_\parallel,0\right)\right]. \quad (20)$$

A perturbative solution, up to first order in ϵ, leads to

$$\beta_{nm}\left(k_\parallel,t\right)\mathrm{e}^{i\omega_n\left(k_\parallel\right)t}=-\epsilon\, C_{nm}\left(k_\parallel\right)f_{nm}\left(k_\parallel,t\right), \quad (21)$$

where we defined

$$C_{nm}\left(k_\parallel\right)=\frac{A_n(0)A_m(0)}{\sqrt{\omega_m\left(k_\parallel\right)\omega_n\left(k_\parallel\right)}}\frac{k_n(0)k_m(0)}{\omega_n\left(k_\parallel\right)+\omega_m\left(k_\parallel\right)}, \quad (22)$$

$$f_{nm}\left(k_\parallel,t\right)=\int_0^t \mathrm{d}t'\dot{\xi}(t')\mathrm{e}^{i\left[\omega_m\left(k_\parallel\right)+\omega_n\left(k_\parallel\right)\right]t'} \quad (23)$$

with $\omega_n^2\left(k_\parallel\right)=\omega_n^2\left(k_\parallel,0\right)$. The number of created particles in a given mode with $k_z=k_n$ and with a parallel moment between $\mathbf{k}_\parallel$ and $\mathbf{k}_\parallel+\mathrm{d}^2\mathbf{k}_\parallel$ is

$$\mathcal{N}_n\left(\mathbf{k}_\parallel,t_f\right)\mathrm{d}^2\mathbf{k}_\parallel=\epsilon^2\frac{L^2}{\left(2\pi\right)^2}\sum_{m=1}^{\infty}\left|C_{nm}\left(k_\parallel\right)f_{nm}\left(k_\parallel,t_f\right)\right|^2\mathrm{d}^2\mathbf{k}_\parallel. \quad (24)$$

The total number of created particles inside the cavity takes the form

$$\mathcal{N}\left(t_f\right)=\epsilon^2\frac{L^2}{2\pi}\sum_{n,m=1}^{\infty}\int_0^\infty \mathrm{d}k_\parallel\; k_\parallel\left|C_{nm}\left(k_\parallel\right)f_{nm}\left(k_\parallel,t_f\right)\right|^2, \quad (25)$$

and the total energy is given by

$$\mathcal{E}\left(t_f\right)=\epsilon^2\frac{L^2}{2\pi}\sum_{n,m=1}^{\infty}\int_0^\infty \mathrm{d}k_\parallel\; k_\parallel\omega_m\left(k_\parallel\right)\left|C_{nm}\left(k_\parallel\right)f_{nm}\left(k_\parallel,t_f\right)\right|^2. \quad (26)$$

For the harmonic motion considered before, with $\omega t_f\gg 1$), we get

$$\left|f_{nm}\left(k_\parallel,t_f\right)\right|^2=\frac{\pi\omega^2 t_f}{4k_\parallel}\left(\frac{\omega_n\left(k_\parallel\right)\omega_m\left(k_\parallel\right)}{\omega_n\left(k_\parallel\right)+\omega_m\left(k_\parallel\right)}\right)\delta\left(k_{nm}(\omega)-k_\parallel\right), \quad (27)$$

where $\sqrt{k_{nm}^2(\omega)+k_n^2}+\sqrt{k_{nm}^2(\omega)+k_m^2}-\omega=0$. Using last result for $\left|f_{nm}\left(k_\parallel,t_f\right)\right|^2$, we obtain

$$\mathcal{N}\left(t_f\right)=\epsilon^2\frac{L^2 t_f}{8\omega}\sum_{n,m=1}^{\infty}\left(A_nA_mk_nk_m\right)^2\Theta\left(\omega-k_n-k_m\right) \quad (28)$$

and $\mathcal{E}\left(t_f\right)=\mathcal{N}\left(t_f\right)\omega/2$. Figure 4 shows the behavior of the total number of created particles inside the plates in terms of the frequency ω of the moving plate. We plot $\mathcal{N}\left(t_f\right)$ divided by $\epsilon^2\omega^3L^2t_f$ in terms of $\omega a_0/\pi$.

Solid lines connected by dotted lines correspond to the DD case, while solid lines connected by dashed lines, to a RD case. The discontinuities occur at the resonant values ($\omega=k_n+k_m$). The main difference between

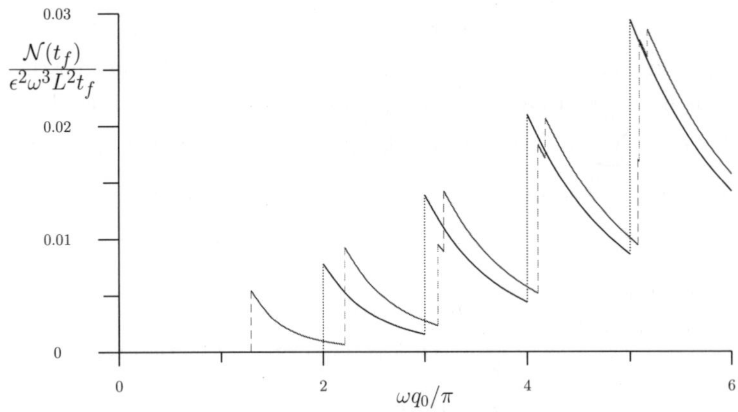

Fig. 4. Total number of created particles for an open three-dimensional cavity formed by two parallel plates as a function of the frequency of the moving plate.

DD and RD cases consists in the fact that the resonances for the former are equally spaced, while for the latter they are not equally spaced, as can be seen from Figure 4. Note the presence of small solid lines for the RD case, a direct consequence of the degeneracy breaking that happens when we use RBC, as discussed previously. It is worth noting the similarity of the graph for the D-D case with that for the electromagnetic field inside two parallel and perfectly conducting plates discussed by Mundarain and Maia Neto.[14]

4. Final comments

In this work we considered RBC in one-dimensional cavities and in a three-dimensional open cavity formed by two parallel plates. Using the instantaneous basis method[11] we computed the number of created particles when the frequency of the oscillating plate was at resonance. As we showed, for one-dimensional cavities, there are more resonances for the RD case than for the DD or ND cases, due to the degeneracy breaking discussed in the text. For the same reason, there are more discontinuities in Figure 4 when a RBC is involved than for the case where both plates impose a DBC on the field. An important difference between the 1+1 and 3+1 cases treated here is that in the former the total number of created particles, $\mathcal{N}\,(t_f)$, is proportional to t_f^2, while in the latter, $\mathcal{N}\,(t_f)$ is proportional to t_f, as shown in equations (17) and (28). This occurs because in the 1+1 case we have a closed cavity, while the system formed by two parallel plates correspond, in fact, to an open cavity.

The possibility of suppression of the DCE[5,6] was not investigated, since we considered here always resonant cavities. It would be interesting to study a massless scalar field in 3+1 dimensions submitted to a RBC at one moving plate and check if suppression of the DCE still occurs. We think that RBC, as well as more realistic BC, should be more investigated in the DCE, whose experimental verification seems imminent[15] (see also the recent proposal of experiment[16]). In this work we were concerned only with the regions inside the cavities, but an analysis involving also the outside regions, including a discussion of the dissipative force on the moving plate and the energy balance, can be made and will appear elsewhere.

References

1. GT Moore, J. Math. Phys. **11** (1970) 2679
2. VV Dodonov, Modern Nonlinear Optics, Advances in Chem. Phys. Series **119**, 309, ed. MW Evans (Wiley, New York, 2001)
3. *Special Issue on the Non-stationary Casimir effect and quantum systems with moving boundaries*, J. Opt. B: Quantum Semiclass. Opt. **7** S3 (2005)
4. A Romeo and AA Saharian, J. Phys. A**35** (2002) 1297
5. B Mintz, C Farina, PA Maia Neto and R Rodrigues, J. Phys. A**39** (2006) 6559
6. B Mintz, C Farina, PA Maia Neto and R Rodrigues, J. Phys. A**39** (2006) 11325
7. VM Mostepanenko and NN Trunov, Sov. J. Nucl. Phys. **45** (1985) 818
8. G Chen and J Zhou, *Vibration and Damping in Distributed Systems*, ol. **1** (Boca Raton, FL:CRC, 1992), pg 15
9. LH Ford and A Vilenkin, Phys. Rev. D**25** (1982) 2569
10. A Lambrecht, MT Jaekel and S Reynaud, Phys. Rev. Let. **77** (1996) 615
11. CK Law, Phys. Rev. A**49** (1994) 433
12. VV Dodonov and AB Klimov, Phys. Rev. A**53** (1996) 2664
13. DT Alves, ER Granhem and C Farina, Phys. Rev. A**73** (2006) 063818
14. D F Mundarain and P A Maia Neto, Phys. Rev. A**57** (1998) 1379
15. C Braggio, G Bressi, G Carugno, C Del Noce, G Galeazzi, A Lombardi A Palmieri, G Ruoso and D Zanello, Europhys. Lett. **70** (2005) 754
16. JR Johansson, G Johansson, CM Wilson and Franco Nori, Phys. Rev. Lett. **103** (2009) 147003

THEORETICAL STUDIES ON DYNAMICAL CASIMIR EFFECT IN A SUPERCONDUCTING ARTIFICIAL ATOM

TOSHIYUKI FUJII, SHIGEMASA MATSUO, KOHJI TAKASHIMA
and NORIYUKI HATAKENAKA

Graduate School of Integrated Arts and Sciences, Hiroshima University, Higashi-hiroshima, 739-8521, Japan

SUSUMU KURIHARA

Department of Physics, Waseda University, Tokyo, 169-8555, Japan

ANTON ZEILINGER

Institute of Experimental Physics, University of Vienna, Boltzmanngasse 5, 1090 Vienna, Austria

A double SQUID consisting of a superconducting ring interrupted by dc SQUID is regarded as a quantum harmonic oscillator with a time-dependent frequency under an appropriate junction parameter. A rapid change of the frequency brought by the magnetic flux modulation in the dc SQUID produces squeezed states of the quantum flux in the double SQUID due to nonadiabatic mixing of quantum states. Here we show that nonadiabatic aspect of the system can be measured by squeezing parameter of a quantum flux. This provides an alternative scheme for verification of dynamical Casimir effect.

Keywords: Dynamical Casimir effect, superconducting artificial atom.

1. Introduction

Dynamical Casimir effect predicts creation of a quantum out of the vacuum[1] essentially from a nonadiabatic distortion of the vacuum states due to the quickly moving boundaries. This can be imitated by using parametric Josephson junctions[2–6] without using real material boundaries. Along this line, we proposed an alternative scheme for exploiting dynamical Casimir effect by creating nonstationary situations using a double SQUID, regarded as a Josephson artificial atom,[7] and numerically confirmed that a vacuum state evolves into squeezed states resulting from a sudden change in the Josephson critical current.[8] Here we show that squeezing parameter can measure the degrees of nonadiabatic effect brought by sudden or sinusoidal changes in frequencies. This might be greatly helpful for testing dynamical Casimir effect in Josephson artificial atoms.

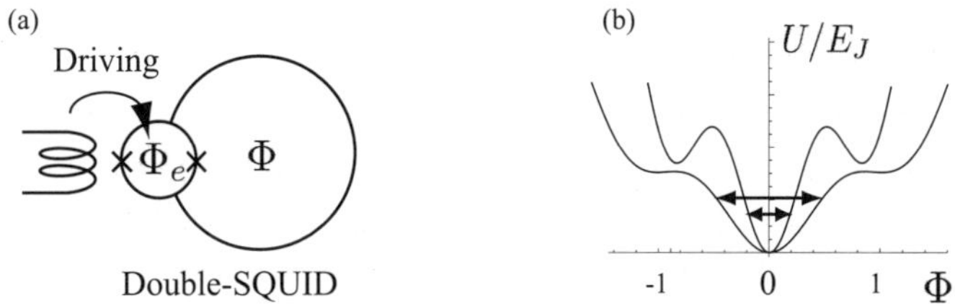

Fig. 1. (a)Schematics of a double SQUID. (b) Potential energy of the double SQUID and its dynamical motion. The potential is normalized by the Josephson energy $E_J = \hbar I_0/2e$.

2. Josephson artificial atoms

The system we consider here consists of a superconducting loop with the inductance L interrupted by a dc-SQUID (double SQUID) as shown in Fig. 1 (a). The double SQUID behaves as an rf-SQUID with tunable Josephson critical current $I_c(t)$. Figure 1 (b) shows the potential profile as a function of the magnetic flux threading the large loop Φ. The motion of the Φ is described by a particle in the potential $U(\Phi, t) = (\hbar I_c(t)/2e)(1 - \cos 2\pi\Phi/\Phi_0) + (\Phi - \Phi_e)^2/2L$, where $I_c(t) = I_{J0} \cos 2\pi\Phi_e(t)/\Phi_0$ with I_{J0} being the Josephson critical current and Φ_0 being the flux quantum defined as $\Phi_0 = h/2e$. For the sufficiently deep potential, this system is considered as a harmonic oscillator with a time-dependent frequency; $\hat{H} = \hat{Q}^2/2C + C\omega(t)^2\hat{\Phi}^2/2$ where the $\omega(t) = (2\pi I_c(t)/\Phi_0 C + 1/LC)^{1/2}$ is time-dependent frequency driven by the external flux bias Φ_e. We define the annihilation ($\hat{a} = (C\omega_0/2\hbar)^{1/2}\hat{\Phi} + i(1/2\hbar C\omega_0)^{1/2}\hat{Q}$) and creation ($\hat{a}^\dagger = (C\omega_0/2\hbar)^{1/2}\hat{\Phi} - i(1/2\hbar C\omega_0)^{1/2}\hat{Q}$) operators at time $t = 0$ where $\omega_0 = \omega(t = 0)$ is an initial frequency. By using $\hat{a}$ and $\hat{a}^\dagger$, Hamiltonian at time t is written as

$$\hat{H}(t) = \frac{\hbar\omega_0}{4}\left[\left\{\left(\frac{\omega(t)}{\omega_0}\right)^2 + 1\right\}(\hat{a}^\dagger\hat{a} + \hat{a}\hat{a}^\dagger) + \left\{\left(\frac{\omega(t)}{\omega_0}\right)^2 - 1\right\}(\hat{a}^{\dagger 2} + \hat{a}^2)\right]. \quad (1)$$

This is diagonalized as $\hat{H}(t) = (\hbar\omega(t)/2)\left\{\hat{b}^\dagger(t)\hat{b}(t) + \hat{b}(t)\hat{b}^\dagger(t)\right\}$ by using the Bogoliubov transformation, $\hat{b}(t) = \hat{V}(t)\hat{a}\hat{V}^\dagger(t)$ and $\hat{b}(t)^\dagger = \hat{V}(t)\hat{a}^\dagger\hat{V}^\dagger(t)$, with $\hat{V}(t) = \exp\left\{-(\mu(t)/2)\left(\hat{a}^2 - \hat{a}^{\dagger 2}\right)\right\}$ and $\mu(t) = -(1/2)\ln(\omega(t)/\omega_0)$.

The number state $|n(t)\rangle$ at time t defined by the snapshot number operator $\hat{b}^{\dagger}(t)\hat{b}(t)$ is generated by the operator $\hat{V}(t)$ from number state $|n\rangle$ at time $t = 0$, i.e., $|n(t)\rangle = \hat{V}(t)|n\rangle$. In other words, the unitary operator $\hat{V}(t)$ is regarded as time-translational operator for basis and it represents that the definition of the eigenstate changes over time.

3. Quantum-state evolutions

Let us consider the time evolution of the state $|\phi(t)\rangle$ which is initially prepared in ground state $|0\rangle$. The state $|\phi(t)\rangle$ at time t is expressed in terms of the snapshot eigenstates $|n(t)\rangle$, i.e., $|\phi(t)\rangle = \sum_{n=0} c_n(t)|n(t)\rangle$ with $c_n(t)$ being the time-dependent expansion coefficient for n-th snapshot number state. In this expansion, both the coefficient and basis are time-dependent. To focus on time evolution of $c_n(t)$ by excluding the time evolution of the basis, we apply the operator $\hat{V}^{\dagger}(t)$ to $|\phi(t)\rangle$ and define the state $|\phi'(t)\rangle \equiv \hat{V}^{\dagger}(t)|\phi(t)\rangle$, which is governed by the Schrödinger equation

$$\begin{aligned} i\hbar\frac{\partial}{\partial t}|\phi'(t)\rangle &= \left\{\hat{V}^{\dagger}(t)\hat{H}(t)\hat{V}(t) - i\hbar\hat{V}^{\dagger}(t)\frac{\partial}{\partial t}\hat{V}(t)\right\}|\phi'(t)\rangle \\ &= \left\{\hbar\omega(t)\left(\hat{a}^{\dagger}\hat{a} + \frac{1}{2}\right) + i\frac{\hbar\dot{\mu}(t)}{2}\left(\hat{a}^2 - \hat{a}^{\dagger 2}\right)\right\}|\phi'(t)\rangle \\ &\equiv \{\hat{H}_0(t) + \hat{H}_1(t)\}|\phi'(t)\rangle \end{aligned} \tag{2}$$

with $\dot{\mu}(t) = -(1/2)\frac{d}{dt}\{\ln(\omega(t)/\omega_0)\} = -\dot{\omega}(t)/2\omega(t)$. The second term $\hat{H}_1(t)$ in the right-hand side of Eq. (2) is proportional to $\dot{\omega}(t)/\omega(t)$ that shows the nonadiabatic effect, leading to exciting the system to the even-numbered levels. Thus the state $|\phi'(t)\rangle$ no longer stays in the ground state when the $H_1(t)$ becomes significant. The Hamiltonian $\hat{H}_1(t)$ is equivalent to the squeezed Hamiltonian in nonlinear quantum optics. Therefore, the squeezed state generation due to nonadiabatic changes of the boundaries is clear evidence for demonstrating dynamical Casimir effect in Josephson artificial atoms.

4. Nonadiabatic effect and squeezing

To investigate the effect of $H_1(t)$, we move to the interaction picture, $|\phi'(t)\rangle = U(t)|\phi''(t)\rangle$ with $U(t) = \exp\left[-\frac{i}{\hbar}\int_0^t \hat{H}_0(t')dt'\right]$. The time evolution of $|\phi''(t)\rangle$ due to nonadiabatic effect is governed by the equation

$$i\hbar\frac{\partial}{\partial t}|\phi''(t)\rangle = \hat{H}_1^{int}(t)|\phi''(t)\rangle, \tag{3}$$

with $\hat{H}_1^{int}(t) = \hat{U}^\dagger(t)\hat{H}_1(t)\hat{U}(t) = i(\hbar\dot{\mu}(t)/2)\left[\hat{a}^2 e^{-2i\int_0^t \omega(t')dt'} - H.C.\right]$. The formal solution of this equation is expressed by $|\phi''(t)\rangle = \hat{T}\exp\left[-\frac{i}{\hbar}\int_0^t \hat{H}_1^{int}(t')dt'\right]|0\rangle$ where $\hat{T}$ is time-ordering operator. However, it is difficult to obtain the analytical solution for equation (3) because Hamiltonian cannot commute itself at different times. Hereafter we consider the situation where $[\hat{H}_1^{int}(t), \hat{H}_1^{int}(t')] = i\hbar^2\dot{\mu}(t)\dot{\mu}(t')\sin\left(\int_{t'}^t \omega(t'')dt''\right)\left(\hat{a}^\dagger\hat{a} + \hat{a}\hat{a}^\dagger\right) \simeq 0$. This seems to be appropriate in an adiabatic case where $\dot{\mu}(t)$ is small and also in a nonadiabatic case where $\dot{\mu}(t)$ is localized in time domain owing to quickly frequency changes. Under this situation, we obtain an approximate solution by neglecting the time-ordering operator, $|\phi''(t)\rangle = \exp\left[-\frac{i}{\hbar}\int_0^t \hat{H}_1^{int}(t')dt'\right]|0\rangle$. In this approximation, the state with time-dependent bases at time t is obtained as

$$\begin{aligned} |\phi(t)\rangle &= \hat{V}(t)e^{-\frac{i}{\hbar}\int_0^t \hat{H}_0(t)dt'}|\phi''(t)\rangle \\ &= e^{-\frac{i}{2}\int_0^t \omega(t')dt'} e^{-\frac{1}{2}\left\{\nu\hat{b}(t)^2 - \nu^*\hat{b}^{\dagger 2}(t)\right\}}|0(t)\rangle. \end{aligned} \tag{4}$$

This is nothing but *the vacuum squeezed state* with the squeezed parameter

$$\nu(t) = \frac{1}{2}\int_0^t \frac{\dot{\omega}(t')}{\omega(t')} e^{2i\int_t^{t'} \omega(t'')dt''} dt'. \tag{5}$$

This is our main result of this paper. The nonadiabatic effect $\dot{\omega}(t')/\omega(t')$ is connected to squeezing parameter $\nu(t)$. Therefore, dynamical counterpart of the Casimir effect can be confirmed by investigating squeezing of the quantum flux in double SQUID.

Figure 2 (a) shows the squeezed parameter at $t = \infty$ as a function of a nonadiabatic parameter c defined by $\omega(t) = \omega_1 + (\omega_0 - \omega_1)(1 - \tanh ct)/2$ for a sudden change. Our approximate solutions agree well with the numerical result obtained by the fitting of the distribution of the number states.[8] In the nonadiabatic limit (for large c), the $|\nu(\infty)|$ approaches asymptotically to $\nu = \mu(t = \infty) = -(1/2)\ln(\omega_1/\omega_0)$ as expected. Even in the opposite limit (for small c), the approximate solution also works well since small $\dot{\mu}$ makes the approximation $[\hat{H}_1^{int}(t), \hat{H}_1^{int}(t')] \simeq 0$ reasonable.

The sudden change scheme requires large frequency ratio of the initial and final frequencies as well as the rapid frequency change for producing the nonadiabatic effect. On the other hand, Figure 2 (b) shows the time evolution of $\nu(t)$ under $\omega(t) = \omega_0 + \omega_1 \sin\omega t$ at $\omega_1/\omega_0 = 0.4$. Squeezing grows linearly in time. Thus, sinusoidal changes allow us to create nonadiabatic situations more effectively using current technologies.

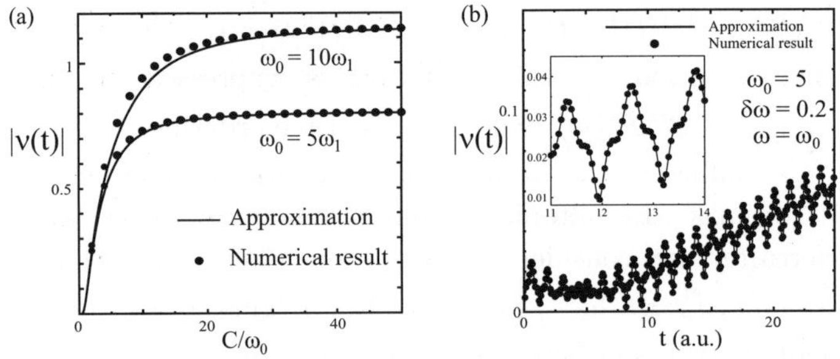

Fig. 2. (a)The squeezed parameter as a function of the nonadiabatic parameter c for sudden frequency changes. (b) Time evolution of the squeezed parameter for sinusoidal change. The dots and solid lines are numerical and approximate solutions, respectively.

5. Summary

We have investigated the nonstationary properties of a quantum flux in the double SQUID system in terms of squeezing. We have confirmed that squeezing parameter can measure the nonstationary aspect of the quantum flux through numerical and analytical calculations in the case of sudden or sinusoidal changes in frequencies. This provides an alternative scheme for verification of dynamical Casimir effect.

Acknowledgments

We would like to thank A. Ustinov and S. Poletto for valuable discussions. This work was supported in part by a grant from Research Foundation for Opto-Science and Technology and by KAKENHI (No. 195836) from MEXT of Japan.

References

1. G. T. Moore, J. Math. Phys. **11**, 2679 (1970).
2. V. V. Dodonov, V. I. Man'ko and O. V. Man'ko, J. Sov. Laser Res. **10**, 413 (1989).
3. V. I. Man'ko, J. Sov. Laser Res. **12**, 383 (1991).
4. O. V. Man'ko, J. Korean Phys. Soc. **27**, 1 (1994).
5. V. V. Dodonov, Adv. Chem. Phys. **119**, 309 (2001).
6. J. R. Johansson, G. Johansson, C. M. Wilson and F. Nori, Phys. Rev. Lett. **103**, 147003 (2009).
7. K. Takashima, N. Hatakenaka, S. Kurihara and A. Zeilinger, J. Phys. A: Math. Theor. **41**, 164036 (2008).
8. K. Takashima, S. Matsuo, T. Fujii, N. Hatakenaka, S. Kurihara and A. Zeilinger, J. Phys. : Conf. Ser. **150**, 052260 (2009).

CONTROLLING AND HARNESSING CRITICAL CASIMIR FORCES

A. GAMBASSI

SISSA – International School for Advanced Studies and INFN, via Beirut 2-4, 34151 Trieste, Italy
E-mail: gambassi@sissa.it

This contribution provides a brief overview of recent advances in the theoretical and experimental study of the critical Casimir effect originating from the confinement of the enhanced thermal fluctuations which occur close to a critical point. The resulting effect turns out to be highly tunable, it influences the behavior of soft matter, from wetting films of fluids to colloidal dispersions, and therefore might find applications for the manipulation of matter at the micrometer and sub-micrometer scale.

Keywords: Critical Casimir effect, wetting films, colloids.

1. The critical Casimir effect

Quite generally, whenever a medium in which fluctuations of some physical nature take place is confined, effective forces arise on the confining boundaries. Perhaps the best-known example is the Casimir effect in quantum electrodynamics.[1] An analogous effect arises in statistical physics when the relevant fluctuations are the thermal ones of the order parameter ϕ of a near-critical medium such as pure ^{4}He close to the normal/superfluid transition or a classical binary liquid mixture close to the demixing critical point.[2] These fluctuations are characterized by a temperature-dependent spatial correlation length $\xi \sim \xi_0|T/T_c - 1|^{-\nu}$ which also sets the range of the resulting force. Here T_c is the temperature of the bulk critical point, ξ_0 a (microscopic) material-dependent length, and $\nu > 0$ a universal critical exponent. Assuming that the medium is confined within a film of thickness L, if L and ξ are much larger than the microscopic molecular scale ξ_0, the temperature dependence of this critical Casimir force F becomes largely independent of the specific material properties of the medium and of the film surfaces but depends only on some of their gross features. In particular, upon approaching the critical point, the specific interaction between the medium and the confining surfaces (provided by, *e.g.*, a solid substrate or the interface between the fluid medium and its vapor), which locally affects the order parameter ϕ, eventually results into effective boundary conditions (BCs) for ϕ. The character of such BCs is again largely independent of the material properties of each surface and of its interaction with the fluid but depends only on the absence (O) or the presence of preferential adsorption with enhancement (+) or depletion (−) of ϕ at the surfaces.[4] All these informations are encoded in the *scaling function* $\vartheta_{\parallel}$ which is universal in the specific sense of critical phenomena and determines the force F per area S acting on the boundaries, as a function of the correlation length ξ and the thermal energy $k_{\mathrm{B}}T$:

$$\frac{F}{S} = \frac{k_{\mathrm{B}}T}{L^3}\vartheta_{\parallel}(L/\xi)\,. \tag{1}$$

Due to the universality which characterizes $\vartheta_{\parallel}$, a theoretical analysis of critical Casimir forces in a certain system can be done by studying suitable representative models which belong to the same universality class as the medium and the boundaries and which lend themselves for an easier analytical or numerical study (primarily Monte Carlo simulations).

Below we briefly discuss how it is possible to detect and use critical Casimir forces in soft matter systems (a more comprehensive account and list of references can be found in Ref. 3).

2. Indirect evidence: Wetting films

In order to realize and control a film of microscopically large and constant thickness L one can expose to a solid substrate a bulk vapor phase which is thermodynamically close to the condensation transition. Indeed, if the substrate is suitably chosen, a liquid wetting film of a certain equilibrium thickness L condenses on it as a result of an effective repulsive Casimir-Lifshitz interaction. Substrates which strongly attract the fluid undergo complete wetting, so that L becomes macroscopically large and diverges as the undersaturation of the vapor phase vanishes. In the resulting wetting film the fluid is naturally confined between the solid/liquid and the liquid/vapor interfaces. If such a confined fluid is driven thermodynamically towards a critical point, the associated critical fluctuations of ϕ – which are confined within the wetting film – give rise to a force F as described by Eq. (1). This force displaces the liquid/vapor interface from the equilibrium position it would have under the effect of dispersion forces alone, *i.e.*, in the absence of critical fluctuations. Accordingly, by monitoring experimentally the temperature dependence of L along a certain thermodynamic path one has *indirect* evidence of critical Casimir forces.[5] This kind of approach was adopted for studying wetting films of pure ^{4}He[6] and ^{4}He/^{3}He mixtures[7] close to the normal-superfluid transition and of classical binary mixtures close to the demixing critical point.[8] These experimental studies turned out to be in a remarkable quantitative agreement with the theoretical predictions based on the Monte Carlo simulations of suitable models.[9,10] Depending on the properties of the confining *surfaces*, F can be either attractive [$(\pm,\pm)$ and (O,O) BCs] or repulsive [$(\pm,\mp)$ and $(\pm,O)$], leading to the observed thinning[6,8] or thickening[7] of the wetting films.

3. Direct evidence and applications: Colloids

Similarly to the case of wetting films, a sphere which is immersed in a near-critical fluid is subject to a critical Casimir force when it approaches a flat substrate as the fluctuations of the order parameter are effectively confined within the gap between the surfaces of the sphere and of the substrate. This effect becomes particularly relevant for micrometer-sized particles, known as *colloids*. Indeed, colloids are typically characterized by an energy scale $k_{\mathrm{B}}T$, *i.e.*, the same as the critical Casimir force [see Eq. (1)]. Accordingly, colloids can be used for a *direct* measurement of such a force and conversely the latter can be used for manipulating colloids.

If the distance z between the surface of the colloid and the substrate is much smaller than the radius R of the colloid, the critical Casimir force F acting on the sphere is

$$F(z \ll R) = k_{\mathrm{B}}T\frac{R}{z^2}\vartheta_{|\circ}(z/\xi)\,. \tag{2}$$

The scaling function $\vartheta_{|\circ}$ can be calculated on the basis of $\vartheta_{\|}$ via the so-called Derjaguin (or proximity-force) approximation, which turns out to be rather accurate already for $z/R \lesssim 0.3$.[11] In the limit of strong preferential adsorption, $\vartheta_{|\circ}$ retains the same qualitative features as $\vartheta_{\|}$, being negative (attractive F) for (sphere, substr.) $= (\pm,\pm)$ BCs and positive (repulsive F) for $(\pm,\mp)$ BCs. The critical Casimir potential $\Phi(z) = \int_z^\infty \mathrm{d}z' F(z')$ can be measured by monitoring the Brownian motion of a single colloidal particle close to a substrate, both immersed in a near-critical fluid.[12,13] A recent experimental measurement of $\Phi(z)$ in a water-lutidine liquid mixture[12] provided the first direct evidence of the critical Casimir effect, in remarkable quantitative agreement with the theoretical predictions based

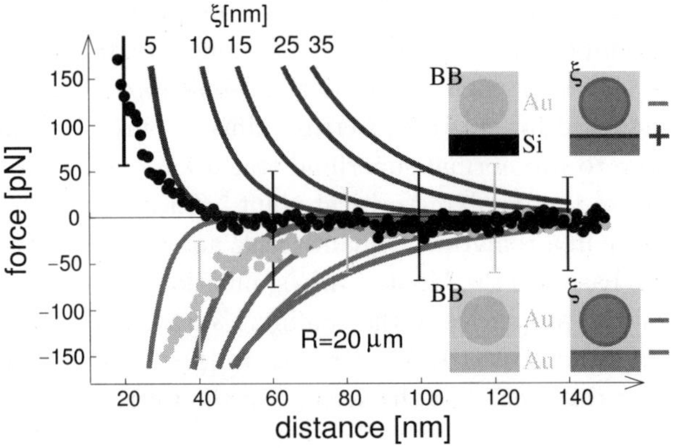

Fig. 1. Quantitative comparison between Casimir-Lifshitz (data points) and critical Casimir forces (solid lines, see the main text for details).

on the Derjaguin approximation and on the Monte Carlo results for $\vartheta_{\parallel}$ in three dimensions.[10] As in the case of wetting films, the sign of F is determined by the preferential adsorptions of the *surfaces* of the substrate and of the colloid, whereas its range, set by ξ, is controlled by minute temperature changes around T_c. Due to this high degree of tunability this force might find concrete applications – together with the omnipresent dispersion forces – for operating parts of micro-electromechanical systems. As an illustration, Fig. 1 compares a recent measurement[14] of the Casimir-Lifshitz force (data points and error bars) acting on a sphere of radius $R = 20\,\mu\text{m}$, close to a substrate and immersed in a fluid at room temperature $T \simeq 300\,\text{K}$, to the critical Casimir force (solid lines) expected within the same setting but in the presence of critical fluctuations with correlation length ξ and of surfaces with strong preferential adsorption for the order parameter. In the former case, turning attraction into repulsion requires a change of the bulk substrate from gold (lower data set) to silica (upper set) and a careful selection of the liquid medium (bromobenzene, BB). The resulting force is not significantly affected by changes of T. In contrast, in the case of the critical Casimir force, the actual material properties of the medium do not play such a crucial role in determining the strength and sign of the force and they can be suitably selected in order to have a critical point (*e.g.*, of demixing) at a convenient temperature. Moreover, the particle and the substrate can be realized with convenient materials (*e.g.*, silica) via well-established technological processes. The adsorption preferences of their surfaces are then controlled by means of suitable *surface* treatments. In addition, upon slightly varying T, *i.e.*, for correlation lengths $\xi \simeq 5$–$35\,\mathrm{nm}$, the range and the strength of the critical Casimir force can be controlled *reversibly* and to a large extent, both in the case of attraction [$(-,-)$ BCs, lower set of solid curves in Fig. 1] and repulsion [$(-,+)$ BCs, upper set].

This pronounced dependence on surface properties makes rather easy to control the spatial direction of the critical Casimir force by exposing the colloid to flat substrates with a spatially varying preferential adsorption.[11,13,15,16] This feature might have important applications to central problems in the physics of colloids such as controlling reversibly the spatial distribution and the rheology of dilute solutions, influencing aggregation, phase separation, and dynamics of colloidal mixtures etc. Most of these aspects, however, still need a proper theoretical understanding.

Acknowledgments

The author acknowledges financial support by the European Science Foundation (ESF) within the activity "New Trends and Applications of the Casimir Effect" (www.casimir-network.com) and by MIUR within the program "Incentivazione alla mobilità di studiosi straneri e italiani residenti all'estero".

References

1. H. B. Casimir, *Proc. K. Ned. Akad. Wet.* **51**, 793 (1948).
2. M. E. Fisher and P. G. de Gennes, *C. R. Acad. Sci. Paris Ser. B* **287**, 207 (1978).
3. A. Gambassi, *J. Phys.: Conf. Ser.* **161**, 012037 (2009).
4. H. W. Diehl, in *Phase Transitions and Critical Phenomena*, eds. C. Domb and J. L. Lebowitz (Academic, London, 1986), vol. 10, p. 76.
5. M. Krech and S. Dietrich, *Phys. Rev. A* **46**, 1922 (1992).
6. R. Garcia and M. H. W. Chan, *Phys. Rev. Lett.* **83**, 1187 (1999); A. Ganshin, S. Scheidemantel, R. Garcia and M. H. W. Chan, *Phys. Rev. Lett.* **97**, 075301 (2006).
7. R. Garcia and M. H. W. Chan, *Phys. Rev. Lett.* **88**, 086101 (2002); A. Maciołek, A. Gambassi and S. Dietrich, *Phys. Rev. E* **76**, 031124 (2007).
8. M. Fukuto, Y. F. Yano and P. S. Pershan, *Phys. Rev. Lett.* **94**, 135702 (2005).
9. A. Hucht, *Phys. Rev. Lett.* **99**, 185301 (2007).
10. O. Vasilyev, A. Gambassi, A. Maciołek and S. Dietrich, *EPL* **80**, 60009 (2007); *Phys. Rev. E* **79** 041142 (2009).
11. M. Tröndle, S. Kondrat, A. Gambassi, L. Harnau and S. Dietrich, *EPL* **88**, 40004 (2009).
12. C. Hertlein, L. Helden, A. Gambassi, S. Dietrich and C. Bechinger, *Nature* **451**, 172 (2008); A. Gambassi, A. Maciołek, C. Hertlein, U. Nellen, L. Helden, C. Bechinger and S. Dietrich, *Phys. Rev. E* **80**, 061143 (2009).

13. F. Soyka, O. Zvyagolskaya, C. Hertlein, L. Helden and C. Bechinger, *Phys. Rev. Lett.* **101**, 208301 (2008).
14. J. N. Munday, F. Capasso and V. A. Parsegian, *Nature* **457**, 170 (2009).
15. U. Nellen, L. Helden and C. Bechinger, *EPL* **88**, 26001 (2009).
16. F. Parisen Toldin and S. Dietrich, in this volume.

CRITICAL CASIMIR FORCES INVOLVING A CHEMICALLY STRUCTURED SUBSTRATE

F. PARISEN TOLDIN* and S. DIETRICH

Max-Planck Institut für Metallforschung,
Heisenbergstr. 3, D-70569, Stuttgart, Germany
and
Institut für Theoretische und Angewandte Physik, Universität Stuttgart,
Pfaffenwaldring 57, D-70569 Stuttgart, Germany
**E-mail: parisen@mf.mpg.de*

Motivated by recent experiments with confined binary liquid mixtures near their continous demixing phase transition we study the critical behavior of a system, which belongs to the Ising universality class, for the film geometry with one planar wall chemically structured such that there is a laterally alternating adsorption preference for the species of the binary liquid mixture. By means of Monte Carlo simulations and finite-size scaling analysis we determine the critical Casimir force and the corresponding universal scaling function.

Keywords: Critical phenomena, Casimir effect, confined fluids, finite-size scaling.

1. Introduction

If a field exhibiting fluctuations with long wavelengths is confined between surfaces, long-ranged effective forces between them arise. The order parameter of a second-order phase transition represents such a field. The resulting effective force is known as the critical Casimir force. This phenomena, first predicted by Fisher and de Gennes[1] is the analogue of the Casimir effect in quantum electrodynamics.[2,3] The critical Casimir force is characterized by a universal scaling function, which is determined by the bulk and surface universality classes (UC) of the confined system. It is independent of the microscopic details of the system, and it depends only on a few global and general properties, such as the number of components of the order parameter, the shape of the confinement, and the boundary conditions (b.c.) there.

In recent years, the critical Casimir effect has attracted numerous experimental and theoretical investigations; see Refs. 3,4 for recent reviews.

Wetting layers of binary liquid mixtures[5] have provided indirect evidence and measurements of critical Casimir forces. Recently, a direct determination of the critical Casimir force has been reported[6] using individual colloidal particles immersed in a binary liquid mixture close to its critical demixing point and exposed to a planar wall.

Quantitatively reliable calculations of critical Casimir forces for laterally homogeneous b.c. have been obtained recently by means of Monte Carlo (MC) simulations. In this context, the Ising UC,[8] which describes the demixing transition in a binary liquid mixture, has been investigated and the critical Casimir force has been determined.

Experiments with binary liquid mixtures have also been used to study critical Casimir forces acting on a colloid in front of a chemically structured substrate,[9] leading to a laterally varying adsorption preference. Such a system has been investigated theoretically for planar geometries within mean-field theory[10] and for a curved geometry within the Derjaguin approximation.[11] Here we consider the Casimir force in the film geometry $L \times L_\parallel \times L_\parallel$, $L_\parallel \gg L$, such that the free energy density $\mathcal{F}$ per volume decomposes as

$$\mathcal{F}(\beta, L, L_\parallel) = f_{\text{bulk}}(\beta) + \frac{1}{L} f_{\text{surf}}(\beta) + \frac{1}{\beta} f_{\text{ex}}(\beta, L, L_\parallel), \tag{1}$$

where $f_{\text{bulk}}(\beta)$ and $f_{\text{surf}}(\beta)$ are the bulk and the surface free energies per volume and area, respectively, in the thermodynamic limit $L_\parallel, L \to \infty$, and f_{ex} is the excess free energy. The Casimir force F_C per area and per $\beta^{-1} = k_B T$ is

$$F_C \equiv - \left. \frac{\partial (L f_{\text{ex}})}{\partial L} \right|_{\beta, L_\parallel}. \tag{2}$$

According to finite-size scaling,[12] close to a continuous phase transition F_C attains the following asymptotic scaling behavior:

$$F_C = \frac{1}{L^3} \theta \left(t (L/\xi_0^+)^{1/\nu}, \rho \right), \qquad t \equiv (T - T_C)/T_C, \tag{3}$$

where T_C is the bulk critical temperature, $\rho \equiv L/L_\parallel$ is the aspect ratio, and ξ_0^+ is the non-universal amplitude of the correlation length $\xi = \xi_0^+ t^{-\nu}$ in the disordered phase. The scaling function $\theta(x, \rho)$ is universal, i.e., it depends only on the UC of the bulk phase transition and on the surface UC, which is determined by the b.c. applied on the two surfaces. In the case of a confined binary liquid mixture, the order parameter ϕ is the difference between the local and the critical bulk concentration of one of the two components.

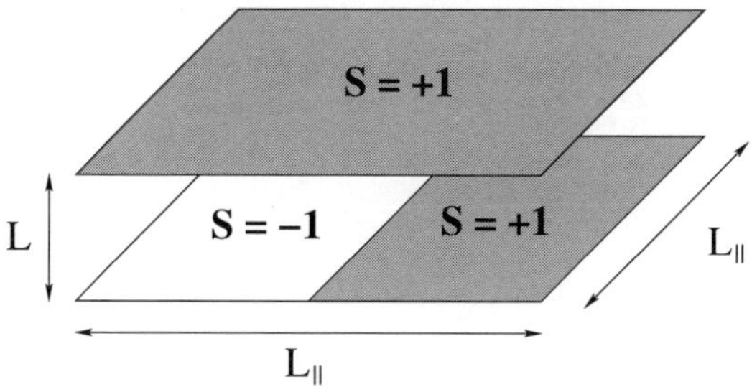

Fig. 1. Film geometry with a single chemical step; S is the spin value at the surface.

The substrates generically prefer one species over the other, so that $|\phi|$ is enhanced near the surface: this corresponds to the so-called extraordinary or normal surface UC[13] which, in a lattice model, can be implemented by external fields acting at the surfaces. For laterally homogeneous substrates, one has to distinguish between the case of surfaces with the same $(++)$ and opposite $(+-)$ adsorption preferences. For these systems laterally varying adsorption preferences can be realized.[9]

Motivated by corresponding experimental results,[9] we have performed MC simulations of a representative spin model on a lattice belonging to the Ising UC.[14] As illustrated in Fig. 1, the spins on the upper surface are fixed to $+1$. The lower surface is divided into two halves, one with spins fixed to -1 and the other with spins fixed to $+1$; the lattice constant is set to 1. This mimics a single chemical step as the simplest lateral heterogeneity.

2. Simulation method

In order to calculate the Casimir force, we use the so-called coupling parameter approach.[8] Given two systems with Hamiltonians $\mathcal{H}_0$ and $\mathcal{H}_1$ and the same configuration space, the free energy difference ΔF between the two systems can be computed as

$$\Delta F = \int_0^1 d\lambda \langle \Delta\mathcal{H} \rangle_\lambda, \qquad \Delta\mathcal{H} \equiv \mathcal{H}_1 - \mathcal{H}_0, \tag{4}$$

where $\langle \ldots \rangle_\lambda$ denotes the thermal average over the ensemble described by the crossover Hamiltonian $\mathcal{H}_\lambda \equiv (1-\lambda)\mathcal{H}_0 + \lambda\mathcal{H}_1$, with $\lambda \in [0,1]$. This average can be evaluated by standard MC simulations, followed by numerically carrying out the integral in Eq. (4). We have used Eq. (4) by considering $\mathcal{H}_0$ as the Hamiltonian of the system shown in Fig. 1 and $\mathcal{H}_1$ as that of the corresponding film but with thickness $L-1$ and an additional, decoupled

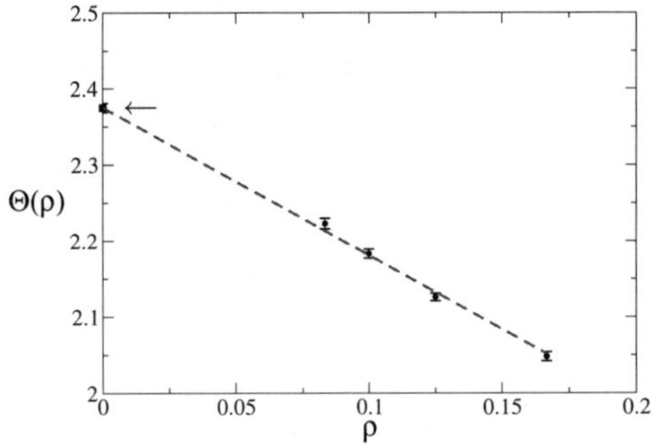

Fig. 2. The amplitude $\Theta(\rho)$ of the critical Casimir force at T_c for the system shown in Fig. 1 as a function of the aspect ratio $\rho = L/L_{\parallel}$.

two-dimensional (2D) layer. One obtains

$$\frac{\Delta F}{L_{\parallel}^2} = -f_{\rm bulk}(\beta) + f_{\rm 2D}(\beta) + \frac{1}{\beta}\frac{1}{L^3}\theta\left(t(L/\xi_0^+)^{1/\nu}, \rho\right), \tag{5}$$

where $f_{\rm 2D}(\beta)$ is the free energy density of the decoupled 2D layer and we have set the lattice constant to 1, so that each term is dimensionless. We note that Eq. (5) is correct only up to corrections to scaling.

3. Results

At the bulk critical temperature the expression in Eq. (5) reduces to

$$\frac{\Delta F}{L_{\parallel}^2} = -f_{\rm bulk}(\beta_c) + f_{\rm 2D}(\beta_c) + \frac{1}{\beta_c}\frac{1}{L^3}\Theta(\rho), \tag{6}$$

where $\Theta(\rho) \equiv \theta(0, \rho)$ is the critical Casimir amplitude. By fitting the MC data to this equation and by adding the expected additional scaling corrections, one can determine the critical Casimir amplitude.[14]

As shown in Fig. 2 the Casimir amplitude varies linearly for small ρ. We find $\Theta(1/6) = 2.048(6)$, $\Theta(1/8) = 2.126(5)$, $\Theta(1/10) = 2.183(6)$, $\Theta(1/12) = 2.223(7)$, and $\Theta(\rho \to 0) = 2.386(4)$. In the limit $\rho \to 0$ the force is expected to be the average of the force for laterally homogeneous $(++)$ and $(+-)$ b.c. At criticality[8] $(\Theta_{++} + \Theta_{+-})/2 = 2.33(4)$, in agreement with our results. The calculation of the full scaling function $\theta(x, \rho)$ requires the subtraction of the $L-$independent terms in Eq. (5) (see Ref. 14 for details). In Fig. 3 we report the resulting scaling function for various aspect ratios. We also show a comparison with the scaling function obtained as the average of those for homogeneous $(++)$ and $(+-)$ b.c., based on the data of Ref. 8. For $\rho \to 0$ we find good agreement with our results.

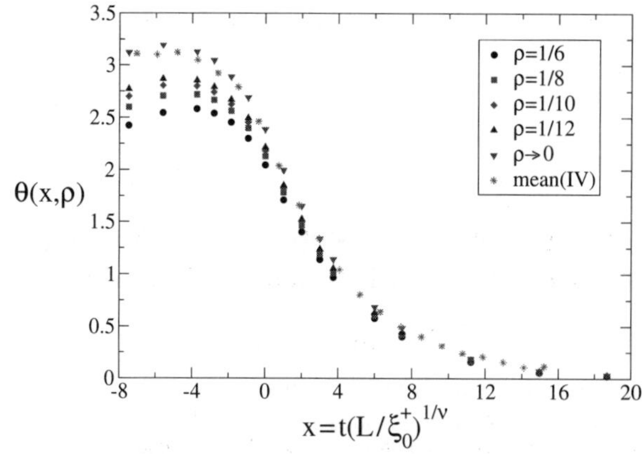

Fig. 3. The universal scaling function of the critical Casimir force for the system shown in Fig. 1 and for various aspect ratios $\rho = L/L_{\parallel}$. We also show the scaling function obtained as the mean value of those for laterally homogeneous (++) and (+−) b.c. based on the approximant IV presented in Ref. 8. Error bars are smaller than the symbol size.

References

1. M. E. Fisher, P. G. de Gennes, *C. R. Acad. Sci. Paris Ser. B* **287**, 207 (1978).
2. H. B. Casimir, *Proc. K. Ned. Akad. Wet.* **51**, 783 (1948).
3. A. Gambassi, *J. Phys.: Conf. Ser.* **161**, 012037 (2009).
4. A. Gambassi, *see these proceedings.*
5. M. Fukuto, Y. F. Yano, P. S. Pershan, *Phys. Rev. Lett.* **94**, 135702 (2005); S. Rafaï, D. Bonn, J. Meunier, *Physica A* **386**, 31 (2007); R. Garcia, M. H. W. Chan, *Phys. Rev. Lett.* **88**, 086101 (2002).
6. C. Hertlein, L. Helden, A. Gambassi, S. Dietrich, C. Bechinger, *Nature* **451**, 172 (2008); A. Gambassi, A. Maciołek, C. Hertlein, U. Nellen, L. Helden, C. Bechinger, S. Dietrich, *Phys. Rev. E* **80**, 061143 (2009).
7. A. Hucht, *Phys. Rev. Lett.* **99**, 186301 (2007).
8. O. Vasilyev, A. Gambassi, A. Maciołek, S. Dietrich, *EPL* **80**, 60009 (2007); O. Vasilyev, A. Gambassi, A. Maciołek, S. Dietrich, *Phys. Rev. E* **79**, 041142 (2009).
9. F. Soyka, O. Zvyagolskaya, C. Hertlein, L. Helden, C. Bechinger, *Phys. Rev. Lett.* **101**, 208301 (2008).
10. M. Sprenger, F. Schlesener, S. Dietrich, *J. Chem. Phys.* **124**, 134703 (2006).
11. M. Tröndle, S. Kondrat, A. Gambassi, L. Harnau, S. Dietrich, *EPL* **88**, 40004 (2009).
12. V. Privman, in *Finite-size scaling and numerical simulations of statistical system*, V. Privman ed. (World Scientific, Singapore, 1990), p. 1.
13. H. W. Diehl, in *Phase transitions and critical phenomena*, C. Domb and J. Lebowitz eds. (Academic, London, 1986), Vol. 10, p. 75.
14. F. Parisen Toldin, S. Dietrich, *in preparation.*

TWISTOR-BEAMS AND HOLOGRAPHIC STRUCTURE OF A PRE-QUANTUM KERR-SCHILD GEOMETRY

A. BURINSKII

Laboratory of Theoretical Physics, NSI Russian Academy of Sciences,
B.Tulskaya 52, Moscow, 115191, Russia
E-mail: bur@ibrae.ac.ru, www.ibrae.ac.ru

The exact Kerr-Schild (KS) solutions for electromagnetic excitations of black-holes have the form of singular beams supported on twistor lines. The beams have very strong back reaction on metric and horizon and produce a fluctuating KS geometry. Its holographic structure forms a pre-quantum spacetime taking intermediate position between the Classical and Quantum gravity.

Keywords: Kerr-Schild solutions; pp-waves; beams; Kerr theorem; twistors; holography; quantum gravity.

1. Introduction

Singular pp-waves,[1] playing the role of plane waves in gravity, differ drastically from plane waves in flat spacetime, which may be the origin of incompatibility of the Quantum theory and Gravity. Quantum theory works basically in momentum space, while Gravity demands explicit representation in configuration space. Twistor theory removes this incompatibility. Geometrically, twistor is a null line formed by a pair $(x^\mu,\ \theta^\alpha)$, where θ^α is a two-component spinor joined to the point $x^\mu \in M^4$. This spinor fixes a null direction $\sigma^\mu_{\dot\alpha\alpha}\bar\theta^{\dot\alpha}\theta^\alpha$ corresponding to momentum p^μ of a massless particle. A plane wave in twistor coordinates $T^I = \{\theta^\alpha,\ \mu_{\dot\alpha}\}$, $\mu_{\dot\alpha} = x_\nu\sigma^\nu_{\dot\alpha\alpha}\theta^\alpha$ has the form $\exp\{ix_\mu\sigma^\mu_{\dot\alpha\alpha}\tilde\theta^{\dot\alpha}\theta^\alpha\}$ which may be transformed to twistor space by a 'twistor' Fourier transform. The result,[2] (p.2.5), corresponds to a singular beam in direction p^μ supported on the null line with twistor coordinates $\{\theta^\alpha, \mu_{\dot\alpha}\}$. Such twistor-beams are sources of singular pp-wave solutions in gravity, and similar twistor-beams are obtained for typical exact electromagnetic (em) excitations of black-holes (BH)[3] obtained in the KS class[4] which covers a wide range of the rotating and non-rotating BH and cosmological solutions. Appearance of twistor-beams in the exact KS solutions is not accidental, since twistors form a skeleton of the KS spacetime.

Twosheeted holographic structure of the KS geometry is perfectly adapted to quantum treatment, as suggested in,[5,6] and the considered in[3] fluctuating twistor-beam solutions form a pre-quantum geometry taking intermediate position between the Classical and Quantum gravity.

2. Twistor structure of the KS geometry

The KS solutions are based on the KS form of metric

$$g_{\mu\nu} = \eta_{\mu\nu} + 2Hk_\mu k_\nu, \tag{1}$$

in which $\eta_{\mu\nu}$ is metric of auxiliary Minkowski space-time M^4 and k_μ is a field of null directions, forming a principal null congruence (PNC) $\mathcal{K}$. The BH solutions may be represented in two different forms $g^\pm_{\mu\nu} = \eta_{\mu\nu} + 2Hk^\pm_\mu k^\pm_\nu$, were $k^{\mu\pm}(x)$, $(x = x^\mu \in M^4)$ correspond to two different vector fields tangent to different congruences $\mathcal{K}^\pm$. The directions $k^\pm_\mu$ are determined in the null coordinates $u = (z-t)/\sqrt{2}$, $v = (z+t)/\sqrt{2}$, $\zeta = (x+iy)/\sqrt{2}$, $\bar\zeta = (x-iy)/\sqrt{2}$, by the form

$$k^{(\pm)}_\mu dx^\mu = P^{-1}(du + \bar Y^\pm d\zeta + Y^\pm d\bar\zeta - Y^\pm \bar Y^{(\pm)} dv), \tag{2}$$

which depends on the complex angular coordinate $Y = e^{i\phi}\tan\frac{\theta}{2}$. Two solutions $Y^\pm(x)$ are determined by **Kerr Theorem**[1,4,7,8] which states that the (necessary for solutions of type D) geodesic and shear-free null congruences in M^4 are generated by algebraic equation $F = 0$, where $F(Z^p)$ is arbitrary holomorphic function of the projective twistor coordinates.[a]

The Kerr congruence represents a vortex of null lines which cover the spacetime twice: in the form of ingoing ($k^{\mu -} \in \mathcal{K}^-$) and outgoing field ($k^{\mu +} \in \mathcal{K}^+$). It forms two sheets with two different metrics $g^\pm_{\mu\nu}$ on the same spacetime M^4. The em field has to be aligned with the rays of PNC and turns out to be different on the in- and out- sheets, and these fields should not be mixed. This twosheetedness is ignored in perturbative approaches, causing drastic discrepancy in the form of fundamental solutions. The typical exact em solutions on the Kerr background have the form of singular beams propagating along the rays of PNC, contrary to the smooth angular dependence of the wave solutions obtained perturbatively!

3. Holographic KS structure

For a long time this twosheetedness was a mystery of the Kerr BH (see refs in[9]). It was suggested to truncate the second sheet and replace it by a

[a]Projective twistor coordinates Z^p, $p = 1, 2, 3$ are related with twistor coordinates $\{\theta^\alpha,\ \mu_{\dot\alpha}\}$ as follows $T^I = \theta^1(1, Z^p)$.

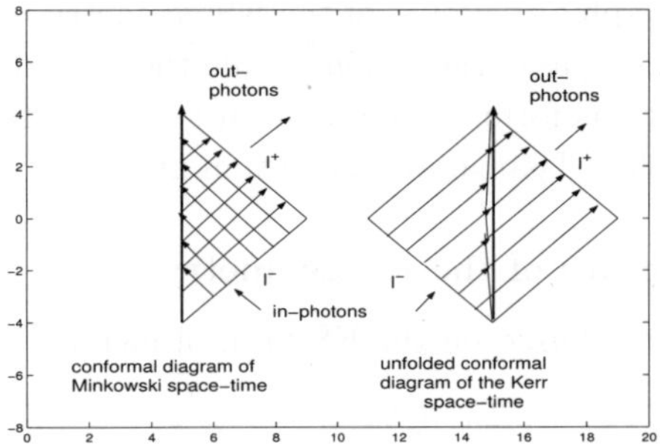

Fig. 1. Penrose conformal diagrams. Unfolding of the auxiliary M^4 space of the Kerr spacetime yields twosheeted structure of a pre-quantum BH spacetime.

rotating disk-like bubble covering the Kerr singular ring. Alternatively, the Kerr singular ring was considered as a closed 'Alice' string forming a gate to 'Alice' mirror world of the advanced fields.[10] Holographic approach resolves this problem unifying the both points of view. The source of Kerr solution is considered as a membrane separating the in- and out- parts of the KS space. The both sheets of the KS space are necessary for description of quantum fields in gravity. In particular, Gibbons states in[5] that the curved spacetime $\mathcal{M}$ should be separated into two time-ordered regions $\mathcal{M}_-$ and $\mathcal{M}_+$ which are associated with ingoing and outgoing vacuum states $|0_- >$ and $|0_+ >$. If the source is absent, the KS basic solutions may be extended analytically from in- to out-sheet. Presence of the source breaks this analyticity, separating the retarded and advanced fields, and allows one to consider the BH evaporation as a scattering of the in-vacuum on the BH-source. Similarly, 't Hooft et.al.[6] consider a pre-quantum BH spacetime with separated the in- and out- sheets which correspond to a holographic correspondence: the source forms a holographically dual boundary (membrane) separating the in- and out- regions. In the modern version,[11] this is a domain wall boundary of a superconducting bubble. The usual Penrose conformal diagram, containing the in- and out-fields on the same of M^4, has to be unfolded to split twosheetedness, as it is shown on the Fig.2. The unfolded KS diagram displays explicitly realization of the holographic principle in the KS space. The null rays of Kerr congruence perform the lightlike projection of the past null infinity I^-, so the Kerr source appears as a holographic image of the data on the past null infinity I^-.

Conformal structure of the KN solution is determined by complex function $Y(x)$. Tetrad derivatives of function $Y(x)$ determine principal parameters of the KS holographic projection. The functions $Y(x)$ obeying the conditions $Y_{,2} = Y_{,4} = 0$ define the shear-free and geodesic congruences providing conformal properties of the KS holographic projection.

4. Twistor-beam KS solutions

For the Kerr congruence function $F(Y,x)$ of the Kerr theorem is quadratic in Y, and the solution $Y(x)$ of the eq. $F=0$ determines congruence via (2), and function H. In particular, for the Kerr geometry at rest

$$H = \frac{mr - |\psi|^2/2}{r^2 + a^2\cos^2\theta}, \tag{3}$$

where r, θ are oblate spheroidal coordinates. Electromagnetic field is determined by vector potential

$$\alpha = \alpha_\mu dx^\mu = -\frac{1}{2} Re\ [(\frac{\psi}{r + ia\cos\theta})e^3 + \chi d\bar{Y}], \tag{4}$$

where $\chi = 2\int(1+Y\bar{Y})^{-2}\psi dY$, which obeys the alignment condition $\alpha_\mu k^\mu = 0$. For the Kerr solution m is mass of BH and $\psi = 0$.

The Kerr-Newman solution corresponds to $\psi = q = const.$. However, any nonconstant holomorphic function $\psi(Y)$ yields also an exact KS solution,[4] On the other hand, any nonconstant holomorphic functions on sphere acquire at least one pole. A single pole at $Y = Y_i$, $\psi_i(Y) = q_i/(Y - Y_i)$ produces the beam in angular directions $Y_i = e^{i\phi_i}\tan\frac{\theta_i}{2}$. The function $\psi(Y)$ acts immediately on the function H which determines the metric and the position of the horizon. The given in[12] analysis showed that electromagnetic beams have very strong back reaction to metric and deform topologically the horizon, forming the holes which allows matter to escape interior. For $\psi(Y) = \sum_i \frac{q_i}{Y - Y_i}$, the exact solution has the beams in angular directions $Y_i = e^{i\phi_i}\tan\frac{\theta_i}{2}$, leading to the horizon with many holes. In far zone the twistor beams tend to the known exact singular pp-wave solutions.[1]

5. Pre-quantum fluctuating KS geometry

The stationary KS beamlike solutions may be generalized to time-dependent wave pulses,[3] Since the horizon is extra sensitive to electromagnetic excitations, it may also be sensitive to the vacuum electromagnetic field, and the vacuum beam pulses may produce a fine-grained structure of fluctuating microholes in the horizon, allowing radiation to escape interior of black-hole, see details and refs. in.[3] Twistor-beam pulses depend on retarded time τ and obey to the nonstationary Debney-Kerr-Schild (DKS) equations.[4] The corresponding solutions acquire an extra radiative term $\gamma(Y,\tau)$. The long-term attack on the DKS equations,[3] has led to the obtained time-dependent solutions of the fluctuating Kerr-Schild spacetimes and showed explicitly that electromagnetic radiation from a black-hole interacting with vacuum contains two components:

a) a set of the singular beam pulses (determined by function $\psi(Y, \tau)$) propagating along the Kerr PNC and breaking the topology and stability of the horizon; and

b)the regularized radiative component (determined by $\gamma_{reg}(Y, \tau)$) which is smooth and determines evaporation of the black-hole.

The obtained solutions describe excitations of electromagnetic beams on the Kerr-Schild background, the fine-grained fluctuations of the black-hole horizon, and the consistent back reaction of the beams to the metric. The holographic space-time forms a fluctuating twosheeted pre-geometry which reflects the dynamics of singular beam pulses. This pre-geometry is classical, but has to be still regularized to get the usual smooth classical space-time. In this sense, it takes an intermediate position between the classical and quantum gravity. Note also that use of the Kerr theorem with functions $F(Y)$ of higher degrees in Y leads to multi-particle KS solutions[13] with complicate networks of twistor-beams.

Acknowledgments

Author thanks Organizers and especially Kim Milton for invitation to attend this extremely interesting conference and for financial support.

References

1. D.Kramer, H.Stephani, E. Herlt, M.MacCallum, *Exact Solutions of Einstein's Field Equations* , Cambridge Univ. Press, 1980.
2. E. Witten, *Comm. Math. Phys.* **252** 189 (2004).
3. A. Burinskii, First Award of Gravity Research Foundation, 2009, *Gen. Rel. Grav.* **41** 2281 (2009), arXiv: 0903.3162[gr-qc]; details in 0903.2365[hep-th].
4. G.C. Debney, R.P. Kerr and A.Schild, *J. Math. Phys.* **10**, 1842 (1969).
5. G. W. Gibbons, *Comm. Math. Phys.* **45** 191 (1975).
6. C.R. Stephens, G. 't Hooft and B.F. Whiting, *Class. Quant. Grav.* **11** 621 (1994).
7. R. Penrose, *J. Math. Phys.* **8** 345 (1967);
8. A. Burinskii, *Phys. Rev.* **D 70**, 086006 (2004).
9. A. Burinskii, E.Elizalde, S. R. Hildebrandt and G. Magli, *Phys. Rev.* **D65** 064039 (2002), arXiv:gr-qc/0109085.
10. A. Burinskii, *Sov.Phys.JETP*, **39** 193 (1974); *Russian Phys.J.* **17**, 1068 (1974), DOI:10.1007/BF00901591; *Grav. Cosmol.* **14** 109 (2008).
11. A. Burinskii,*Superconducting Source of the Kerr-Newman Electron*, arXiv: 0910.5388[hep-th].
12. A. Burinskii, E. Elizalde, S.R. Hildebrandt and G. Magli, *Phys. Rev.* **D 74** 021502 (2006); *Grav. Cosm.* 12 119 (2006).
13. A. Burinskii, *Grav. Cosmol.* **11**, 301 (2005; *IJGMMP*, **4**, n.3, 437 (2007).

REPULSIVE CASIMIR FORCES AND ALTERNATIVES TO EINSTEINIAN GRAVITY

E. ELIZALDE

Consejo Superior de Investigaciones Científicas ICE/CSIC-IEEC,
Campus UAB, Facultat de Ciències, Torre C5-Parell-2a pl,
E-08193 Bellaterra (Barcelona), Spain
E-mail: elizalde@ieec.uab.es, elizalde@math.mit.edu
www.ieec.fcr.es/english/recerca/ftc/eli/eli.htm

A basic property of the Casimir force is that, in most cases, it is attractive. The quest for workable configurations which exhibit repulsion is, at the same time, important and difficult, both in the nanophysics laboratory (e.g. to avoid stickness) and at cosmological scales (to account for the accelerated expansion of our universe). We shall here report on some configurations which exhibit the two signs of the Casimir force in function of the distance, and also on some alternatives to Einsteinian gravitation with the eventual aim to try to get around the cosmological constant problem.

Keywords: Casimir effect; repulsive Casimir force; boundary conditions; topological configurations; cosmological constant; alternatives to GR.

1. Introduction

A new generation of precision data has brought up evidence which does not fit within the previously accepted model of the universe evolution. In particular, its expansion rate seems to have been accelerating since many years ago. This evidence has led to the concept of dark energy whose nature could conceptually be understood as a cosmological constant, in its modern version as a contribution of quantum vacuum fluctuations. But, fitting the precise value of the cosmic acceleration has severe difficulties (e.g. the *new* cosmological constant problem). Regularization of vacuum contributions was never easy but, on the other hand, it may be possible to obtain, from quite natural arguments, some vacuum contributions that have the appropriate order of magnitude corresponding to the observed acceleration.[1] Moreover, in braneworld models Casimir energy contributions may lead to the stabilization of interesting configurations, what also proves the relevance of vacuum fluctuations at a cosmological scale.[2] In those and related problems, although in a seemingly distant domain, associated with MEM and NEM devices, the search for a repulsive Casimir force is of paramount importance.

We will here evaluate the Casimir energy and force for a massive scalar field with general curvature coupling parameter, subject to Robin (BCs) on two codimension-one parallel plates, located on a (D+1)-dimensional background spacetime with an arbitrary internal space.[3] Robin BCs can be made conformally invariant, while purely-Neumann ones cannot, therefore, they are needed when one wishes to preserve this invariance. They are an extension of those imposed on perfectly conducting boundaries and may, in some geometries, be useful for modeling the finite penetration of the field through the boundary, the skin-depth parameter being related to the Robin coefficient. They are most suitable to describe physically realistic situations. They naturally arise for scalar and fermion bulk fields in the Randall-Sundrum model, where the Robin coefficients are related to the curvature scale and to the boundary mass terms of the field. At laboratory level, they appear in the study of Casimir forces between the boundary planes of films.[4] Also interesting is to note that boundary problems with non-local BCs can be reduced to corresponding ones with Robin BCs, with the coefficients depending on the wave vector components along the plates.

With independence of the geometry of the internal space, Casimir forces are generically attractive for Dirichlet or Neumann BCs on both plates, and repulsive for Dirichlet on one and Neumann on the other (see e.g. Ref. 5). For Robin BCs, they can be either attractive or repulsive, depending on the Robin coefficients and also on the separation between the plates, what is both remarkable and useful for the different applications envisaged. An equilibrium point for the interplate distance, which is stabilized due to the Casimir force, appears and its stability improves in presence of extra dimensions. Applications of these properties in braneworld models are discussed.

As advanced, a different approach to the subject in cosmology is to consider alternatives to GR, as the Horăva-Lifshitz (HL) gravity. In this respect we summarize here our main results on a characterization of the cosmological phase space of this theory. The analysis allows to compare physical consequences of the imposition (or not) of the detailed balance condition. An interesting result of the investigation is that one of the attractors in the ensuing theory corresponds to an oscillatory behavior that can be associated to a bouncing universe (as previously described by Brandenberger) and prevents, in the detailed balance case, a possible evolution towards a de

Sitter universe. Other results show that the cosmological models generated by HL gravity without the detailed balance assumption have the potential to describe the transition between the Friedmann and the dark energy eras. The whole analysis leads to the plausible conclusion that a cosmology compatible with the present observations of the universe can be achieved only if the detailed balance condition is broken in those models.

2. Casimir Energy for Two Plates with Robin BCs

Consider a scalar field $\varphi(x)$, with arbitrary curvature coupling parameter ζ, satisfying the equation of motion

$$\left(g^{MN}\nabla_M\nabla_N + m^2 + \zeta R\right)\varphi(x) = 0, \tag{1}$$

$M, N = 0, 1, \ldots, D$, with R being the scalar curvature for a $(D+1)$-dimensional background spacetime. For the special cases of minimally and of conformally coupled scalars one has, respectively, $\zeta = 0$ and $\zeta = \zeta_D \equiv (D-1)/4D$. We will assume that the background spacetime has a topology $R^{(D_1,1)} \times \Sigma$, where $R^{(D_1,1)}$ is (D_1+1)-dimensional Minkowski spacetime and Σ a D_2-dimensional internal manifold, $D = D_1 + D_2$. The corresponding line element has the form

$$ds^2 = g_{MN}dx^M dx^N = \eta_{\mu\nu}dx^\mu dx^\nu - \gamma_{il}dX^i dX^l, \tag{2}$$

with $\eta_{\mu\nu} = \mathrm{diag}(1, -1, \ldots, -1)$ being the metric for the (D_1+1)-dimensional Minkowski spacetime and the coordinates X^i cover the manifold Σ. Here and below $\mu, \nu = 0, 1, \ldots, D_1$ and $i, l = 1, \ldots, D_2$. For the scalar curvature of the metric tensor, from (2) one has $R = -R_{(\gamma)}$, where $R_{(\gamma)}$ is the scalar curvature for the metric tensor γ_{il}.

We want to calculate the Casimir energy density and force for two parallel plates of codimension one, located at $x^{D_1} = a_1$ and $x^{D_1} = a_2$, $a_1 < a_2$. On the boundaries the scalar field obeys Robin BCs

$$\left(1 + \beta_j n^M \nabla_M\right)\varphi(x) = [1 + \beta_j(-1)^{j-1}\partial_{D_1}]\varphi(x) = 0, \quad x^{D_1} = a_j,\ j = 1, 2, \tag{3}$$

with constant coefficients β_j. For $\beta_j = 0$ these BCs are reduced to Dirichlet and for $\beta_j = \infty$ to Neumann BCs.

Between the plates, $a_1 < x^{D_1} < a_2$, the eigenfunctions satisfying the BC on the plate at $x^{D_1} = a_j$, are

$$\varphi_\alpha(x^M) = C_\alpha \exp\left(-i\sum_{\mu,\nu=0}^{D_1-1}\eta_{\mu\nu}k^\mu x^\nu\right)\cos\left[k^{D_1}|x^{D_1} - a_j| + \alpha_j\right]\ \psi_\beta(X), \tag{4}$$

where α denotes a set of quantum numbers specifying the solution and

$$k^0 = \omega = \sqrt{k^2 + (k^{D_1})^2 + m_\beta^2}, \quad m_\beta^2 = \lambda_\beta^2 + m^2,$$
$$k = |\mathbf{k}|, \ \mathbf{k} = (k^1, \ldots, k^{D_1-1}). \tag{5}$$

In Eq. (4), the α_j, $j = 1, 2$, are defined by the relations

$$\sin \alpha_j = \frac{1}{\sqrt{(k^{D_1})^2 \beta_j^2 + 1}}, \ \cos \alpha_j = \frac{k^{D_1}\beta_j}{\sqrt{(k^{D_1})^2 \beta_j^2 + 1}}. \tag{6}$$

The modes $\psi_\beta(X)$ are the eigenfunctions of the operator $\Delta_{(\gamma)} + \zeta R_{(\gamma)}$:

$$\left[\Delta_{(\gamma)} + \zeta R_{(\gamma)}\right] \psi_\beta(X) = -\lambda_\beta^2 \psi_\beta(X), \tag{7}$$

with the eigenvalues λ_β^2, and fulfill the normalization condition

$$\int d^{D_2}X \sqrt{\gamma}\psi_\beta(X)\psi_{\beta'}^*(X) = \delta_{\beta\beta'}. \tag{8}$$

$\Delta_{(\gamma)}$ the Laplace-Beltrami operator for γ_{il}. We assume $\lambda_\beta \geqslant 0$.

From the BC on the second plate, the eigenvalues for k^{D_1} are solutions of

$$F(z) = (1 - b_1 b_2 z^2) \sin z - (b_1 + b_2) z \cos z = 0,$$
$$z = ak^{D_1}, \ a = a_2 - a_1, \ b_j = \beta_j / a. \tag{9}$$

We denote by $z = z_n$, $n = 1, 2, \ldots$, the zeros of the function $F(z)$ in the right half-plane of the complex variable z, arranged in ascending order, $z_n < z_{n+1}$. In the discussion below we will assume that all these zeros are real. This is the case for the conditions $\{b_1 + b_2 \geqslant 1, b_1 b_2 \leqslant 0\} \cup \{b_{1,2} \leqslant 0\}$. The coefficient C_α in (4) is determined from the orthonormality condition for the eigenfunctions, and is equal to

$$C_\alpha^2 = \frac{(2\pi)^{1-D_1}}{\omega(z_n)a} \left[1 + \frac{1}{z_n} \sin(z_n) \cos(z_n + 2\alpha_j)\right]^{-1}, \tag{10}$$

being $\omega(z_n) = \sqrt{k^2 + z_n^2/a^2 + m_\beta^2}$ the eigenfrequencies.

The vacuum energy in the region between the plates (per unit volume along the directions $x^1, \ldots, x^{D_1-1}$) is given by the expression

$$E_{[a_1,a_2]} = \frac{1}{2} \int \frac{d\mathbf{k}}{(2\pi)^{D_1-1}} \sum_\beta \sum_{n=1}^{\infty} \sqrt{k^2 + z_n^2/a^2 + m_\beta^2}. \tag{11}$$

In order to be meaningful, this needs to be regularized, and we can use either a cut-off function or zeta function regularization. There is here no place to describe these procedures which, in the case of zeta regularization,[6] starts from the expression

$$\zeta_{12}(s) = \frac{1}{2}\int \frac{d\mathbf{k}}{(2\pi)^{D_1-1}} \sum_{\beta}\sum_{n=1}^{\infty} \left(k^2 + z_n^2/a^2 + m_\beta^2\right)^{-s}, \quad E_{[a_1,a_2]} = \zeta_{12}(-\frac{1}{2}), \tag{12}$$

and we will just give the final result.

3. The Varying Sign of the Casimir Force

For the different cases above, after a rather involved calculation we obtain for the force by unit area acting on each plate (in the original parallel plate configuration):

$$P = -\frac{2(4\pi)^{-D_1/2}}{V_\Sigma \Gamma(D_1/2) a^{D_1+1}} \sum_\beta \int_{am_\beta}^{\infty} dx \, \frac{x^2 (x^2 - a^2 m_\beta^2)^{D_1/2-1}}{\frac{(b_1 x-1)(b_2 x-1)}{(b_1 x+1)(b_2 x+1)} e^{2x} - 1}. \tag{13}$$

This force is attractive when its sign is negative, and repulsive when it is positive. As seen from Eq. (13), the sign of the vacuum stress on the plate is determined by the sign of the integral here. In particular, the Casimir force between the plates is always attractive for symmetric BCs with $\beta_1 = \beta_2 < 0$. This result is a special case of the general theorem,[7,8] which dictates attraction between bodies with the same properties.

For the special case of Dirichlet and Neumann BCs, using the recurrence relations for Bessel's $K_\nu(z)$, the Casimir forces can be written as

$$P^{(\mathrm{J},\mathrm{J})} = -\frac{2(4\pi)^{-D_1/2}}{V_\Sigma \Gamma(D_1/2)} \sum_\beta \int_{m_\beta}^{\infty} dx \, x^2 \frac{(x^2 - m_\beta^2)^{D_1/2-1}}{e^{2ax} - 1} \tag{14}$$

$$= \frac{2a^{-D_1-1}}{(8\pi)^{(D_1+1)/2} V_\Sigma} \sum_\beta \sum_{n=1}^{\infty} \frac{1}{n^{D_1+1}} \left[f_{(D_1+1)/2}(2nam_\beta) - f_{(D_1+3)/2}(2nam_\beta) \right],$$

being $f_\nu(z) = z^\nu K_\nu(z)$ and $\mathrm{J} = \mathrm{D}, \mathrm{N}$. In the case $D_1 = 3$ and for a massless scalar field this result reduces to the one obtained in Ref. 9. The forces described by Eq. (14) are attractive for all distances between the plates, irrespective of the geometry of the internal subspace. For Dirichlet BCs on one plate and Neumann BCs on the other, the expression for the Casimir force is obtained from Eq. (14) after introducing an additional factor $(-1)^{n+1}$ in the summation over n. In this case, Casimir forces are repulsive. For Robin BCs the Casimir forces can be either attractive or repulsive, depending on the coefficients present in the definition of the BCs and, what is even more important, on the distance between plates.

Let us now consider the asymptotic behavior of the Casimir force as a function of the size of the internal space. Note that if the size of the internal space is of the order L, then for nonzero modes one has $\lambda_\beta \sim 1/L$. For small values of L and for the nonzero modes, λ_β is large. The contribution of these modes is exponentially suppressed and the main contribution comes from the zero mode. We then recover the Casimir force for two parallel plates in (D_1+1)-dimensional Minkowskian spacetime, namely

$$V_\Sigma P \approx -\frac{2(4\pi)^{-D_1/2}}{\Gamma(D_1/2)} \int_m^\infty dx \, \frac{x^2(x^2-m^2)^{D_1/2-1}}{\frac{(\beta_1 x-1)(\beta_2 x-1)}{(\beta_1 x+1)(\beta_2 x+1)} e^{2ax} - 1}. \tag{15}$$

For the case of a degenerated zero eigenstate the corresponding degeneracy factor must be included on the right-hand side. In some models of compactification the zero mode is absent (for example, in models with twisted BCs along the compactified dimensions, see below). In such cases, for small values of L the main contribution to the Casimir force comes from the lowest mode $\lambda_\beta = \lambda_0$ and, to leading order, one gets

$$V_\Sigma P \approx -\frac{m_0^{D_1+1}}{(4\pi)^{D_1/2}} \frac{(\beta_1 m_0 x+1)(\beta_2 m_0 x+1)}{(\beta_1 m_0 x-1)(\beta_2 m_0 x-1)} \frac{e^{-2am_0}}{(am_0)^{D_1/2}}, \tag{16}$$

where $m_0 = \sqrt{\lambda_0^2 + m^2}$. Hence, here the Casimir forces are exponentially suppressed for small sizes of the internal space.

For small values of the inter-plate distance, $a/|\beta_j| \ll 1$, the main contribution into the integral comes from larger values of x and, to leading order,

$$P \approx -\frac{2(4\pi)^{-D_1/2}}{V_\Sigma \Gamma(D_1/2)} \sum_\beta \int_{m_\beta}^\infty dx \, x^2 \frac{(x^2-m_\beta^2)^{D_1/2-1}}{e^{2ax}-1}, \tag{17}$$

except for the case of Dirichlet BCs on one plate and non-Dirichlet on the other. Note in this limit (17) the Casimir force is attractive. However, for of Dirichlet BCs on one plate and non-Dirichlet on the other it is repulsive, at small distances, what coincides with known results in ordinary dimensions.

Particular cases of this general configuration have been analyzed in Ref. 3, as the special case where $\Sigma = S^1$ and the case of a D_2-dimensional torus as internal space, $\Sigma = (S^1)^{D_2}$, and also the issue of embedding the model in string theory, together with the discussion there of the holographic principle, in which case the presence of the internal space $\Sigma = S^{D_2}$ in the analysis carried out here is of very special interest.

4. Hořava-Lifshitz Cosmology in the Detailed Balance Case

Among the alternatives to Einsteinian gravitation which have been mentioned in the introduction, there is one, about one year old, that has attracted a lot of interest. Hořava[10] has made a proposal for an ultraviolet completion of general relativity (GR) that is renormalizable at the level of power counting. Such ultraviolet behavior has been obtained by introducing irrelevant operators which improve the ultraviolet behavior of the theory, at the price of explicitly breaking Lorentz invariance which, however, is expected to be recovered at low energies, as an accidental symmetry of the theory. The theory is reminiscent of the Lifshitz theory (this is the reason for his name being associated with it), which is indeed a very interesting connection in the context of these proceedings. In order to reduce the number of independent coupling constants, Hořava imposed on the theory additional symmetries, namely the *projectability condition* and the *detailed balance condition*. The reason for these restriction is purely pragmatic (as Hořava himself recognizes): to limit the proliferation of independent couplings. The projectability condition does not affect the cosmological analysis to follow, since the associated shift variable which follows from it is only time dependent.[10] On the other hand, the name detailed balance comes from the way this second particular restriction is implemented, which is specially reminiscent of methods used in non-equilibrium critical phenomena and quantum critical systems. Borrowing terminology from non-equilibrium dynamics, Hořava said that theories whose potential has a specific form, quadratic in a term which is contracted through the inverse De Witt metric, which term itself follows from a variational principle from some potential, satisfy the detailed balance condition.[10] The analysis below will allow to compare some key physical consequences of the imposition of detailed balance and we will go here a bit further concerning the discussion of the introduction of matter fields and of the generalization of a recently proposed extension of the theory.

The potential for the detailed balance case turns out to be

$$\mathcal{L}_{potential-detail} = \beta C_{ij}C^{ij} + \gamma\epsilon^{ijk}R_{il}\nabla_j R^l_k + \zeta R_{ij}R^{ij} + \eta R^2 + \delta R + \sigma\,, \tag{18}$$

where $\sqrt{g}C^{ij} = \epsilon^{ikl}\nabla_k(R^j_l - 1/4Rg^j_l)$ and the Greek letters define the different coupling constants (we follow the notation of Ref. 12). Only three out of the whole set of coupling constants above are independent, due to the detailed balance constraint. The field equations for this theory, on a

FRW ansatz can be written as[11]

$$\alpha(3\lambda-1)\dot{H}+\alpha(3\lambda-1)H^2=-\frac{1}{6}(1+3w)\rho+\mathbb{A}\alpha(1-3\lambda)\frac{k^2}{a^4}+\mathbb{A}\alpha(3\lambda-1)\Lambda^2,$$

$$3\alpha(3\lambda-1)H^2+6\mathbb{A}\alpha(3\lambda-1)\Lambda\frac{k}{a^2}=\rho+3\mathbb{A}\alpha(3\lambda-1)\frac{k^2}{a^4}+3\mathbb{A}\alpha(3\lambda-1)\Lambda^2,$$

$$\dot{\rho}+3(w+1)H\rho=0\,, \tag{19}$$

where we leave the parameters α, λ, Λ explicit because of their cosmological relevance. Here we have defined the variable $\mathbb{A}=\frac{-\zeta}{\alpha(1-3\lambda)^2}$ which is always positive, since $\zeta\leq 0$ owing to the aforementioned detailed balance constraint. At this level, only a subset of all the initially defined couplings $(\alpha, \lambda, \Lambda, \mathbb{A})$, plays a role in the cosmology. In order to analyze the phase space of this cosmological model, we define the variables

$$\Omega=\frac{\rho}{3\alpha H^2}\,,\qquad z=\frac{\mathbb{A}\Lambda^2}{H^2}\,,\qquad K=2\Lambda\frac{k\mathbb{A}}{a^2H^2}\,,\qquad C=\frac{k^2\mathbb{A}}{a^4H^2}\,, \tag{20}$$

with the cosmic time $\mathcal{N}=\ln[a(t)]$.

The cosmological equations (19) are then equivalent to the system

$$\begin{aligned}\Omega'&=\tfrac{1+3w}{3\lambda-1}\Omega^2+\Omega(2C-2z-3w-1),\\ z'&=-2z^2+z\left[2+2C+\tfrac{1+3w}{3\lambda-1}\Omega\right],\\ K'&=K\left[2C-2z+\tfrac{1+3w}{3\lambda-1}\Omega\right],\\ C'&=2C^2-C\left[2+2z-\tfrac{1+3w}{3\lambda-1}\Omega\right],\end{aligned} \tag{21}$$

with the Gauss constraint

$$1+K-z-C+\frac{\Omega}{1-3\lambda}=0\,, \tag{22}$$

where the prime denotes the derivative with respect to $\mathcal{N}$. The Gauss constraint (22) allows to eliminate one of the equations. If we choose to eliminate the one for Ω, we obtain

$$\begin{aligned}z'&=z[3+C+K-3z-3w(C-K+z-1)]\,,\\ C'&=C\left[C+K-3z-3w(C-K+z-1)-1\right]\,,\\ K'&=K\left[1+C+K-3z-3w(C-K+z-1)\right]\,.\end{aligned} \tag{23}$$

This system presents three invariant submanifolds $z=0$, $K=0$, $C=0$ which, by definition, cannot be crossed by any orbit. This implies that no global attractor can exist in this type of HL cosmology. Also, the structure of (22) reveals that the system is non-compact and asymptotic analysis is required in order to complete the study of the phase space. Due to its dimensionality it is not easy to draw a plot of the phase space. Since invariant submanifolds are present, one can think to give a plot of orbits belonging to them, however the physical meaning of such manifold is not clear. For example, the invariant submanifold $C = 0$ might be associated at first glance to flat cosmology, however from the definitions (20), one can see that this would be the case only for $K = 0$, so that the rest of $C = 0$ has no real physical meaning. Similar arguments can be given for the $z = 0$ and $K = 0$ submanifolds. As a consequence, only in orbits in the phase space bulk correspond to physical evolutions for the system and we can describe them without the aid of graphics. Actually, we could consider the $\Omega = 0$ invariant submanifold which could be of interest, but in our setting this would require the derivation of a new system, in which another variable is eliminated.

5. Hořava-Lifshitz Cosmology with No Detailed Balance

If one does not impose detailed balance to hold, the cosmological equations in presence of matter (assumed to be a barotropic fluid) can be written as (following the notation of Ref. 13)

$$\begin{aligned} \left(1-\tfrac{3\xi}{2}\right)\left(\dot{H}+H^2\right)+\tfrac{1}{2}\kappa^2\rho(1+3w)-\tfrac{\chi_1}{6}+\tfrac{\chi_3 k^2}{6a^4}+\tfrac{\chi_4 k}{3a^6}=0\,,\\ \left(1-\tfrac{3\xi}{2}\right)H^2-\tfrac{\chi_2 k}{6a^2}-\kappa^2\rho-\tfrac{\chi_1}{6}-\tfrac{\chi_3 k^2}{6a^4}-\tfrac{\chi_4 k}{6a^6}=0\,,\\ \dot{\rho}+3(w+1)H\rho=0\,, \end{aligned} \tag{24}$$

where

$$\begin{aligned} \kappa^2=\tfrac{1}{6\alpha},\ \chi_1=\tfrac{g_0\alpha^3}{6},\ \chi_2=-6g_1\alpha^2>0,\\ \chi_3=12\alpha\left(3g_2+g_3\right),\ \chi_4=24\left(9g_4+3g_5+g_6\right), \end{aligned} \tag{25}$$

and where we have defined $\xi = 1 - \lambda$ assuming also that $\xi \neq 2/3$. Taking $\xi > 2/3$ would imply that the energy density in the Friedmann equation has a negative sign. This range of the parameter corresponds to spin zero modes of the theory, which can be excluded at the cosmological level and are related to unwanted ghost modes. The sign of the term χ_1 determines the sign of an (effective) cosmological constant in the model and χ_2 can be always taken to be negative. Comparing the system above with the one in the former case, we note that, as pointed out in Ref. 13, there is not much difference between the two cases.[11]

After carrying out the analysis of the two cases in detail, the results are the following. In the first case, the phase space exhibits four finite fixed points, three of which represent physical solutions of the system. Although one of these points is associated to an unstable classical Friedmann solution that could be certainly useful to model the nucleosynthesis and the structure formation periods, our analysis does not reveal any useful fixed point which could model an inflationary or dark energy phase. It has been proposed that, because of the changes in the value of the speed of light contained in the theory, the absence of an explicit inflationary phase might not be such a serious shortcoming of the theory,[12] although at first glance it is difficult in this setting to produce a dark energy era. A more conclusive analysis of this issue, however, requires a complete numerical study.

An interesting result of the investigation is that one of the attractors in the theory corresponds to an oscillatory behavior. Other attractors are associated with a recollapsing solution in the form of a Gauss function. These solutions, which also appear in the context of GR plus a scalar field, higher-order and scalar-tensor gravities, could—if proven to be unstable—model some kind of superinflationary phase whose properties ought to be carefully investigated. However, most of the orbits that contain these points will either evolve towards recollapse or to the oscillatory attractor. In other words, when detailed balance is imposed, although one is able to prove that the corresponding cosmology could, given suitable initial conditions, behave like a Friedmann one, it turns out that it will eventually evolve towards an oscillatory behavior or either recollapse, leaving no space for a dark-energy era. Another interesting results is related to the HL cosmology without detailed balance. In this case, in fact, the additional freedom in the values of the parameters allows for the existence of cosmic histories which contain a Friedmann era and evolve towards a dark energy one. This follows because the phase space contains a fixed point associated to the standard Friedmann solution which is unstable and another one which can be associated to a de Sitter solution which is an attractor. The last point can then model a dark energy era. However, the existence of these fixed points is only a necessary condition: because of the presence of invariant manifolds and the constraints on the parameters only a subset of the phase space and the parameter space will realize this scenario. On the other hand the fact that the fixed points all lie on invariant submanifolds guarantee that such orbits can exist. Unfortunately the high dimension of the phase space makes it quite hard to perform any qualitative analysis. Therefore only numerical methods will allow the investigation of the details of these orbits. Notwithstanding these problems, it seems safe to conclude that a cosmology compatible with the present observations can be obtained, in the HL framework, only if the detailed balance is broken.[11] Such result makes this type of HL gravity a promising phenomenological model for both the study of dark energy and quantum gravity. Hints that point out to the same conclusion that detailed balance cannot be maintained have come from other directions (see e.g., Ref. 14), which go further to prescribe a particular extension of the theory in order to make it consistent. We are now working on the most general extension along these lines and on the problem of the introduction of massive fields.[15]

Acknowledgments

I am much indebted with my several collaborators in this research, in special S. Carloni, S.D. Odintsov, A.A. Saharian and P.J. Silva. The author acknowledges financial support by the European Science Foundation (ESF) within the activity "New Trends and Applications of the Casimir Effect" (www.casimir-network.com). This investigation has been also supported in part by MICINN (Spain), project FIS2006-02842 and grant PR2009-0314, and by AGAUR (Generalitat de Catalunya), contract 2005SGR-00790. It was partly done while the author was on leave at the Department of Physics and Astronomy, Dartmouth College, 6127 Wilder Laboratory, Hanover, NH 03755, USA.

References

1. E. Elizalde, J. Phys. **A39**, 6299 (2006); J. Phys. **A40**, 6647 (2007).
2. E. Elizalde, S. Nojiri, S.D. Odintsov and S. Ogushi, Phys. Rev. **D67** 063515 (2003); S. Capozziello, V.F. Cardone, E. Elizalde, S. Nojiri and S.D. Odintsov, Phys. Rev. **D73**, 043512 (2006).
3. E. Elizalde, S.D. Odintsov, and A.A. Saharian. Phys. Rev. **D79**, 065023 (2009).
4. F.M. Schmidt and H.W. Diehl. Phys. Rev. Lett. **101**, 100601 (2008).
5. A. Romeo and A.A. Saharian, J. Phys. **A35**, 1297 (2002).
6. E. Elizalde, S.D. Odintsov, A. Romeo, A.A. Bytsenko and S. Zerbini, *Zeta regularization techniques with applications* (World Scientific, Singapore, 1994); K. Kirsten, *Spectral functions in mathematics and physics* (Chapman & Hall, London, 2001); E. Elizalde, *Ten physical applications of spectral zeta functions*, Lecture Notes in Physics (Springer-Verlag, Berlin, 1995).
7. O. Kenneth and I. Klich, Phys. Rev. Lett. **97**, 160401 (2006).
8. C.P. Bachas, J. Phys. A: Math. Gen. **40**, 9089 (2007).
9. K. Kirsten and S.A. Fulling, Phys. Rev. **D79**, 065019 (2009).
10. P. Hořava, Phys. Rev. **D79**, 084008 (2009).
11. S. Carloni, E. Elizalde and P.J. Silva, *An analysis of the phase space of Horava-Lifshitz cosmologies*, arXiv:0909.2219 [hep-th].

12. E. Kiritsis and G. Kofinas, *Hořava-Lifshitz Cosmology*, arXiv:0904.1334 [hep-th].
13. T.P. Sotiriou, M. Visser and S. Weinfurtner, *Quantum gravity without Lorentz invariance*, arXiv:0905.2798 [hep-th].
14. D. Blas, O. Pujolas and S. Sibiryakov, *On the extra mode and inconsistency of Hořava Gravity*, arXiv:0906.3046 [hep-th].
15. S. Carloni, E. Elizalde and P.J. Silva, to be issued.

QUANTUM BACK REACTION TO ASYMPTOTICALLY ADS BLACK HOLES

A. FLACHI, K. KASHIYAMA, N. TANAHASHI, T. TANAKA

Kyoto University, Kyoto 606-8502, Japan

In recent work, we have analyzed the effects of the quantum back reaction due to a conformal field theory (CFT) on a black hole spacetime with negative cosmological constant. We have found a sequence of configurations without horizon in thermal equilibrium (*CFT stars*), followed by a sequence of configurations with horizon. We studied the thermodynamic properties of the system and discussed how back reaction effects alter the space-time structure. We also provided an interpretation of the above sequence of solutions in terms of the AdS/CFT correspondence. The dual five-dimensional description is given by the Karch-Randall model, in which a sequence of five-dimensional floating black holes followed by a sequence of brane localized black holes correspond to the above solutions. This note is a brief summary of our recent work.

In this brief note, we will report on our study of quantum back reaction effects for black holes in AdS space. For brevity, we limit ourselves to present the main results and invite the interested reader to consult Ref. 1 for details.

The main issue that triggered our attention in the problem is that in the Randall-Sundrum model[2] (RS) no *large*, stable, static black hole solution localized on the brane or floating in the bulk have so far been found, whereas *small* localized solutions have been constructed numerically[3] for black holes with size smaller than the curvature scale ℓ. The above situation is in tune with the prediction of the AdS/CFT correspondence[4] that relates 5D classical gravity in the RS model to a dual 4D gravity coupled to a CFT (with cut-off ℓ), and suggests the absence of static 4D black holes that would evaporate due to the presence of the CFT. The absence of 4D static solutions, according to the correspondence, would imply the existence of a 5D classical dynamical process analogous to the 4D quantum evaporation, clarifying why 5D solutions cannot be found.[5,6]

It is clearly difficult to provide any relevant (numerical) proof, and it makes sense to take an indirect approach by considering modifications to the RS model. A natural thing to do is to relax the asymptotic flatness of the branes by detuning their tension. This leads to asymptotically AdS branes, *i.e.* Karch-Randall model (KR).[7]

Contrary to the RS model, in the KR model two small black hole solutions are expected to exist: an unstable one close to the brane, located at $y = y_b$, and a stable one near $y = 0$, far from the brane[a]. The latter is infinitely far from the brane in the RS limit of the KR model.[8] As in the RS model, the AdS/CFT correspondence works also in the KR model, but with a very different outcome. As we have mentioned, in the RS model the CFT back reaction causes the black hole to evaporate preventing the staticity of the solution[b]. In the KR model, the branes are asymptotically AdS, and the presence of a non-zero cosmological constant changes the situation dramatically. Since the lapse function in AdS behaves at large distances as r/L (L is the 4D AdS curvature scale), the temperature and hence the energy density of (thermal) CFT decrease rapidly for $r \gg L$, reducing the effects of the back reaction. If the black hole size is large, the energy density due to CFT will stay negligiblly small at any radius. Whilst, if the size of the black hole is small, the back reaction becomes important and a static black hole solution becomes non-trivial. Roughly speaking, such a small black hole will be unstable against the CFT back reaction and should 'evaporate' into a CFT star of the same mass.

The sequence of the CFT stars can be tagged by the central density, and the end-point of the sequence corresponds to a star with singular central density and lapse vanishing at the center. Thus, this sequence of the CFT stars will naturally flow into the sequence of quantum corrected black holes, whose starting-point corresponds to a small black hole in the limit of zero horizon radius.

According to the AdS/CFT correspondence, we may expect that a five-dimensional black hole in the KR model will be dual to the above sequence of 4D CFT stars and quantum black holes.[8] Naive expectation is that a brane-localized black hole and a floating black hole in the KR model are, respectively, dual to a four-dimensional black hole with back reaction of CFT halo and a star composed of CFT, which we refer to as quantum black hole and CFT star. If it is really the case, we can examine black

[a]The coordinate y parametrizes the extra dimension. The junction conditions across the brane determine y_b and relate it to the brane tension σ: $\sigma = -3\tanh(y_b/\ell)/(4\pi G_5 \ell)$.

[b]For clarity, we mention that, since we are interested in equilibrium configurations, the vacuum state that we consider throughout the paper refers to the Hartle-Hawking vacuum state.

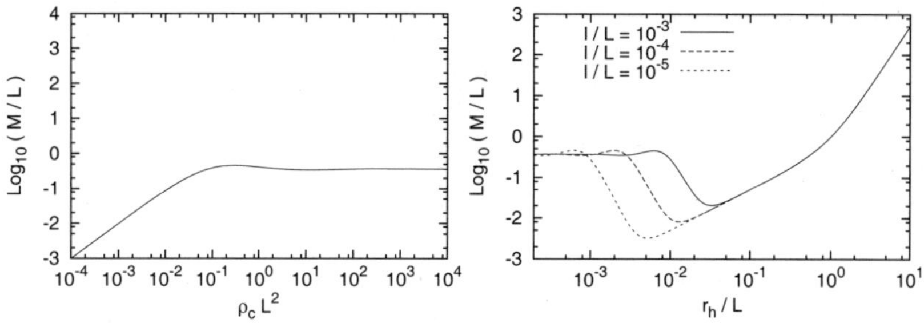

(a) CFT star

(b) Black hole + CFT

Fig. 1. Total mass of CFT stars (left panel) and quantum black holes (right panel) with respect to the central density of the star and black hole horizon radius, respectively. In the right panel we set the parameter $\ell^2/L^2 = 10^{-6}, 10^{-8}10^{-10}$. The transition between the two sequences occurs at $M/L = 0.36$.

holes in the KR model by analyzing the four-dimensional system. Thus, we can interpret the sequence of four-dimensional quantum black holes and CFT stars from a five-dimensional view point as follows. At the transition point of the sequence, the lapse vanishes at the center of the system. This four-dimensional configuration corresponds to a five-dimensional black hole floating in the bulk and just touching the brane, since the lapse vanishes at the touching point for this five-dimensional configuration too. In this way, we may speculate that the sequence of floating black holes corresponds to the sequence of CFT stars, while the sequence of brane-localized black holes corresponds to the sequence of quantum black holes.

In Ref. 1 we have studied the relevant quantum back reaction problem and illustrated explicitly the above picture. Here, for brevity, we will only report the results for the thermodynamical quantities. Figs. (1), (2), and (3) show the total mass, temperature and entropy. The results for the CFT star and for the quantum black holes are shown next to each other, illustrating the smooth transition from one to the other. The two sequences are connected in the limit of infinite central density for the star configuration sequence, and in the limit of zero mass and zero horizon radius for the black hole sequence. A CFT star with large central density 'becomes' a small mass black hole at the connection point. The transition occurs at

$$M/L = 0.36\ ,\quad \ell^{1/2}L^{1/2}T = 0.21\ ,\quad \ell^{-1/2}L^{-3/2}S = 2.0\ ,$$

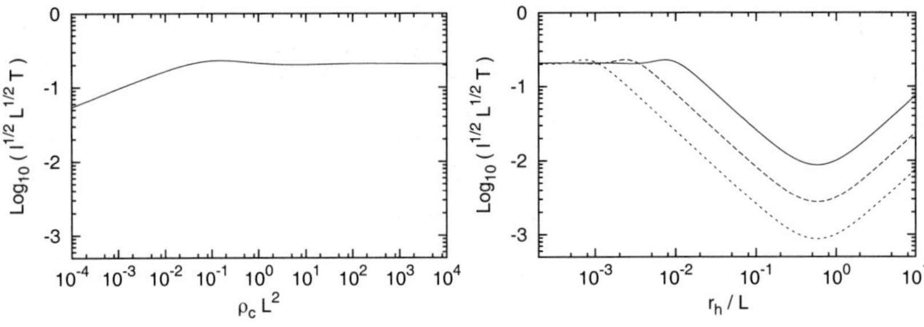

(a) CFT star

(b) Black hole + CFT

Fig. 2. Plots for the temperature in the same way as in Fig. 1. The transition between the two sequences occurs at $\ell^{1/2}L^{1/2}T = 0.21$.

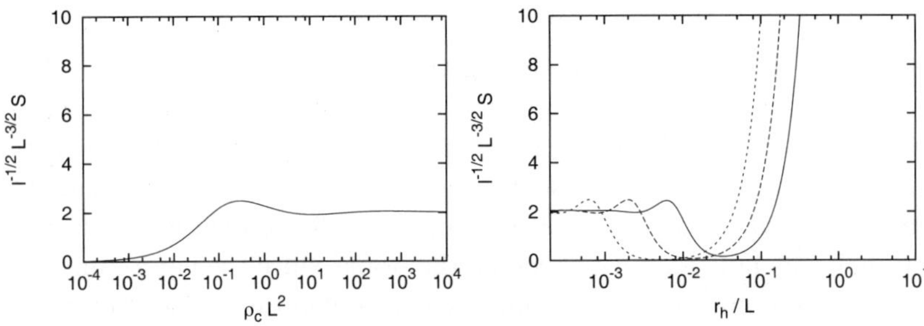

(a) CFT star

(b) Black hole + CFT

Fig. 3. Plots for the temperature in the same way as in Fig. 1. The transition between the two sequences occurs at $\ell^{-1/2}L^{-3/2}S = 2.0$.

and these critical values do not depend on the ratio ℓ/L. The various features of the thermodynamical quantities and the transition point can be analyzed analytically and are in agreement with the numerical results. Fig. (4) shows the relation between M and T. The dotted line refers to the star sequence, while the solid line to the quantum black hole sequence. In order to clarify the back reaction effects, two additional reference curves are also shown in the same figure. The dashed line refers to the purely Schwarzschild AdS black hole case, and the dotted-dashed line refers to the sum of the black hole mass and the energy due to the CFT without taking

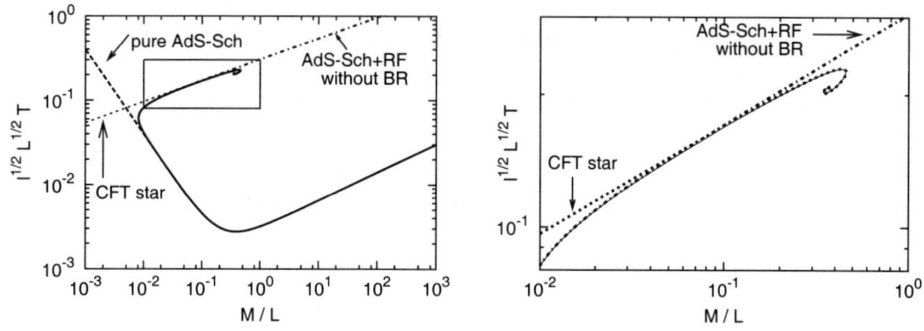

(a) Phase diagram of quantum corrected black hole and CFT star

(b) Closeup of the transition point

Fig. 4. Relation between M and T for CFT stars (dotted line) and quantum black holes (solid line). For the black hole system, we set $l^2/L^2 = 10^{-8}$. To understand the back reaction effect more clearly we add temperature-energy relation for Schwarzschild AdS space with (dotted-dashed line) and without (dashed line) the contribution of the radiation fluids. The right panel shows the closeup around the transition point.

into account the back reaction to the geometry. Once again, the smooth transition between the sequences of CFT stars and quantum black holes is evident. In addition to the thermodynamic functions, we also analyzed how the CFT back reaction alters the geometry. Details can be found in Ref. 1. The AdS/CFT correspondence relates the above sequence of 4D CFT stars and quantum black holes, to a sequence of classical 5D floating and localized black holes in the KR model. For instance, one can estimate the expected size of the classical black hole on the 5D side and the transition points, by relating the entropy to the area of black hole horizon. Here, we spare the details of the discussion that can be found in Ref. 1, and conclude by summarizing the expected phase diagram of black hole solutions in the KR model in Fig. (5).

Acknowledgments

We acknowledge the support of JSPS through Grants Nos. 19540285, 19GS0219, 2056381, 20740133, 21244033 and the support of the Global COE Program "The Next Generation of Physics, Spun from Universality and Emergence" from the Ministry of Education, Culture, Sports, Science and Technology (MEXT).

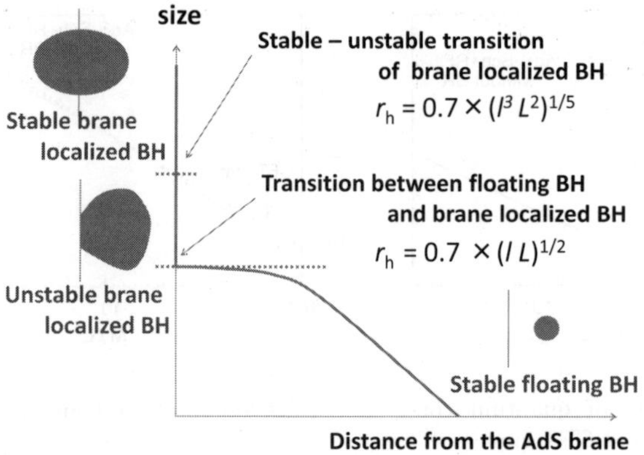

Fig. 5. Phase diagram of BH solutions in the KR model.

References

1. K. Kashiyama *et al.*, '*Quantum back reaction to asymptotically AdS black holes*', arXiv:0910.5376 [gr-qc].
2. L. Randall, R. Sundrum, Phys. Rev. Lett. **83** 4690 (1999).
3. H. Kudoh, T. Tanaka, N. Nakamura, Phys. Rev. D **68** 024035 (2003).
4. S. W. Hawking, T. Hertog, H. S. Reall, Phys. Rev. D **62** 043501 (2000).
5. T. Tanaka, Prog. Theor. Phys. Suppl. **148** 307 (2003).
6. R. Emparan, A. Fabbri, N. Kaloper, J. High Energy Phys. **0208** 043 (2002).
7. A. Karch, L. Randall, Phys. Rev. Lett. **87** 061601 (2001).
8. T. Tanaka, Prog. Theor. Phys. Suppl. **121** 1133 (2009).

TOPOLOGICAL CASIMIR EFFECT IN POWER-LAW FRW COSMOLOGIES

A. L. MKHITARYAN and A. A. SAHARIAN*

Department of Physics, Yerevan State University,
1 Alex Manoogian Street, 0025 Yerevan, Armenia
**E-mail: saharian@ictp.it*

We investigate the vacuum expectation values of the field squared and the energy-momentum tensor for a massless scalar field with general curvature coupling parameter in spatially flat Friedmann-Robertson-Walker universes with an arbitrary number of toroidally compactified dimensions. When the comoving lengths of the compact dimensions are short compared to the Hubble length, the topological parts coincide with those for a conformal coupling. This limit corresponds to the adiabatic approximation. In the opposite limit of large comoving lengths of the compact dimensions, in dependence of the curvature coupling parameter, two regimes are realized with monotonic or oscillatory behavior of the vacuum expectation values.

Keywords: Topological Casimir effect; Friedmann-Robertson-Walker cosmology.

1. Introduction

In the present talk, based on Ref. 1, we consider an exactly soluble problem for the topological Casimir effect on background of Friedmann-Robertson-Walker (FRW) universes with a power-law scale factor. The vacuum polarization and the particle creation in the FRW cosmological models with trivial topology have been considered in a large number of papers (see Refs. 2,3). In particular, the vacuum expectation values of the field squared and the energy-momentum tensor in models with power law scale factors have been discussed in Refs. 4.

In most work on the topological Casimir effect in cosmological backgrounds, the results for the corresponding static counterparts were used replacing the static length scales by comoving lengths in the cosmological bulk. This procedure is valid in conformally invariant situations or under the assumption of a quasi-adiabatic approximation. For non-conformal fields the calculations should be done directly within the framework of quantum field theory on time-dependent backgrounds (for the topological Casimir effect in toroidally compactified de Sitter spacetime see Refs. 5).

The paper is organized as follows. In the next section we evaluate the vacuum expectation value (VEV) of the field squared in spatially flat FRW model with topology $R^p \times (S^1)^q$. In Sec. 3 we consider the VEV of the energy-momentum tensor. The main results are summarized in Sec. 4.

2. VEV of the field squared

We consider a scalar field with curvature coupling parameter ξ evolving on background of the $(D+1)$-dimensional spatially flat FRW spacetime with power law scale factor $a(t) = \alpha t^c$. In addition to the synchronous time coordinate t it is convenient to introduce the conformal time τ in accordance with $t = [\alpha(1-c)\tau]^{1/(1-c)}$. Here we assume that $c \neq 1$. Note that one has $0 \leqslant \tau < \infty$ for $0 < c < 1$ and $-\infty < \tau \leqslant 0$ for $c > 1$.

We will assume that the spatial coordinates z^l, $l = p+1, \ldots, D$, are compactified to S^1: $0 \leqslant z^l \leqslant L_l$, and for the other coordinates we have $-\infty < z^l < +\infty$, $l = 1, \ldots, p$. Hence, we consider the spatial topology $R^p \times (S^1)^q$ with $p + q = D$. Along the compact dimensions we will consider the boundary conditions $\varphi(\tau, \mathbf{z}_p, \mathbf{z}_q + \mathbf{e}_l L_l) = e^{2\pi i \alpha_l} \varphi(\tau, \mathbf{z}_p, \mathbf{z}_q)$ with constant phases α_l, where $\mathbf{z}_p = (z^1, \ldots, z^p)$, $\mathbf{z}_q = (z^{p+1}, \ldots, z^D)$, and $\mathbf{e}_l$, $l = p+1, \ldots, D$, is the unit vector along the direction z^l.

For the VEV of the field squared we have the following decomposition:

$$\langle \varphi^2 \rangle_{p,q} = \langle \varphi^2 \rangle_{\rm FRW} + \langle \varphi^2 \rangle_{p,q}^{(\rm t)}, \; \langle \varphi^2 \rangle_{p,q}^{(\rm t)} = \sum_{j=p}^{D-1} \Delta_{j+1} \langle \varphi^2 \rangle_{j,D-j}, \tag{1}$$

where $\langle \varphi^2 \rangle_{\rm FRW} = \langle \varphi^2 \rangle_{D,0}$ is the VEV in the spatial topology R^D and the part $\langle \varphi^2 \rangle_{p,q}^{(\rm t)}$ is induced by the nontrivial topology. For the topological part induced due to the compactness of the z^{p+1} - direction we have

$$\Delta_{p+1} \langle \varphi^2 \rangle_{p,q} = \frac{4A\eta^{2b}}{(2\pi)^{(p+3)/2} V_{q-1}} \sum_{\mathbf{n}_{q-1} \in \mathbf{Z}^{q-1}} \int_0^\infty dy y \, [I_{-\nu}(y\eta) + I_\nu(y\eta)]$$
$$\times K_\nu(y\eta) \sum_{n=1}^\infty \frac{\cos(2\pi n \alpha_{p+1})}{(nL_{p+1})^{p-1}} f_{(p-1)/2}(nL_{p+1}\sqrt{y^2 + \mathbf{k}_{\mathbf{n}_q - 1}^2}), \tag{2}$$

where $I_\nu(z)$ and $K_\nu(z)$ are the modified Bessel functions, $f_\nu(x) = x^\nu K_\nu(x)$, and $A = \alpha^{1-D}[\alpha|1-c|]^{(D-1)c/(c-1)}$. Here the notations $\eta = |\tau|$, $b = (cD - 1)/[2(c-1)]$, $V_{q-1} = L_{p+2}, \ldots, L_D$ and $\mathbf{n}_{q-1} = (n_{p+2}, \ldots, n_D)$, $\mathbf{k}_{\mathbf{n}_{q-1}}^2 =$

$\sum_{l=p+2}^{D}(2\pi n_l/L_l)^2$, are introduced. The parameter ν is defined as

$$\nu = \frac{1}{2|1-c|}\sqrt{(cD-1)^2 - 4\xi Dc\,[(D+1)c-2]}. \tag{3}$$

In figure Fig. 1, we have plotted the ratio $\langle\varphi^2\rangle^{(\mathrm{t})}_{D-1,1}/\langle\varphi^2\rangle^{(\mathrm{t,c})}_{D-1,1}$ for the special case of topology $R^{D-1}\times S^1$ as a function of L/η, with $L=L_D$ being the length of the compact dimension, for untwisted $D=3$ scalar field ($\alpha_D=0$) and for various values of the parameter ν. Note that the ratio L/η is related to the comoving length of the compact dimension, measured in units of the Hubble length, by $L_l/\eta = (|1-c|/c)L_l^{(\mathrm{c})}/r_{\mathrm{H}}$. Figure 1 clearly shows that the adiabatic approximation for the topological part is valid only for small values of the ratio $L^{(\mathrm{c})}/r_{\mathrm{H}}$.

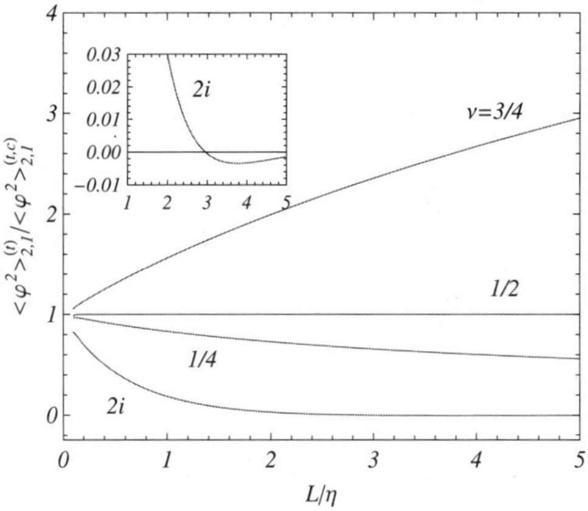

Fig. 1. The ratio $\langle\varphi^2\rangle^{(\mathrm{t})}_{D-1,1}/\langle\varphi^2\rangle^{(\mathrm{t,c})}_{D-1,1}$ versus L/η for different values of the parameter ν (numbers near the curves) and for $D=3$ scalar field with the periodicity condition along the compact dimension.

In the limit $L_l^{(\mathrm{c})} \ll r_{\mathrm{H}}$ the topological part coincides with that for a conformal coupling and behaves like $\langle\varphi^2\rangle^{(\mathrm{t})}_{p,q} \propto t^{c(1-D)}$. This limit corresponds to the adiabatic approximation. In the opposite limit the behavior of the VEV is qualitatively different for real and imaginary values of the parameter ν. For real values, the topological part behaves as $\langle\varphi^2\rangle^{(\mathrm{t})}_{p,q} \propto t^{2(c-1)\nu-cD+1}$. In the limit $L_l^{(\mathrm{c})} \gg r_{\mathrm{H}}$ and for imaginary values ν the asymptotic behavior is oscillatory: $\langle\varphi^2\rangle^{(\mathrm{t})}_{p,q} \propto t^{1-cD}\cos[2|\nu|(c-1)\ln(t/t_0)+\psi]$.

3. VEV of the energy-momentum tensor

For the the VEV of the energy-momentum tensor we have the formula

$$\langle T_i^k \rangle_{p,q} = \langle T_i^k \rangle_{\mathrm{FRW}} + \langle T_i^k \rangle_{p,q}^{(\mathrm{t})}, \ \langle T_i^k \rangle_{p,q}^{(\mathrm{t})} = \sum_{j=p}^{D-1} \Delta_{j+1} \langle T_i^k \rangle_{j,D-j}, \tag{4}$$

where $\langle T_i^k \rangle_{\mathrm{FRW}}$ is the part corresponding to the uncompactified FRW spacetime and $\langle T_i^k \rangle_{p,q}^{(\mathrm{t})}$ is induced by the the nontrivial topology. The first term is well investigated in the literature. Here for the topological part we have (no summation over i)

$$\Delta_{p+1} \langle T_i^i \rangle_{p,q} = \frac{4A\Omega^{-2}}{(2\pi)^{(p+3)/2} V_{q-1}} \sum_{\mathbf{n}_{q-1} \in \mathbf{Z}^{q-1}} \sum_{n=1}^{\infty} \frac{\cos(2\pi n \alpha_{p+1})}{(nL_{p+1})^{p-1}} \int_0^{\infty} dy \, y^{3-2b}$$
$$\times \left[f_{(p-1)/2}(z) F^{(i)}(\eta y) - f_p^{(i)}(z) \frac{\tilde{I}_\nu(\eta y) \tilde{K}_\nu(\eta y)}{(nL_{p+1} y)^2} \right]_{z = nL_{p+1}\sqrt{y^2 + \mathbf{k}_{\mathbf{n}_{q-1}}^2}}, \tag{5}$$

with $\tilde{K}_\nu(z) = z^b K_\nu(z)$, $\tilde{I}_\nu(z) = z^b \left[I_\nu(z) + I_{-\nu}(z) \right]$, and

$$F^{(0)}(z) = \frac{1}{2} \tilde{I}'_\nu \tilde{K}'_\nu + \frac{D\xi c}{z(1-c)} (\tilde{I}_\nu \tilde{K}_\nu)' - \frac{1}{2} \left[1 - \frac{\xi D(D-1)c^2}{z^2(1-c)^2} \right] \tilde{I}_\nu \tilde{K}_\nu,$$
$$F^{(l)}(z) = 2\left(\xi - \frac{1}{4}\right) \tilde{I}'_\nu \tilde{K}'_\nu - \frac{1}{z} \frac{c\xi}{1-c} (\tilde{I}_\nu \tilde{K}_\nu)'$$
$$+2\left[\xi - \frac{1}{4} - Dc\xi \frac{(Dc-2)(\xi - \xi_D) + \xi c}{z^2(1-c)^2} \right] \tilde{I}_\nu \tilde{K}_\nu, \tag{6}$$

where $l = 1, \ldots, D$. In Eq. (5) we have used the notations

$$f_p^{(0)}(z) = 0, \ f_p^{(i)}(z) = f_{(p+1)/2}(z), \ f_p^{(l)}(z) = (nL_{p+1} k_l)^2 f_{(p-1)/2}(z),$$
$$f_p^{(p+1)}(z) = -p f_{(p+1)/2}(z) - z^2 f_{(p-1)/2}(z), \tag{7}$$

where $i = 1, 2, \ldots p$ and $l = p+2, \ldots D$.

For $L_l^{(\mathrm{c})} \ll r_{\mathrm{H}}$ the topological part behaves like $\langle T_i^k \rangle_{p,q}^{(\mathrm{t})} \propto t^{-c(D+1)}$ and to the leading order the stresses along the uncompactified dimensions are equal to the vacuum energy density. In the limit $L_l^{(\mathrm{c})} \gg r_{\mathrm{H}}$ and for real values ν the asymptotic has the form $\langle T_i^k \rangle_{p,q}^{(\mathrm{t})} \propto t^{2(c-1)\nu - cD-1}$. The corresponding vacuum stresses are isotropic and the equation of state for the topological parts in the vacuum energy density and pressures is of the barotropic type. For $L_l^{(\mathrm{c})} \gg r_{\mathrm{H}}$ and for imaginary values ν we have the asymptotic behavior $\langle T_i^k \rangle_{p,q}^{(\mathrm{t})} \propto t^{-cD-1} \cos[2|\nu|(c-1)\ln(t/t_0) + \psi_i]$.

4. Conclusion

We have investigated one-loop quantum effects for a scalar field with general curvature coupling, induced by toroidal compactification of spatial dimensions in spatially flat FRW cosmological models with power law scale factor. General boundary conditions with arbitrary phases are considered along compact dimensions. The boundary conditions imposed on possible field configurations change the spectrum of vacuum fluctuations. Among the most important characteristics of the vacuum state are the expectation values of the field squared and the energy-momentum tensor. Though the corresponding operators are local, due to the global nature of the vacuum these VEVs carry an important information on the global structure of the background spacetime. We present the VEVs as the sum of the function for topologically trivial FRW model and the topological part. The latter is finite in the coincidence limit and in this way the renormalization of the VEVs is reduced to that for the FRW universe with trivial topology. The topological parts are given by formulae (1) and (2) for the field squared and by formulae (4), (5) for the energy density and the stresses. A.L.M. gratefully acknowledges the organizers of the conference QFEXT09 for the opportunity to present this paper.

References

1. A. A. Saharian and A. L. Mkhitaryan, arXiv:0908.3291.
2. N. D. Birrel and P. C. W. Davies, *Quantum Fields in Curved Space* (Cambridge University Press, Cambridge, 1982); A. A. Grib, S. G. Mamayev, and V. M. Mostepanenko, *Vacuum Quantum Effects in Strong Fields* (Friedmann Laboratory Publishing, St. Petersburg, 1994).
3. V. M. Mostepanenko and N.N. Trunov, *The Casimir Effect and Its Applications* (Clarendon, Oxford, 1997); K. A. Milton, *The Casimir Effect: Physical Manifestation of Zero-Point Energy* (World Scientific, Singapore, 2002); M. Bordag, G. L. Klimchitskaya, U. Mohideen and V. M. Mostepanenko, *Advances in the Casimir Effect* (Oxford University Press, Oxford, 2009).
4. M. Bordag, J. Lindig, V. M. Mostepanenko, and Yu. V. Pavlov, *Int. J. Mod. Phys. D* **6**, 449 (1997); M. Bordag, J. Lindig, and V. M. Mostepanenko, *Class. Quantum Grav.* **15**, 581 (1998).
5. A. A. Saharian and M. R. Setare, Phys. Lett. B **659**, 367 (2008); S. Bellucci and A. A. Saharian, Phys. Rev. D **77**, 124010 (2008); A. A. Saharian, Class. Quantum Grav. **25**, 165012 (2008); E. R. Bezerra de Mello and A. A. Saharian, JHEP **12**, 081 (2008).

THE TRACE ANOMALY AND DYNAMICAL VACUUM ENERGY IN COSMOLOGY

EMIL MOTTOLA

Theoretical Division, Los Alamos National Laboratory, Los Alamos, NM 87545, USA
and
Theoretical Physics Group, PH-TH, CERN, CH-1211, Geneva 23, Switzerland
E-mail: emil@lanl.gov and emil.mottola@cern.ch

The trace anomaly of conformal matter implies the existence of massless scalar poles in physical amplitudes involving the stress-energy tensor. These poles may be described by a local effective action with massless scalar fields, which couple to classical sources, contribute to gravitational scattering processes, and can have long range gravitational effects at macroscopic scales. In an effective field theory approach, the effective action of the anomaly is an infrared relevant term that should be added to the Einstein-Hilbert action of classical General Relativity to take account of macroscopic quantum effects. The additional scalar degrees of freedom contained in this effective action may be understood as responsible for both the Casimir effect in flat spacetime and large quantum backreaction effects at the horizon scale of cosmological spacetimes. These effects of the trace anomaly imply that the cosmological vacuum energy is dynamical, and its value depends on macroscopic boundary conditions at the cosmological horizon scale, rather than sensitivity to the extreme ultraviolet Planck scale.

Keywords: Vacuum Energy, Trace Anomaly, Cosmological Constant, Dark Energy, Casimir Effect.

1. The Cosmological Constant and Vacuum Energy

In classical General Relativity, the requirement that the field eqs. involve no more than two derivatives of the metric tensor allows for the possible addition of a constant term, the cosmological term Λ, to Einstein's eqs.

$$R^{\mu}{}_{\nu} - \frac{R}{2}\,\delta^{\mu}{}_{\nu} + \Lambda\,\delta^{\mu}{}_{\nu} = \frac{8\pi G}{c^4}\,T^{\mu}{}_{\nu}. \tag{1}$$

If transposed to the right side of this relation, the Λ term corresponds to a constant energy density $\rho_\Lambda = c^4\Lambda/8\pi G$ and isotropic pressure $p_\Lambda =$

$-c^4\Lambda/8\pi G$ permeating all of space uniformly, and independently of any localized matter sources. Hence, even if the matter stress tensor $T_{\mu}^{\ \nu} = 0$, a cosmological term causes spacetime to become curved with a radius of curvature of order $|\Lambda|^{-\frac{1}{2}}$.

In purely classical physics there is no natural scale for Λ. Indeed if $\hbar = 0$ and $\Lambda = 0$, there is no fixed length scale at all in the vacuum Einstein equations, G/c^4 being simply a conversion factor between the units of energy and those of length. Hence Λ may take on any value whatsoever with no difficulty (and with no explanation) in classical General Relativity.

As soon as we allow $\hbar \neq 0$, there is a quantity with the dimensions of length that can be formed from $\hbar, G$, and c, namely the Planck length

$$L_{pl} \equiv \left(\frac{\hbar G}{c^3}\right)^{\frac{1}{2}} = 1.616 \times 10^{-33}\,\text{cm}. \tag{2}$$

Hence when quantum theory is considered in a general relativistic setting, the quantity,

$$\lambda \equiv \Lambda L_{pl}^2 = \frac{\hbar G \Lambda}{c^3} \tag{3}$$

becomes a dimensionless pure number, whose value one might expect a theory of gravity incorporating quantum effects to address.

Some eighty years ago W. Pauli was apparently the first to consider the question of the effects of quantum vacuum fluctuations on the the curvature of space.[1] Pauli recognized that the sum of zero point energies of the two transverse electromagnetic field modes *in vacuo*

$$\rho_\Lambda = 2\int^{L_{min}^{-1}} \frac{d^3\vec{k}}{(2\pi)^3}\frac{\hbar\omega_k}{2} = \frac{1}{8\pi^2}\frac{\hbar c}{L_{min}^4} = -p_\Lambda \tag{4}$$

contribute to the stress-energy tensor of Einstein's theory as would an effective cosmological term $\Lambda > 0$. Since the integral (4) is quartically divergent, an ultraviolet cutoff L_{min}^{-1} of (4) at large $|\vec{k}|$ is needed. Taking this short distance cutoff L_{min} to be of the order of the classical electron radius e^2/mc^2, Pauli concluded that if his estimate were correct, Einstein's theory with this large a Λ would lead to a universe so curved that its total size "could not even reach to the moon." If instead of the classical electron radius, the apparently natural but much shorter length scale of $L_{min} \sim L_{pl}$ is used to cut off the frequency sum in (4), then the estimate for the cosmological term in Einstein's equations becomes vastly larger, and the entire universe would be limited in size to the microscopic scale of L_{pl} (2) itself, in even more striking disagreement with observation.

Clearly the naive estimate of the contribution of short distance modes of the electromagnetic field to the curvature of space, by using (4) as a source for Einstein's eqs. (1) is not correct. The question is why. Here the Casimir effect may have something to teach us. The vacuum zero point fluctuations being considered in (4) are the same ones that contribute to the Casimir effect, but this estimate of the scale of vacuum zero point energy, quartically dependent on a short distance cutoff L_{min}, is certainly *not* relevant for the effect observed in the laboratory.[2] In calculations of the Casimir force between conductors, one subtracts the zero point energy of the electromagnetic field in an infinitely extended vacuum (with the conductors absent) from the modified zero point energies in the presence of the conductors. It is this *subtracted* zero point energy of the electromagnetic vacuum, depending upon the *boundary conditions* imposed by the conducting surfaces, which leads to experimentally well verified results for the force between the conductors. In this renormalization procedure the ultraviolet cutoff L_{min}^{-1} drops out, and the distance scale of quantum fluctuations that determine the magnitude of the Casimir effect is not the microscopic classical electron radius, as in Pauli's original estimate, much less the even more microscopic Planck length L_{pl}, but rather the relatively *macroscopic* distance d between the conducting boundary surfaces. The resulting subtracted energy density of the vacuum between the conductors is

$$\rho_v = -\frac{\pi^2}{720}\frac{\hbar c}{d^4}\,. \tag{5}$$

This energy density is of the opposite sign as (4), leading to an attractive force per unit area between the plates of 0.013 dyne/cm^2 $(\mu m/d)^4$, a value which is both independent of the ultraviolet cutoff L_{min}^{-1}, and the microscopic details of the atomic constituents of the conductors. This is a clear indication, confirmed by experiment, that the *measurable* effects associated with vacuum fluctuations are *infrared* phenomena, dependent upon macroscopic boundary conditions, which have little or nothing to do with the extreme ultraviolet modes or cutoff of the integral in (4).

By the Principle of Equivalence, local short distance behavior in a mildly curved spacetime is essentially equivalent to that in flat spacetime. Hence on physical grounds we should not expect the ultraviolet cutoff dependence of (4) to affect the universe in the large any more than it affects the force between metallic conductors in the laboratory.

In the case of the Casimir effect a constant zero point energy of the vacuum, no matter how large, does not affect the force between the plates. In the case of cosmology it is usually taken for granted that any effects of boundary conditions can be neglected. It is not obvious then what should play the role of the conducting plates in determining the magnitude of ρ_v in the universe, and the magnitude of any effect of quantum zero point energy on the curvature of space has remained unclear from Pauli's original estimate down to the present. In recent years this has evolved from a question of fundamental importance in theoretical physics to a central one of observational cosmology as well. Observations of type Ia supernovae at moderately large redshifts ($z \sim 0.5$ to 1) have led to the conclusion that the Hubble expansion of the universe is *accelerating*.[3] This is consistent also with microwave background measurements.[4] According to Einstein's equations accelerated expansion is possible if and only if the energy density and pressure of the dominant component of the universe satisfies the inequality,

$$\rho + 3p = \rho\,(1 + 3w) < 0\,. \tag{6}$$

A vacuum energy with $\rho > 0$ and $w \equiv p_v/\rho_v = -1$ leads to an accelerated expansion, a kind of "repulsive" gravity in which the relativistic effects of a negative pressure can overcome a positive energy density in (6). Taken at face value, the observations imply that some 74% of the energy in the universe is of this hitherto undetected $w = -1$ dark variety.[3,4] This leads to a non-zero inferred cosmological term in Einstein's equations of

$$\Lambda_{\rm meas} \simeq (0.74)\,\frac{3H_0^2}{c^2} \simeq 1.4 \times 10^{-56}\ {\rm cm}^{-2} \simeq 3.6 \times 10^{-122}\ \frac{c^3}{\hbar G}\,. \tag{7}$$

Here H_0 is the present value of the Hubble parameter, approximately $73\,{\rm km/sec/Mpc} \simeq 2.4\times 10^{-18}\,{\rm sec}^{-1}$. Thus the value of the cosmological dark energy inferred from the SN Ia data in terms of Planck units, $L_{\rm pl}^{-2} = \frac{c^3}{\hbar G}$, gives the dimensionless number in (3) the extremely small but finite value,

$$\lambda \simeq 3.6 \times 10^{-122}\,. \tag{8}$$

Explaining the value of this smallest number in all of physics is the basic form of the *cosmological constant problem.*

As we have already noted, if the universe were purely classical, L_{pl} would vanish and Λ, like the overall size or total age of the universe, could take on any value whatsoever without any technical problem of naturalness. Likewise as the Casimir effect makes clear, if $G = 0$ and there are also no boundary effects to be concerned with, then the cutoff dependent zero point energy of flat space (4) could simply be subtracted, with no observable consequences. A naturalness problem arises only when the effects of quantum zero point energy on the large scale curvature of spacetime are considered. This is a problem of the gravitational energy of the quantum vacuum or ground state of the system at *macroscopic* distance scales, very much greater than L_{pl}, when both $\hbar \neq 0$ and $G \neq 0$.

2. Effective Field Theory and Anomalies

The treatment of quantum effects at distances much larger than any ultraviolet cutoff is precisely the context in which effective field theory (EFT) techniques should be applicable. This means that we assume that we do not need to know every detail of physics at extremely short distance scales of 10^{-33} cm or even 10^{-13} cm in order to discuss cosmology at 10^{28} cm scales. In EFT one assumes some organizing principle or symmetry of low energy dynamics, expresses degrees of freedom in terms of local fields having well-defined covariant transformation properties under the symmetry, and expands the effective action in local invariants of increasing number of derivatives of the fields. The dimensionful parameters multiplying the terms in the action determine the scale at which the derivative expansion is expected to break down.

The organizing principle in gravity is the Principle of Equivalence, *i.e.* invariance under general coordinate transformations, which greatly restricts the form of any EFT of gravity. In his search for field equations (1) for a metric theory with universal matter couplings, which incorporates the Equivalence Principle automatically, but which is no higher than second order in derivatives of the metric, Einstein was using what we would now recognize as EFT reasoning. In an EFT quantum effects and any ultraviolet (UV) divergences they generate at very short distance scales are absorbed into a few, finite low energy effective parameters, such as G and Λ.

In extending Einstein's classical theory to take account of the quantum properties of matter, the classical stress-energy tensor of matter $T_{\mu}^{\ \nu}$ must be replaced by a quantum expectation value $\langle T_{\mu}^{\ \nu}\rangle$. In this *semi-classical* theory with both $\hbar$ and G different from zero, the quantum zero-point and vacuum energy effects first appear, while the spacetime geometry can still be treated classically. This is clearly an approximation to a more exact treatment, which can be formally justified by taking the number N of matter degrees of freedom to infinity.

Since the expectation value $\langle T_{\mu}^{\ \nu}\rangle$ suffers from the quartic divergence (4), a regularization and renormalization procedure is necessary in order to define the semi-classical EFT. General coordinate invariance requires a more careful renormalization procedure than the simple subtraction of (4) which suffices for the original Casimir calculations in flat space. The UV divergent terms of the stress tensor contain subleading quadratic and logarithmic dependence upon the cutoff L_{min} which must be isolated and removed in a way consistent with the Equivalence Principle to extract physical effects correctly. These more general renormalization procedures, involving *e.g.* proper time, covariant point splitting or dimensional regularization have been developed in the context of quantum field theory in curved spacetime.[5] The non-renormalizability of the classical Einstein theory poses no particular obstacle for this semi-classical EFT approach. It requires only that certain additional terms be added to the effective action to take account of UV divergences which are not of the form of a renormalization of G or Λ. One such set of terms that arise from a consistent covariant renormalization scheme are those associated with the trace anomaly of $\langle T_{\mu}{}^{\nu}\rangle$ whose effects we discuss in succeeding sections. The result of the renormalization program for quantum fields and their vacuum energy in curved space is that General Relativity can be viewed as a low energy quantum EFT of gravity, provided that the classical Einstein-Hilbert classical action is augmented by these additional terms required when $\hbar \neq 0$.

The essential physical assumption in any EFT approach is the hypothesis of *decoupling*, namely that low energy physics is independent of very short distance degrees of freedom and the details of their interactions. All of the effects of these short distance degrees of freedom is subsumed into a few phenomenological coefficients of the infrared relevant terms of the EFT. Notice that this will not be the case if the low energy Λ relevant for dark energy and cosmology depends upon the quantum zero point energies of all fields up to some UV cutoff, as in (4). Taken seriously this would indicate *quartic* power sensitivity of extreme infrared physics to the ultraviolet cutoff. In addition to violating any intuitive notion of decoupling, this is clearly not how the Casimir effect works. The hierarchy between the scale of electroweak symmetry breaking in the Standard Model compared to the Planck scale also suggests that low energy physics does not have even quadratic power law sensitivity to the extreme UV cutoff scale L_{Pl}.

Power law sensitivity to ultraviolet cutoffs may well be an artificial problem of a poor regularization technique, since for example it does not occur in dimensional regularization. On the other hand *logarithmic* scale sensitivity is the basis of renormalization group analyses, and in the case of the Standard Model has been verified experimentally.[6] The argument of a logarithm necessarily involves the ratio of a UV scale to an IR scale. This distinguishes logarithms from simple additive UV contributions to dimensionful quantities such as the Higgs mass, or the cosomological vacuum energy in (4). One of these logarithmic corrections to classical gravity arises from the conformal or trace anomaly.[5,7]

Anomalies violate strict decoupling of UV from IR degrees of freedom in the sense that the effective action that describes them is necessarily non-local in terms of the original local field degrees of freedom. Because it is non-local in terms of the original fields, the usual EFT approach of expanding in local invariants with higher numbers of derivatives of those fields will miss the anomaly. Instead the non-local effective action of the anomaly must be added explicitly to the local EFT action. Alternatively, an anomaly generally implies massless poles signifying additional massless degrees of freedom which do not decouple, and these new degrees of freedom need to be added to the action to complete the low energy effective theory.

3. The Axial Anomaly and Its Massless Pole

The best known example of a quantum anomaly is the chiral or axial anomaly of QED,[8] later extended to QCD. Despite this there are some features of the QED axial anomaly that remain somewhat under appreciated, which are directly relevant to the gravity case. Therefore it is well to review the IR features of the QED axial anomaly before proceeding to gravitational applications.

In QED the Dirac equation

$$-i\gamma^\mu(\partial_\mu - ieA_\mu)\psi + m\psi = 0\,, \tag{9}$$

implies that the vector current $J^\mu = \bar{\psi}\gamma^\mu\psi$ is conserved:

$$\partial_\mu J^\mu = 0\,. \tag{10}$$

The axial current $J_5^\mu = \bar{\psi}\gamma^\mu\gamma^5\psi$ (with $\gamma^5 \equiv i\gamma^0\gamma^1\gamma^2\gamma^3$) apparently obeys

$$\partial_\mu J_5^\mu = 2im\,\bar{\psi}\gamma^5\psi \qquad \text{(classically)}. \tag{11}$$

In the limit of vanishing fermion mass $m \to 0$, the classical Lagrangian has a $U_{ch}(1)$ global symmetry under $\psi \to e^{i\alpha\gamma^5}\psi$, in addition to $U(1)$ local gauge invariance, and J_5^μ is the Noether current corresponding to this chiral symmetry. As is well known, both symmetries cannot be maintained simultaneously at the quantum level.[8] Let us denote by $\langle J_5^\mu(z)\rangle_A$ the expectation value of the chiral current in the presence of a background electromagnetic potential A_μ. Enforcing $U(1)$ gauge invariance (10) on the full quantum theory leads necessarily to a finite axial current anomaly,

$$\partial_\mu \langle J_5^\mu\rangle_A\Big|_{m=0} = \frac{e^2}{16\pi^2}\,\epsilon^{\mu\nu\rho\sigma}F_{\mu\nu}F_{\rho\sigma} = \frac{e^2}{2\pi^2}\,\vec{E}\cdot\vec{B}\,, \tag{12}$$

in an external electromagnetic field. Varying this expression twice with respect to the external A field and Fourier transforming, we see that the anomaly must appear in the amplitude,

$$\Gamma^{\mu\alpha\beta}(p,q) \equiv -i \int d^4x \int d^4y\, e^{ip\cdot x+iq\cdot y}\, \frac{\delta^2 \langle J_5^\mu(0)\rangle_A}{\delta A_\alpha(x)\delta A_\beta(y)}\Bigg|_{A=0}$$
$$= ie^2 \int d^4x \int d^4y\, e^{ip\cdot x+iq\cdot y}\, \langle T J_5^\mu(0) J^\alpha(x) J^\beta(y)\rangle\big|_{A=0}\,. \qquad (13)$$

At the lowest one-loop order it is given by a triangle diagram with the axial current J_5^μ at one vertex with four-momentum k^μ, and the vector currents J^α and J^β at the other two vertices, coupling to photons with four-momenta p^μ and q^μ respectively. Momentum conservation requires $k^\mu = p^\mu + q^\mu$.

The anomaly may be regarded as a clash of symmetries. Either but not both of the two classically valid relations (10) and (11) can be maintained at the quantum level. There are various ways to see this, but intuitively, defining the quantum amplitude (13) at one-loop order introduces implicitly or explicitly an additional scale into the amplitude, which violates naive identities dependent upon global scale and chiral invariance. Although this is usually presented as a UV scale, necessary to regulate the triangle amplitude, in fact it is not difficult to show that the axial anomaly (12) is determined by its UV *finite*, *i.e.* cutoff independent parts, together with Lorentz invariance, vector current conservation (10), and Bose symmetry of exchanging the two external photon lines.[9,10] Thus, the anomaly can be equally well thought of as due to the logarithmic sensitivity of the amplitude (13) to the *infrared* cutoff at large distances when $m = 0$ and $k = p+q = 0$.

Consistent with this infrared interpretation of the axial anomaly, one can demonstrate explicitly the existence of a massless pole in the amplitude (13), and in the matrix element of the axial current between the vacuum and the physical two-photon state, $|p,q\rangle$, giving[10]

$$\langle 0|J_5^\mu(0)|p,q\rangle = i\Gamma^{\mu\alpha\beta}(p,q)\tilde{A}_\alpha(p)\tilde{A}_\beta(q)\big|_{m^2=p^2=q^2=0}$$
$$= \frac{ie^2}{2\pi^2 k^2}\, k^\mu \epsilon^{\alpha\beta\rho\sigma} p_\rho q_\sigma \tilde{A}_\alpha(p)\tilde{A}_\beta(q)\,, \qquad (14)$$

When contracted with k_μ the pole cancels and we recover (12) in momentum space. The existence of the pole at $k^2 = 0$ in a physical amplitude (14) implies the existence of a propagating massless pseudoscalar 0^- intermediate state, with low energy long range effects. The discontinuity of the amplitude contains a δ function in the corresponding spectral representation, the signal of a massless intermediate state.[10] In fact, because of the off-diagonal nature of the pole appearing in the two-particle intermediate state between J_5^μ and $J^\alpha J^\beta$, there are actually *two* massless degrees of freedom in the EFT description of the low energy amplitude, one with positive norm, the other with negative norm. These new states appear in the two-particle correlations of massless fermions moving collinearly at the speed of light, and are a genuine quantum effect. The effective action for the anomaly is non-local in terms of the original fields but can be rendered local by the introduction of two new massless pseudoscalar fields whose propagator gives rise to the $1/k^2$ pole in (14). The negative norm state cannot be present in the S matrix and indicates instead some non-trivial infrared vacuum structure for zero mass fermions.

In real QED these infrared effects are suppressed by the non-zero physical electron mass $m > 0$, and the additional fact that macroscopic chirality violating sources for J_5^μ which would be sensitive to the anomaly are difficult to create. In QCD the propagating pseudoscalar state of positive norm in the low energy EFT is identified with the π^0, whose decay to two photons, $\pi^0 \to 2\gamma$ is correctly given by the triangle amplitude.[11] In fact, it was the experimental agreement between the measured decay rate to that predicted by the axial anomaly computed in the UV theory of 3 colors of fractionally charged quarks that gave one of the strongest early confirmations of QCD. It is the fact that the anomaly may be computed in the UV theory but gives rise to a low energy amplitude, $\pi^0 \to 2\gamma$ that led to the principle of anomaly *matching*.[12] The lesson to be taken away from this example is that anomalies are a unique window which the low energy EFT provides to short distance physics. As such their long distance effects must be taken into account by explicitly adding the IR relevant terms they induce in the low energy effective action.[11]

4. The Trace Anomaly in Two Dimensions

Consider next gravity in $D = 2$ dimensions. The local action,

$$S_{cl}[g] = \int d^2x \sqrt{-g}\,(\gamma R - 2\Lambda) \tag{15}$$

actually contains *no* local degrees of freedom at all, since all metrics in $D = 2$ are locally conformally flat, and hence may be expressed in the form

$$g_{\mu\nu} = \exp(2\sigma)\,\bar{g}_{\mu\nu} \tag{16}$$

for some $\sigma(x)$ and a fixed metric $\bar{g}_{\mu\nu}$ which may be taken to be flat. Owing to the identity,

$$R\sqrt{-g} = \bar{R}\sqrt{-\bar{g}} - 2\sqrt{-\bar{g}}\,\bar{\Box}\sigma\,, \qquad (D = 2) \tag{17}$$

the σ dependence of the Einstein-Hilbert term in (15) is a total derivative and gives no metric variation. Hence the theory described by the local action (15) has no local dynamical degrees of freedom at all.

When massless conformal matter is coupled to the geometry, this situation changes due to the conformal trace anomaly,[5]

$$\langle T^{\mu}_{\ \mu}\rangle = \frac{N}{24\pi} R\,, \qquad (D=2) \tag{18}$$

where $N = N_S + N_F$ is the total number of massless fields, either scalar (N_S) or fermionic (N_F). It is not difficult to show that the amplitude $\int d^2x\, e^{ik\cdot x}\, \langle T^{\mu\nu}(x) T^{\alpha\beta}(0)\rangle$ at one-loop order develops a pole at $k^2 = 0$ when the mass of the quantum field vanishes, whose residue is just proportional to the coefficient of the anomaly (18).[13] Accordingly no local coordinate invariant action exists whose metric variation leads to (18).

A non-local action corresponding to (18) can easily be found by using the relation (17) in (18), which implies that the conformal variation, $\delta\Gamma/\delta\sigma = \sqrt{-g}\,\langle T_{\mu}^{\ \mu}\rangle$ of the effective action Γ reproducing the anomaly is linear in σ. Hence this Wess-Zumino effective action[14] in two dimensions is

$$\Gamma^{(2)}_{WZ}[\bar{g};\sigma] = \frac{N}{24\pi}\int d^2x\,\sqrt{-\bar{g}}\left(-\sigma\bar{\Box}\sigma + \bar{R}\,\sigma\right)\,. \tag{19}$$

By solving (17) for σ it is now straightforward to find a non-local scalar functional $S_{anom}[g]$ of the full metric in (16) such that

$$\Gamma^{(2)}_{WZ}[\bar{g};\sigma] = S^{(2)}_{anom}[g = e^{2\sigma}\bar{g}] - S^{(2)}_{anom}[\bar{g}]\,, \tag{20}$$

upon also using the fact that $\sqrt{-g}\Box = \sqrt{-\bar{g}}\bar{\Box}$ is conformally invariant in two dimensions. In this way we find

$$S^{(2)}_{anom}[g] = -\frac{N}{96\pi}\int d^2x\,\sqrt{-g}\int d^2x'\,\sqrt{-g'}\,R(x)\,\Box^{-1}(x,x')\,R(x')\,, \tag{21}$$

with $\Box^{-1}(x,x')$ denoting the Green's function inverse of the scalar differential operator $\Box$. The parameter N is replaced by $N-25$ if account is taken of the contribution to the anomaly coefficient of the metric fluctuations themselves in addition to those of the N matter fields.

The anomalous effective action (21) is a scalar under coordinate transformations and therefore fully covariant and geometric in character, as required by the Equivalence Principle. However since it involves the Green's function $\Box^{-1}(x,x')$, which requires boundary conditions for its unique specification, it is quite non-local, and dependent upon more than just the local curvature invariants of spacetime. In this important respect it is quite different from the classical action (15), and describes rather different physics. In order to expose that physics it is most convenient to recast the non-local and non-single valued functional of the metric, $S^{(2)}_{anom}$ into a local form by introducing a scalar auxiliary field φ satisfying

$$-\Box\varphi = R\,. \tag{22}$$

Then one may check that varying

$$S^{(2)}_{anom}[g;\varphi] \equiv -\frac{N}{96\pi}\int d^2x\,\sqrt{-g}\,\left(g^{\mu\nu}\,\nabla_\mu\varphi\,\nabla_\nu\varphi - 2R\,\varphi\right) \tag{23}$$

with respect to φ gives the eq. of motion (22) for the auxiliary field, which when solved formally by $\varphi = -\,\Box^{-1}R$ and substituted back into $S^{(2)}_{anom}[g;\varphi]$ returns the non-local form of the anomalous action (21), up to a surface term. The non-local information in addition to the local geometry which was previously contained in the specification of the Green's function $\Box^{-1}(x,x')$ now resides in the local auxiliary field $\varphi(x)$, and the freedom to add to it homogeneous solutions of (22).

In the local form (23), we see that a new local scalar degree of freedom has appeared in the form of the auxiliary field φ, which was not present in the original classical action (15). This field is associated with the scalar conformal deformations of the metric, which now fluctuates freely thanks to the kinetic term in (23), whereas the metric was constrained in (15). The massless pole in the intermediate state of $\int d^2x\, e^{ik\cdot x}\,\langle T^{\mu\nu}(x)T^{\alpha\beta}(0)\rangle$ is exactly the φ propagator. Its fluctuations lead to the gravitational "dressing" of the critical exponents of conformal matter in a gravitational background.[15]

Since critical exponents are characteristic of long range fluctuations in a second order phase transition, this shows that the effective action of the anomaly (23) is definitely a relevant operator in the infrared. The anomalous action is also responsible for the infrared *running* of Λ. In other words, the bare parameter of the classical action is renormalized by the quantum fluctuations of the φ field, and becomes scale dependent. For all of these reasons the action (23) and the additional scalar degree of freedom contained in it must be added to the classical action (15), to get a complete low energy EFT of two-dimensional gravity.

5. Massless Scalar Degrees of Freedom in 4*D* Gravity

With the examples of the QED/QCD axial anomaly and the conformal anomaly in two dimensions, we consider finally the conformal or trace anomaly in $D = 4$ dimensions,

$$\langle T^\mu_{\ \mu}\rangle = bF + b'\left(E - \frac{2}{3}\Box R\right) + b''\Box R + \sum_i c_i H_i\,, \tag{24}$$

where

$$E \equiv {}^*\!R_{\mu\nu\alpha\beta}\, {}^*\!R^{\mu\nu\alpha\beta} = R_{\mu\nu\alpha\beta}R^{\mu\nu\alpha\beta} - 4R_{\mu\nu}R^{\mu\nu} + R^2\,, \tag{25a}$$

$$F \equiv C_{\mu\nu\alpha\beta}C^{\mu\nu\alpha\beta} = R_{\mu\nu\alpha\beta}R^{\mu\nu\alpha\beta} - 2R_{\mu\nu}R^{\mu\nu} + \frac{R^2}{3}\,. \tag{25b}$$

and H_i denotes any number of dimension 4 conformally invariant scalars constructed from the gauge fields externally coupled to the matter in question. For example in QED in *flat* space with $E = F = 0$, with massless fermions coupled to electromagnetism, $H = F_{\mu\nu}F^{\mu\nu}$ and one finds

$$\langle T^\mu_{\ \mu}\rangle_A\Big|_{m=0,flat} = c\,H = -\frac{e^2}{24\pi^2}\,F_{\mu\nu}F^{\mu\nu}\,, \tag{26}$$

in complete analogy with (12). The coefficients b and b' in (24), like the coefficient in (12) do not depend on any ultraviolet short distance cutoff, but instead are determined only by the number and spin of massless fields,[5,7]

$$b = \frac{\hbar}{120(4\pi)^2}\,(N_S + 6N_F + 12N_V)\,, \tag{27a}$$

$$b' = -\frac{\hbar}{360(4\pi)^2}\,(N_S + \frac{11}{2}N_F + 62N_V)\,, \tag{27b}$$

with (N_S, N_F, N_V) the number of fields of spin $(0, \frac{1}{2}, 1)$ respectively.

For the flat space trace anomaly (26) one can perform a full analysis of the one-loop triangle contribution to the amplitude,

$$\Gamma^{\mu\nu\alpha\beta}(p,q) = \int d^4x \int d^4y\; e^{ip\cdot x + iq\cdot y}\; = \frac{\delta^2\langle T^{\mu\nu}(0)\rangle_A}{\delta A_\alpha(x)\delta A_\beta(y)}\Big|_{A=0}\,, \tag{28}$$

analogous to (13). This triangle amplitude also develops a pole at $k^2 = 0$ when the combined limits $p^2 = q^2 = m^2 = 0$ are taken. The 0^+ scalar pole survives in matrix elements of the stress-energy tensor of massless fermions to physical two-photon states, analogous to (14). The residue of the pole is proportional to the coefficient of the anomaly. The discontinuity of the $\langle TJJ\rangle$ triangle diagram also exhibits a δ function in the intermediate two-electron state with total spin zero, which signifies that a new massless scalar degree of freedom with gravitational coupling is required by the anomaly.[10]

The effective action of the massless degree(s) of freedom is again non-local in terms of the original local metric and electromagnetic field strengths. However it can be rewritten in a completely local form by the introduction of the *two* local scalar fields whose quanta are responsible for the massless poles in anomalous amplitudes such as the $\langle TJJ\rangle$ and $\langle TTT\rangle$ triangle diagrams. By following steps analogous to those in the two dimensional case of the previous section, explicitly integrating the anomaly eq.

(24) and introducing two auxiliary fields φ and ψ to account for the two independent invariants $b'E$ and bF in the trace anomaly (24) one finds[16]

$$S_{anom}[g;\varphi,\psi] = \frac{b'}{2}\int d^4x\,\sqrt{-g}\,\left\{-\varphi\Delta_4\varphi + \left(E - \frac{2}{3}\Box R\right)\varphi\right\}$$
$$+\frac{b}{2}\int d^4x\,\sqrt{-g}\,\left\{-2\varphi\Delta_4\psi + \left(F + \frac{c}{b}\,H\right)\varphi + \left(E - \frac{2}{3}\Box R\right)\psi\right\} \quad (29)$$

where in QED, $c = -e^2/24\pi^2$, $H = F_{\mu\nu}F^{\mu\nu}$ and in general,

$$\Delta_4 \equiv \Box^2 + 2R^{\mu\nu}\nabla_\mu\nabla_\nu - \frac{2}{3}R\Box + \frac{1}{3}(\nabla^\mu R)\nabla_\mu\,. \quad (30)$$

By variation of (29) the auxiliary scalar fields satisfy the linear fourth order eqs. of motion,

$$\Delta_4\,\varphi = \frac{1}{2}E - \frac{1}{3}\Box R\,, \quad (31a)$$
$$\Delta_4\,\psi = \frac{1}{2}F + \frac{c}{2b}\,H\,. \quad (31b)$$

The effective action (29) analogous to (23) in $D = 2$ generates all the anomalous amplitudes, by successive variations with respect to background metric and/or external gauge potentials, including diagrams with multiple stress-energy tensor insertions, such as $\langle TTT...JJ\rangle$ and $\langle TTT...\rangle$.

Since from the free variation of the effective action (29) φ and ψ obey non-trivial massless wave eqs., they are additional massless scalar degrees of freedom in low energy gravity, over and above the usual transverse, trace-free gravitational waves of the Einstein theory, and can have long range, macroscopic effects. The poles in the amplitude (28) survive in low energy scattering processes involving two photons with a gravitational strength.[10]

For QED the effects of the trace anomaly $\langle TJJ\rangle$ in flat space are again screened at distances greater than $\hbar/mc$ by the finite mass of the electron. However the trace anomalies in $\langle TTT...\rangle$ amplitudes which couple to purely gravitational sources and scattering processes are mediated by truly massless particles, such as the photon itself. Unlike the axial case, where non-trivial sources for J_5^μ may not be easy to come by, the sources here are gravitational which are omnipresent in the Universe. Since the coupling is through the stress-energy, it is universal but weak. However gravity is unscreened by any other interaction and its effects are cumulative over large distances. The long range nature of massless fluctuations make the scalar degrees of freedom contained in the effective action (29) relevant at the even the very largest macroscopic distance scales of cosmology.

6. Dynamical Vacuum Energy

The example of the axial anomaly, well-tested in QCD, and the logarithmic scaling of the anomaly action (29) imply that the full effective action of low energy gravity should be the sum of the classical Einstein-Hilbert action of classical General Relativity together with the effective action (29). In this EFT the additional long range scalar modes have some interesting effects.

With a covariant action functional (29) one may compute a covariantly conserved stress-energy tensor[16]

$$T_{\mu\nu}^{(anom)} \equiv -\frac{2}{\sqrt{-g}}\frac{\delta S_{anom}}{\delta g^{\mu\nu}} \tag{32}$$

the b' term of which gives the tensor,

$$\begin{aligned} E_{\mu\nu} = & -\frac{2}{3}\partial_\mu\partial_\nu\Box\varphi - 2(\partial_{(\mu}\varphi)\partial_{\nu)}\Box\varphi + 2(\partial_\mu\partial_\nu\varphi)\Box\varphi + \frac{2}{3}(\partial_\alpha\varphi)(\partial^\alpha\partial_\mu\partial_\nu\varphi) \\ & -\frac{4}{3}(\partial_\mu\partial_\alpha\varphi)(\partial_\nu\partial^\alpha\varphi) + \frac{1}{6}\eta_{\mu\nu}\left\{-3(\Box\varphi)^2 + \Box\left[(\partial_\alpha\varphi)(\partial^\alpha\varphi)\right]\right\} \end{aligned} \tag{33}$$

in flat space, which is conserved by use of the eqs. of motion (31). A particular solution of (31a) in flat space with $E = \Box R = 0$ is

$$\varphi = a\,\frac{z^2}{d^2}\,, \tag{34}$$

(with the second auxiliary field $\psi = 0$) which leads to

$$T_{\mu\nu}^{(anom)} = b'E_{\mu\nu} = -\frac{2b'a^2}{3d^4}\left(\eta_{\mu\nu} - 4\delta_\mu^z\delta_\nu^z\right) = \frac{C}{d^4}\,\mathrm{diag}(-1,1,1,-3)\,. \tag{35}$$

This is exactly the form of the Casimir vacuum stress-energy between two infinite parallel conducting plates a distance d apart in (5). The constant a and therefore $C = -2b'a^2/3$ depends upon the boundary conditions imposed on the conductors. Notice from (27b) that $b' < 0$ so that the constant $C > 0$, corresponding to an attractive force between the plates for any real a. It is remarkable that the auxiliary field φ and its stress tensor, obtained from the local form of *quantum* anomaly in the *trace* of the stress-energy tensor in *curved* space may be regarded through a particular homogeneous solution (34) of the *classical* eqs. (31) as responsible for the Casimir stress tensor (35) in *flat* space, where it is *tracefree.*

Several other examples of the auxiliary fields and stress tensor (32) in curved black hole and cosmological spacetimes have now been studied and used to compute vacuum polarization, illuminating their physical meaning.[16,17] Solving classical differential eqs. for the auxiliary fields in a given fixed spacetime background allows one to survey a great number of physical states of the underlying quantum field theory, taking account the spin of the field through the b, b' coefficients (27), rather than having to decompose the solutions of the field eqs. for each spin into normal modes, impose boundary conditions on those modes, construct the stress tensor and regularize it and renormalize it in each quantum state. The stress-energy (32) is particularly important in the vicinity of black hole and cosmological horizons, which it can dominate even the classical curvature terms, and lead to large quantum vacuum polarization effects there.[16–18]

In addition, when the scalar fields φ, ψ in (29) are treated as dynamical fields and quantized in their own right, they lead to infrared renormalization and finite volume dependence of the effective cosmological term (3). In other words the quantity $\lambda = \hbar G\Lambda/c^3$ becomes a dynamical quantity in the EFT, running with IR renormalization scale as every other coupling affected by light degrees of freedom. A one-loop calculation with the auxiliary field propagator given by the inverse of (30) gives the volume scaling relation[19]

$$V\frac{d\lambda}{dV} = 4\,(2\delta - 1)\,\lambda\,, \tag{36}$$

with the anomalous dimension,

$$2\delta - 1 = \frac{\sqrt{1-\frac{8}{Q^2}} - \sqrt{1-\frac{4}{Q^2}}}{1+\sqrt{1-\frac{4}{Q^2}}} \leq 0\,, \tag{37}$$

and

$$Q^2 \equiv -32\pi^2 b' = \frac{1}{180}\left(N_S + \frac{11}{2}N_F + 62N_V\right) + Q^2_{grav} \tag{38}$$

in terms of the anomaly coefficient b' in (24), and Q^2_{grav}, the contribution of graviton fluctuations to the anomaly coefficient (approximately 7.9). The anomalous scaling dimension (37) is negative for all $Q^2 \geq 8$. This implies that the dimensionless cosmological term λ has an infrared fixed point at zero as the volume $V \to \infty$. Thus the cosmological term is dynamically driven to zero as $V \to \infty$ by infrared fluctuations of the conformal part of the metric described by (29). There is no fine tuning involved here and no free parameters enter except Q^2, which is determined by the trace anomaly coefficient b' by (38). Once Q^2 is assumed to be positive, then $2\delta - 1$ is negative, and λ is driven to zero at large volumes or large distances by the conformal fluctuations of the metric. This identifies a mechanism for the dynamical screening of the vacuum energy at large distances, relying only on the four dimensional quantum physics of the trace anomaly with no additional assumptions.

Thus, the fluctuations of the new scalar degrees of freedom in the effective action (29) of the anomaly allow the cosmological "constant" vacuum energy of classical General Relativity to vary dynamically. This is qualitatively similar to the effect of the φ conformal degree of freedom in $2D$ gravity (23) at second order critical points.[15] The fixed point of $\lambda = 0$ is stable to marginal deformations by the Einstein-Hilbert terms, and describes a quantum conformal phase of $4D$ gravity. To take account of this mechanism and understand the role of the conformal phase where the fluctuations of the new dynamical scalar degrees of freedom are important in a consistent cosmological theory of vacuum dark energy is the remaining task.

7. Linear Response in de Sitter Space and Cosmological Horizon Modes

To this end we have recently studied the effect of the scalar fluctuations in the linear response of coupled matter-geometry perturbations around de Sitter spacetime, relevant for both inflationary and present day dark energy cosmology.[18] Here only the main results are summarized.

Linear response in gravity means solving the linear eqs. for small perturbations of the metric and matter stress-energy renormalized expectation value $\langle T^{\mu}_{\ \nu}\rangle_R$ expanded around a self-consistent solution of the semi-classical Einstein equations. A self-consistent solution of these eqs. is de Sitter space,

$$ds^2 = -c^2 d\tau^2 + a^2(\tau)(dx^2 + dy^2 + dz^2) = -d\tau^2 + e^{2H\tau} d\vec{x}\cdot d\vec{x} \qquad (39)$$

with $a(\tau) = e^{H\tau}$ the de Sitter Robertson-Walker (RW) scale factor, and conformal matter fields in their de Sitter invariant Bunch-Davies state.[5] The linear variation of the semi-classical Einstein eqs. is:

$$\delta\left\{R^{\mu}_{\ \nu} - \frac{R}{2}\delta^{\mu}_{\ \nu} + \Lambda\delta^{\mu}_{\ \nu}\right\} = \frac{8\pi G}{c^4}\,\delta\,\langle T^{\mu}_{\ \nu}\rangle_R\,, \qquad (40)$$

The variation on the left side is purely geometrical, obtained by varying the metric from its de Sitter value $g_{\mu\nu}$ given in RW coordinates (39) to $g_{\mu\nu} + \delta g_{\mu\nu}$. The variation of the expectation value on the right side contains two kinds of terms. The first kind are also proportional to the metric variation $h_{\mu\nu} = \delta g_{\mu\nu}$ and involve the retarded response function of stress-energy fluctuations, namely $\theta(t-t')\langle[T^{\mu}_{\ \nu}(x), T^{\alpha\beta}(x')]\rangle$ integrated over all points x' in the causal past of the point x. It turns out that this first kind of variation leads to solutions of linear response eqs. (40) which have spacetime dependence only on the Planck scale (2). Since this is the ultraviolet cutoff scale at which the semi-classical EFT breaks down, one cannot trust any physical conclusion obtained this kind of solutions of (40) at this scale.

The second kind of term in the stress-energy variation $\delta \langle T^{\mu}_{\ \nu}\rangle_R$ in (40) arise from the possibility of varying the quantum state of the field in which the expectation value $\langle T^{\mu}_{\ \nu}\rangle_R$ is evaluated, independently of the variation of the metric. The quantum state is specified by boundary conditions on the cosmological horizon scale c/H, having nothing to do with the microscopic Planck scale L_{Pl}. The scalar auxiliary field eqs. (31) in de Sitter space parametrize additional state dependent contributions to $\delta \langle T^{\mu}_{\ \nu}\rangle_R$ from the variation of $\delta T^{(anom)}_{\mu\nu}$ of (32). The gauge invariant combination of auxiliary field $\delta\varphi$ and metric perturbation $h_{\tau\tau}$ given by

$$u = \frac{1}{H^2}\left(\frac{\partial^2}{\partial\tau^2} + H\frac{\partial}{\partial\tau} - \frac{\vec{\nabla}^2}{a^2}\right)\delta\varphi - 2\,h_{\tau\tau} \tag{41}$$

in the gauge $g^{ij}h_{ij} = 0 = \nabla^i h_{i\tau} - \frac{1}{2}h_{\tau\tau}$ satisfies the second order homogeneous eq.,[18]

$$\left(\frac{\partial^2}{\partial\tau^2} + 5H\frac{\partial}{\partial\tau} + 6H^2 - \frac{\vec{\nabla}^2}{a^2}\right) u = 0\,, \tag{42}$$

the general solution of which is a linear combination of

$$u_{\vec{k},\pm}(\tau,\vec{x}) = v_{\vec{k},\pm}(\tau,\vec{x}) = \frac{1}{a^2}\,\exp\left(\pm\frac{ik}{Ha} + i\vec{k}\cdot\vec{x}\right)\,. \tag{43}$$

in Fourier space. These modes and those of the same form arising from the second auxiliary field ψ (called v), give rise in the linear response eq. (40) to perturbations of the Ricci tensor,

$$\delta R^{\tau}_{\ \tau} = -\delta R^{i}_{\ i} = \frac{\varepsilon'}{2a^2}\,\vec{\nabla}^2 u - \frac{\varepsilon}{6a^2}\,\vec{\nabla}^2 v\,, \tag{44}$$

with $\delta R = 0$ and

$$\varepsilon \equiv 32\pi G H^2 b\,, \tag{45a}$$

$$\varepsilon' \equiv -\frac{32\pi}{3} G H^2 b'\,. \tag{45b}$$

Thus the auxiliary fields of the anomaly action yield non-trivial gauge invariant solutions for the stress tensor and corresponding linearized Ricci tensor perturbations (44). Being solutions of (42) which itself is independent of the Planck scale, these solutions vary instead on arbitrary scales determined by the wavevector $\vec{k}$, and are therefore genuine low energy modes of the semi-classical effective theory. The Newtonian gravitational constant G and the Planck scale enter (44) only through the small coupling parameters ε and ε' between the auxiliary fields and the metric perturbation. In the limit of either flat space, or arbitrarily weak coupling $GH^2 \to 0$ these modes decouple from the metric perturbations at linear order.

These infrared scalar u and v modes are associated with the cosmological horizon scale c/H in de Sitter space and for that reason may be called *cosmological horizon modes.* To show this connection, one may introduce the static coordinates of de Sitter space, *viz.*

$$ds^2 = -(c^2 - H^2r^2)dt^2 + \frac{dr^2}{1 - H^2r^2/c^2} + r^2(d\theta^2 + \sin^2\theta d\phi^2)\,, \qquad (46)$$

related to the RW coordinates (39) by the coordinate transformation,

$$r = |\vec{x}|\, e^{H\tau}\,, \qquad (47a)$$

$$t = \tau - \frac{1}{2H}\ln\left(1 - \frac{H^2|\vec{x}|^2}{c^2}e^{2H\tau}\right). \qquad (47b)$$

In these static coordinates the eqs. of motion (42) possess the time independent solution,

$$u = v = \frac{1}{1 - H^2r^2/c^2} \qquad (48)$$

which diverge on the cosmological horizon $r = c/H$ centered at the origin. The corresponding stress-energy tensor perturbation in the static frame is

$$\delta\langle T^r_{\ r}\rangle_R = \delta\langle T^\theta_{\ \theta}\rangle_R = \delta\langle T^\phi_{\ \phi}\rangle_R = -\frac{1}{3}\,\delta\langle T^t_{\ t}\rangle_R = C\,\frac{\hbar\, H^4/c^3}{(1 - H^2r^2/c^2)^2}\,, \qquad (49)$$

with a quadratically *divergent* value on the cosmological horizon. This form of the stress tensor perturbation is also the form of a finite temperature fluctuation away from the Hawking-de Sitter temperature $T_H = \hbar H/2\pi k_B$ of the Bunch-Davies state in static coordinates.[20] It corresponds therefore to a change of the boundary conditions on the state of the underlying quantum fields on the horizon, with a corresponding change in the vacuum polarization effects of the fields near the horizon.

That these Casimir like effects are dynamical and functions of the boundary conditions is not surprising given our experience with the Casimir effect in flat space. It is also known that the Casimir stress-energy can diverge as a curved surface of a perfect conductor is approached.[21] In electromagnetism we know that perfect conductors do not exist and this mathematical divergence is cut off by the finite conductivity and skin depth of a metallic surface. A diverging stress tensor on the cosmological horizon signals the breakdown of boundary conditions there as well, except that in classical General Relativity the horizon is supposed to be purely a mathematical boundary with no physical stress tensor.

In the semi-classical EFT with S_{anom} included in the effective action there are additional scalar degrees of freedom in gravity that become important in the vicinity of geometries with horizons and there their fluctuations must be taken into account, just as charged matter fluctuations must be taken into account in an imperfect conductor. The stress-energy (49) of these degrees of freedom in the vicinity of the cosmological horizon can cause large backreaction on the classical geometry. Because the horizon is a null surface, the massless propagator pole associated with the anomalous amplitudes as in (14) can lead to large quantum correlations in multi-stress tensor amplitudes $\langle TTT... \rangle$ on the horizon as well. Such large amplitudes in fluctuations from the mean $\langle T^{\mu}_{\ \nu} \rangle$ is characteristic of a *phase transition* in which the semi-classical mean field theory breaks down, and where the rigid cosmological constant term of the classical theory can change.

Thus the fluctuations of the scalar degrees of freedom determined by the anomaly may lead to a phase transition to precisely the conformally invariant phase of gravity described by the fixed point $\lambda = 0$ of (36) in the near vicinity of the horizon. This suggests a rather different cosmological model than the standard one, in which we live inside a kind of "bubble" of vacuum energy condensate, with a preferred origin and a physical surface at the cosmological horizon.[19,22] At the horizon the quantum fluctuations of the scalar degrees of freedom contained in the anomaly lead to a phase transition in which the spacetime condensate Λ "melts." The value of the cosmological dark energy in the interior is then dynamical and fixed by the boundary conditions on the surface at the infrared Hubble scale c/H, much as the Casimir effect is, with no regard to the ultraviolet sensitive and clearly incorrect estimate of (4).

References

1. W. Pauli (unpublished); N. Straumann, e-print arxiv: gr-qc/0208027; See also W. Nernst, *Verh. Dtsch. Phys. Ges.* **18**, 83 (1916); C. P. Enz and A. Thellung, *Helv. Phys. Acta* **33**, 839 (1960); S. E. Rugh and H. Zinkernagel, *Stud. Hist. Philos. Mod. Phys.* **33**, 663 (2002).
2. K. A. Milton, *The Casimir effect: physical manifestations of zero-point energy* (World Scientific, Singapore, 2001); M. Bordag, *Advances In the Casimir Effect* (Oxford Univ. Press, Oxford, 2009), and references therein.
3. A. G. Riess *et. al.*, *Astron. J.* **116**, 1009 (1998); *Astron. J.* **607** 665 (2004); S. Perlmutter *et. al.*, *Astrophys. J.* **517** 565 (1999); J. L. Tonry *et. al.*, *Astrophys. J.* **594**, 1 (2003).
4. C. L. Bennett *et. al*, WMAP team, Astrophys. J. **583**, 1 (2003); e-print arXiv: 1001.4758 (2010).

5. N. D. Birrell and P. C. W. Davies, *Quantum Fields in Curved Space* (Cambridge Univ. Press, Cambridge, 1982), and references therein.
6. See *e.g.* P. Achard *et. al.*, L3 Collaboration, *Phys. Lett. B* **623**, 26 (2005).
7. D. Capper and M. Duff, *N. Cimento A* **23**, 173 (1974); *Phys. Lett. A* **53**, 361 (1975);
   S. Deser, M. Duff and C. J. Isham, *Nucl. Phys. B* **111**, 45 (1976);
   M. Duff, *Nucl. Phys. B* **125**, 334 (1977); *Nucl. Phys. B* **234**, 269 (1984); *Class. Quant. Grav.* **11**, 1387 (1994).
8. S. L. Adler, *Phys. Rev.* **177**, 2426 (1969);
   J. S. Bell and R. Jackiw, *N. Cimento A***60**, 47 (1969).
9. A. D. Dolgov and V. I. Zakharov, *Nucl. Phys. B* **27**, 525 (1971);
   J. Hořejší, *Czech. J. Phys. B* **35**, 820 (1985); *Phys. Rev. D* **32**, 1029 (1985); *J. Phys. G: Nucl. Phys.* **12**, L7 (1986); *Czech. J. Phys.* **42**, 345 (1992).
10. M. Giannotti and E. Mottola, *Phys. Rev. D* **79**, 045014 (2009).
11. W. A. Bardeen, H. Fritzsch and M. Gell-Mann, in *Scale and Conformal Symmetry in Hadron Physics*, R. Gatto, ed. (Wiley, New York, 1973);
    T. P. Cheng and L. F. Li, *Gauge Theory of Elementary Particle Physics* (Oxford Univ. Press, Oxford, 1984);
    S. Weinberg, *The Quantum Theory of Fields*, Vol. 2 (Cambridge Univ. Press, Cambridge, 1995).
12. G. 't Hooft in *Recent Developments in Gauge Theories, Procedings, NATO Advanced Study Institute, Cargese*, Nato Advanced Study Institutes Series B, Physics, Vol. 59 (Plenum, New York,1980).
13. R. A. Bertlmann and E. Kohlprath, *Ann. Phys.* **288** 137 (2001).
14. J. Wess and B. Zumino, *Phys. Lett. B* **37** 95 (1971).
15. V. G. Knizhnik, A. M. Polyakov, and A. B. Zamolodchikov, *Mod. Phys. Lett. A* **3** 819 (1988);
    F. David, *Mod. Phys. Lett. A* **3** 1651 (1988).
16. E. Mottola and R. Vaulin, *Phys. Rev. D* **74**, 064004 (2006).
17. P. R. Anderson, E. Mottola, and R. Vaulin, *Phys. Rev. D* **76**, 124028 (2007).
18. P. R. Anderson, C. Molina-Pariís and E. Mottola, *Phys. Rev. D* **80**, 084005 (2009).
19. I. Antoniadis, P. O. Mazur, and E. Mottola, *N. Jour. Phys.* **9** 11 (2007).
20. P. R. Anderson, W. A. Hiscock, and D. A. Samuel, *Phys. Rev. D* **51**, 4337 (1995).
21. D. Deutsch and P. Candelas, *Phys. Rev. D* **20**, 3063 (1979).
22. P. O. Mazur and E. Mottola, arXiv:gr-qc/0109035; *Proc. Nat. Acad. Sci.* **101**, 9545 (2004).

QUASI-BLACK HOLES AND LORENTZ-ABRAHAM ELECTRON IN GENERAL RELATIVITY

O. B. ZASLAVSKII

Astronomical Institute of Kharkov V.N. Karazin National University,
35 Sumskaya St., Kharkov, 61022, Ukraine
E-mail: ozaslav@kharkov.ua

Usually, a pure field model of the classical analogue of an "elementary particle" implies that 1) the mass has pure electromagnetic origin, 2) bare non-electromagnetic stresses vanish. Meanwhile, if effects of general relativity are taken into account properly, it opens a new possibility. Namely, it is shown that for extremal quasi-black (when the size of a system approaches its gravitational radius but the horizon does not form) the surface non-zero stresses can exist but, nonetheless, they do not contribute to the total mass.. As a result, we obtain a mass of pure electromagnetic origin in spite of presence of bare stresses.

1. Introduction

In recent years, the new class of objects was introduced in general relativity - so-called quasiblack holes (QBHs). The general concepts were developed in Refs. 1, 2 where a reader can also find further details. The purpose of my talk is to review briefly some aspects of QBHs with the main emphasis on the problem of the analogue of classical electron of the Lorentz-Abraham Electron in general relativity. This concept implies that we want to examine the possibility of self-consistent analogue of an elementary particle in general relativity having a mass of pure electromagnetic origin.

In flat space-time Coulomb repulsion prevents such a construction, so one needs Poincaré stresses for such a construction. But, by including gravitation, one may possibly dispense with those stresses, the attractive force of gravitation making the question reasonable within the theory of general relativity. On the face of it, it would seem natural that as we want to have electromagnetic and gravitational forces alone, we must require the absence of a bare tension on the surface. Otherwise, this would mean that apart from electromagnetism and gravitation there were also external forces of different nature, of Poincaré type, needed to keep the system in equilibrium. An attempt of this kind was made,[3] where it was argued that a charged shell with empty space inside obeys this criteria in the extremal case, $m = q$. This was criticized in Ref. 4 where it was shown that, actually, the surface stresses do not vanish in such a model even in the extremal limit. Instead, another model was suggested in Ref. 4, where the external extremal Reissner-Nordström metric was glued to the Bertotti-Robinson tubelike geometry inside. Then, it turned out that the surface stresses vanish in the limit when the surface of gluing approaches its own horizon.

However, it turns out that two issues "mass of pure electromagnetic origin" and "absence of bare stresses" in general relativity may be different in one exceptional situation that is connected just with the presence of a QBH. Namely, if the surface of the charged body approaches the quasihorizon, the contribution of the bare tension on the surface to the total mass in the extremal case completely vanishes, although these stresses by themselves remain finite. As a result, we obtain a model in which a distant observer measures a mass as having purely electromagnetic origin, although locally on the surface there are extraneous additional forces. Moreover, one can even allow nonelectromagnetic fields inside in the bulk, since their contribution to the total mass vanishes in the quasihorizon limit. All the region beyond the quasihorizon including the quasihorizon itself turns out to be frozen and gives no contribution to the mass (for the nonextremal case the inner region also is frozen but the boundary is not).

One reservation is in order. It is well known that for a real electron the value of mass $m \ll e$ (in geometric units) where e is a charge and, thus, the corresponding Reissner-Nordström metric corresponds to naked singularity, not to a black hole. Therefore, we do not pretend to make the literal description of a electron that, obviously, needs essentially quantum treatment. Rather, it demonstrates some non-trivial features of space-time due to general relativity combined with classical electrodynamics and which can (or should) be taken into account on a quantum level too. An additional motivation for such a pure classical treatment comes from the interesting correspondence between the problem of self-energy in classical and quantum electrodynamics[3] (Sec. 6).[5]

2. Definition of QBH

Below I outline briefly derivation of this rather unexpected result. First of all, let me give a definition of QBH. Roughly speaking one can say that a QBH is an object on the verge of becoming a black hole but actually is distinct from it in many ways. To give more rigorous definition, let me start, for simplicity, from the spherically-symmetrical systems. Let us consider the metric

$$ds^2 = -B(r)\,dt^2 + V^{-1}(r)dr^2 + r^2\,d\Omega^2\,, \tag{1}$$

where r is the Schwarzschild radial coordinate, $d\Omega^2 = d\theta^2 + \sin^2\theta\,d\phi^2$, and $V(r)$ and $A(r)$ are metric potentials.

Let an inner matter configuration, with an asymptotic flat exterior region, exist with the properties (a) the function $V(r)$ attains a minimum at some $r^* \neq 0$, such that $V(r^*) = \varepsilon$, with $\varepsilon << 1$, this minimum being achieved either from both sides of r^* or from $r > r^*$ alone, (b) for such a small but nonzero ε the configuration is regular everywhere with a nonvanishing metric function B, at most the metric contains only delta-function like shells, and (c) in the limit $\varepsilon \to 0$ the metric coefficient $B \to 0$ for all $r \leq r^*$. These three features define a QBH. Note that although the above definition of QBHs relies on the coordinate system and metric coefficient V, actually, this definition can be done in a form invariant under the choice of the radial coordinate. Indeed, it is sufficient to replace V by $(\nabla r)^2$. In the Schwarzschild coordinates one has $(\nabla r)^2 = V$.

In turn, these three features entail some nontrivial consequences: (i) there are infinite redshift whole regions, (ii) when $\varepsilon \to 0$, a free-falling observer finds in his own frame infinitely large tidal forces in the whole inner region, showing some form of degeneracy, although the spacetime curvature invariants remain perfectly regular everywhere, (iii) in the limit, outer and inner regions become mutually impenetrable and disjoint, and (iv) for external far away observers the space-time is virtually indistinguishable from that of black holes. The QBH is on the verge of forming an event horizon, but it never forms one, instead, a quasihorizon appears. For a QBH the metric is well defined and everywhere regular. However, properties, such as when $\varepsilon = 0$, QBH space-time become degenerate, almost singular, have to be examined with care.

3. Relationship between regular and singular features

3.1. *Structure of space-time*

A very simple example can be invoked, where gluing between an inner flat metric and an external Reissner-Nordström metric is performed. Such a construction was discussed in Ref. 3 as an example of a classical model of an elementary particle where the Reissner-Nordström metric was chosen to be extremal. For definiteness and bearing in mind a distinguished role of the extremal case (see below), I also restrict myself by the extremal case. Consider an external space-time given by equation (1) with $B = V = \left(1 - \frac{m}{r}\right)^2$ for $r \geq r_0$, and an inner space-time given by the Minkowski metric,

$$ds^2 = -dT^2 + dr^2 + r^2 d\Omega^2 \tag{2}$$

where $0 < r \leq r_0$. On the border, the condition of matching both parts of the space-time leads to

$$t = \frac{T r_0}{r_0 - m}, \tag{3}$$

so that the time part of the metric (2) can be written as $-dT^2 = -\frac{(r_0-m)^2}{r_0^2}\, dt^2$. Then, if time t is used, the metric coefficient $g_{00} \to 0$ in the limit $r_0 \to m$. This is the reason why this construction can be considered as an example of a QBH. We obtain an infinite redshift due to the mismatch in time rescaling in equation (3). Also, we cannot achieve the continuous matching if T is considered as a legitimate coordinate inside since the surface $r = m$ is time-like in the metric (2) but lightlike in the outer metric where the coordinate t is used. One may try to repair this by considering inside the same time t as outside. However, the term $-\frac{(r_0-m)^2}{r_0^2}\, dt^2$ disappears in this limit and the space-time becomes degenerate. This example clearly shows the entanglement of regular and singular features in the space-time of QBH. This reveals not in the divergences of curvature invariants - moreover, in the present case the gravitational field inside is absent at all!

3.2. *Infinite redshift*

In terms of the time t, the outer time, the product $\omega\sqrt{B(r)} = \omega_{\rm c}$ remains constant on each ray during the propagation of light in a static gravitational field, where here $\omega_{\rm c}$ is some constant frequency. Note that $\omega_{\rm c}$ is a frequency measured with respect to time t, ω is the frequency measured with respect to the proper time at a given point r. Since at infinity $B = 1$, one obtains that $\omega_{\rm c}$ is $\omega(r \to \infty) \equiv \omega_\infty$, so that one can write $\omega\sqrt{B(r)} = \omega_\infty$. A distant observer would register an infinite redshift ($\omega_{\rm c} \to 0$) if an emitted particle had a finite ω inside the matter since $B \to 0$ there in the QBH limit. Only high-frequency photons with infinite ω inside the quasihorizon but finite ω_∞ can escape to infinity. This occurs for any Bonnor star whose boundary gets arbitrarily close to the horizon ($B^{\rm RN}(r_0) \to 0$), this property being model-independent.

3.3. *Naked behavior in free-falling frame*

It was observed in Ref. 6 that some components of curvature tensor which vanish on the horizon of a black hole become nonzero in the free-falling frame. It was further observed[7,9] that, moreover, they can become infinite in spite of the finiteness of Kretschmann scalar Kr and other curvature invariants. This is due to the fact that while in the static frame Kr represents the sum of squared components of the tensor, so that its finiteness entails the finiteness of each component, in the free-falling frame separate components enter Kr with different signs, so that they can cancel each other. Returning from the black hole case to the QBH one, we can observe that now the same phenomenon occurs in the whole region $r \leq r^*$. Thus, typically, the inner region becomes degenerate not only in what concerns space-time properties but also dynamically As result, it disjoints from the outer space since an external observer cannot penetrate inside. In the example discussed above there is no gravitational field inside. But, instead, there are surface stresses and one can show that they remain finite in the static frame but diverge in the free-falling frame. The issue of surface stresses is important by itself, and below I outline briefly what happens in the static frame in a general case. I also demonstrate that this results in the mass formula which is the exact counterpart of the mass formula for black hole.

4. Stresses in the static frame and mass formula

Consider the static configuration (not necessarily spherically-symmetric) with the metric in written in Gauss normal

$$ds^2 = -N^2 dt^2 + dl^2 + g_{ab} dx^a dx^b \,, \tag{4}$$

The Tolman formula gives us[8]

$$M = \int (-T_0^0 + T_k^k) \sqrt{-g}\, d^3x \,. \tag{5}$$

Then, one can split it to three contributions from the inside, boundary and outside: $M_{\text{tot}} = M_{\text{in}} + M_{\text{surf}} + M_{\text{out}}$. The quantity M_{in} contains the factor N and vanishes since $N \to 0$ inside by the definition of a QBH. The quantity M_{out} is the standard Tolman mass M_{tol} calculated between the quasihorizon and infinity. The most interesting contribution comes from M_{surf}. It follows from the Israel formalism of thin shells or directly form the Einstein equations that

$$M_{\text{surf}} = \frac{1}{4\pi} \int_{\text{surf}} \left[\left(\frac{\partial N}{\partial l} \right)_+ - \left(\frac{\partial N}{\partial l} \right)_- \right] d\sigma \,. \tag{6}$$

where integration is taken over the boundary surface, "+" and "-" refer to the outer and inner sides, respectively. Inside $\left(\frac{\partial N}{\partial l}\right)_{-} \to 0$ since $N \to 0$ uniformly there by the definition of QBH. Outside, $\left(\frac{\partial N}{\partial l}\right)_{+} \to \kappa$ where κ is the surface gravity. As a result, we obtain $M_{\rm surf} = \frac{\kappa A}{4\pi}$ where A is the horizon surface in accordance with the mass formula $M = \frac{\kappa A}{4\pi} + M_{\rm tol}$.[10] It is instructive to trace back the origination of nonzero terms in $M_{\rm surf}$. They come from S_{μ}^{ν}, the delta-like part of T_{μ}^{ν} so that $S_{\mu}^{\nu} = \int T_{\mu}^{\nu}\, dl$. Then, it is easy to show that the transverse stresses contain the term $N^{-1}\left(\frac{\partial N}{\partial l}\right)_{+}$ which is to be multiplied by the factor N that stems in the integrand of $M_{\rm surf}$ from $\sqrt{-g}$. As a result, in the nonextremal case ($\kappa \neq 0$) the surface stresses are infinite but give the finite nonzero contribution. In the nionextremal one, by definition, $N \sim \exp(-Bl)$ where B is a constant and $l \to \infty$, so that stresses are finite but give zero contribution to the mass.

5. Conclusion

It follows from the results outlined briefly in my talk that the two issues "mass of pure electromagnetic origin" and "absence of bare stresses" in general relativity may be different in one exceptional situation. If the surface of the charged body approaches the quasihorizon, the contribution of the bare tension on the surface to the total mass in the extremal case completely vanishes, although these stresses by themselves can remain nonzero. As a result, we obtain a model in which a distant observer measures a mass as having purely electromagnetic origin, although locally, on the surface there are extraneous additional forces. Moreover, one can even allow non-electromagnetic fields inside in the bulk, since anyway, their contribution to the total mass vanishes in the quasihorizon limit. All the region beyond the quasihorizon including the quasihorizon itself turns out to be frozen and gives no contribution to the mass (for the non-extremal case the inner region also is frozen but the boundary is not).

It is also worth noting that the general statement of Ref. 3 about the distinguished role of extremal black holes (now we would rephrase it as "black and quasi-black holes") turns out to be correct. They are suitable candidates for the role of classical models of elementary particles since only in this case the mass can have pure electromagnetic origin. Thus we have obtained that an extremal quasi-black hole (in contrast to the non-extremal one) can serve as a physically reasonable classical model of an Abraham-Lorentz electron in that both the inner and surface contribution of forces with non-electromagnetic origin vanish. In doing so, we showed that one may weaken the requirement of vanishing surface stresses since the finite stresses have zero contribution to the total mass.

In the present talk the static case was considered. However, as the charge is in a sense produces the effects similar to those from rotation, the present results are expected to be useful for corresponding analysis.[11] In the first place, it concerns the problem of constructive a regular source for the Kerr metric.[12–15] This is a separate interesting subject for future researches.

Acknowledgments

I thank organizers and especially Kim Milton for support and a nice conference.

References

1. J. P. S. Lemos and O. B. Zaslavskii, *Phys. Rev.* D **76**, 084030 (2007).
2. J. P. S. Lemos and O. B. Zaslavskii, *Phys. Rev. D* **78**, 124013 (2008)..
3. A. V. Vilenkin and P. I. Fomin, *Nuovo Cimento* A **45**, 59 (1978).
4. O. B. Zaslavskii, *Phys. Rev.* **70**, 104017 (2004).
5. A. V. Vilenkin and P. I. Fomin, *ZETF* **67,** 12 (1974) (Engl. Transl.: *Sov. Phys. JETP* **40 6** (1975).
6. G. T. Horowitz and S. F. Ross, *Phys. Rev.* D **56**, 2180 (1997); ibid., **57**, 1098 (1998).
7. V. Pravda, O. B. Zaslavskii, *Class. Quant. Grav.* **22**, 5053 (2005).
8. R. Tolman. *Relativity, Thermodynamics and Cosmology*, (Dover, New York, 1987), section 92 (first published in 1934, by Clarendon Press).
9. O. B. Zaslavskii, *Phys. Rev.* **76**, 024015 (2007).
10. J. M. Bardeen, B. Carter, and S. W. Hawking, *Commun. Math. Phys.* **31**, 161 (1973).
11. J. P. S. Lemos and O. B. Zaslavskii, *Phys. Rev. D* **79**, 044020 (2009).
12. W. Israel, *Phys. Rev. D* **2**, 641 (1970).
13. M. Gürses and F. Gürsey, *J. Math. Phys.* **16**, 2385 (1975).
14. C.A. López, *Phys. Rev.* **30**, 313 (1984).
15. A. Burinskii, E. Elizalde, S.R. Hildebrandt and G. Magli *Phys. Rev.* D **65** 064039 (2002).

WORLDLINE APPROACH TO QFT ON MANIFOLDS WITH BOUNDARY

F. BASTIANELLI* and O. CORRADINI†

Dipartimento di Fisica, Università di Bologna and INFN, Sezione di Bologna, Via Irnerio 46, I-40126 Bologna, Italy
** bastianelli@bo.infn.it*
† corradini@bo.infn.it

P. A. G. PISANI

IFLP (CONICET), Departamento de Física de la Universidad Nacional de La Plata, c.c. 67, 1900 La Plata, Argentina
E-mail: pisani@obelix.fisica.unlp.edu.ar

C. SCHUBERT

Instituto de Física y Matemáticas, Universidad Michoacana de San Nicolás de Hidalgo, Edificio C-3, Apdo. Postal 2-82, C.P. 58040, Morelia, Michoacán, México
E-mail: schubert@ifm.umich.mx

We use the image charge method to compute the trace of the heat kernel for a scalar field on a flat manifold with boundary, representing the trace by means of a worldline path integral and obtain useful non-iterative master formulae for n insertions of the scalar potential. We discuss possible extensions of the method.

Keywords: Worldline formalism; manifolds with boundary.

1. Worldline formalism on manifolds with boundary

The worldline formalism (see Ref. 1 for a review) is an alternative method to compute effective actions, amplitudes and anomalies in quantum field theory. For example, the one-loop effective action can be written as a "trace log" of a differential operator which be exponentiated using a Schwinger proper time integral and the trace can be written in terms of a quantum mechanical path integral. For the simplest case of a real massless scalar field with self-interaction $U(\phi)$, propagating in a flat boundaryless space, the one-loop effective action formally reads $\Gamma[\varphi] = -\frac{1}{2}\int_0^\infty \frac{dT}{T}\mathrm{Tr}\ e^{-TH}$ where

$H = -\Box + U''(\phi)$ and the partition trace reads

$$\mathrm{Tr}\; e^{-TH} = \int_{PBC} Dx\; \exp\left[-\int_0^1 d\tau \left(\frac{1}{4T}\dot{x}^2 + T\,V(x(\tau)) \right) \right] \tag{1}$$

where $V(x) \equiv U''(\varphi(x))$ and the path integral is over the space of all closed paths on the unit circle and the Teichmuller parameter T is the proper length of the circle. The short-time expansion of the operator $\mathrm{Tr}\; e^{-TH}$, known as heat kernel expansion, takes the form

$$\mathrm{Tr}\; e^{-TH} = \int d^D x\; K(T; x, x) = \frac{1}{(4\pi T)^{D/2}} \sum_{n=0}^{\infty} a_n\; T^n \tag{2}$$

and the integrated heat kernel coefficients a_n can be straightforwardly obtained as a short-time expansion of the path integral (1) by Wick contracting the Taylor expansion of the potential.[2]

The worldline path integral approach to QFT on manifolds with boundary has been carried out using Monte Carlo simulations[3] and many interesting results are obtained with this method.[4] However, a serious difficulty one has to face with such a method concerns boundary conditions different than Dirichlet and an alternative method that might help overcoming this difficulty would be quite welcome.

In a manifold with boundary a heat kernel expansion for the trace, cfr. Eq. (2), still holds but the sum involves half-integer powers and the coefficients include boundary contributions as well as bulk contributions.[5] We have developed[6] a method that generalizes analytic worldline techniques to flat manifolds with boundary $M = \mathbb{R}_+ \times \mathbb{R}^{D-1}$ using the image charge method to map the path integral on a half space to the combination of two path integrals on the whole space

$$\mathrm{Tr}_M\; e^{-TH} = \int_M d^D x\; K(T; x, x) \mp \int_M d^D x\; K(T; \tilde{x}, x) \tag{3}$$
$$x = (y, \vec{z})\,, \quad y \in \mathbb{R}_+\,,\; \vec{z} \in \mathbb{R}^{D-1}$$

where $\tilde{x} = (-y, \vec{z})$ is the image charge of x and the upper (lower) sign corresponds to Dirichlet (Neumann) boundary conditions (in Ref. 7 an extension of the method to Robin boundary conditions was considered). The above kernels are whole space kernels computed with an evenly extended potential

$$V(x) \;\rightarrow\; \tilde{V}(x) = \theta(y)V(x) + \theta(-y)V(\tilde{x}) = V_+(x) + \epsilon(y)V_-(x)\;. \tag{4}$$

The reflection property of the potential also allows to extend the overall Riemannian integral to the whole space, so that the above two contributions can be written as

$$\begin{aligned} K^{dir}(T) &\equiv \int_M d^D x\ K(T;x,x) = \frac{1}{2}\int d^D x\ K(T;x,x) \\ &= \frac{1}{2}\int_{PBC} Dx\ \exp\left[-\int_0^1 d\tau \left(\frac{1}{4T}\dot{x}^2 + T\tilde{V}(x(\tau))\right)\right] \end{aligned} \tag{5}$$

$$\begin{aligned} K^{ind}_{\partial M}(T) &\equiv \int_M d^D x\ K(T;\tilde{x},x) = \frac{1}{2}\int d^D x\ K(T;\tilde{x},x) \\ &= \frac{1}{2}\int_{(A)PBC} Dx\ \exp\left[-\int_0^1 d\tau \left(\frac{1}{4T}\dot{x}^2 + T\tilde{V}(x(\tau))\right)\right] \end{aligned} \tag{6}$$

and will be referred to as the "direct contribution" and the "indirect contribution" respectively. Above, the suffix $_{(A)PBC}$ indicates that the coordinate $y(\tau)$ satisfies (anti)-periodic boundary conditions, whereas coordinates $\vec{z}(\tau)$ satisfy periodic boundary conditions. The potential $\tilde{V}$ includes a distribution $\epsilon(y)$ and the naive application of the Wick theorem results nontrivial.[6,7] However, we[8] demonstrated that (i) upon Fourier representing the sign function $\epsilon(y) = \int \frac{dp}{\pi p}\sin(py) = \int_{ev}\frac{dp}{i\pi p}\, e^{ipy}$ and (ii) upon carefully separating out bulk contributions from boundary contributions, one can safely use it: two new coefficients for the half-space, a_4 and $a_{9/2}$, were computed.

1.1. *Indirect contribution to the heat kernel trace*

For this contribution the coordinate $y(\tau)$ is antiperiodic and therefore its kinetic action has no zero mode. Hence, $\int_{ABC} Dy\ e^{-\frac{1}{4T}\int_0^1 d\tau \dot{y}^2} = \frac{1}{2}$ and we can safely Taylor expand the potential about the boundary $(0,\vec{z})$

$$\tilde{V}(y(\tau),\vec{z}+\vec{z}(\tau)) = e^{\vec{z}(\tau)\cdot\vec{\partial}}\left[e^{y(\tau)\partial_y}V_+(0,\vec{z}) + \int_{ev}\frac{dp}{i\pi p}e^{y(\tau)D^0(p)}V_-(0,\vec{z})\right] \tag{7}$$

with $D^0(p) = \partial_y + ip$. Inserting the latter into (6) one obtains

$$\begin{aligned} K^{ind}_{\partial M}(T) &= \frac{1}{4(4\pi T)^{\frac{D-1}{2}}}\sum_{n=0}^{\infty}\frac{(-T)^n}{n!}\int_0^1 d\tau_1\cdots\int_0^1 d\tau_n\int_{\partial M} d^{D-1}z \\ &\times \exp\left[-\frac{T}{2}\sum_{i,j=1}^{n}\left(G_P(\tau_i,\tau_j)\vec{\partial}_i\cdot\vec{\partial}_j + G_A(\tau_i,\tau_j)D^0_i(p)D^0_j(p)\right)\right] \\ &\times \prod_{k=1}^{n}\left[V_+^{(k)}(0,\vec{z}) + \int_{ev}\frac{dp_k}{i\pi p_k}V_-^{(k)}(0,\vec{z})\right] \end{aligned} \tag{8}$$

and the suffix i on the derivative means that it acts on the term of potential labelled accordingly. The ABC propagator appearing above is given by $G_A(\tau,\sigma) = |\tau-\sigma| - \frac{1}{2}$, whereas the expression for the PBC propagator depends on the prescription one adopts for factoring out the zero mode $\vec{z}$. For example, in the "String Inspired" method we have $G_P(\tau,\sigma) = |\tau-\sigma| - (\tau-\sigma)^2$, and using (worldline) DBC we instead have $G_P(\tau,\sigma) = |\tau-\sigma| + \frac{1}{2}(1-2\tau)(1-2\sigma) - \frac{1}{2}$. The two methods yield different unintegrated heat kernel expansions and their difference resides on total derivative terms. However, since here these terms are boundary total derivatives their integrals vanish and the integrated expression (8) is scheme-independent. Scheme-independence will be slightly more subtle for the direct contribution that we describe next.

1.2. *Direct contribution to the heat kernel trace*

Here all the coordinates have periodic boundary conditions and we Taylor expand the potential insertions about the zero modes $(y,\vec{z})$ and get

$$K^{\text{dir}}(T) = \frac{1}{2(4\pi T)^{\frac{D}{2}}} \sum_{n=0}^{\infty} \frac{(-T)^n}{n!} \int_0^1 d\tau_1 \cdots \int_0^1 d\tau_n \int_{-\infty}^{\infty} dy \int_{\partial M} d^{D-1}z$$
$$\times \exp\left[-\frac{T}{2}\sum_{i,j=1}^{n} G_P(\tau_i,\tau_j)\partial_i\cdot\partial_j\right] \prod_{k=1}^{n}\left[V_+^{(k)}(x) + (\epsilon(y)V_-(x))^{(k)}\right] \quad (9)$$

where the notation is meant to convey that derivatives may act on $\epsilon(y)$ as well as on $V_\pm$. When derivatives act on $\epsilon(y)$, δ functions or derivatives thereof are generated giving rise to boundary terms. Contributions where no derivatives act on $\epsilon(y)$ are bulk terms. Namely

$$K_M^{\text{dir}}(T) = \frac{1}{2(4\pi T)^{\frac{D}{2}}} \sum_{n=0}^{\infty} \frac{(-T)^n}{n!} \int_0^1 d\tau_1 \cdots \int_0^1 d\tau_n \int_{-\infty}^{\infty} dy \int_{\partial M} d^{D-1}z$$
$$\times \exp\left[-\frac{T}{2}\sum_{i,j=1}^{n} G_{D,S}(\tau_i,\tau_j)\partial_i\cdot\partial_j\right] \prod_{k=1}^{n}\left[V_+^{(k)}(x) + \epsilon(y)V_-^{(k)}(x)\right] . \quad (10)$$

Subtracting (10) from (9) yields the boundary terms associated to the direct contribution, namely

$$K^{\rm dir}_{\partial M}(T) = \frac{1}{2(4\pi T)^{\frac{D}{2}}} \sum_{n=0}^{\infty} \frac{(-T)^n}{n!} \int_0^1 d\tau_1 \cdots \int_0^1 d\tau_n \int_{-\infty}^{\infty} dy \int_{\partial M} d^{D-1}z$$
$$\times \int_0^1 dw \frac{\partial}{\partial w} \exp\left[-\frac{T}{2} \sum_{i,j=1}^{n} G_P(\tau_i,\tau_j) D_i(wp)\cdot D_j(wp)\right]$$
$$\times \prod_{k=1}^{n} \left[V_+^{(k)}(x) + \int_{ev} \frac{dp_k}{i\pi p_k}\, e^{ip_k y} V_-^{(k)}(x)\right] \qquad (11)$$

where $D(wp) = (D^0(wp), \vec{\partial})$ and the total derivative on w takes care of the aforementioned subtraction. The evaluation of (11) is done as follows:

(i) Taylor expand potentials about the boundary: it is safe as by construction all terms in (11) are boundary terms;
(ii) integrate over y: it yields a δ function involving various p's and ∂'s;
(iii) w derivative cancels one (spurious) pole;
(iv) integrate over all p's, then over w and finally over τ_i.

Expressions (10) and (11) separately are scheme-dependent but the scheme-dependent terms cancel out in the sum (9). In fact as mentioned above scheme-dependence of the bulk part is encoded in a set of total derivative terms that upon integration yield boundary terms.

2. Outlook

We discussed a path integral method to compute the heat kernel trace for a self-interacting scalar field on a flat manifold with boundary. A natural generalization is the inclusion of interaction with external fields. This is clearly feasible by evenly extending to the whole space the coupling $A_M(y,\vec{z})x^M$ or $h_{MN}(y,\vec{z})x^M x^N$ for the spin-one and spin-two case. Another possible generalization involves the inclusion of particles with spin in the loop, e.g. by representing the effective action in terms of a spinning particle action on the circle.

Acknowledgments

The work of F.B. and O.C. was partly supported by the Italian MIUR-PRIN contract 20075ATT78. The work of P.P. was partly supported by PIP 6160, UNLP proj. 11/X381 and DAAD.

References

1. C. Schubert, *Phys. Rept.* **355**, 73 (2001).
2. D. Fliegner, M. G. Schmidt and C. Schubert, *Z. Phys.* **C64**, 111 (1994).
3. H. Gies, K. Langfeld and L. Moyaerts, *JHEP* **06**, 018 (2003).
4. H. Gies, *Geothermal Casimir phenomena*, plenary talk presented at QFexT09.
5. H. P. McKean and I. M. Singer, *J. Diff. Geom.* **1**, 43 (1967).
6. F. Bastianelli, O. Corradini and P. A. G. Pisani, *JHEP* **02**, 059 (2007).
7. F. Bastianelli, O. Corradini and P. A. G. Pisani, *J. Phys.* **A41**, 164010 (2008).
8. F. Bastianelli, O. Corradini, P. A. G. Pisani and C. Schubert, *JHEP* **10**, 095 (2008).

VACUUM ENERGY, SPECTRAL DETERMINANT AND HEAT KERNEL ASYMPTOTICS OF GRAPH LAPLACIANS WITH GENERAL VERTEX MATCHING CONDITIONS

J. M. HARRISON* and K. KIRSTEN†

Department of Mathematics, Baylor University, Waco, TX 76798, USA

**E-mail: jon_harrison@baylor.edu*

†*E-mail: klaus_kirsten@baylor.edu*

We consider Laplace operators on metric graphs, networks of one-dimensional line segments (bonds), with matching conditions at the vertices that make the operator self-adjoint. Such quantum graphs provide a simple model of quantum mechanics in a classically chaotic system with multiple scales corresponding to the lengths of the bonds. For graph Laplacians we briefly report results for the spectral determinant, vacuum energy and heat kernel asymptotics of general graphs in terms of the vertex matching conditions.

1. Introduction

Quantum graphs provide a simple, analytically tractable, model in which to investigate quantum phenomena. This is of particular interest for the study of vacuum energy and related spectral quantities as the spectrum of a quantum graph generically incorporates the features associated with quantum mechanical systems that are classically chaotic.[1] In particular the eigenvalues of the Laplace operator on a graph are typically distributed as the eigenvalues of large random matrices where the appropriate random matrix ensemble for comparison is determined by the symmetries of the quantum system; this correspondence is known as the Bohigas-Giannoni-Schmit conjecture.[2]

Quantum graphs were introduced as a model of quantum chaos by Kottos and Smilansky.[3] Graph models are also of current interest in many areas of mesoscopic physics like Anderson localization, photonic crystals, microelectronics, nanotechnology and the theory of wave guides; see Ref. 4 for a review. In this article we present results for the vacuum energy,[5,6] spectral determinant[7–9] and heat kernel asymptotics[10,11] of graph Laplacians. The unified approach we adopt, based on the spectral zeta function of the graph, provides new results in each case. The full description of the construction of zeta functions of quantum graphs will appear in Ref. 12.

2. Graph Laplacian

Let $G = (\mathcal{V}, \mathcal{B})$ be a graph where $\mathcal{V}$ is the set of vertices and $\mathcal{B}$ is the set of bonds (or edges); for an example see Fig. 1. Each bond $b \in \mathcal{B}$ joins a pair of vertices and for notational convenience we assume that bonds are directed with an initial and terminal vertex specified via the functions $o : \mathcal{B} \to \mathcal{V}$ and $t : \mathcal{B} \to \mathcal{V}$ respectively. We insist that the set $\mathcal{B}$ is symmetric so that $b \in \mathcal{B}$ if and only if there is another bond $\overline{b} \in \mathcal{B}$ the reversal of b such that $o(b) = t(\overline{b})$ and $t(b) = o(\overline{b})$. The total number of bonds $B = |\mathcal{B}|$ is therefore twice the physical number of connections in the network. Each bond b is associated with an interval $[0, L_b]$ so that the local coordinate $x_b = 0$ at $o(b)$ and $x_b = L_b$ at $t(b)$. We naturally require that $L_{\overline{b}} = L_b$ so there is a single length associated with a physical connection in the network, consequently $x_{\overline{b}} = L_b - x_b$. The total length of the graph G is denoted by $\mathcal{L} = \frac{1}{2}\sum_{b=1}^{B} L_b$.

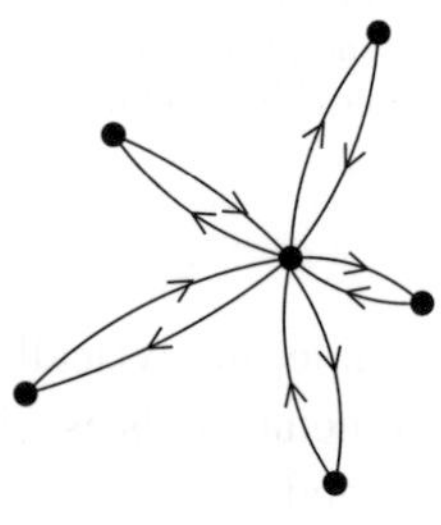

Fig. 1. A star (or hydra) graph.

A function ψ on G is defined by the collection of functions $\{\psi_b(x_b)\}_{b\in\mathcal{B}}$ on the bonds of the graph. We require $\psi_b(x_b) = \psi_{\overline{b}}(L_b - x_b)$ in order that ψ be physically meaningful. The Hilbert space on the graph is

$$L^2(G) = \bigoplus_{b=1}^{B} L^2([0, L_b]) \ . \tag{1}$$

To define a Laplacian on G we consider the operator $-\frac{\mathrm{d}^2}{\mathrm{d}x_b^2}$ on the bonds with matching conditions between functions ψ_b meeting at vertices of G to make the graph Laplacian self-adjoint: for example Neumann like matching conditions require ψ to be continuous at the vertices and $\sum_{b|o(b)=v} \psi_b'(0) =$

0 for all vertices $v \in \mathcal{V}$. All such self-adjoint matching conditions were classified by Kostrykin and Schrader[13] and we follow their scheme. The matching conditions on the whole graph are specified by the pair of $B \times B$ matrices $\mathbb{A}, \mathbb{B}$ via

$$\mathbb{A}\boldsymbol{\psi} + \mathbb{B}\boldsymbol{\psi}' = \mathbf{0} \; , \tag{2}$$

where $\boldsymbol{\psi} = \left(\psi_1(0), \ldots, \psi_B(0)\right)^T$ and $\boldsymbol{\psi}' = \left(\psi_1'(0), \ldots, \psi_B'(0)\right)^T$. These matching conditions define the domain of a self-adjoint Laplace operator iff $\mathrm{rank}(\mathbb{A}, \mathbb{B}) = B$ and $\mathbb{A}\mathbb{B}^\dagger = \mathbb{B}\mathbb{A}^\dagger$.

To find the graph spectrum we want to solve the Laplace eq. on the bonds

$$-\frac{\mathrm{d}^2}{\mathrm{d}x_b^2}\psi_b(x_b) = k^2\psi_b(x_b) \; . \tag{3}$$

In the case of a star graph, Fig. 1, solving Eq. (3) on the bonds and substituting in the matching condition at the center one obtains the well known secular eq. of the star graph $\sum_{b=1}^{B} \tan kL_b = 0$, whose solutions $\{k_j | k_j \geqslant 0\}$ are square roots of the eigenvalues of the Laplacian on the star.

If instead one substitutes solutions of (3) in the general form of matching conditions (2) we find the sequence $\{k_j\}$ are the solutions of a secular eq. $f(k) = 0$ where,

$$f(k) = \det\left(\mathbb{A} + k\mathbb{B}\begin{pmatrix} -\cot(k\mathbf{L}) & \csc(k\mathbf{L}) \\ \csc(k\mathbf{L}) & -\cot(k\mathbf{L}) \end{pmatrix}\right) \; , \tag{4}$$

and $\mathbf{L} = \mathrm{diag}\{L_1, \ldots, L_B\}$. This can be compared to the scattering matrix formulation introduced by Kottos and Smilansky.[3,14]

3. Results

The zeta function of the quantum graph can be formulated as a contour integral applying the argument principle to (4) following the technique introduced in Refs. 15,16, namely

$$\zeta(s) = \sum_{j=0}^{\infty}{}' k_j^{-2s} = \frac{1}{2\pi \mathrm{i}}\int_c z^{-2s}\frac{f'(z)}{f(z)}\,\mathrm{d}z = \frac{1}{2\pi \mathrm{i}}\int_c z^{-2s}\frac{\mathrm{d}}{\mathrm{d}z}\log f(z)\,\mathrm{d}z \; . \tag{5}$$

The prime on the sum denotes the exclusion of zero modes if present. Generically the poles of f are the whole of the set $\{m\pi/L_b | m \in \mathbb{Z}, b = 1, \ldots, B\}$ and this is the case we consider.[a] The contour c is then chosen so as to

[a] It is possible for particular choices of matching conditions $\mathbb{A}, \mathbb{B}$ and bond lengths $\{L_b\}$ that individual poles cancel in the determinant, however, a small perturbation of the bond lengths removes this.

include the zeros of f on the positive real axis and exclude the poles. Making a contour transformation to an integral along the imaginary axis closed with a semicircular arc on the right we obtain an expression for $\zeta(s)$ valid in the strip $0 < \Re s < 1$,

$$\zeta(s) = \zeta_R(2s) \sum_{b=1}^{B} \left(\frac{\pi}{L_b}\right)^{-2s} + \frac{\sin \pi s}{\pi} \int_0^\infty t^{-2s} \frac{\mathrm{d}}{\mathrm{d}t} \log \hat{f}(t)\, \mathrm{d}t \ , \tag{6}$$

$$\hat{f}(t) = \det\left(\mathbb{A} - t\mathbb{B}\begin{pmatrix} \coth(t\mathbf{L}) & -\mathrm{csch}(t\mathbf{L}) \\ -\mathrm{csch}(t\mathbf{L}) & \coth(t\mathbf{L}) \end{pmatrix}\right) , \tag{7}$$

where ζ_R is the Riemann zeta function.

Regularized formulations of the vacuum energy, and the spectral determinant as well as the heat kernel asymptotics are obtained directly from the zeta function. The vacuum energy, formally $E = \frac{1}{2}\sum_{j=0}^{\infty}{}' k_j$ where k_j^2 is an eigenvalue of the Laplacian, is defined as $E_c = \frac{1}{2}\zeta(-1/2)$. Differentiating with respect to the L_β, the length of the bond β, we obtain the Casimir force on the bond,

$$F_c^\beta = \frac{\pi}{24L_\beta^2} + \frac{1}{\pi}\int_0^\infty \frac{\partial}{\partial L_\beta} \log \hat{f}(t)\, \mathrm{d}t \ . \tag{8}$$

The spectral determinant is formally the product of the eigenvalues $\prod_{j=0}^{\infty}{}' k_j^2$. The zeta function representation (6) allows a direct evaluation of the regularized spectral determinant, as

$$\det{}'(-\triangle) = \exp\big(-\zeta'(0)\big) = \frac{2^B \hat{f}(0)}{c_N \prod_{b=1}^{B} L_b} \ . \tag{9}$$

Here, c_N denotes the first non-zero coefficient in the $t \to \infty$ asymptotic expansion of

$$\hat{f}(t) \sim \det(\mathbb{A} - t\mathbb{B}) = \det \mathbb{B}\, t^{2B} + c_{2B-1} t^{2B-1} + \cdots + c_1 t + \det \mathbb{A} \ . \tag{10}$$

So $c_N = \det \mathbb{B}$ when $\det \mathbb{B} \neq 0$. The factor of c_N in (9) is introduced by the analytic continuation of $\zeta(s)$ to $s = 0$.

The heat kernel for a quasi one-dimensional system has an expansion for $t \to 0$ of the form, $K(t) = \sum_{j=1}^{\infty} \mathrm{e}^{-k_j^2 t} \sim \sum_{\ell=0,1/2,1,\ldots}^{\infty} a_\ell t^{\ell-1/2}$. The heat kernel coefficients are related to the zeta function[17] by $a_\ell = \mathrm{Res}(\zeta(s)\Gamma(s))|_{s=1/2-\ell}$. Subtracting sufficiently many terms in the $t \to \infty$ expansion of $\hat{f}$, the zeta function (6) can be extended as far left of the imaginary axis as required and we obtain the full asymptotic expansion of the heat kernel of the general graph Laplacian,

$$K(t) \sim \frac{\mathcal{L}}{\sqrt{4\pi t}} + \frac{N}{2} - \sum_{k=1,3/2,2,\dots}^{\infty} \frac{b_{2k-1}}{\Gamma\left(k-\frac{1}{2}\right)} t^{k-1/2} . \tag{11}$$

The coefficients b_n are obtained from the expansion of $\log \hat{f}(t)$ at infinity,

$$\log \hat{f}(t) \sim N \log t + \log c_N + \sum_{n=1}^{\infty} \frac{b_n}{t^n} . \tag{12}$$

For a given set of matching conditions defined by matrices $\mathbb{A}, \mathbb{B}$ the coefficients b_n can be easily obtained with a computer algebra package.

Acknowledgments

The authors would like to thank G Berkolaiko, JP Keating, P Kuchment, J Marklof, R Piziak and B Winn for helpful suggestions. KK is supported by National Science Foundation grant PHY–0554849 and JMH is supported by National Science Foundation grant DMS–0604859.

References

1. S. Gnutzmann and U. Smilansky, *Adv. Phys.* **55**, 527–625 (2006).
2. O. Bohigas, M.-J. Giannoni and C. Schmit, *Phys. Rev. Lett.* **52**, 1-4 (1984).
3. T. Kottos and U. Smilansky, *Phys. Rev. Lett.* **79**, 4794–4797 (1997).
4. P. Kuchment, *Waves Random Media* **12**, R1–24 (2002).
5. S. Fulling, L. Kaplan and J. H. Wilson, *Phys. Rev. A* **76**, 012118 (2007).
6. G. Berkolaiko, J. M. Harrison and J. H. Wilson, *J. Phys. A: Math. Theor.* **42**, 025204 (2009).
7. J. Desbois, *J. Phys. A: Math. Gen.* **33**, L63–L67 (2000).
8. J. Desbois, *Eur. Phys. J. B* **24**, 261–266 (2001).
9. L. Friedlander, in: G. Berkolaiko, R. Carlson, S. A. Fulling, and P. Kuchment (Eds) *Quantum Graphs and Their Applications*, Contemporary Mathematics, **415**, 151–160 (AMS, 2006).
10. V. Kostrykin, J. Potthoff and R. Schrader, in F. Germinet and P. D. Hislop (Eds), *Adventures in mathematics,* Contemporary Mathematics **447**, 175–198 (AMS, 2007).
11. J. Bolte and S. Endres, *Ann. Henri Poincare* **10**, 189–223 (2009).
12. J. M. Harrison and K. Kirsten, "Zeta functions of quantum graphs," (Preprint) `arXiv:0911.2509`.
13. V. Kostrykin and R. Schrader, *J. Phys. A: Math. Gen.* **32**, 595–630 (1999).
14. T. Kottos and U. Smilansky, *Ann. Phys.* **274**, 76–124 (1999).
15. K. Kirsten and A. J. McKane, *Ann. Phys.* **308**, 502–527 (2003).
16. K. Kirsten and A. J. McKane, *J. Phys. A: Math. Gen.* **37**, 4649–4670 (2004).
17. R.T. Seeley, in *Singular Integrals, Chicago. Proc. Sympos. Pure Math.* **10**, 288–307 (1966), (AMS, 1968).

EFFICIENT EVALUATION OF EFFECTIVE ACTION IN RADIAL BACKGROUNDS

HYUNSOO MIN

Department of Physics, University of Seoul,
Seoul, 130-743, Korea
E-mail: hsmin@dirac.uos.ac.kr

Recently a new caculational scheme for effective actions in radial background fields was developed. The effective action is expressed as an infinite sum of partial-wave contributions, using the rotational symmetry of the system. The sum becomes convergent after proper regularization and renormalization, but the rate of convergence is rather slow. We introduce a systematic way of accelerating the rate of convergence. This method is based on a radial WKB series in the angular momentum cut-off. We demonstrate the power of this scheme by applying it to the calculation of instanton determinant in QCD.

Keywords: Effective action, partial wave, functional determinant, WKB, renormalization.

1. Introduction

The one-loop effective action plays a central role in quantum field theories. However, the explicit evaluation of it is usually very difficult and analytic results are known only in limited cases.

Recently there has been a significant progress in this problem when background fields have radial symmetry, using the partial wave analysis.[1,2] Introduction of a cut-off in the partial wave sum and separation of the sum into two parts is a key of the idea. A radial WKB expansion which is uniformly valid for the large angular momentum part is developed. Proper renormalization counterterms are taken into account there. Combination of the leading terms of this WKB expansion and the contributions from the low angular momentum part provides us a finite renormalized value in the limit of large cutoff value. It also turns out that the inclusion of higher order terms in the uniform radial WKB expansion greatly improves the rate of convergence in the infinite sum.[3]

In this work, we present a general scheme of this method and some results for the one-loop effective action in the case of QCD instanton.

2. Partial Wave Method

The one-loop effective action for a complex scalar field is defined in terms of functional determinants formally as

$$\Gamma = \ln\left(\frac{\det[\mathcal{M}+m^2]}{\det[\mathcal{M}^{\text{free}}+m^2]}\right) \tag{1}$$

where $\mathcal{M} = -\partial^2 + V(r)$, $\mathcal{M}^{\text{free}} = -\partial^2$. Here $V(r)$ is a radial potential vanishing sufficiently fast at infinity. Using the partial wave analysis, the effective action can be written as

$$\Gamma = \sum_{l=0}^{\infty} g_l \Omega_l, \quad \Omega_l = \ln\left(\frac{\det[\mathcal{M}_l+m^2]}{\det[\mathcal{M}_l^{\text{free}}+m^2]}\right). \tag{2}$$

Here l denotes the angular momentum quantum number appropriate to each partial wave and $g_l = (2l+d-2)(l+d-3)!/l!(d-2)!$ is the degeneracy factor.[4] The associated radial differential operator$\mathcal{M}_l$ is given by $\mathcal{M}_l = -\partial^2 + \mathcal{V}_l$ with the effective potential

$$\mathcal{V}_l = \frac{(l+\frac{d-3}{2})(l+\frac{d-1}{2})}{r^2} + V(r), \tag{3}$$

and $\mathcal{M}_l^{\text{free}} = \mathcal{M}_l$ with $V=0$.

The individual radial determinant ratio Ω_l in (2) can be evaluated easily by using the Gel'fand-Yaglom method[5]

$$\Omega_l = \lim_{r\to\infty} \ln \frac{\psi_l(r)}{\psi_l^{\text{free}}(r)}, \tag{4}$$

where the wave functions $\psi_l(r)$ and $\psi_l^{\text{free}}(r)$ are the solutions of $(\mathcal{M}_l + m^2)\psi_l = 0$ and $(\mathcal{M}_l^{\text{free}} + m^2)\psi_l^{\text{free}} = 0$ respectively and both of them behave as $r^{l+(d-2)/2}$ near $r=0$.

The infinite sum in (2) is formally divergent. This problem is related to renormalization. There exists an elegant way to extract the renormalized quantity Γ_{ren} from Γ.[1,4] Another problem is the slow rate of convergence of the l-sum. Both of these problems can be solved by splitting the sum into two pieces: the low angular momentum part Γ_{L} and the high angular momentum part Γ_{H} as

$$\Gamma_{\text{ren}} = \Gamma_{\text{L}} + \Gamma_{\text{H}} = \sum_{l=0}^{L} g_l\Omega_l + (\sum_{l>L}^{\infty} g_l\Omega_l + \delta\Gamma). \tag{5}$$

Here we introduce a cut-off L and $\delta\Gamma$ denotes the 'conventional' renormalization counterterm. Each Ω_l in the part $\Gamma_{\rm L}$ can be evaluated using the above Gel'fand-Yaglom method. Since Ω_l behaves like $\sim 1/l$ for large l and g_l increases as l^{d-2}, $\Gamma_{\rm L}$ behaves like L^{d-2} for $d \geq 2$ in the large L limit. (This reveals the divergent structures in the formal expression in (2)). As for the part $\Gamma_{\rm H}$, we can evaluate it *analytically* in a uniform asymptotic series of the form

$$\Gamma_{\rm H} = \int_0^\infty dr \left(Q_{\log} + \sum_{n=2-d}^{\infty} Q_{-n} L^{-n} \right), \tag{6}$$

where Q_{-n}'s may have an implicit L dependence of $O(L^0)$ and $Q_{\log}$ behaves as $O(\ln L)$ in the large L limit. One can find explicit forms of the Q's with derivation in a recent work.[3,6]

Now let us consider the case with very large value of L. Note that, as $L \to \infty$, unsuppressed terms in the expansion (6) may grow but they match precisely the divergences coming from $\Gamma_{\rm L}$ with the opposite sign. Hence, combining these two and taking the $L \to \infty$ limit yields $\Gamma_{\rm ren}$:

$$\Gamma_{\rm ren} = \lim_{L\to\infty} \left[\Gamma_{\rm L} + \Gamma_{\rm H}^{(1)} \right], \quad \Gamma_{\rm H}^{(1)} = \int_0^\infty dr \left(Q_{\log} + \sum_{n=0}^{d-2} Q_n L^n \right). \tag{7}$$

In principle, one can use this expression to obtain the renormalized effective action. But we still have a practical problem related with the convergence rate.

With a finite value of L, we can write the following formula for $\Gamma_{\rm ren}$:

$$\Gamma_{\rm ren} = \Gamma_{\rm L} + \Gamma_{\rm H}^{(1)} + \int_0^\infty dr \sum_{n=1}^{N} Q_{-n} \frac{1}{L^n} + O\left(\frac{1}{L^{N+1}} \right), \tag{8}$$

where N refers to the order of truncation. In this formula the error is indicated by the last term and it is totally under control. It is apparent that we get a more accurate value of $\Gamma_{\rm ren}$ by taking into account more $\frac{1}{L}$-suppressed terms for a given value of the cutoff L. In the subsequent sections, by applying this method to the evaluation of the instanton determinant, we demonstrate the power of our method.

3. Instanton determinant with $m = 0$

We apply our method to the case of the instanton determinant with $m = 0$ where an exact computation is possible.[7] Now the partial derivative ∂_μ in $\mathcal{M}$ should be replaced with $D_\mu = \partial_\mu - iA_\mu$. Here the background field $A_\mu =$

$A^a_\mu \tau^a/2$ is an SU(2) single instanton solution and $A^a_\mu(x) = 2\eta_{\mu\nu a}x_\nu f(r)$ with $f(r) = 1/(1+r^2)$, taking the size parameter $\rho = 1$. It is convenient to introduce the total angular momentum $\vec{J} = \vec{L} + \vec{T}$. Then each partial wave is labeled by (l, j) with $j = l \pm 1/2$ and l takes half integral values. The effective potential has the form:

$$\mathcal{V}_{l,j} = \frac{(2l+\frac{1}{2})(2l+\frac{3}{2})}{r^2} + 4f(r)[j(j+1) - l(l+1) - 3/4] + 3r^2 f(r)^2. \quad (9)$$

We may exactly solve the radial ODE, $\mathcal{M}_{(l,j)}\psi_{l,j} = 0$ and get a master formula of Ω_l:

$$\Omega_l = \Omega_{l,l+\frac{1}{2}} + \Omega_{l+\frac{1}{2},l} = \ln\frac{2l+1}{2l+2}. \quad (10)$$

The WKB large-L expansion of $\Gamma_{\rm H}$ is evaluated as

$$\Gamma_{\rm H} = \Gamma_{\rm H}^{(1)} - \frac{1}{6L} + +\frac{119}{1440L^2} - \frac{13}{240L^3} + \frac{1597}{40320L^4} - \frac{103}{3360L^5} + O(L^{-6}),$$
$$\Gamma_{\rm H}^{(1)} = \frac{127}{72} - \frac{1}{3}\ln 2 - \frac{1}{6}\ln L + 4L + 2L^2. \quad (11)$$

Then the renormalized effective action is (setting the renormalization parameter $\mu = 1$)

$$\Gamma_{\rm ren} = \lim_{L\to\infty}\Big(\sum_{l=0,1/2,\cdots,L} g_l \ln\frac{2l+1}{2l+2} + \Gamma_{\rm H}^{(1)}\Big) \equiv \alpha, \quad (12)$$

with $\alpha = -(5/72) - 1/6\ln 2 - 2\zeta'(-1) = 0.145873312863\ldots$. Taking a finite value of L, we get the approximate value of the effective action:

$$\Gamma_{\rm ren}(L) = \Gamma_{\rm L} + \Gamma_{\rm H}(\text{truncated}). \quad (13)$$

When $L = 20$, comparing the exact and approximate values we find that the difference is just 3.7×10^{-10}. This result clearly shows the elegance of our method.

4. Numerical values and asymptotic expansions: $m \neq 0$

When $m \neq 0$ it is no longer possible to find a master formula for Ω_l. Related ODE's must be solved in a numerical way. We have found exact numerical values of the effective action for various values of m.[1]

There are other approximate but analytic expressions for the effective action. The large mass expansion is directly obtained from the Schwinger-DeWitt expansion as described in the work[8]

$$\Gamma_{\rm ren} = -\frac{\ln m}{6} - \frac{1}{75m^2} - \frac{17}{735m^4} + \frac{232}{2835m^6} - \frac{7916}{148225m^8} + \cdots \quad (14)$$

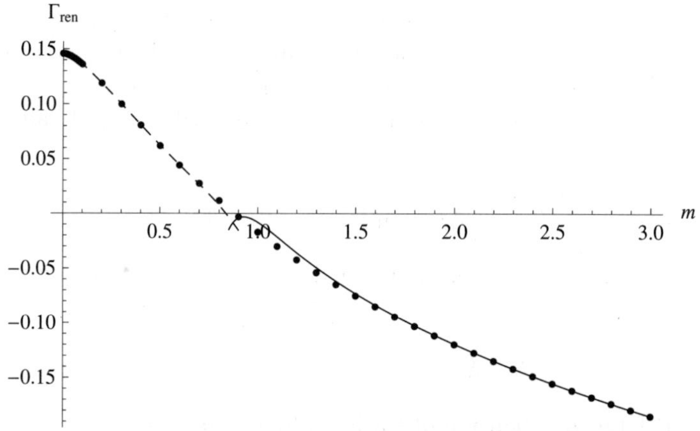

Fig. 1. Plot of our small mass expansion (dashed line) and the large mass expansion (solid line) together with the exact numerical result (dots).

In a recent paper,[9] the small mass expansion is reported as

$$\Gamma_{\text{ren}} = \alpha(1/2) + \frac{m^2}{2}(\ln m + \gamma + 1/2 - \ln 2) - \frac{1}{4}m^4 \ln^2 m$$
$$+m^4(\frac{\ln m}{2}[1/2 - \gamma + \ln 2] - 0.382727) + O(m^6) \tag{15}$$

In Fig. 1, we have plotted these two expansions and the exact numerical values of the effective action Γ_{ren} as a function of m.

Acknowledgments

This work is supported in part by University of Seoul 2007 Research Funding Program.

References

1. G. V. Dunne, J. Hur, C. Lee and H. Min, Phys. Rev. Lett. **94**, 072001 (2005); Phys. Rev. D **71**, 085019 (2005).
2. G. V. Dunne, J. Hur and C. Lee, Phys. Rev. D **74**, 085025 (2006).
3. G. V. Dunne, J. Hur, C. Lee and H. Min, Phys. Rev. D **77**, 045004 (2008).
4. G. V. Dunne and K. Kirsten, J. Phys. A. **39**, 11915 (2006).
5. I. M. Gelfand and A. M. Yaglom, J. Math. Phys. **1**, 48 (1960); K. Kirsten and A. J. McKane, Annals Phys. **308**, 502 (2003).
6. J. Hur and H. Min, Phys. Rev. D **77**, 125033 (2008).
7. G. 't Hooft, Phys. Rev. D **14**, 3432 (1976); [Errata *ibid.* D **18**, 2199 (1978)].
8. O. K. Kwon, C. Lee and H. Min, Phys. Rev. D **62**, 114022 (2000)
9. J. Hur, C. Lee and H. Min, Arxiv:hep-th/0909.5515 (2009).

LOCAL MEASURES OF CONVEX SURFACES INDUCED BY THE WIENER MEASURE OF PATHS

M. SCHADEN

Department of Physics, Rutgers University, 101 Warren Street, Newark, NJ 07102, USA
E-mail: mschaden@rutgers.edu

The Wiener measure induces a measure of closed, convex, $d-1$-dimensional, Euclidean (hyper-)surfaces that are the convex hulls of closed d-dimensional Brownian bridges. I present arguments and numerical evidence that this measure, for odd d, is generated by a local classical action of length dimension 2 that depends on geometric invariants of the $d-1$-dimensional surface only.

Keywords: Probability measures; convex surfaces; Wiener measure.

1. Introduction

It was recently[1] observed that the spectral function $\phi_{\mathfrak{D}}(\beta)$, or trace of the heat kernel $\mathfrak{K}_{\mathfrak{D}}(\beta)$, of a free massless scalar field that vanishes on the boundary $\partial\mathfrak{D}$ of a d-dimensional convex domain $\mathfrak{D}$,

$$\phi_{\mathfrak{D}}(\beta) = \sum_{n\in\mathbb{N}} e^{-\beta\lambda_n/2} = \int_{\mathfrak{D}} d\mathbf{x}\mathfrak{K}_{\mathfrak{D}}(\mathbf{x},\mathbf{x};\beta), \tag{1}$$

naturally induces a measure on convex surfaces that are the convex hulls of closed Brownian bridges. Here $\{\lambda_n, n\in\mathbb{N}\}$ is the spectrum of the (negative) Laplacian Δ with Dirichlet boundary conditions on $\partial\mathfrak{D}$ and the heat kernel $\mathfrak{K}_{\mathfrak{D}}(\mathbf{x},\mathbf{y};\beta)$ solves the diffusion equation, $\partial_\beta\mathfrak{K}_{\mathfrak{D}}(\mathbf{x},\mathbf{y};\beta) = \frac{1}{2}\Delta_x\mathfrak{K}_{\mathfrak{D}}(\mathbf{x},\mathbf{y};\beta)$, with initial condition $\mathfrak{K}_{\mathfrak{D}}(\mathbf{x},\mathbf{y};0) = \delta(\mathbf{x}-\mathbf{y})$ and vanishes on $\partial\mathfrak{D}$.

The Feynman-Kac theorem[2] implies that $\phi_{\mathfrak{D}}(\beta)$ can be expressed by the probability $\mathcal{P}[\ell_\beta(\mathbf{x})\subset\mathfrak{D}]$ that a standard Brownian bridge (SBB) $\ell_\beta(\mathbf{x}) = \{\mathbf{x}+\mathbf{B}_\tau, 0\le\tau\le\beta; \mathbf{B}_0 = \mathbf{B}_\beta = \mathbf{0}\}$ starting at $\mathbf{x}$ and returning to $\mathbf{x}$ in "proper time" β does not exit $\mathfrak{D}$,

$$\phi_{\mathfrak{D}}(\beta) = \int_{\mathfrak{D}} \frac{d\mathbf{x}}{(2\pi\beta)^{d/2}}\mathcal{P}[\ell_\beta(\mathbf{x})\subset\mathfrak{D}] \ . \tag{2}$$

A numerical approximation to a SBB derives from a discrete random walk $\{\mathbf{X}_n, n = 0,\ldots,N; \mathbf{X}_0 = \mathbf{0}\}$ of N independent steps with average displacement $\langle \mathbf{X}_{n+1} - \mathbf{X}_n \rangle = \mathbf{0}$ and variance $\langle (\mathbf{X}_{n+1} - \mathbf{X}_n)^2 \rangle = \beta d/N$. $\ell_\beta(\mathbf{x})$ then is approximated by $\{\mathbf{Y}_n, n = 0,\ldots,N; \mathbf{Y}_n = \mathbf{X}_n + \mathbf{x} - n\mathbf{X}_N/N\}$. $\mathbf{Y}_N = \mathbf{Y}_0 = \mathbf{x}$ and the steps remain uncorrelated and of variance $d\beta/N$ by construction. The continuum limit is obtained by letting $N \to \infty$.

To ascertain whether a SBB is entirely within a bounded domain generally requires examining all of its N points. However, in the case of *convex* bounded domains $\mathfrak{D}$, it suffices to check whether the smallest convex surface enclosing all points of the SBB, its *convex hull*, lies entirely within $\mathfrak{D}$. Of advantage is that the convex hull of a SBB has exponentially fewer vertices (see Fig.1) than the original loop has points and that its construction is independent of the convex domain $\mathfrak{D}$. One thus can generate and store a large number of convex hulls of SBBs once and for all and only has to check whether the few vertices of the hulls lie within a particular convex domain $\mathfrak{D}$. The computer time required to construct the hull of a set of N points is $\mathcal{O}(N \ln N)$ only[3] and thus is comparable to the time required to examine the N points of a loop – but the convex hull can be used for any convex domain. The ensemble of convex hulls of SBBs furthermore shares all global symmetries of the Wiener measure.

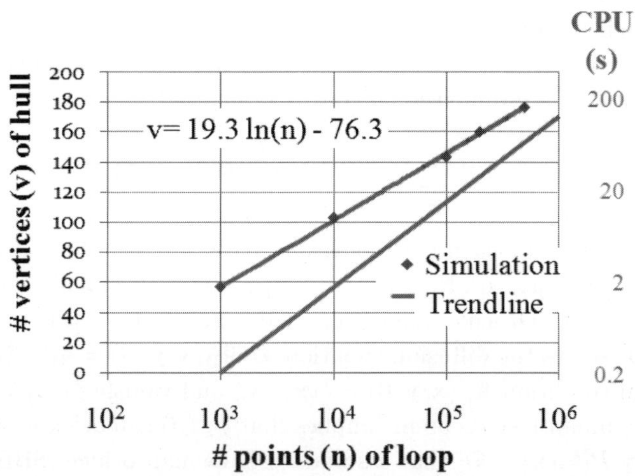

Fig. 1. Average number of vertices v (left axis) of the hull of a 3-dimensional SBB with n points (red). The trendline is given by the inset formula. The right axis and lower (green) line give the average CPU-time required to compute the hulls on a laptop with a 2GHz processor.

2. Measuring convex surfaces with the Wiener measure

One can define open sets of convex domains $\mathfrak{C} = \{C \text{ is convex}\}$ that include all convex domains that, much like russian dolls, fit into a particular one. One can similarly define open sets of the boundaries of convex domains,

$$\mathcal{M}(\mathfrak{D}) := \{\partial C; C \subset \mathfrak{D} \in \mathfrak{C}\} = \{\mathcal{H}[\ell_\beta(\mathbf{x})], \ell_\beta(\mathbf{x}) \subset \mathfrak{D}\} , \tag{3}$$

where $\mathcal{H}[\ell_\beta(\mathbf{x})]$ is the convex hull of the SBB $\ell_\beta(\mathbf{x})$ based at $\mathbf{x} \in \mathfrak{D}$. In view of the correspondence in Eq.(3) it is natural to assign the set $\mathcal{M}(\mathfrak{D})$ the translation, $O(d)$ and scale- invariant measure provided by the spectral function,

$$\mu_\beta[\mathcal{M}(\mathfrak{D})] := \phi_{\mathfrak{D}}(\beta) = \int_{\mathfrak{D}} \frac{d\mathbf{x}}{(2\pi\beta)^{d/2}} \mathcal{P}[\ell_\beta(\mathbf{x}) \subset \mathfrak{D}] . \tag{4}$$

Although $\tilde{\mathcal{M}}(\mathfrak{D})$ does not refer to a scale, the induced probability measures on convex surfaces defined by Eq.(4) depend on the parameter β, of length dimension 2, that characterizes the SBBs. The measures μ_β may be extended to the entire Borel set by using the definition of Eq.(4) and properties of the probability space of Brownian loops[4] as well as the fact that $\bigcap_{i>0} \mathcal{M}(\mathfrak{D}_i) = \mathcal{M}(\cap_{i>0}\mathfrak{D}_i)$.

3. Locality

These measures on convex surfaces are *local* in the sense that they can be generated by a Metropolis-like algorithm[5] involving only local updates of convex surfaces. This perhaps is less surprising upon recalling that convexity is equivalent to demanding that the scalar curvature of the surface is positive everywhere. The latter constraint is *local.* The following construction of the ensemble of convex surfaces with a measure proportional to Eq.(4) is *not* the most efficient, but does imply that these measures on convex surfaces in $d > 1$ are generated by a local action.

Note first that the Wiener measure is local in this sense, because the relative frequency of two SBBs, ℓ'_β and ℓ_β, that differ only in the immediate vicinity of $\mathbf{Q} = \mathbf{x}_k$ is,

$$\begin{aligned} \frac{d\mathcal{P}[\ell'_\beta]}{d\mathcal{P}[\ell_\beta]} &\sim 1 + N(\mathbf{x}'_k - \mathbf{x}_k) \cdot (\mathbf{x}_{k+1} - 2\mathbf{x}_k + \mathbf{x}_{k-1})/\beta \\ &= 1 + \delta_Q S \text{ with } S = \int_0^\beta \dot{\mathbf{x}}^2 d\tau; \quad \mathbf{x}(\beta) = \mathbf{x}(0) \end{aligned} \tag{5}$$

The local classical action S in this case may be reconstructed explicitly, and a Metropolis algorithm[5] may be designed that proposes local updates of the SBB with precisely this probability, i.e. the algorithm accepts every proposal.

Now consider the corresponding two hulls $h = \mathcal{H}[\ell_\beta]$ and $h' = \mathcal{H}[\ell'_\beta]$. There are four distinct possibilities:(i) $\mathbf{x}_k$ is a vertex of h and $\mathbf{x}'_k$ is a vertex of h'; (ii) $\mathbf{x}_k$ is a vertex of h but $\mathbf{x}'_k$ is not a vertex of h'; (iii) $\mathbf{x}_k$ is not a vertex of h but $\mathbf{x}'_k$ is a vertex of h' and (iv) $\mathbf{x}_k$ is not a vertex of h and $\mathbf{x}'_k$ is not a vertex of h'. Since only a single point of the underlying loop is moved, (iv) implies that $h' = h$. In all other cases, $h' \neq h$ and at least one vertex has changed. The update is almost local in that the average distance δ from the point $\mathbf{Q} = (\mathbf{x}_k + \mathbf{x}'_k)/2$ on the convex surface that is affected is of order $\delta \sim \sqrt{2\bar{R}\Delta} = \mathcal{O}(\sqrt{\beta}N^{-1/4})$. [The average radius of curvature of the convex surface $\bar{R} = \mathcal{O}(\sqrt{\beta})$ does not depend on N and the average displacement of the updated point is $\Delta\mathbf{x} = \sqrt{\langle(\mathbf{x}' - \mathbf{x})^2\rangle} = \mathcal{O}(\sqrt{\beta/N})$.] This region thus decreases with increasing refinement N of the SBB like $N^{(1-d)/4}$ and the update becomes local in the continuum limit. As can be seen in Fig. 1 the number of vertices v of the hull of a SBB is proportional to $\ln(N)$ and thus, for large N, almost always at most one vertex of the hull is updated. For large N case (iv) occurs most frequently and an algorithm that updates large sections of the underlying loop is far more efficient than this local one.

However, I argue that the possibility of constructing a local algorithm implies that rare events are generated by a local classical action. Since the probability for generating a SBB of large extent s decreases exponentially $\sim \exp[\mathcal{O}(s^2/\beta)]$, the probability for generating a convex hull much larger than average should also asymptotically decrease as

$$\mathcal{P}[\mathfrak{D} \subset \mathfrak{H}[\ell_\beta]] \underset{\beta\sim 0}{\longrightarrow} \propto e^{-S_d(\mathfrak{D})/\beta}\,, \tag{6}$$

where $S_d(\mathfrak{D})$ is a classical action of length dimension 2 that can only depend on geometrical characteristics of $\mathfrak{D}$. Locality of the measure in the continuum limit implies that $S_d(\mathfrak{D})$ is an integrated density that depends on the metric and its derivatives only. The local constraint that $\mathfrak{D}$ have positive curvature everywhere is not holomorphic and therefore is not reflected in the form of $S_d(\mathfrak{D})$, much as the lower bound for the motion of a bouncing ball does not appear in its action.

4. Numerical results

d = 1 The convex hull of a SBB in 1 dimension consists of two points, the maximum and minimum values of the SBB. The only translation invariant action of length dimension two is the square of the distance between these.

Explicit calculation gives,[6]

$$\mathcal{P}[|\mathfrak{H}[\ell_\beta]| < s] = 1 - 2\sum_{n=1}^{\infty}((2sn)^2/\beta - 1)e^{-(2sn)^2/(2\beta)} \underset{\beta\sim 0}{\longrightarrow} 1 - 8s^2/\beta e^{-2s^2/\beta}, \tag{7}$$

where $|\mathfrak{H}[\ell_\beta]|$ denotes the extent of the convex hull, that is the distance between its two points. The asymptotic behavior of Eq.(7) has the expected form with an action $S_1([a,b]) = 2(a-b)^2$ that depends on the two points of the hull only. However, $S_1(\mathfrak{D})$ is not local and the correlation matrix is constant due to the failure of the previous geometrical argument for a discrete set of points. As shown in the talk, the agreement of numerical simulations with Eq.(7) is excellent.

d = 2 The convex hull of a SBB in two dimensions numerically is a piecewise linear, closed, one-dimensional curve, whose length may be called the perimeter of the SBB. It is a local geometrical invariant of dimension 1. Eq.(6) suggests that the measure for rare events will not scale with the perimeter (see Fig. 2), but rather with the area enclosed by it, $S_2(\mathfrak{D}) \propto A = \int_{\mathfrak{D}} d^2x$. The measure also scales with the local quantity of dimension 2 that is the sum of the squares of the lengths of the piecewise linear sections of the curve $P^2 = \sum_i s_i^2$, but $P^2 \to 0$ as $N \to \infty$ and does not survive the continuum limit. The numerical evidence shown in Fig 2a supports this identification.

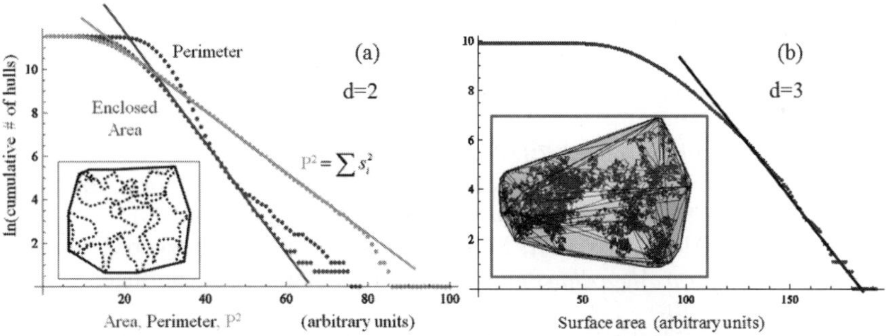

Fig. 2. Logarithm of the cumulative statistics (a) for the enclosed area (red), the perimeter (blue) and the invariant P^2 (green) of the convex hulls of 10^5 SBBs with $N = 10^4$ points in $d = 2$ dimensions and (b) the area of the convex hulls of 10^4 SBBs with $N = 10^4$ steps in $d = 3$ dimensions. The exponential decrease in rare events with the enclosed area and with P^2 in $d = 2$ and with the surface area in $d = 3$ is indicated. The insets show sketches of SBBs and their convex hulls in $d = 2$ and $d = 3$ dimensions.

d = 3 The convex hull of a SBB in $d = 3$ is a triangulated 2-dimensional surface. The only local action of length dimension 2 is its area $S_3(\mathfrak{D}) \propto \int_{\partial D} d^2x \sqrt{\det g_{ik}(\mathbf{x})}$, where $g_{ik}(\mathbf{x})$ is the metric *on* $\partial\mathfrak{D}$ (see Fig. 2b).

5. Outlook and a conjecture for $d > 3$ dimensions

Numerical evidence in four and higher dimensions were not presented, because the construction of hulls and local actions in this case is considerably more involved and has no direct application to Casimir effects. A dimensional argument, however, suggests that the local classical action for *even*-dimensional convex surfaces is determined by the metric on the surface and its (tangential) derivatives only. Local classical actions that asymptotically describe the measures for *odd*-dimensional convex surfaces on the other hand necessarily depend on the bulk metric (see Fig. 2a), since the number of derivatives in local geometric invariants is always even. One thus is led to conjecture that $S_5(\mathfrak{D}) \propto \int_{\partial D} d^4x \sqrt{\det g_{ik}(\mathbf{x})} R(\mathbf{x})$, may be intrinsically defined in terms of the metric and associated Ricci curvature scalar $R(\mathbf{x})$ of the 4-dimensional convex hyper-surface embedded in 5-dimensional flat Euclidean space. The nontrivial intrinsic local classical action in odd $d = 7, 9, \ldots$ are sums of higher derivative terms that are not uniquely determined by dimensional considerations alone.

Acknowledgments

I would like to thank the organizers of QFEXT09 for superbly managing a very engaging conference. This work was supported by the National Science Foundation with Grant No. 0902054.

References

1. M. Schaden, *Phys. Rev. A* **79**, 052105 (2009).
2. R.P. Feynman, *Rev. Mod. Phys.***20**, 367 (1948); M. Kac, *Trans. Am. Math. Soc.***65**, 1 (1949); *ibid Amer. Math. Monthly* **73** Part II, 1 (1966).
3. M. de Berg, M. van Kreveld, M. Overmars, and O. Schwarzkopf, *Computational Geometry: Algorithms and Applications*, 2nd edition (Springer-Verlang, New York, 2000).
4. B. Øksendahl, *Stochastic Differential Equations* 5th edition (Springer-Verlag, New York, 2000)
5. N. Metropolis, A.W. Rosenbluth, M.N. Rosenbluth, A.H. Teller and E. Teller,*J. Chem. Phys.***21**, 1087 (1953); W.K. Hastings, *Biometrika***57**, 97 (1970).
6. M. Schaden, *Phys. Rev. Lett.***102**, 060402 (2009).

NEGATIVE ENERGY DENSITIES IN QUANTUM FIELD THEORY

L. H. FORD

Institute of Cosmology,
Department of Physics and Astronomy,
Tufts University, Medford, MA 02155, USA
E-mail: ford@cosmos.phy.tufts.edu

Quantum field theory allows for the suppression of vacuum fluctuations, leading to sub-vacuum phenomena. One of these is the appearance of local negative energy density. Selected aspects of negative energy will be reviewed, including the quantum inequalities which limit its magnitude and duration. However, these inequalities allow the possibility that negative energy and related effects might be observable. Some recent proposals for experiments to search for sub-vacuum phenomena will be discussed. Fluctuations of the energy density around its mean value will also be considered, and some recent results on a probability distribution for the energy density in two dimensional spacetime are summarized.

1. Introduction

Although the local energy density for the electromagnetic and other known fields is positive at the classical level, this need not be the case at the quantum level. When defining the expectation value of the stress tensor in quantum field theory, a subtraction is needed, after which a negative value can result. In the case of empty Minkowski spacetime, we take the vacuum state as the zero of energy density. Negative values for the mean energy density are examples of sub-vacuum phenomena, where the effects of vacuum fluctuations have effectively been suppressed. We will review some examples of this effect in the Casimir effect and in non-classical quantum states, and also review the quantum inequalities which limit sub-vacuum phenomena. In Sect. 5, some proposed laboratory experiments to detect sub-vacuum effects will be discussed. Finally we turn to the topic of stress tensor fluctuations, whereby negative energy can arise by a quantum fluctuation even when the expectation value of the energy is positive or zero.

Some recent results on the probability distribution for such fluctuations will be discussed.

2. Negative Energy Density in the Casimir Effect

Simply from Casimir's result for the force per unit area[1] between a pair of parallel perfectly reflecting plates, one can construct the entire stress tensor, using conservation, tracelessness and symmetry arguments.[2,3] The result, in units where $\hbar = c = 1$, is

$$T_{\mu\nu} = \begin{pmatrix} T_{00} & 0 & 0 & 0 \\ 0 & T_{xx} & 0 & 0 \\ 0 & 0 & T_{yy} & 0 \\ 0 & 0 & 0 & T_{zz} \end{pmatrix} = \frac{\pi^2}{720\, a^4} \begin{pmatrix} -1 & 0 & 0 & 0 \\ 0 & 1 & 0 & 0 \\ 0 & 0 & 1 & 0 \\ 0 & 0 & 0 & -3 \end{pmatrix} \tag{1}$$

where the plates are separated by a distance a in the z-direction. Thus there is a constant negative energy density between the plates.

The case of plates of finite reflectivity is more complicated, and the local stress tensor can no longer be recovered simply from knowledge of the force per unit area. The reason for this is that symmetry under Lorentz boosts parallel to the plates, a key ingredient in the argument leading to Eq. (1), no longer holds. Helfer and Lang[4] have noted that now there could be a positive self-energy density associated with each plate, even when the plates are widely separated. In this case, an attractive force is no guarantee of negative energy density. Consider for example the case of classical electrostatics, where opposite charges attract, but the local energy density, proportional to the squared electric field, in non-negative. In the case of plates described by a plasma model dielectric function, the local energy density between the plates was calculated in Ref. 5, with the results illustrated in Fig. 1.

The basic result is that the local energy density at the center of the region between the plates will become negative when the reflectivity of the plates is sufficiently large, specifically when $\omega_p a > 100$, where ω_p is the plasma frequency. Thus local negative energy density is possible in the Casimir effect, but is not inevitable.

3. Negative Energy from Quantum Coherence Effects

Another way to create local negative energy densities is with non-classical states of the quantized electromagnetic field, or indeed any quantum field. It was proven many years ago by Epstein *et al* [6] that all quantum field theories contain quantum states in which the local energy density at a

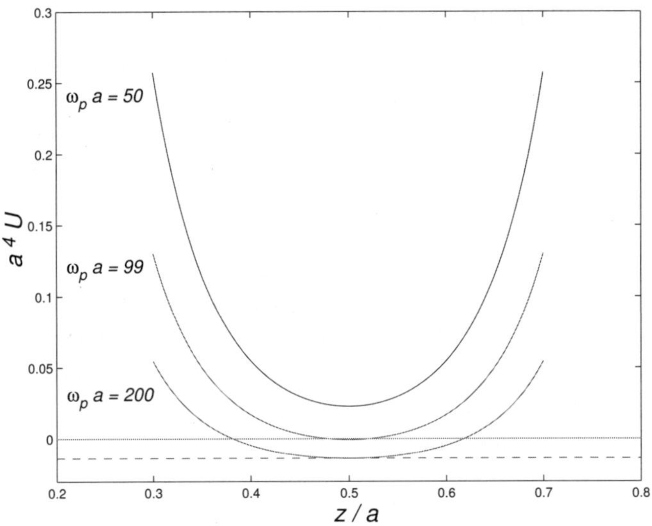

Fig. 1. The energy density in the vacuum region between two dielectric half-spaces is illustrated for three values of the parameter $\omega_p a$. The dashed horizontal line is the energy density for the perfectly conducting limit.

given point may be made negative. In fact, it can be arbitrarily negative.[7] Explicit examples of this phenomenon are rather easy to construct, and include the moving mirror models of Fulling and Davies[8] and the squeezed vacuum states.[9] The energy density in several non-classical states, including squeezed vacua, is discussed, for example, in Refs. 10,11.

In the case of a single monochromatic mode, the energy density will be oscillatory, with both negative and positive energy density intervals, but with the time-averaged energy being positive. By making the frequency of the mode arbitrarily high, one can make the peak values of both the negative and positive energy densities arbitrarily large, but the duration of the period of negative energy will become arbitrarily short. As we will see in Sect. 4, this is a very general feature of negative energy from quantum coherence effects.

Unrestricted negative energy would have dramatic and disturbing consequences, including violations of the second law of thermodynamics,[12] formation of naked singularities in general relativity,[13,14] traversable wormholes,[15,16] and faster than light travel.[17] The later two possibilities could allow the creation of a time machine, with all of the logical problems inherent in time travel.

4. Quantum Inequalities

Unrestricted negative energy would cause serious problems for physics as it is presently formulated. This leads to the suspicion that the laws of physics do place restrictions on negative energy and related effects. This is indeed the case. In the case of free fields in Minkowski spacetime, it is possible to prove "quantum inequalities" which greatly constrain the magnitude and duration of negative energy fluxes or periods of negative energy density. The first versions[12,18] of the quantum inequalities were limits on the negative energy fluxes, which showed that the duration of such a flux has an inverse relation to its magnitude. This limit is sufficient to show that macroscopic violations of the second law are not possible. Later authors[19–23] proved quantum inequalities for the expectation value of the energy density in arbitrary quantum states. Let

$$\rho(t) = \langle T_{tt}(t) \rangle \tag{2}$$

be the expectation value of the energy density operator in an arbitrary state, evaluated on the worldline of an inertial observer at time t. Further, let $g(t, \tau)$ be a sampling function in t with a characteristic width of τ. The quantum inequalities for a massless field take the general form

$$\int \rho(t)\, g(t, \tau)\, dt \geq -\frac{C}{\tau^d}\,, \tag{3}$$

where d is the spacetime dimension, and C is a positive constant. The basic physical content of Eq. (3) is that a observer who sees negative energy lasting for a time of order τ will measure the magnitude of this negative energy density to be bounded by about C/τ^d.

In the case of a massless scalar field in two-dimensional spacetime, $d = 2$, Flanagan[21] has obtained the optimum bound to be given by

$$C = \frac{\tau^2}{24\pi} \int_{-\infty}^{\infty} dt \frac{\dot{g}^2}{g}\,, \tag{4}$$

where $\dot{g} = \partial g / \partial t$. For example, for the case of a Gaussian sampling function, $g(t, \tau) = (\sqrt{\pi}\, \tau)^{-1}\, \mathrm{e}^{-t^2/\tau^2}$, one finds $C = 1/(12\pi)$. This bound will play a key role in the results on energy density fluctuations to be discussed in Sect. 6. Flanagan's bound, Eq. (4), is optimal in the sense that one can construct a quantum state for which this relation is an equality, thus proving that there cannot be a more restrictive bound for arbitrary states. In four dimensional spacetime, the optimal bound is not known, but Fewster and Eveson[22] have proven an inequality for a general sampling function. In this case, typical values of C are of order 10^{-3} for functions such Lorentzians or Gaussians.

It may be shown[12–14,20,25,26] that these inequalities, and their curved spacetime analogs, greatly restrict the effects of negative energy, making macroscopic violations of the second law or of cosmic censorship unlikely, and preventing the construction of macroscopic wormholes or warpdrives.

5. Possible Experiments to Detect Negative Energy?

Although negative energy and related sub-vacuum effects are constrained by quantum inequalities, this does not mean that they are unobservable in principle. Unfortunately, the gravitational effects of negative energy are extremely small. However, there is still a possibility of a laboratory experiment using non-gravitational effects. The role of quantum inequalities in quantum optics has been discussed in Refs. 27,28.

An early attempt to devise an experiment was made in Ref. 29. These authors considered a spin system in a magnetic field interacting with the quantized electromagnetic field in a non-classical state, such as a squeezed vacuum, for a single plane wave mode. Normally, photons will flip spins and cause a decrease in the magnetization of the system. However, in a non-classical photon state, the instantaneous magnetization can actually increase above its value in the vacuum. In the model of Ref. 29, this increase occurs during the interval when the expectation value of energy density of the quantized electromagnetic field is negative. One can interpret this result as follows: quantum vacuum fluctuations cause some de-alignment of the spins, compared to what would occur in a world without vacuum fluctuations. One cannot turn off these fluctuations, but they can be momentarily suppressed, resulting in "repolarization". This effect seems too small for a realistic experiment. It should also be noted that it is really measuring the mean squared magnetic field $\langle B^2 \rangle$ in the non-classical state, rather than the energy density. It is only in the special case of a plane wave mode, where $\langle B^2 \rangle = \langle E^2 \rangle$, that this effect is a measure of negative energy density.

Another proposal to measure sub-vacuum effects was made in Ref. 30, where the effect of squeezed states of the photon field on electron interference was discussed. Normally, the scattering of photons by the electrons will lead to dephasing, decreasing the contrast of the interference pattern. However, when the photons are in a squeezed vacuum state, and the interference pattern is formed only from electrons which pass through the interferometer at selected times in the cycle of the excited mode, then the contrast can increase compared to the vacuum case. This effect of "recoherence" is similar to the repolarization effect for spins. Vacuum fluctuations of the electromagnetic field cause fluctuations of the Aharonov-Bohm phase of the electrons, which in turn lead to a loss of contrast, compared to what would be seen in a world without vacuum fluctuations. Again, the best that we can do is to momentarily suppress these fluctuations, and cause a small increase in contrast. This effect can be expressed as a change in a double surface integral of $\langle F_{\mu\nu}(x)\, F_{\alpha\beta}(x')\rangle$, the electromagnetic field strength correlation function, with the integration taken over a surface bounded by the electron paths. This is not a local quantity like the energy density, but suppression of the Aharonov-Bohm phase fluctuation is a sub-vacuum effect, just as is negative energy density. The effect of recoherence is in principle observable, but probably not with current technology.

Another potentially observable effect of sub-vacuum phenomena arises from the spontaneous decay rates of atoms. It is well known that vacuum fluctuations are essential for spontaneous decay, because atomic energy levels would be eigenstates of the Hamiltonian, and hence stable, were it not for the coupling to the quantized electromagnetic field. It was recently proposed in Ref. 31 that non-classical states of the photon field might lead to observable suppression of atomic decay rates. The proposed experiment involves sending a beam of atoms in an excited state through a cavity, in which one mode is excited in a non-classical state, such as a squeezed vacuum. On average, the effect of the excited state will be to increase the decay rate, as would be expected from stimulated emission effects. However, if the atoms pass through the cavity at certain times in the cycle of the excited mode, then the decay probability during the transit period can be reduced compared to the case when the electromagnetic field in the cavity is in the vacuum state. As in the effects discussed in the previous two paragraphs, this can be interpreted as suppression of the usual vacuum fluctuation effects. Under certain conditions, this effect can serve to measure $\langle E^2\rangle$, the shift in mean squared electric field due to the non-classical state (but not including Casimir effects due solely to the cavity). In the case where the frequency of the cavity field is near the atomic transition frequency, the transit time is small compared to the associated period, and the mode function is approximately constant along the atom's path, then the decay probability is

$$\frac{P}{P(0)} = 1 + \frac{1}{f^2(\mathbf{x_0})}\, \langle E^2(\mathbf{x_0}, t)\rangle\,. \tag{5}$$

Here $f^2(\mathbf{x_0})$ is the squared mode function for the excited mode, evaluated at a point $\mathbf{x_0}$ on the atom's path, P is the decay probability for the non-classical state, and $P(0)$ is that for the vacuum state. When $\langle E^2(\mathbf{x_0}, t)\rangle < 0$, then $P < P(0)$, and the decay rate has been suppressed. In Ref. 31, a quantum inequality is proven which states that

$$\langle E^2(\mathbf{x_0}, t)\rangle \geq -f^2(\mathbf{x_0}), \tag{6}$$

thus guaranteeing that $P \geq 0$, as required. However, a state which comes close to saturating the quantum inequality bound will lead to a significant fractional decrease in decay probability when the atom is in the cavity. Note that this is quite different from the more familiar suppression of atomic decay rates in cavities compared to empty space, which can be interpreted in terms of a lack of available modes in to which to atom can decay. The decay probability during transit is small to begin with, of order 10^{-8} in some examples treated in Ref. 31, so a large number of atom needs to be used to produce a statistically significant result. However, a realistic experiment might be feasible, and is currently under study.

6. Quantum Stress Tensor Fluctuations

So far, we have been discussing situations where the expectation value of the energy density can be negative. However, there is another sense in which negative energy density can arise in quantum field theory. This is when quantum fluctuations momentarily create a region of negative energy. This can occur even when the expectation value is non-negative. A simple example is the Minkowski vacuum state, where $\langle T_{\mu\nu}\rangle = 0$, but the state is not an eigenstate of $T_{\mu\nu}$. This means that there must be both positive and negative fluctuations and an associated probability distribution function. To find this, we need the probability distribution for a stress tensor operator averaged over a sampling function in time, or spacetime.

In general, this is still an unsolved problem. However, it has recently been solved for the case of two-dimensional conformal field theory with a Gaussian average in time.[32] Here we simply quote the results, which will be derived in Ref. 32, and discuss some of their physical implications. Let T_{tt} be the energy density operator, and define the averaged energy density by

$$u = \frac{1}{\sqrt{\pi}\,\tau}\int_{-\infty}^{\infty} T_{tt}(t,x)\, \mathrm{e}^{-t^2/\tau^2}\, dt\,. \tag{7}$$

The associated probability distribution is a Gamma distribution given by

$$P(x) = \frac{\pi^{c/12}}{\Gamma(c/12)}\,(x+x_0)^{c/12-1}\, \mathrm{e}^{-\pi(x+x_0)}\,, \qquad x > -x_0\,, \tag{8}$$

and $P(x) = 0$ for $x < -x_0$. Here $x = u\,\tau^2$, c is the central charge, and $-x_0/\tau^2$ is the quantum inequality bound on expectation values of T_{tt} in arbitrary quantum states. The probability distribution is plotted in Fig. 2 for the case of a free massless scalar field in two-dimensional Minkowski spacetime, for which $c = 1$.

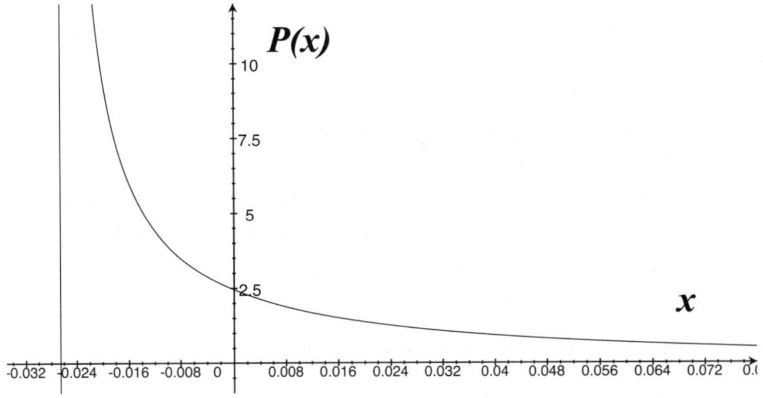

Fig. 2. The probability distribution $P(x)$ for the smeared energy density of a massless scalar field in two-dimensional Minkowski spacetime is plotted. Here $x = u\,\tau^2$, where u is the energy density operator averaged in time with a Gaussian function of width τ. The lower limit of $P(x)$ occurs at $x = -x_0 = -1/(12\pi)$, illustrated by the vertical line.

The lower cutoff at the quantum inequality bound is expected to be a general feature of the probability distribution for vacuum fluctuations of an averaged energy density. It is of interest to note that 0.84 of the area of the graph in Fig. 2 lies to the left of the origin. This means that a measurement of the averaged energy density will find a negative result 84% of the time. However, when a positive value is found, it is typically larger in magnitude.

The probability distribution for the energy density in four-dimensional theories has not yet been found, but is of considerable interest. One application is to inflationary cosmology, where quantum stress tensor fluctuations might contribute a potentially observable component to the cosmological density fluctuations.[33] This component would be non-Gaussian in a way which is associated with the skewness of the quantum stress tensor probability distribution. A distinct, more exotic, application to cosmology arises in models which employ anthropic reasoning to compute probabilities of various observables. These models require a counting of observers, usually assumed to be beings like ourselves in the sense of having arisen from biological evolution on earth-like planets. However, another possibility are "Boltzmann brains" which have nucleated from the vacuum in deSitter or even Minkowski spacetime. Quantum energy fluctuations can sometimes produce large concentrations of energy, which very occasionally might be capable of conscious thought. The probability per unit volume of this is very tiny, but the available volume is vastly larger than that for biological systems. If "Boltzmann brains" are the more prevalent type of observer, it would greatly complicate attempts at anthropic prediction. (For further discussion and references, see, for example Refs. 34,35.) The key to studying this question is in the details of the long positive tail of the probability distribution, which can tell us how likely the appearance of a "Boltzmann brain" in a given region might be.

7. Summary

We have seen that quantum field theory allows local negative energy and other sub-vacuum effects. These effects are strongly restricted by quantum inequalities, but are nonetheless potentially observable. We have reviewed some proposals to measure these effects in laboratory experiments, the most promising of which involves changes in the decay rates of atoms, and could conceivably lead to measurements of negative mean squared electric fields. We have also discussed some new results on the probability distribution for vacuum stress tensor fluctuations. This distribution has a lower cutoff at the quantum inequality bound on the expectation value in an arbitrary state, but has a tail in the positive direction. Thus a typical fluctuation in the local energy density is negative but bounded below, while rare but extremely large positive fluctuations are possible. Stress tensor fluctuations effects could be important in inflationary cosmology.

Acknowledgments

I would like to thank Chris Fewster and Tom Roman for valuable discussions. This research was supported in part by the US National Science Foundation under Grant No. PHY-0855360.

References

1. H. B. G. Casimir, Proc. Kon. Ned. Akad. Wet. **51**, 793 (1948).
2. L. S. Brown, G. J. Maclay, Phys. Rev. **184**, 1272 (1969).
3. B.S. DeWitt, Phys. Rep. **19C**, 295 (1975).
4. A. D. Helfer and A. S. Lang, J. Phys. A: Math. Gen. **32**, 1937 (1999).
5. V. Sopova and L.H. Ford, Phys. Rev. D **66**, 045026 (2002), quant-ph/0204125.

6. H. Epstein, V. Glaser, and A. Jaffe, Nuovo Cim. **36**, 1016 (1965).
7. C.J. Fewster, "Energy inequalities in quantum field theory", in *Proceedings of the XIVth International Congress on Mathematical Physics*, edited by J.C. Zambrini (World Scientific, Singapore, 2005), math-ph/0501073
8. S.A. Fulling and P.C.W. Davies, Proc. R. Soc. London **A348**, 393 (1976); P.C.W. Davies and S.A. Fulling, Proc. R. Soc. London **A356**, 237 (1977).
9. C. M. Caves, Phys. Rev. D **23**, 1693 (1981).
10. A. Borde, L.H. Ford, and T.A. Roman, Phys. Rev. D **65**, 084002 (2002), qr-qc/0109061.
11. L.H. Ford and T.A. Roman, Phys. Rev. D **77**, 045018 (2008), arXiv:0705.3003.
12. L. H. Ford, Proc. Roy. Soc. Lond. **A364**, 227 (1978).
13. L.H. Ford and T.A. Roman, Phys. Rev. D **41**, 3662 (1990).
14. L.H. Ford and T.A. Roman, Phys. Rev. D **46**,1328 (1992).
15. M. Morris, K. Thorne, and U. Yurtsever, Phys. Rev. Lett. **61**, 1446 (1988).
16. M. Visser, *Lorentzian Wormholes; From Einstein to Hawking,* (AIP, Woodburry, N. Y., 1995).
17. M. Alcubierre, Class. Quantum Grav. **11**, L73 (1994).
18. L. H. Ford, Phys. Rev. **D43**, 3972 (1991).
19. L.H. Ford and T.A. Roman, Phys. Rev. D **51**, 4277 (1995), gr-qc/9410043.
20. L.H. Ford and T.A. Roman, Phys. Rev. D **55**, 2082 (1997), gr-qc/9607003.
21. E.E. Flanagan, Phys. Rev. D, **56**, 4922 (1997), gr-qc/9706006.
22. C.J. Fewster and S.P. Eveson, Phys. Rev. D **58**, 084010 (1998), gr-qc/9805024.
23. C.J. Fewster, Class. Quantum Grav. **17**, 1897 (2000), gr-qc/9910060.
24. L.H. Ford and T.A. Roman, Phys. Rev. D **53**, 5496 (1996), gr-qc/9510071.
25. M.J. Pfenning and L.H. Ford, Class. Quantum Grav., **14**, 1743 (1997), gr-qc/9702026.
26. T.A. Roman, "Some Thoughts on Energy Conditions and Wormholes", in *Proceedings of the Tenth Marcel Grossmann Meeting on General Relativity*, edited by S.P. Bergliaffa and M. Novello, (World Scientific, Singapore, 2006), gr-qc/0409090.
27. P. Marecki, Phys. Rev. **A 66**, 053801 (2002), quant-ph/0203027.
28. P. Marecki and N. Szpak, Ann. Phys. (Leipzig) 14, 428 (2005), quant-ph/0407186.
29. L.H. Ford, P.G. Grove, and A.C. Ottewill, Phys. Rev. D **46**, 4566 (1992).
30. J.T. Hsiang and L.H. Ford, Phys. Rev. D **78**, 065012 (2008), rXiv:0806.1512.
31. L.H. Ford and T.A. Roman, arXiv:0907.1638.
32. C.J. Fewster, L.H. Ford, and T.A. Roman, manuscript in preparation.
33. C.-H. Wu, K.-W. Ng, and L.H. Ford, Phys. Rev. D **75**, 103502 (2007), gr-qc/0608002.
34. G. Garriga and A. Vilenkin, Phys. Rev. D **77**, 043526 (2008), arXiv:0711.2559.
35. A. De Simone, A.H. Guth, A. Linde, M. Noorbala, M.P. Salem, and A. Vilenkin, arXiv:0808.3778.

VACUUM ENERGY DENSITY AND PRESSURE NEAR BOUNDARIES

S. A. FULLING*

Mathematics Department, Texas A&M University,
College Station, TX 77843-3368, USA
** E-mail: fulling@math.tamu.edu*
www.math.tamu.edu/˜fulling/

In vacuum energy calculations with an ultraviolet cutoff, divergences arise that clearly are related to the physics of boundaries. We point out how to find the cut-off energy density and pressure most directly from a Green function. Then we discuss three apparent paradoxes arising in this work that are in various stages of resolution.

Keywords: Vacuum energy; Stress tensor; Cylinder kernel; Optical approximation; Multiple scattering.

1. Introduction

The main principles of our working philosophy toward vacuum energy and its divergences are the following.[1]

Local energy density and pressure, the components of the stress tensor $T^{\mu\nu}(\mathbf{r})$, are important for understanding the physics. Semiclassical path analysis (reducing in the simplest cases to exact image solutions) is illuminating. Idealized models with perfect reflection boundary conditions (and a scalar field, often in lower dimensions) can still be instructive, provided one is honest and sensible about their limitations.

To understand the physics of divergences, it is best to use an ultraviolet cutoff. Zeta and dimensional methods of regularization hide the divergences in an ad hoc way and yield global energies that are inconsistent with the local $T^{\mu\nu}(\mathbf{r})$ and hence with general relativity.[2] At the other extreme, detailed modeling of real materials is hard, and also too specific: real metals have little to do with hadron bags, cosmological branes, thermal fluctuations in biological membranes, and other situations where Casimir-type calculations have been conducted. An ultraviolet cutoff in quantum vacuum calculations is like the repulsive core in the Lennard–Jones potential. A more accurate atomic potential would be based on the electronic structure of atoms — but would not apply to nucleon-nucleon scattering. Similarly, a simple frequency cutoff is *universally applicable*, mathematically convenient, and qualitatively physically plausible — but can't be expected to model real metals at atomic-scale distances.

Our mathematically preferred cutoff is the factor $e^{-t\omega_j}$, but others are possible and may be physically superior. (Note that the ω_j^{-2} of dielectric theory is *not sufficient*. Further account of the granular structure of matter is needed.[3,4]) Thus we study a regularized total energy

$$E(t) \equiv \frac{1}{2}\sum_{j=1}^{\infty} \omega_j e^{-t\omega_j} \quad (t \downarrow 0) \tag{1}$$

and a regularized energy density

$$\rho(\mathbf{r}, t) \equiv \langle T^{00} \rangle \equiv \frac{1}{2}\sum_{j=1}^{\infty} \omega_j |\phi_j(\mathbf{r})|^2 e^{-t\omega_j} \tag{2}$$

with similar formulas for the other components of $T^{\mu\nu}$. (Equation (2) applies to the value $\xi = \frac{1}{4}$ for the curvature coupling parameter. Other cases are similar but more complicated.[1])

2. Integral Kernels in Billiards and Cavities

Associated with any self-adjoint second-order linear differential operator H is a dizzying array of integral kernels (also called Green functions or spectral functions), most of which implement the solution of some partial differential equation involving H. They include the resolvent kernel (energy-domain Green function) $G(k, \mathbf{r}, \mathbf{r}') = \langle \mathbf{r}|(H - k^2)^{-1}|\mathbf{r}'\rangle$, the heat kernel $K(t, \mathbf{r}, \mathbf{r}') = \langle \mathbf{r}|e^{-tH}|\mathbf{r}'\rangle$, the quantum kernel $U(t, \mathbf{r}, \mathbf{r}') = \langle \mathbf{r}|e^{-itH}|\mathbf{r}'\rangle$, the local zeta function $Z(s, \mathbf{r}, \mathbf{r}') = \langle \mathbf{r}|H^{-s}|\mathbf{r}'\rangle$, and the *cylinder kernel*

$$\overline{T}(t, \mathbf{r}, \mathbf{r}') = \left\langle \mathbf{r} \left| \frac{e^{-t\sqrt{H}}}{-\sqrt{H}} \right| \mathbf{r}' \right\rangle. \tag{3}$$

Of foremost importance to us is that derivatives of the cylinder kernel comprise the regularized stress tensor:

$$\rho(\mathbf{r}, t) = -\frac{1}{2}\frac{\partial^2}{\partial t^2}\overline{T}(t, \mathbf{r}, \mathbf{r}) \quad (\text{if } \xi = \tfrac{1}{4}), \quad \text{etc.} \tag{4}$$

Note that $U(t) = K(it)$; similarly, the analytic continuation of $\overline{T}$ to imaginary time is one of the Green functions associated with the wave equation. In a compact system the eigenvalue density is an integral of the imaginary part of G, and this fact has a local counterpart:

$$\sigma(\mathbf{r},k)\,dk = \frac{1}{\pi}\operatorname{Im} G(\mathbf{r},\mathbf{r},k)\,d(k^2). \tag{5}$$

A Laplace transform with respect to k^2 leads from σ to K, and from K to G. A Laplace transform in k leads directly from σ to $\overline{T}$, which is also related to K by

$$\int_0^\infty t^{-1/2} e^{-\tau^2/4t} K(t)\,dt = -\sqrt{\pi}\,\overline{T}(\tau). \tag{6}$$

Because of all these interrelations, each spectral function in some sense contains all the information in any of the others, though perhaps in a less (or more) accessible form.

In the asymptotic construction of Green functions it is useful to use what is known in general relativity as the *Synge–DeWitt formalism*[5,6] and used in classical optics in different notations.[7] Let $L(\mathbf{r},\mathbf{r}')$ denote the distance from $\mathbf{r}'$ to $\mathbf{r}$ along (in the present context) a straight path with specular reflections, and let $\sigma(\mathbf{r},\mathbf{r}') \equiv \frac{1}{2}L(\mathbf{r},\mathbf{r}')^2$. Then $(\nabla \equiv \nabla_{\mathbf{r}})$

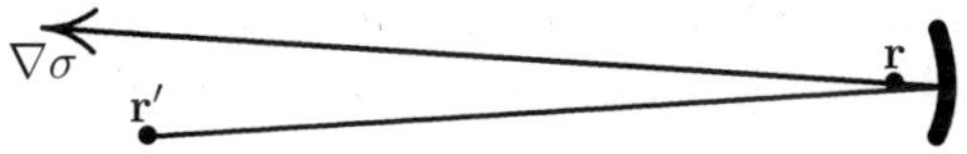

(i) $\nabla\sigma = L\nabla L = L\hat{\mathbf{n}}$, where $\hat{\mathbf{n}}$ is the unit vector at $\mathbf{r}$ in the direction of the path;
(ii) $(\nabla\sigma)^2 = L^2 = 2\sigma$;
(iii) $\nabla^2\sigma = 1 + L\nabla\cdot\hat{\mathbf{n}} = d + O(L^2)$ (where d is the spatial dimension, usually 3);
(iv) For the *direct* path (no reflections), $\sigma = \frac{1}{2}|\mathbf{r}-\mathbf{r}'|^2$, $\nabla\sigma = \mathbf{r}-\mathbf{r}'$, and $\nabla^2\sigma = d$.

For $H = -\nabla^2$ in a cavity, we can now construct any kernel by the *optical approximation.* Let G stand for the function in question (not necessarily the resolvent kernel). We seek it in the form of a sum over specularly reflecting paths of terms

$$G_j = (-1)^j D_j(\mathbf{r},\mathbf{r}')F(\sigma(\mathbf{r},\mathbf{r}')), \tag{7}$$

where $(-1)^j$ means the parity of the number of reflections (so that the Dirichlet condition is satisfied by the sum), and D does *not* depend on the parameter t or k (though F does). The central observation of this section is that *D comes out the same for all the kernels.* (This is true for billiards and cavities, not for Schrödinger operators with potentials, whose kernels do not factor as nicely as in Eq. (7).) Thus we have the option of studying $\overline{T}$ directly, instead of starting with G or U and passing through σ, as usually done.

Proof and construction. Plug ansatz (7) into the partial differential equation defining G. Group the terms by order of singularity; it will be seen that the leading term vanishes if $F(\frac{1}{2}|\mathbf{r}-\mathbf{r}'|) = G_0(\cdot, \mathbf{r}, \mathbf{r}')$, the corresponding kernel in $\mathbf{R}^3$. For the most familiar kernels, the singularity occurs in the limit $t \downarrow 0$ or $k \to +\infty$. For $\overline{T}$, "singularity" refers to powers of $(t^2 + 2\sigma)^{-1}$, and the conclusion thus far is that

$$F = \overline{T}_0 = -\frac{1}{2\pi^2} \frac{1}{t^2 + |\mathbf{r}-\mathbf{r}'|^2} \,. \tag{8}$$

Then the next term vanishes if $D \equiv |\det M|^{1/2}$, where $M_{jk} \equiv \partial^2\sigma/\partial r_j\, \partial r'_k$. In other terms, $\det M = L^{d-1}\delta$ where $\delta(\mathbf{r}, \mathbf{r}')$ is the *enlargement factor* $d(\text{angle})/d(\text{area})$ exploited by Scardicchio and Jaffe.[7] □

3. Paradox Past

Why is the "renormalized" total energy finite (and independent of ξ) when the "renormalized" energy density has a nonintegrable, ξ-dependent singularity at a boundary? (Here "renormalization" refers to the "unspeakable act" of discarding the divergent terms in $E(t)$ and $\rho(\mathbf{r}, t)$ as $t \to 0$; this is not the place to discuss its physical justification, or lack thereof.) This question was essentially raised by Deutsch and Candelas[2] and essentially answered by Ford and Svaiter.[8] I have dwelt upon it at previous QFExt meetings, so I'll dismiss it quickly here: $\lim_{t\downarrow 0} \rho(\mathbf{r}, t)$ is a limit nonuniform in $\mathbf{r}$, so taking the limit outside the integral over space is not the same as taking it inside. (See figure in Sec. 5.) In fact, the integral $E(t)$ for small, nonzero t may have the same or the opposite sign as $\lim \rho(\mathbf{r}, t)$, depending on ξ and other features of the configuration studied.[1] When t is finite, so is $E(t) = \iint dS \int \rho\, dz$ ($z \equiv$ distance from boundary). Thus the cutoff theory is internally consistent (so far).

4. Paradox Present

The cutoff-dependent surface energy of a sphere depends on radius. Where is the corresponding pressure? The regularized energy near a spherical surface of radius R, on either the inside or the outside, is

$$E(t) = -\frac{R^2}{2t^3} + O\left(\frac{R}{t^2}\right). \tag{9}$$

(This result can be derived from either the heat-kernel expansion or the leading term in a multiple-reflection expansion.) It follows that there must be a generalized force and corresponding radial pressure

$$F = -\frac{\partial E}{\partial R} = \frac{R}{t^3}, \qquad p_r = \frac{F}{4\pi R^2} = \frac{1}{4\pi R t^3}. \tag{10}$$

For a thin shell these quantities should be multiplied by 2 to account for both inside and outside. (This discussion refers to a scalar field. It is well known that the corresponding quantities vanish for electromagnetism because of cancellation between two classes of modes.)

However, one seeks in vain for such a term in previous direct studies of $T_{\mu\nu}$ near a sphere.[9–11] Furthermore, in [1] we said that in a rectangle, there is a divergent pressure on a particular side, but it is unrelated to the divergent energy density in the region adjacent to that side (which is independent of the position of the side). Instead, it corresponds to the length dependence of the divergent energy in the edge regions along the perpendicular sides. (In a piston it is cancelled by the exterior.) But in a sphere, there is no perpendicular side! The sphere pressure manifestly comes from p_r, not the pressure $p_\perp$ in the tangential directions, and it must somehow arise from the curvature.

Note that here we are *not* talking about an aspect of the "first-order curvature term" in the heat kernel, spectral density, and energy. *That* effect has *opposite signs for inside and outside* and *same sign for Dirichlet and Neumann conditions. This* effect must have *same sign for inside and outside* and *opposite signs for Dirichlet and Neumann*, like the curvature-independent surface term in the energy from which it arises.

We first investigate the situation in the optical approximation, using the $\overline{T}$ formalism from above and the enlargement factor D for a sphere calculated by Scardicchio and Jaffe.[7] With $2\sigma = 4(r-R)^2$, we find

$$\rho = -\frac{2t^2 R}{\pi^2 r}(t^2+2\sigma)^{-3} + \frac{R}{2\pi^2 r}(t^2+2\sigma)^{-2}, \tag{11}$$

$$p_\perp = -\left(\frac{R}{2\pi^2 r}\right)(t^2+2\sigma)^{-2} - \frac{R}{16\pi^2 r^3}(t^2+2\sigma)^{-1}, \tag{12}$$

$$p_r = 0 - \frac{rR}{8\pi^2 (R-r) r^3}(t^2+2\sigma)^{-1}. \tag{13}$$

The vanishing leading term in Eq. (13) emphasizes that the radial pressure is in a sense less singular than the other components, which removes any contradiction with [10]. Very surprising, however, is the remaining term in

Eq. (13), which displays a divergence at the surface, $r = R$. This divergence is purely geometrical, not ultraviolet; it exists although the cutoff t is still finite! It constitutes a failure of the single-reflection optical approximation. (It arises in terms involving 2 derivatives of D, and those are of same order as terms omitted from $\overline{T} \approx D(t+2\sigma)^{-1}$. This is a generic limitation of WKB-type methods.)

Attempts to construct an optical approximation of higher order in stationary phase (routine in curved manifolds without boundary[5]) break down at boundaries. (See [12] for the analogous problem with the heat kernel.) Such expansions can't naturally yield arctangent terms, which are almost certainly present, as we'll see presently. Instead, we shall turn to the full *multiple-reflection* expansion, integrating over paths whose reflection need not be specular.

First, however, consider the implications of the conservation law, $\nabla_\mu T^{\mu\nu} = 0$. Under the symmetry conditions of our problem,[10] one of its components is

$$\frac{\partial p_r}{\partial r} + \frac{2p_r}{r} - \frac{2p_\perp}{r} = 0 \Rightarrow p_r = \frac{2}{r^2}\int^r p_\perp \, r\, dr. \tag{14}$$

Then Eq. (12) suggests that p_r must contain a term

$$-\frac{R^3}{\pi^2 t^3 r^2}\frac{4R^2+3t^2}{(4R^2+t^2)^2}\tan^{-1}\left(\frac{t}{2(R-r)}\right) \sim -\frac{1}{8\pi R t^3} \tag{15}$$

(for $r = R$, $t \ll R$). Of all the terms in the integral (14), this one seem uniquely capable of supplying the expected pressure — but is off by a factor $-\frac{1}{2}$, to which we'll return shortly.

Now we consider the first term in the multiple-scattering expansion of Balian and Bloch:[13]

$$G(k,\mathbf{r},\mathbf{r}') - G_0 = -2\int_S dS_{\mathbf{q}} \frac{\partial G_0(k,\mathbf{r},\mathbf{q})}{\hat{n}_{\mathbf{q}}} G_0(k,\mathbf{q},\mathbf{r}') + \cdots . \tag{16}$$

Liu[14] observed that because

$$\overline{T}(t,\mathbf{r},\mathbf{r}') = \frac{2}{\pi}\int_0^\infty e^{-tk}\mathrm{Im}G(k,\mathbf{r},\mathbf{r}')\, dk, \tag{17}$$

it follows that

$$\overline{T} - \overline{T}_0 = \frac{1}{4\pi^3}\int dS_{\mathbf{q}} \cos\theta_1 \left(\frac{1}{r_1{}^2(L^2+t^2)} + \frac{2L^2}{r_1 r_2 (L^2+t^2)^2}\right) + \cdots \tag{18}$$

where $\cos\theta_1 \equiv (\mathbf{q}-\mathbf{r})\cdot\hat{n}_{\mathbf{q}}$, $r_1 \equiv |\mathbf{r}-\mathbf{q}|$, $r_2 \equiv |\mathbf{q}-\mathbf{r}'|$, and $L \equiv r_1 + r_2$. This formula applies to *any* surface, not just a sphere. (It can also be obtained directly from the single-scattering term in $\overline{T}$, in accordance with the philosophy of Sec. 2.)

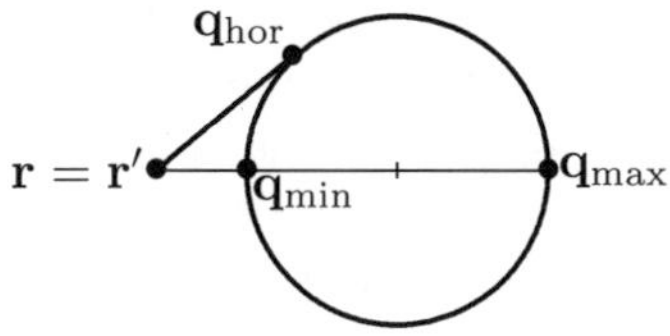

Now specialize to the (exterior of the) sphere.[14] There are three critical lengths of paths,

$$L_{\rm min} = 2(r - R), \quad L_{\rm max} = 2(r + R), \quad L_{\rm hor} = 2\sqrt{r^2 - R^2}\,. \tag{19}$$

Paths intersecting the sphere beyond the point of tangency at $L = L_{\rm hor}$ are expected to cancel approximately against paths with three reflections, at least in the optical limit.[15]

The integral from the visible region $(L_{\rm min}, L_{\rm hor})$ is

$$\frac{1}{4\pi^2 a}\left[\frac{2R}{t^2 + L_{\rm min}^2} + \frac{1}{t}\left(\tan^{-1}\frac{L_{\rm min}}{t} - \tan^{-1}\frac{L_{\rm hor}}{t}\right)\right]. \tag{20}$$

The first term here is recognized as the optical approximation to $\overline{T}$. The second term is the predicted arctangent. (It gives a correction of order $O(L_{\rm min}^{-3})$ to the leading term in $\rho(\mathbf{r}, 0)$, which is $O(L_{\rm min}^{-4})$ with $L_{\rm min} = 2(r - R) \to 0$ at the boundary.)

The integral from the shadow region $(L_{\rm hor}, L_{\rm max})$ is

$$\frac{1}{4\pi^2 a}\left[\frac{2R}{t^2 + L_{\rm max}^2} + \frac{1}{t}\left(\tan^{-1}\frac{L_{\rm hor}}{t} - \tan^{-1}\frac{L_{\rm max}}{t}\right)\right]. \tag{21}$$

The first term is the optical term from reflection on the far side of the sphere. Although one expects this term to be compensated by terms with more reflections, in the full multiple-reflection integral a consistent truncation at $L_{\rm hor}$ is not possible, because contributions from $L \approx L_{\rm min}$ are entangled with those from comparably short 3-reflection paths. Better calculational technology is needed to get a good handle on these low-frequency details.

To get $\langle T^{\mu\nu}\rangle$ we need derivatives of $\overline{T}$ at $\mathbf{r}' = \mathbf{r}$. (Evaluating the integrals for $\overline{T}$ at $\mathbf{r}' \neq \mathbf{r}$ appears hopeless.) After much *Mathematica* work we reproduce in the limit of small t

$$E \sim -\frac{R^2}{2t^3} + \frac{5}{192\pi R} + O(t^2/R^3) \tag{22}$$

(cf. Eq. (9), whose $O(R/t^2)$ term comes from two reflections) and

$$p_r \sim -\frac{1}{8\pi Rt^3} + \frac{1}{64\pi R^3 t} + O(t^0/R^4)\,; \tag{23}$$

the (cut-off) radial pressure is finite at the boundary and originates in the subdominant arctangent term. (So there is no contradiction with [9].)

This is a qualitative resolution of the sphere pressure paradox — but only qualitative. We found $p_r \sim -(8\pi Rt^3)^{-1}$ but were expecting $p_r \sim +(4\pi Rt^3)^{-1}$. This prompted us to look at *flat* plates intersecting at right angles in dimension 3, where we find the same discrepancy $-\frac{1}{2}$! That in turn forced a reexamination of the 2-dimensional calculation in [1], mentioned above. It turns out that there the numerical factor is correct *but the sign is wrong.* (In dimension n the discrepant factor (for flat plates) is $-1/(n-1)$.)

5. Paradox Future

So, *why is the energy-balance equation*

$$\frac{\partial E}{\partial h} = -\int_S p_h \tag{24}$$

(for general parameter h) violated in the cutoff theory? (Note that a similar, nonparadoxical, result was found in [16].)

Let us take a critical look at the leading, bulk term in the vacuum $\langle T^{\mu\nu}(\mathbf{r})\rangle$, which we usually simply throw away. The ultraviolet cutoff yields

$$T^{\mu\nu} = \frac{1}{2\pi^2 t^4}\operatorname{diag}(3,1,1,1) \equiv \frac{1}{2\pi^2 t^4} T_{\text{rad}}\,, \tag{25}$$

proportional to the stress tensor of massless radiation. If we were to take it seriously, we would note that it is (i) ambiguous, (ii) not Lorentz-invariant, and (iii) inconsistent with Eq. (24).

As concerns ambiguity, note that elsewhere[17] we argued that

$$T^{\mu\nu} \propto \operatorname{diag}(-1,1,1,-3) \equiv T_{\text{Cas}}\,, \tag{26}$$

the tensor characteristic of conformally invariant Casimir energy[18] (which happens to be invariant under Lorentz transformations parallel to the x–y plane). In that calculation the three space-time directions (t, x, y) were treated alike, but differently from z. Effectively, we used a k_z cutoff instead of a frequency cutoff.

This problem was confronted in the early years of quantum field theory in curved space-time.[5,19–21] In particular, Christensen[5] calculated the bulk vacuum stress tensor

$$T^{\mu\nu} = \frac{1}{2\pi^2 t^4}\left(g^{\mu\nu} - 4\, u^\mu u^\nu \operatorname{sgn}(u_\rho u^\rho)\right), \tag{27}$$

where u^μ was the unit 4-vector in the direction of "point separation". $T_{\rm rad}$ corresponds to point-splitting in the time direction, $T_{\rm Cas}$ to point-splitting in the direction of a certain spatial coordinate. In the gravitational context it was taken as self-evident that a physically acceptable term of zeroth order in curvature must be covariant and hence proportional to

$$T_{\rm dark} \equiv \mathrm{diag}(-1,1,1,1) \tag{28}$$

(i.e., to the metric tensor). It can then be absorbed by renormalization of the cosmological constant. Various rationalizations of this step appeared in the literature. Zel'dovich[19] advocated Pauli–Villars regularization, which maintains Lorentz invariance at all times. Adler et al. and Christensen[20,21] proposed to average over u (getting 0 for massless fields). Others just "renormalized" direction-dependent terms, resolving any ambiguity by requiring $\nabla_\mu T^{\mu\nu} = 0$ (conservation) for the remainder.

Unlike $T_{\rm rad}$ and $T_{\rm Cas}$, $T_{\rm dark}$ satisfies Eq. (24). Note that the latter does not automatically follow from $\nabla_\mu T^{\mu\nu} = 0$ for the material inside a container; it requires also an equation of state that is consistent with what actually happens to the "matter" when the volume of the container changes. For bulk vacuum energy, *nothing* happens, so only the dark-energy equation of state, $p = -\rho$, is allowed.

The lesson seems to be that a cutoff model is not a complete, consistent dynamical system, so formal properties such as covariance, conservation, and energy balance may be disrupted and need to be restored ad hoc.

In standard static Casimir problems, we have a rest frame, so this direction dependence has always been ignored. Now we see that the issue reemerges in the cutoff-dependent terms near boundaries. There is no apparent physical reason why point-splitting in the time direction should be preferred. Modeling of a real conductor suggests that a wave-number cutoff in directions *parallel* to the boundary may be more appropriate,[4] and what is appropriate for the energy may not be so for the pressure.[22]

Numerically, the energy-balance equation (in particular, the sign) for rectangular boxes is disrupted because, for timelike point-splitting, the graph of the cut-off energy density displays a dip ($*$) at the boundary that does not occur in the pressure function:

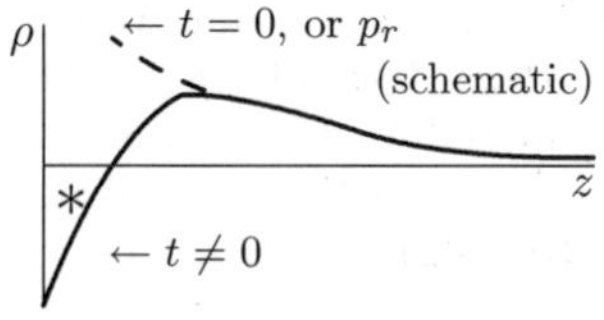

But it was precisely that dip that restored consistency with the integrated energy, resolving the first of the three paradoxes! Therefore, it may be that the famous formula

$$E(t) \sim c_0 V t^{-4} + c_1 S t^{-3} + \cdots \tag{29}$$

is totally meaningless as far as cutoff-dependent terms are concerned. In future work the *full variety* of possible cutoffs needs to be investigated in search of formal consistency of stress tensor components as well as physical plausibility in comparison with particular real boundary interactions.

Acknowledgments

This work was supported by National Science Foundation Grant PHY-0554849. It incorporates input from Martin Schaden, my students Zhonghai Liu, Phuongmai Truong, and Kevin Resil, and the coauthors of [1].

References

1. S. A. Fulling, L. Kaplan, K. Kirsten, Z. H. Liu and K. A. Milton, *J. Phys. A* **42**, 155402 (2009).
2. D. Deutsch and P. Candelas, *Phys. Rev. D* **20**, 3063 (1979).
3. G. Barton, *J. Phys. A* **34**, 4083 (2001).
4. G. Barton, *J. Phys. A* **37**, 1011 (2004); *ibid.* **38**, 2997 (2005).
5. S. M. Christensen, *Phys. Rev. D* **14**, 2490 (1976).
6. F. H. Molzahn, T. A. Osborn and S. A. Fulling, *Ann. Phys.* **204** 64 (1990).
7. A. Scardicchio and R. L. Jaffe, *Nucl. Phys. B* **704**, 552 (2005).
8. L. H. Ford and N. F. Svaiter, *Phys. Rev. D* **58**, 065007 (1998).
9. A. Scardicchio and R. L. Jaffe, *Nucl. Phys. B* **743**, 249 (2006).
10. D. Schwarz-Perlov and K. Olum, *Phys. Rev. D* **72**, 065013 (2005).
11. K. A. Milton, *Phys. Rev. D* **68**, 065020 (2003).
12. D. M. McAvity and H. Osborn, *Class. Quantum Grav.* **8**, 603 (1991).
13. R. Balian and C. Bloch, *Ann. Phys.* **60**, 401 (1970).
14. Z. H. Liu, Ph.D. dissertation, Texas A&M University, 2009, `http://www.math.tamu.edu/~fulling/qvac09/liudiss.pdf`.
15. R. Balian and C. Bloch, *Ann. Phys.* **69**, 76 (1972).
16. G. Barton, *J. Phys. A* **37**, 3725 (2004).
17. K. A. Milton, *The Casimir Effect*, (World Scientific, Singapore, 2001); K. A. Milton, S. A. Fulling, P. Parashar, A. Romeo, K. V. Shajesh and J. A. Wagner, *J. Phys. A* **41**, 164052 (2008).
18. L. S. Brown and G. J. Maclay, *Phys. Rev.* **184**, 1272 (1969).
19. Ya. B. Zel'dovich, *Sov. Phys. Usp.* **11**, 381 (1968).
20. S. L. Adler, J. Lieberman and Y. J. Ng, *Ann. Phys.* **106**, 279 (1977).
21. S. M. Christensen, *Phys. Rev. D* **17**, 946 (1978).
22. G. Barton, *J. Phys. A* **38**, 3021 (2005).

LOCAL ANALYSIS OF THE SINE-GORDON KINK QUANTUM FLUCTUATIONS

INÉS CAVERO-PELÁEZ
Departamento de Física Teórica, Universidad de Zaragoza, Spain
E-mail:cavero@unizar.es, cavero@nhn.ou.edu

JUAN MATEOS GUILARTE
Departamento de Física Fundamental y IUFFyM, Universidad de Salamanca, Spain
E-mail:guilarte@usal.es

We develop a local study of the Green's function and energy density for mesons fluctuating around the sine-Gordon kink background.

Keywords: Sine-Gordon kink, quantum fluctuations, Green's function, local energy density.

1. Introduction

Topological defects are solutions of classical field equations that break translational and other symmetries. Finite energy solutions that conserve their shape at all times, even after collisions, are called solitons. In this paper, we shall study the fluctuations around a well known soliton, the sine-Gordon kink.

We consider a scalar field $\bar{\Phi}(t,\bar{z})$ in (1+1) dimensions and Lagrangian dynamics given by

$$\bar{\mathcal{L}} = -\frac{1}{2}\partial_\mu\partial^\mu\bar{\Phi}^2 - \bar{U}[\bar{\Phi}], \tag{1}$$

with potential energy density,

$$\bar{U}[\bar{\Phi}] = \frac{m^4}{\lambda}(1-\cos\frac{\sqrt{\lambda}}{m}\bar{\Phi}) \quad \rightarrow \quad U[\Phi] = (1-\cos\Phi), \tag{2}$$

where we have performed a change in scale to deal with non-dimensional fields, space-time coordinates, and Lagrangian density

$$\Phi = \frac{\sqrt{\lambda}}{m}\bar{\Phi}, \qquad z = m\bar{z}, \qquad \mathcal{L} = \frac{\lambda}{m^4}\bar{\mathcal{L}}. \tag{3}$$

In this system the classical vacua associated to the minima of $U(\Phi)$ are infinitely degenerated, $\Phi_n = 2\pi n$, $n \in \mathbb{Z}$. The choice of vacuum breaks the internal dihedral symmetry $\mathbb{D}_\infty = \mathbb{Z} \times \mathbb{Z}_2 = \{\Phi \to \Phi + 2\pi n, \Phi \to -\Phi\}$ to the $\mathbb{Z}_2$ subgroup and there is also room for kink-solutions of the field equations connecting two of such a vacua,

$$\Phi_K(t,z) = \pm 4 \arctan e^z + 2\pi n. \tag{4}$$

Studying quantum fluctuations around these kinks is a problem closely related to the analysis of vacuum fluctuations in the Casimir effect. In particular, the one-loop correction to the classical mass of the kink due to the kink quantum fluctuations can be seen as the kink Casimir effect, where the kink replaces the plates or other geometries constraining the dynamics of the fluctuations. Dashen, Hasslacher and Neveu[1] (DHN) computed the sine-Gordon kink Casimir effect as long back as the mid seventies. The classical kink energy at the center of mass is $E[\Phi_K] = 8m^3/\lambda$ (versus that of the vacua $E[\Phi_n] = 0$) and DHN found the quantum correction to be $E_C[\Phi_K] = -m/\pi + \mathcal{O}(\lambda/m)$.

They controlled the ultraviolet divergences arising in this calculation by a combined regularization/renormalization procedure: 1) the kink and vacuum zero point energy were regularized using a cut-off in the number of normal fluctuation modes and subsequently the vacuum zero point energy was subtracted from the kink zero point energy. 2) The result was still divergent and it was necessary to implement a mass renormalization process in both sectors of the field theoretical system.

Other spectral zeta function regularization processes closer in spirit to the techniques used in the theoretical study of the ideal Casimir effect were developed[2] and successfully applied to the calculation of the one-loop mass correction to the sine-Gordon and other relativistic kinks[3].

The above scenario offers a global knowledge of the total amount of energy taken from the kink by the quantum fluctuations. Our aim is to attack the problem by studying the Green's function of the small fluctuations in the kink background by importing the Green's function method as it is used in the Casimir effect.[4] This point of view has not been explored as extensively as the zeta function approach. We address here a concrete example in a formulation within the general framework developed in Bordag&Lindig[5] for generic field backgrounds, a paper focusing in local properties of vacuum fluctuations. Our purpose is to benefit from the local information that the time-space Green's function can give through the computation of the energy momentum tensor and the subsequent knowledge of the one-loop energy density and radiation pressure.

2. Kink fluctuation dynamics

By expanding the classical field around the kink background $\Phi(t,z) = \Phi(z)_K + \phi(t,z)$, the Lagrangian up to quadratic terms in the quantum fluctuations $\phi(t,z)$ becomes,

$$\mathcal{L} = -\frac{1}{2}\partial_\mu\partial^\mu\phi^2 - \frac{1}{2}V(z)\phi^2 - \frac{1}{2}\phi^2 + \mathcal{O}(\phi^3), \tag{5}$$

where the "space dependent" mass is,

$$V(z) = \begin{cases} 0, & z < -a, \\ -\frac{2}{\cosh^2 z}, & -a < z < a, \\ 0, & z > a. \end{cases} \tag{6}$$

Here a is a point on the z-axis where the effect of the kink is almost negligible, that means $\cosh^2 a \to \infty$, which can as well be used as infrared cut-off. We are allowed to split the potential in this way thanks to the localization of the classical kink energy density.

Therefore, we face a three region problem where we will impose matching and boundary conditions. In a final step, we shall make a tend to infinity. The four dimensional Green's function $G(x,x')$ is the solution of the Euler-Lagrange PDE of (5) with unit source,

$$\left[\partial_\mu\partial^\mu - V(z) - 1\right] G(x,x') = \delta(x-x'). \tag{7}$$

Notice that even though we deal with the $(1+1)$-dimensional sine-Gordon model, it can be easily embedded in the analogous system in $(3+1)$-dimensional Minkowski space by transforming the kink in a domain wall invariant on the x and y axes. Thus, we can Fourier-transform the Green's function in the perpendicular coordinates to the direction of the kink and on time,

$$G(x,x') = \int \frac{d\omega}{2\pi} \int \frac{d^2k_\perp}{(2\pi)^2}\, e^{-i\omega(t-t')}\, e^{i\mathbf{k}_\perp(\mathbf{x}_\perp - \mathbf{x}'_\perp)}\, g(z,z'). \tag{8}$$

The reduced Green's function $g(z,z')$ satisfies,

$$\left[-\partial_z^2 + \lambda^2 + V(z)\right] g(z,z') = \delta(z-z'), \tag{9}$$

and the constant λ is[a], $\lambda^2 = k_\perp^2 - \omega^2 + 1$.

We next solve the differential equation (9) in the three regions determined by the potential in (6) for appropriate boundary conditions.

[a]In this paper, whenever we need to use the value of λ, we assume that $\mathbf{k}_\perp = \mathbf{0}$ but the generalization is interesting since it is compulsory in the membrane or domain wall interpretation.

3. Decaying boundary conditions

We assume continuity of the Green's function and its derivatives in $-a$ and a and exponentially decaying solutions at infinity.
For $z < -a$ and $z > a$ we find,

$$g_1(z,z') = -\frac{1}{2\lambda}\, e^{\lambda(z_< - z_>)} + a_1\, e^{\lambda(z+z')}, \tag{10}$$

where $z_<$ and $z_>$ are the lesser or the greater of z and z'. The coefficient is given by

$$a_1 = \frac{e^{2\lambda a}}{2\lambda\Delta}\{e^{-2\lambda a}\left[2\lambda(\lambda+\tanh a)\cosh^2 a - 1\right] - e^{2\lambda a}\left[2\lambda(\lambda-\tanh a)\cosh^2 a - 1\right]\}, \tag{11}$$

with

$$\Delta = e^{-2\lambda a} - e^{2\lambda a}\,\delta^2, \qquad \delta = \left[2\lambda(\lambda - \tanh a)\cosh^2 a - 1\right]. \tag{12}$$

In the region of the kink, where $-a < z < a$, we find the following Green's function,

$$g_2(z,z') = -\frac{1}{2\lambda}\frac{1}{(-\lambda^2+1)}\Bigg\{ f_1^-(z_>) f_1^+(z_<) + \frac{\delta}{\Delta}\Big\{\delta\, e^{2\lambda a}\Big[f_1^+(z')\, f_1^-(z) + f_1^-(z')\, f_1^+(z)\Big] + \Big[f_1^+(z')\, f_1^+(z) + f_1^-(z')\, f_1^-(z)\Big]\Big\}\Bigg\}, \tag{13}$$

where

$$f_1^-(z) = e^{\lambda z}(-\lambda + \tanh z) \tag{14a}$$
$$f_1^+(z) = e^{-\lambda z}(\lambda + \tanh z). \tag{14b}$$

The poles of $g_2(z,z')$ lay on $\omega = 0$ and $\omega = \pm 1$ as they should since, in the limit $a \to \infty$, the potential in (6) is a transparent Pösch-Teller well. When it is shifted by 1, there is a bound state of zero energy $\omega = 0$, the so-called translational mode, and a "half-bound" state lying in the threshold of the continuous spectrum $\omega^2 = 1$; see Barton[6] and Graham & Jaffe.[7] Notice that $g_2(z,z')$ captures all these states whereas $g_1(z,z')$ does not include the kink translational mode. The decaying boundary conditions exclude the states in the continuous spectrum.
We can now construct the energy-momentum tensor and get the zero-zero component corresponding to the local density energy. In the area where the kink lives it has the form,

$$\langle T_{00}(z)\rangle = \frac{1}{2i}\left[\partial_z\partial_{z'} + \lambda^2 + 2\omega^2 - \frac{2}{\cosh^z}\right] g_2(z,z')\Big|_{z=z'}. \tag{15}$$

Using equation (13) and after subtracting the bulk term we find,

$$\langle T_{00}(z)\rangle = \frac{-1}{2i\lambda(-\lambda^2+1)}\frac{\delta}{\Delta}\Bigg\{\delta e^{2\lambda a}\left[2\omega^2(\tanh^2 z-\lambda^2)+\frac{1}{\cosh^2 z}-3\frac{\tanh^2 z}{\cosh^2 z}\right]$$

$$+\cosh(2\lambda z)\left[2(\omega^2+\lambda^2)(\lambda^2+\tanh^2 z)+\frac{1}{\cosh^2 z}(1-4\lambda^2-3\tanh^2 z)\right]$$

$$+\sinh(2\lambda z)\left[-4\lambda(\lambda^2+\omega^2)\tanh z+6\lambda\frac{\tanh z}{\cosh^2 z}\right]\Bigg\}. \tag{16}$$

4. Periodic boundary conditions

Here, we only consider the region where the potential is different from zero and impose periodicity on the Green's function and its first derivative at the borders $-a$ and a. The resulting Green's function is then,

$$g(z,z') = \frac{1}{2\lambda(-\lambda^2+1)}\Bigg\{f_1^-(z_<)f_1^+(z_>) - \frac{1}{\Delta^C}\Bigg\{\frac{\tanh a}{\Delta^0}$$

$$\times\left\{\left[f_1^+(z')\,f_1^+(z)+f_1^-(z')\,f_1^-(z)\right]+\left[f_1^+(z')\,f_1^-(z)\right.\right.$$

$$\left.\left.+\,f_1^-(z')\,f_1^+(z)\right]\right\}\Bigg\}+e^{-\lambda a}\Big[-\lambda(\lambda+\tanh a)\cosh^2 a+1\Big]$$

$$\times\left[f_1^+(z')\,f_1^-(z)+f_1^-(z')\,f_1^+(z)\right]\Bigg\}, \tag{17}$$

where the deltas in the denominators are,

$$\Delta^0 = e^{-\lambda a}(\lambda+\tanh a)-e^{\lambda a}(\lambda-\tanh a). \tag{18a}$$

$$\Delta^C = e^{\lambda a}\Big[-\lambda(\lambda-\tanh a)\cosh^2 a+1\Big]-e^{-\lambda a}\Big[-\lambda(\lambda+\tanh a)\cosh^2 a+1\Big], \tag{18b}$$

The zeros of Δ^0 are at $\lambda=\pm 1$ whereas the zeroes of Δ^C occur for $\lambda = i\frac{\pi n}{a}-\frac{1}{a}\text{arcth}(1/\lambda)$. Thus, we find the whole spectrum of meson fluctuations around the kink at the poles of the Green's function; $\omega = 0$ and $\omega^2 = \frac{\pi^2}{a^2}n^2+1$ plus a term which is suppressed for large a. A very subtle point is the fact that the periodic boundary conditions allow a second zero mode, besides the normalizable one, that will disappear in the limit of $a\to\infty$.

The local density energy for a fixed ω is

$$\begin{aligned}
\langle T_{00}(z)\rangle = {} & \frac{1}{4i\lambda(-\lambda^2+1)\Delta^C}\Bigg\{ -\frac{\tanh a}{\Delta^0}\Big\{e^{-2\lambda z}\Big[2(\lambda^2+\omega^2)(\lambda+\tanh z)^2 \\
& + \frac{1}{\cosh^2 z}(-4\lambda^2+1-6\lambda\tanh z-3\tanh^2 z)\Big] + e^{2\lambda z}\Big[2(\lambda^2+\omega^2) \\
& \times(-\lambda+\tanh z)^2 + \frac{1}{\cosh^2 z}(-4\lambda^2+1+6\lambda\tanh z-3\tanh^2 z)\Big] \\
& + 2\Big[2\omega^2(-\lambda^2+\tanh^2 z) + \frac{2}{\cosh^2 z}(1-3\tanh^2 z)\Big]\Big\} \\
& + e^{-\lambda a}\Big[-\lambda(\lambda+\tanh a)\cosh^2 a+1\Big]\Bigg[4\omega^2(-\lambda^2+\tanh^2 z) \\
& + \frac{2}{\cosh^2 z}(1-3\tanh^2 z)\Bigg]\Bigg\}. \qquad (19)
\end{aligned}$$

5. Comparison and prospects

In order to calculate the energy, we need to integrate the above expressions over the whole region as well as over frequencies. After integration we are going to make $a \to \infty$, therefore we need to consider only the region where the kink lives, that means the interval $[-a, a]$. The total energy would be,

$$T_{00} = \frac{1}{2\pi}\int_{-\infty}^{\infty}\int_{-a}^{a} d\zeta\, dz \langle T_{00}(z,\zeta)\rangle, \qquad (20)$$

where we have rotated to the imaginary axes, $\omega \to i\zeta$. Since we are interested in studying local effects, we focus on the integrand of the above expression and plot it for a large value of a keeping ω constant (or equivalently ζ) in each graph. In figure (1) we plot the local density energy around

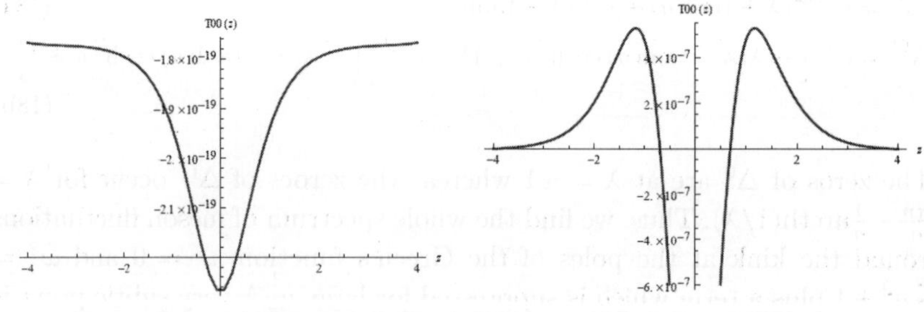

Fig. 1. Local density energy for periodic boundary conditions. On the left $\omega = 2$ and on the right $\omega = 0.2$. The minimum in the right plot is $T_{00} = -1.87 \times 10^{-6}$.

the kink for periodic boundary conditions for frequencies of 2 on the left and 0.2 on the right. The case of decaying boundary conditions at that frequency is qualitatively the same as for periodic, it differs on the value of the energy which is roughly the same as the plotted one except for the powers of ten which do not show for decaying boundary conditions. We observe by looking at the figure (1) that the behavior at low frequencies is quite different than that at large frequencies. Clearly the effect of the kink over the passing mesons is stronger when they come with low frequency (graph on the right). The scale on the plots indicate that high frequency mesons passing by the kink background (graph on the left) are almost ignoring its presence and the bend on energy in that region is very small (notice that the plots do not have the same scale). Graphs shown in figure (2) describe the behavior for large frequencies. We observe that for decaying boundary conditions (on the top), the value of the density energy tends to a constant value, while for periodic boundary conditions (on the bottom) the plot gets distorted as the frequency gets larger, and it diverges.

In a coming publication we hope to expand this analysis by regularizing and renormalizing the ultraviolet divergences following the heat kernel regularization method.[5]

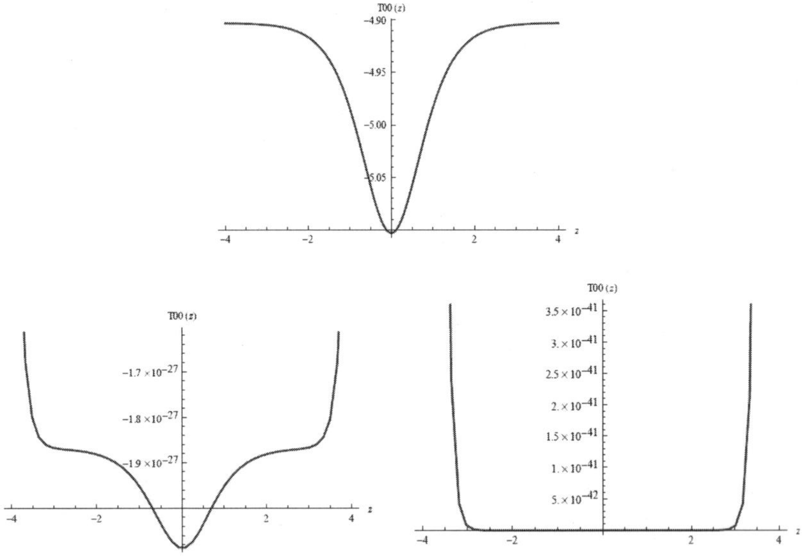

Fig. 2. Density energy for high frequencies. Decaying boundary conditions on the top ($\omega = 5$) and periodic boundary conditions on the bottom ($\omega = 3$ left and $\omega = 5$ right).

Acknowledgments

We thank the European Science Foundation(ESF) within the activity 'New Trends and Applications of the Casimir Effect' exchange grant 2301, the Spanish Ministerio de Educación y Ciencia and the Junta de Castilla y León under grants FIS2006-09417 and GR224 for partial support of this research. We especially thank Kimball A. Milton for an excellent organization and a successful workshop. IC acknowledges travel support by the ESF.

References

1. R. Dashen, B. Hasslacher, A. Neveu, Phys. Rev. **D10** (1974) 4130.
2. M. Bordag, U. Mohideen, V. M. Mostepanenko, Phys. Rept.**353**:1-2005,2001.
3. A. A. Izquierdo, W. G. Fuertes, M. A. G. Leon, J. M. Guilarte, Nucl. Phys. **B 635**(2002) 525-557.
4. K. A. Milton, J. of Phys **A37**(2004)R209.
5. M. Bordag and J. Lindig, J. of Phys **A29** (1996) 4481.
6. G. Barton, J. of Phys **A18**(1985)479.
7. N. Graham and R. Jaffe, Nucl. Phys, **B544**(1999)432-447.

QUANTUM FLUCTUATIONS OF TOPOLOGICAL $\mathbb{S}^3$-KINKS

A. ALONSO IZQUIERDO[(a)], M. A. GONZALEZ LEON[(a)],
J. MATEOS GUILARTE[(b)], M. J. SENOSIAIN[(c)]

[(a)] *Departamento de Matematica Aplicada and IUFFyM*
[(b)] *Departamento de Fisica Fundamental and IUFFyM*
[(c)] *Departamento de Matematicas, Universidad de Salamanca, Spain*

The kink Casimir effect in the massive non-linear S^3-sigma model is analyzed.

Keywords: Kink Casimir effect, spectral zeta function, non-linear sigma model.

1. Introduction

Quantum fluctuations around background kink fields are sophisticated cousins of vacuum fluctuations. Van Nieuwenhuizen et al.[1] reported on the state of the art in this topic in QFEXT03 for susy solitons as pointed out by Milton[2]. More recently, new results have been achieved by (almost) the same Stony Brook/Wien group in the analysis of the quantum fluctuations of susy solitons of non-linear sigma models[3,4]. Almost in parallel, we developed a similar program[5–8] for the kinks of the massive non-linear $\mathbb{S}^2$-sigma model in a purely bosonic framework. Our goal in this work is to describe the quantum fluctuations of the $\mathbb{S}^3$-kinks. The bosonic sector of the non-linear version of the Gell-Mann/Levy σ-model[9] is precisely the system that we are going to address. Being non renormalizable in (3+1)-dimensions, it was conceived as an effective theory describing the low energy interactions of nucleons and pions. In (1+1)-dimensions, however, the pion dynamics can be re-interpreted as the dynamics of a linear chain of $\mathbb{O}(4)$ spin fields, which was renormalized by Brezin et al[10]. We just merely add quadratic terms in the fields to escape from infrared divergences.

2. Massive non-linear $\mathbb{S}^3$-sigma model and topological kinks

Let us consider $\phi_a(t,x)$, $a = 1,2,3,4$, four scalar fields in the $(1+1)$-dimensional Minkowski space-time $\mathbb{R}^{1,1}$. The action of the massive nonlinear $\mathbb{S}^3$-sigma model looks very simple

$$S[\phi_1,\phi_2,\phi_3,\phi_4]=\int dtdx\left\{\frac{1}{2}g^{\mu\nu}\sum_{a=1}^{4}\frac{\partial\phi_a}{\partial x^\mu}\frac{\partial\phi_a}{\partial x^\nu}-\frac{1}{2}\sum_{a=1}^{4}\alpha_a^2\,\phi_a^2(x^\mu)\right\}\,, \quad (1)$$

where $\alpha_1^2>\alpha_2^2>\alpha_3^2>\alpha_4^2$, but the fields are constrained to live in the $\mathbb{S}^3$-sphere, $\phi_1^2+\phi_2^2+\phi_3^2+\phi_4^2=R^2$ forming the infinite dimensional space: Maps$(\mathbb{R}^{1,1},\mathbb{S}^3)$. We take $g_{\mu\nu}=\text{diag}(1,-1,-1,1)$ and the natural system of units $\hbar=c=1$. We select $\sigma_2^2=\frac{\alpha_2^2-\alpha_4^2}{\alpha_1^2-\alpha_4^2}=\frac{\gamma_2^2}{\lambda^2}$, $\sigma_3^2=\frac{\alpha_3^2-\alpha_4^2}{\alpha_1^2-\alpha_4^2}=\frac{\gamma_3^2}{\lambda^2}$, such that $0<\sigma_3^2<\sigma_2^2<\sigma_1^2=1$, and define non-dimensional coordinates: $x^\mu\to\frac{x^\mu}{\lambda}$.

The extremely non-linear dynamics implied by (1) plus the constraint is unveiled if one solves ϕ_4 in favor of ϕ_1,ϕ_2,ϕ_3 in the action and introduce the power expansion of the non-polynomial term. This process shows that: (a) There are an infinite number of vertices determining the interactions between the three pseudo-Nambu-Goldstone bosons. (b) $\frac{1}{R^2}$ is the coupling constant. Vertices with different numbers of legs belong to different orders of perturbation theory: $\frac{1}{R^{2n-2}}$ arises as a factor in the vertices with $2n$ legs. (c) In $(1+1)$-dimensions massless bosons are discarded due to the infrared asymptotics. We consider the situation when the three masses are different. The one-loop self-energy graphs of ϕ_1, ϕ_2 and ϕ_3: $\Sigma_2(\sigma_2,\sigma_3)=\sigma_2^2\Sigma_1(\sigma_2,\sigma_3)$, $\Sigma_3(\sigma_2,\sigma_3)=\sigma_3^2\Sigma_1(\sigma_2,\sigma_3)$ are divergent because $\Sigma_1(\sigma_2,\sigma_3)=\frac{2i}{R^2}\left(I(1)+I(\sigma_2^2)+I(\sigma_3^2)\right)$ with $I(c^2)=\int\frac{dk}{4\pi}\frac{1}{\sqrt{k^2+c^2}}$. To tame these infinities the one-loop mass renormalization counter-terms

$$\mathcal{L}_{CT}=-\frac{1}{R^2}\cdot\Sigma_1(\sigma_2,\sigma_3)\left(\phi_1^2(x^\mu)+\sigma_2^2\phi_2^2(x^\mu)+\sigma_3^2\phi_3^2(x^\mu)\right)$$

must be added to the bare Lagrangian. Searching only for semi-classical effects we do not need to care about other divergent graphs.

The classical minima of the action are the static and homogeneous configurations that annihilate the integrand in (1), i.e., the North and South Poles of $\mathbb{S}^3$. There is the possibility of the existence of topological kinks and to search for them it is convenient to use polar hyper-spherical coordinates: $\phi_1=R\sin\psi\sin\theta\cos\varphi$, $\phi_2=R\sin\psi\sin\theta\sin\varphi$, $\phi_3=R\sin\psi\cos\theta$, $\phi_4=R\cos\psi$, $\psi\in[0,\pi)$,$\theta\in[0,\pi)$, $\varphi\in[0,2\pi)$. There are three types of these kinks: (1) in the meridians on the $\phi_3-\phi_4$ plane, $\theta=0$ or π, the non-trivial field equation is: $\frac{\partial^2\psi}{\partial t^2}-\frac{\partial^2\psi}{\partial x^2}+\frac{\sigma_3^2}{2}\sin 2\psi=0$ and the kink solutions, that we shall denote generically as K_1, can be written in the form $\psi_{K_1}(t,x)=2\arctan e^{\pm\sigma_3\overline{x}}$ where $\overline{x}=\frac{x-x_0-vt}{\sqrt{1-v^2}}$; (2) analogously in the meridians on the $\phi_2-\phi_4$ plane, $\theta=\frac{\pi}{2}$, $\varphi=\frac{\pi}{2}$ or $\frac{3\pi}{2}$, the kink solutions will be referred to as K_2 and are given by $\psi_{K_2}(t,x)=2\arctan e^{\pm\sigma_2\overline{x}}$ and (3)

K_3 kinks, which live in the meridians on the $\phi_1 - \phi_4$ plane, $\theta = \frac{\pi}{2}$, $\varphi = 0$ or π, are $\psi_{K_3}(t,x) = 2\arctan e^{\pm\overline{x}}$. The topological $\mathbb{S}^3$-kink classical energies are: $E(K_1) = 2\lambda R^2\sigma_3 < E(K_2) = 2\lambda R^2\sigma_2 < E(K_3) = 2\lambda R^2$.

Changing slightly the notation by denoting $\psi = \theta^1$, $\theta = \theta^2$, $\varphi = \theta^3$, small fluctuations around the kink solution $\theta(x) = \theta_K(x) + \eta(x) = (\theta^1_K(x), \theta^2_K(x), \theta^3_K(x)) + (\eta^1(x), \eta^2(x), \eta^3(x))$ modify the action as:

$$S[\theta^1,\theta^2,\theta^3] = S[\theta^1_K,\theta^2_K,\theta^3_K] + \frac{R^2}{2}\int dtdx\,\eta(x)\Delta(K)\eta(x) + \mathcal{O}(\eta^3) \quad .$$

The second-order operator governing the kink small fluctuations is the geodesic deviation operator plus the Hessian of the potential: $\Delta(K)\eta = -\left(\nabla_{\theta'_K}\nabla_{\theta'_K}\eta + R(\theta'_K,\eta)\theta'_K + \nabla_\eta \mathrm{grad} V\right)$. Standard geometric calculations allow us to conclude that K_1 small fluctuations are governed by the matrix of Schrödinger operators:

$$\Delta(K_1) = \left(-\frac{d^2}{dx^2} - \frac{2\sigma^2}{\cosh^2\sigma_3 x}\right)\mathbb{I} + \mathrm{diag}\left(\sigma_3^2, 1, \sigma_2^2\right) \tag{2}$$

provided that a "parallel frame" to the kink orbit, i. e., fluctuations of the form $\eta^2(x) = \cosh\sigma_3 x\,\xi^2(x)$, $\eta^3(x) = \cosh\sigma_3 x\,\xi^3(x)$, is chosen.

Therefore, the meson spectrum in the K_1 kink sector has three branches that share a perfectly transmitting Pösch-Teller well but have different thresholds. The first branch corresponds to fluctuations tangent to the kink orbit. There is a bound state, $\eta^1_0(x) = \frac{1}{\cosh\sigma_3 x}$, of zero eigenvalue and one-particle scattering states $\eta^1_k(x) = e^{ik\sigma_3 x}(\tanh\sigma_3 x - ik)$ with frequencies $\omega^2(k) = \sigma_3^2(k^2+1)$. In the orthogonal directions the eigenfunctions are the same but the bound state energies and thresholds of the continuous spectra are shifted respectively to: $1-\sigma_3^2$, $\sigma_2^2 - \sigma_3^2$, 1 and σ_2^2.

3. Spectral zeta function and kink mass quantum correction

We choose a normalization interval of length $l = \lambda L$ and impose periodic boundary conditions on the fluctuations: $\eta(-\frac{l}{2}) = \eta(\frac{l}{2})$. At the end of the computations we will send the length l of the normalization interval to infinity. $\Delta(K)$ acts on the Hilbert space $L^2 = L^2_1(\mathbb{S}^1)\bigoplus L^2_2(\mathbb{S}^1)\bigoplus L^2_3(\mathbb{S}^1)$. The heat trace ($\beta$ is a fictitious inverse temperature or Euclidean time) is:

$$\mathrm{Tr}_{L^2} e^{-\beta\Delta(K_1)} = \frac{l\mathcal{A}}{\sqrt{4\pi\beta}} + \tanh\frac{\sigma_3 l}{2}\left(1 + e^{-(1-\sigma_3^2)\beta} + e^{-(\sigma_2^2-\sigma_3^2)\beta}\right)\mathrm{Erf}[\sigma_3\sqrt{\beta}]$$

where $\mathcal{A} = e^{-\sigma_3^2\beta} + e^{-\beta} + e^{-\sigma_2^2\beta}$. It is interesting also to use the short time asymptotics of the heat trace. Due to the structure of the second-order fluctuations operator (2), a power β expansion of the heat trace is sensible[7]:

$$\mathrm{Tr}_{L^2} e^{-\beta\Delta(K_1)} = \mathrm{Tr}_{L^2} e^{-\beta\Delta_0(K_1)} \sum_{n=0}^{\infty} c_n(K_1)\beta^n = \frac{\mathcal{A}}{\sqrt{4\pi\beta}} \sum_{n=0}^{\infty} c_n(K_1)\beta^n \, ,$$

where the coefficients are: $c_0(K_1) = l$, $c_n(K_1) = \frac{2^{n+1}\sigma_3^{2n-1}}{(2n-1)!!}$.

The Casimir energy $\Delta E^C = \Delta E - \Delta E_0 = \frac{\lambda}{2}\left[\mathrm{Tr}^*_{L^2}\Delta^{\frac{1}{2}}(K_1) - \mathrm{Tr}_{L^2}\Delta_0^{\frac{1}{2}}(K_1)\right]$ is ultra-violet divergent. We shall regularize these divergences by using the zeta function method. The zeta functions are the Mellin transform of the heat traces, $\zeta_\Delta(s) = \frac{1}{\Gamma(s)}\int_0^\infty d\beta \beta^{s-1}\, \mathrm{Tr}_{L^2}\, e^{-\beta\Delta}$ and thus we regularize the divergence by assigning to it the value of the spectral zeta function at a regular point of the s-complex plane: $\Delta E^C(s) = \frac{\mu}{2}\left(\frac{\mu^2}{\lambda^2}\right)^s\left[\zeta^*_{\Delta(K_1)}(s) - \zeta_{\Delta_0(K_1)}(s)\right]$. The behaviour of the kink Casimir energy near the physical pole $s = -\frac{1}{2} + \varepsilon$ is:

$$\Delta E^C = -\frac{\lambda\sigma_3}{2\pi}\left[\frac{3}{\varepsilon} + 3\ln\frac{\mu^2}{\lambda^2} + \ln\frac{2^6}{\sigma_3^2\bar{\sigma}_{13}^2\bar{\sigma}_{23}^2} - 4 + F[-\tfrac{\sigma_3^2}{\bar{\sigma}_{13}^2}] + F[-\tfrac{\sigma_3^2}{\bar{\sigma}_{23}^2}]\right] \quad (3)$$

where we denote $F[x] = {}_2F_1^{(0,1,0,0)}[\frac{1}{2}, 0, \frac{3}{2}; x]$, $\bar{\sigma}_{13}^2 = 1 - \sigma_3^2$ and $\bar{\sigma}_{23}^2 = \sigma_2^2 - \sigma_3^2$.

The kink energy due to the mass renormalization counter-terms that must be added, $\Delta E^{MR} = -\lambda\frac{\sigma_3^2}{R^2}[I(1) + I(\sigma_2^2) + I(\sigma_3^2)]\int dx\, \phi_3^{\mathrm{K}_1}(x)\phi_3^{\mathrm{K}_1}(x) = 2\lambda\sigma_3[I(1) + I(\sigma_2^2) + I(\sigma_3^2)]$ is also ultra-violet divergent. The loop integrals become in the finite length normalization interval divergent series susceptible of being regularized as spectral zeta functions:

$$I(c^2) = \frac{1}{2l}\sum_{n=-\infty}^{\infty}\frac{1}{(\sigma_3^2 n^2 + c^2)^{\frac{1}{2}}} = -\frac{\lambda}{\mu l}\lim_{s\to-\frac{1}{2}}\left(\frac{\mu^2}{\lambda^2}\right)^{s+1}\frac{\Gamma(s+1)}{\Gamma(s)}\zeta_{-\frac{d^2}{dx^2}+c^2}(s)$$

The regularized mass renormalization kink energy

$$\Delta E^{\mathrm{MR}}(s) = -\frac{2\sigma_3\lambda^2}{\mu\sqrt{4\pi}}\left(\frac{\mu^2}{\lambda^2}\right)^{s+1}\frac{\Gamma(s+\frac{1}{2})}{\Gamma(s)}\left(1 + \frac{1}{\sigma_2^{2s+1}} + \frac{1}{\sigma_3^{2s+1}}\right)$$

behaves near the physical pole as:

$$\Delta E^{\mathrm{MR}}(-\frac{1}{2}+\varepsilon) = \frac{\lambda\sigma_3}{2\pi}\left[\frac{3}{\varepsilon} + 3\ln\frac{\mu^2}{\lambda^2} + 3(\ln 4 - 2) - \ln\sigma_2^2\sigma_3^2\right] \quad (4)$$

From the short-time asymptotics of the heat trace we obtain an approximated formula for the kink Casimir energy by means of the partial Mellin transform on the $[0, b]$ integration interval of the truncated to N_0 terms heat trace expansion:

$$\Delta E^C(b, N_0) = -\frac{\lambda}{2\sqrt{\pi b}} - \frac{\lambda}{8\pi}\sum_{n=1}^{N_0} c_n(K)\left[\frac{\sigma_3^2}{\sigma_3^{2n}}\gamma[\sigma_3^2 b] + \gamma[b] + \frac{\sigma_2^2}{\sigma_2^{2n}}\gamma[\sigma_2^2 b]\right]$$

where $\gamma[c]=\gamma[n-1,c]$ and $\gamma[z,c]$ is the incomplete Euler gamma function. The contribution $\Delta E^C_{(1)}$ of the term with $c_1(K_1)=4\sigma_3$ to this approximation to the kink Casimir energy is divergent because $z=0$ is a pole of $\gamma[z,c]$. Fortunately, the divergent mass renormalization kink energy $\Delta E^{\rm MR}$ exactly cancels $\Delta E^C_{(1)}$.

Finally, the K_1 semiclassical mass, $E(K_1)=2\lambda\sigma_3^2R^2+\Delta E+\mathcal{O}(\frac{1}{R^2})$, is obtained by adding (3) and (4):

$$\Delta E=-\frac{\lambda\sigma_3}{2\pi}\left[2+F[-\tfrac{\sigma_3^2}{\bar\sigma_{13}^2}]+F[-\tfrac{\sigma_3^2}{\bar\sigma_{23}^2}]+\ln\frac{\sigma_2^2}{\bar\sigma_{13}^2\bar\sigma_{23}^2}\right]. \tag{5}$$

Because the wells in the second-order fluctuation operator are transparent the Cahill-Comtet-Glauber formula[11], $\Delta E(K_1)=-\frac{\lambda\sigma_3}{\pi}[\sin\nu_1+\frac{1}{\sigma_3}\sin\nu_2+\frac{\sigma_2}{\sigma_3}\sin\nu_3-\nu_1\cos\nu_1-\frac{1}{\sigma_3}\nu_2\cos\nu_2-\frac{\sigma_2}{\sigma_3}\nu_3\cos\nu_3]$, with $\nu_1=\arccos(0)=\frac{\pi}{2}$, $\nu_2=\arccos\bar\sigma_{13}$, $\nu_3=\arccos\frac{\bar\sigma_{23}}{\sigma_2}$, giving the one-loop mass shift in terms only of the bound state eigenvalues and the thresholds of the continuous spectra, can be applied[11]. Despite appearances, the result

$$\Delta E(K_1)=-\frac{\lambda\sigma_3}{\pi}\left[3-\frac{\bar\sigma_{13}}{\sigma_3}\arccos(\bar\sigma_{13})-\frac{\bar\sigma_{23}}{\sigma_3}\arccos(\frac{\bar\sigma_{23}}{\sigma_2})\right] \tag{6}$$

is identical to (5) as one can check by plotting of both expressions. A third (approximate) formula, useful in the cases when the spectral information on the kink fluctuations is unknown, is derived from the asymptotic expansion:

$$\Delta E(b,N_0)=-\frac{\lambda}{2\sqrt{\pi b}}-\frac{\lambda}{8\pi}\sum_{n=2}^{N_0}c_n(K)\left[\frac{\sigma_3^2}{\sigma_3^{2n}}\gamma[\sigma_3^2b]+\gamma[b]+\frac{\sigma_2^2}{\sigma_2^{2n}}\gamma[\sigma_2^2b]\right] \tag{7}$$

References

1. A. Rebhan, P. van Nieuwenhuizen, R. Wimmer, [arXiv:hep-th/0401127]
2. K. Milton, Journal of Physics A37(2004)R209, [arXiv: hep-th/0406024]
3. C. Mayrhofer, A. Rebhan, P. van Nieuwenhuizen, R. Wimmer, JHEP (2007) 0709:069, [arXiv:0706.4476]
4. A. Rebhan, A. Schmitt, P. van Nieuwenhuizen, Phys. Rev. **D80**:045012,2009, [arXiv: 0903.5242]
5. A. Alonso, M.A.G. Leon, J.M. Guilarte, Phys. Rev. Lett. **101**131602,2008.
6. A. Alonso, M.A.G. Leon, J.M. Guilarte, Phys. Rev. **D79**:125003,2009.
7. A. Alonso, M. A.G. Leon, J.M. Guilarte, M. J. Senosiain, Jour. Phys. A: Math. Theor. **42** (2009)385403,
8. A. Alonso, W.G. Fuertes, J.M. Guilarte, M. de la Torre Mayado, M. J. Senosiain, PoS(ISFTG)013,2009, [arXiv:0909.2107]
9. M. Gell-Mann, M. Levy, Il Nuovo Cimento XVI (1960) 705
10. E. Brezin, J. Zinn-Justin, J. C. Le Guillou, Phys. Rev. **D14** (1976) 2615
11. K. Cahill, A. Comtet, R. Glauber, Phys. Lett. **64B**(1976) 283-285

TOPOLOGICAL CASIMIR EFFECT IN NANOTUBES AND NANOLOOPS

A. A. SAHARIAN*

Department of Physics, Yerevan State University,
1 Alex Manoogian Street, 0025 Yerevan, Armenia
**E-mail: saharian@ictp.it*

The Casimir effect is investigated in cylindrical and toroidal carbon nanotubes within the framework of the Dirac-like model for the electronic states. The topological Casimir energy is positive for metallic cylindrical nanotubes and is negative for semiconducting ones. The toroidal compactification of a cylindrical nanotube along its axis increases the Casimir energy for metallic-type (periodic) boundary conditions along its axis and decreases the Casimir energy for the semiconducting-type compactifications. For finite length metallic nanotubes the Casimir forces acting on the tube edges are always attractive, whereas for semiconducting-type ones they are attractive for small lengths of the nanotube and repulsive for large lengths.

Keywords: Casimir effect; Carbon nanotubes.

1. Introduction

An interesting application of the quantum field theoretical models with non-trivial topology recently appeared in nanophysics. The long-wavelength description of the electronic states in graphene can be formulated in terms of the Dirac-like theory with the Fermi velocity playing the role of speed of light (see, e.g., Ref. [1]). Single-walled carbon nanotubes are generated by rolling up a graphene sheet to form a cylinder and the background spacetime for the corresponding Dirac-like theory has topology $R^2 \times S^1$. Compactifying the direction along the cylinder axis we obtain another class of graphene made structures called toroidal carbon nanotubes with the background topology $R^1 \times (S^1)^2$.

The boundary conditions imposed on the fermionic field in these nanostructures give rise to the modification of the spectrum for vacuum fluctuations and, as a result, to the Casimir-type contributions in vacuum expectation value (VEV) of the energy-momentum tensor (for the topological

Casimir effect see [2]). In the present paper, based on Refs. [3,4], we investigate the fermionic Casimir effect in nanotubes. The paper is organized as follows. In the next section the topological Casimir effect is considered in cylindrical nanotubes. The VEV of the energy-momentum tensor for a fermionic field in toroidal nanotubes is discussed in Sec. 3. The Casimir forces acting on the edges of finite-length carbon nanotubes are investigated in Sec. 4. The main results are summarized in Sec. 5.

2. Cylindrical nanotubes

A single wall cylindrical nanotube is a graphene sheet rolled into a cylindrical shape. For this case we have spatial topology $R^1 \times S^1$ with the compactified dimension of the length L. The carbon nanotube is characterized by its chiral vector $\mathbf{C}_h = n_w \mathbf{a}_1 + m_w \mathbf{a}_2$, with n_w, m_w being integers, and $L = |\mathbf{C}_h| = a\sqrt{n_w^2 + m_w^2 + n_w m_w}$, $a = |\mathbf{a}_1| = |\mathbf{a}_2| = 2.46\mathring{A}$. A zigzag nanotube corresponds to the special case $\mathbf{C}_h = (n_w, 0)$, and a armchair nanotube corresponds to the case $\mathbf{C}_h = (n_w, n_w)$. All other cases correspond to chiral nanotubes. In the case $n_w - m_w = 3q_w$, $q_w \in Z$, the nanotube will be metallic and in the case $n_w - m_w \neq 3q_w$ the nanotube will be semiconductor with an energy gap inversely proportional to the diameter.

The electronic band structure of a graphene sheet close to the Dirac points shows a conical dispersion $E(\mathbf{k}) = v_F|\mathbf{k}|$, where $\mathbf{k}$ is the momentum measured relatively to the Dirac points and v_F represents the Fermi velocity. The corresponding low-energy excitations can be described by a pair of two-component Weyl spinors. By taking into account that in the presence of an external magnetic field an effective mass term is generated for the fermionic excitations, we consider the general case of massive spinor field ψ on background of 3-dimensional flat spacetime with spatial topology $R^1 \times S^1$. The corresponding line element has the form $ds^2 = dt^2 - (dz^1)^2 - (dz^2)^2$, where $-\infty < z^1 < \infty$ and $0 \leqslant z^2 \leqslant L$.

We assume that the field obeys the boundary condition

$$\psi(t, z^1, z^2 + L) = e^{i\beta}\psi(t, z^1, z^2), \tag{1}$$

with a constant phase β. For metallic nanotubes $\beta = 0$ and for semiconductor nanotubes, depending on the chiral vector, we have two classes of inequivalent boundary conditions corresponding to $\beta = \pm 2\pi/3$. In the expression for the Casimir densities the phase β appears in the form $\cos(n\beta)$ and, hence, the Casimir energy density and stresses are the same for these two cases. The VEV of the energy-momentum tensor for a cylindrical nanotube has the form (no summation over l)

$$\langle T_l^k \rangle_{(\mathrm{cyl})}(L) = \delta_l^k \sum_{n=1}^{\infty} \cos(n\beta) G^{(l)}(nmL) \frac{e^{-nmL}}{\pi L^3 n^3}, \tag{2}$$

with the notations $G^{(0)}(z) = G^{(1)}(z) = 1 + z$, $G^{(2)}(z) = -(2 + 2z + z^2)$. In particular, the Casimir energy density is positive for metallic nanotubes and negative for semiconducting ones.

In the massless case we have $\langle T_0^0 \rangle_{(\mathrm{cyl})} = \langle T_1^1 \rangle_{(\mathrm{cyl})} = -\langle T_2^2 \rangle_{(\mathrm{cyl})}/2 = S_\beta/(\pi L^3)$, with the notation $S_\beta = \sum_{n=1}^{\infty} \cos(n\beta)/n^3$. In particular, $S_0 = 1.202$ and $S_{2\pi/3} = -0.534$. In carbon nanotubes we have two sublattices and each of them gives the contribution to the Casimir densities given by (2). So, for the Casimir energy density on a carbon nanotube with radius L one has $\langle T_0^0 \rangle_{1,1}^{(\mathrm{cn})} = 2\hbar v_F S_\beta/(\pi L^3)$, where the standard units are restored.

3. Toroidal nanotubes

For the geometry of a toroidal nanotube we have the spatial topology $(S^1)^2$. The boundary conditions have the form

$$\psi(t, z^1 + L_1, z^2) = e^{i\beta_1} \psi(t, z^1, z^2),\ \psi(t, z^1, z^2 + L_2) = e^{i\beta_2} \psi(t, z^1, z^2). \tag{3}$$

The corresponding energy density and the vacuum stresses are given by the expressions (no summation over l)

$$\begin{aligned} \langle T_0^0 \rangle_{(\mathrm{tor})} = {} & \sum_{j=1,2} \langle T_0^0 \rangle_{(\mathrm{cyl})}(L_j) + \frac{2}{\pi} \sum_{m_1=1}^{\infty} \sum_{m_2=1}^{+\infty} \frac{1 + mg(\mathbf{L}_2, \mathbf{m}_2)}{g^3(\mathbf{L}_2, \mathbf{m}_2)} \\ & \times \frac{\cos(m_1\beta_1)\cos(m_2\beta_2)}{\exp(mg(\mathbf{L}_2, \mathbf{m}_2))}, \end{aligned} \tag{4}$$

$$\begin{aligned} \langle T_l^l \rangle_{(\mathrm{tor})} = {} & \langle T_0^0 \rangle_{(\mathrm{tor})} - \frac{m^5}{\pi} \sum_{j=1,2} \sum_{m_j=1}^{+\infty} \cos(m_j\beta_j) L_l^2 m_l^2 G(mL_j m_j) \\ & - \frac{2m^5}{\pi} \sum_{m_1, m_2=1}^{+\infty} \cos(m_1\beta_1)\cos(m_2\beta_2) L_l^2 m_l^2 G(mg(\mathbf{L}_2, \mathbf{m}_2)), \end{aligned} \tag{5}$$

with $G(x) = (3 + 3x + x^2)e^{-x}/x^5$, $l = 1, 2$, and $g(\mathbf{L}_2, \mathbf{m}_2) = \sqrt{m_1^2 L_1^2 + m_2^2 L_2^2}$. The corresponding formulae for the Casimir densities in toroidal nanotubes are obtained with an additional factor 2. In standard units the factor $\hbar v_F$ appears as well. From the formulae given above it follows that the toroidal compactification of a cylindrical nanotube along its axis increases the Casimir energy for periodic conditions ($\beta_1 = 0$) and decreases the Casimir energy for the semiconducting-type compactifications.

4. Finite-length nanotubes

In this section we will assume that the nanotube has finite length a. We assume the periodicity condition (1) along the compact dimension. As the Dirac field lives on the cylinder surface it is natural to impose bag boundary conditions $(1 + i\gamma^l n_l)\psi = 0$, $z^1 = 0, a$, on the cylinder edges, with γ^l being the Dirac matrices and n_l is the normal to the boundaries. The additional confinement of the electrons along the tube axis leads to the change of the ground state energy. The corresponding Casimir energy is decomposed as

$$E = aL\langle T_0^0\rangle_{(\mathrm{cyl})} + 2E^{(1)} - \frac{1}{\pi}\sum_{l=-\infty}^{+\infty}\int_{m_l}^{\infty}\frac{x dx}{\sqrt{x^2 - m_l^2}}\ln\left(1 + \frac{x-m}{x+m}e^{-2ax}\right), \tag{6}$$

with $m_l^2 = [(2\pi l + \beta)/L]^2 + m^2$. The part $E^{(1)}$ is the Casimir energy for a single edge (when the other edge is absent) in the half-space. The last term in Eq. (6) is the interaction part. The single edge part of the Casimir energy does not depend on the length of the tube and will not contribute to the Casimir force.

For the Casimir force acting on the edges of the tube we have

$$P = -\langle T_0^0\rangle_{(\mathrm{cyl})} - \frac{2}{\pi L}\sum_{l=-\infty}^{+\infty}\int_{m_l}^{\infty} dx\, \frac{x}{\frac{x+m}{x-m}e^{2ax} + 1}. \tag{7}$$

The corresponding expressions for the Casimir energy and force in finite length cylindrical nanotubes are obtained with an additional factor 2. So, for the Casimir force acting per unit length of the edge of a carbon nanotube one has: $P^{(\mathrm{CN})} = 2\hbar v_{\mathrm{F}} P$. For long tubes, $a/L \gg 1$, the first term on the right of (7) is dominant and we have $P^{(\mathrm{CN})} \approx -0.765\hbar v_{\mathrm{F}}/L^3$ for metallic nanotubes and $P^{(\mathrm{CN})} \approx 0.34\hbar v_{\mathrm{F}}/L^3$ for semiconducting ones. In the limit $a/L \ll 1$ the Casimir force does not depend on the chirality and one has $P^{(\mathrm{CN})} \approx -0.144\hbar v_{\mathrm{F}}/a^3$. In Fig. 1 we have plotted the Casimir forces acting on the edges of metallic (full curves) and semiconducting-type (dashed curves) carbon nanotube as functions of the tube length for different values of the fermion mass. As it is seen, for metallic nanotubes these forces are always attractive, whereas for semiconducting-type ones they are attractive for small lengths and repulsive for large lengths.

5. Conclusion

We have investigated the Casimir effect for cylindrical and toroidal nanotubes within the framework of the Dirac-like model for electrons. The VEV

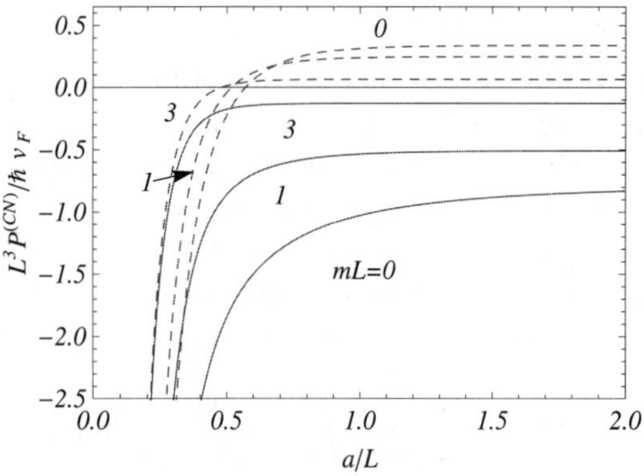

Fig. 1. The fermionic Casimir forces acting on the edges of the metallic (full curves) and semiconducting-type (dashed curves) nanotubes as functions of the tube length.

of the energy-momentum tensor is given by formula (2) for cylindrical nanotubes and by (4) and (5) (with an additional factor 2 which takes into account the presence of two sublattices) for toroidal nanotubes. The topological Casimir energy is positive for metallic cylindrical nanotubes and is negative for semiconducting ones. We have shown that the toroidal compactification of a cylindrical nanotube along its axis increases the Casimir energy for periodic boundary conditions and decreases the Casimir energy for the semiconducting-type compactifications. For finite-length carbon nanotubes the Casimir forces acting on the tube edges are always attractive for metallic nanotubes, whereas for semiconducting-type ones they are attractive for small lengths and repulsive for large lengths. The Casimir energy should be taken into account in the determination of the parameters of nanotubes, such as the diameter and the length. It will also influence the relative number of metallic and semiconducting nanotubes in their formation process.

References

1. A. H. Castro Neto, F. Guinea, N. M. R. Peres, K. S. Novoselov and A. K. Geim, *Rev. Mod. Phys.* **81**, 109 (2009).
2. V. M. Mostepanenko and N. N. Trunov, *The Casimir Effect and Its Applications* (Clarendon, Oxford, 1997); K. A. Milton, *The Casimir Effect: Physical Manifestation of Zero-Point Energy* (World Scientific, Singapore, 2002); M. Bordag, G.L. Klimchitskaya, U. Mohideen and V.M. Mostepanenko, *Advances in the Casimir Effect* (Oxford University Press, Oxford, 2009).
3. S. Bellucci and A. A. Saharian, *Phys. Rev. D* **79**, 085019 (2009).
4. S. Bellucci and A. A. Saharian, *Phys. Rev. D* **80**, 105003 (2009).

THE SEARCH FOR THE SCHWINGER EFFECT: NONPERTURBATIVE VACUUM PAIR PRODUCTION

GERALD V. DUNNE

Department of Physics, University of Connecticut, Storrs, CT 06269, USA

The Schwinger effect is the non-perturbative production of electron-positron pairs when an external electric field is applied to the quantum electrodynamical (QED) vacuum. The inherent instability of the vacuum in an electric field was one of the first non-trivial predictions of QED, but the effect is so weak that it has not yet been directly observed. However, there are exciting new developments in ultra-high intensity lasers, which may bring us to the verge of this extreme ultra-relativistic regime. This necessitates a fresh look at both experimental and theoretical aspects of the Schwinger effect. I describe some new theoretical ideas aimed at making this elusive effect observable, by careful shaping of the laser pulses, much as is done [in a different intensity regime] in the fast-developing field of atto-science.

Keywords: Schwinger effect; non-perturbative; semiclassical.

1. Introduction

The experimental observation of the Schwinger effect, the non-perturbative production of electron-positron pairs from vacuum subjected to an electric field, would open a new window into the largely unexplored regime of non-perturbative quantum field theory, a regime in which we can study matter in extreme environments in a controllable way. This has significant implications beyond QED, for example in particle physics, nuclear physics, plasma physics and gravitational physics. It is well known that quantum vacuum fluctuations mean that the QED vacuum behaves like a polarizable medium that modifies classical behavior, leading to novel quantum effects.[1–4] Some of these effects, such as the Casimir effect or the vacuum birefringence effect, are perturbative and can be well described by perturbative quantum field theory. The Schwinger effect is a non-perturbative effect that cannot be described by any single Feynman diagram; its essence is a truly non-perturbative process, which makes it both fascinating and difficult. The process can be viewed pictorially as in Figure 1: a virtual electron-positron

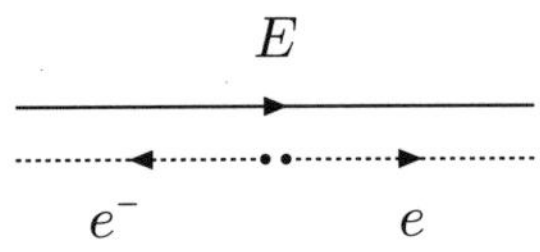

Fig. 1. Pair production as the separation of a virtual vacuum dipole pair under the influence of an external electric field.

pair in vacuum is accelerated apart by an external electric field, becoming a real asymptotic $e^+ e^-$ pair if they gain the binding energy of $2mc^2$ from the external field. This sets the basic scale at which we might expect this process to be significant: when the work done separating the pair by a Compton wavelength matches $2mc^2$. This is a huge scale:

$$\mathcal{E}_{\rm c} = \frac{m^2 c^3}{e\,\hbar} \approx 10^{16}\ {\rm V/cm} \quad ; \quad I_c = \frac{c}{8\pi}\,\mathcal{E}_c^2 \approx 4\times 10^{29}\ {\rm W/cm}^2 \tag{1}$$

Correspondingly, the probability of pair production is exponentially suppressed by the factor (as computed by Heisenberg and Euler[3]):

$$P_{\rm pair\ prod.} \sim \exp\left[-\frac{\pi\, m^2\, c^3}{e\,\mathcal{E}\,\hbar}\right] \tag{2}$$

An analogous estimate for atomic ionization [e.g., for H], again using the approximation of a constant electric field, leads to

$$P_{\rm ionization} \sim \exp\left[-\frac{2}{3}\frac{m^2\, e^5}{\mathcal{E}\,\hbar^4}\right] \quad , \tag{3}$$

setting the basic scale of field strength and intensity near which we expect to observe nonperturbative ionization effects in atomic systems:

$$\begin{aligned}\mathcal{E}_{\rm c}^{\rm ionization} &= \frac{m^2 e^5}{\hbar^4} = \alpha^3 \mathcal{E}_{\rm c} \approx 4\times 10^9\ {\rm V/cm} \\ I_{\rm c}^{\rm ionization} &= \alpha^6 I_c \approx 6\times 10^{16}\ {\rm W/cm}^2 \end{aligned} \tag{4}$$

Indeed, this is close to the scale of atomic ionization experiments, but in fact intensities three orders of magnitude lower are routinely used. This is because the electric field in a laser is not constant, and careful shaping of the laser pulses makes ionization experiments possible at much lower intensities. This simple observation, together with the fact that there are plans at large laser facilities such as ELI,[5] HiPER at Rutherford Laboratory, and the XFEL projects at SLAC and DESY, to approach the $10^{26} W/cm^2$ intensity regime, motivates us to ask:

how critical is the Schwinger critical field (1)?

2. The QED Effective Action

In quantum field theory, the quantum corrections to classical Maxwell electrodynamics are encoded in the "effective action" $\Gamma[A]$.[1,6,7] For example, the polarization tensor $\Pi_{\mu\nu} = \frac{\delta^2\Gamma}{\delta A_\mu \delta A_\nu}$ contains the electric permittivity ϵ_{ij} and the magnetic permeability μ_{ij} of the quantum vacuum, obtained by varying the effective action $\Gamma[A]$ with respect to the external probe $A_\mu(x)$. $\Gamma[A]$ is defined in terms of the vacuum-vacuum persistence amplitude

$$\langle 0_{\rm out} \,|\, 0_{\rm in} \rangle = \exp\left[\frac{i}{\hbar}\left\{{\rm Re}(\Gamma) + i\,{\rm Im}(\Gamma)\right\}\right] \tag{5}$$

$\mathrm{Re}(\Gamma[A])$ describes dispersive effects, such as vacuum birefringence, while $\mathrm{Im}(\Gamma[A])$ describes absorptive effects, such as vacuum pair production. The imaginary part encodes the probability of vacuum par production as[6]

$$P_{\rm production} = 1 - |\langle 0_{\rm out} \,|\, 0_{\rm in} \rangle|^2 = 1 - \exp\left[-\frac{2}{\hbar}\,{\rm Im}\,\Gamma\right] \approx \frac{2}{\hbar}\,{\rm Im}\,\Gamma \tag{6}$$

From a computational perspective, the effective action is defined as[6,7]

$$\Gamma[A] = \hbar \,\ln\,\det\left[i\not{D} - m\right] = \hbar\,{\rm tr}\,\ln\left[i\not{D} - m\right] \tag{7}$$

Here, $\not{D} \equiv \gamma^\mu D_\mu$, where the covariant derivative operator, $D_\mu = \partial_\mu - i\frac{e}{\hbar c}A_\mu$, defines the coupling between electrons and the electromagnetic field A_μ. When the gauge field A_μ is such that the field strength, $F_{\mu\nu} = \partial_\mu A_\nu - \partial_\nu A_\mu$, is constant, this effective action was computed [non-perturbatively!] by Heisenberg and Euler.[3] For example, for a constant electric field $\mathcal{E}$:

$$\Gamma^{\rm HE}[\mathcal{E}] = -\hbar\,{\rm Vol}_4 \frac{e^2\mathcal{E}^2}{8\pi^2}\int_0^\infty \frac{ds}{s^2}\, e^{-\frac{m^2}{e\mathcal{E}}s}\left(\cot(s) - \frac{1}{s} + \frac{s}{3}\right) \tag{8}$$

The leading imaginary part comes from the first pole of the cot(s) function:

$$\frac{{\rm Im}\,\Gamma^{\rm HE}}{{\rm Vol}_4} \sim \hbar \frac{e^2\,\mathcal{E}^2}{8\pi^3}\exp\left[-\frac{\pi\, m^2}{e\,\mathcal{E}}\right] \tag{9}$$

3. The Effective Action in Inhomogeneous Background Fields

It is essential to understand how this fundamental constant field result is modified for more realistic inhomogeneous fields $A_\mu(x)$, such as those describing ultra-short pulse focussed lasers. This is not an easy task, as standard perturbative effective field theory techniques do not apply. The first step in this direction is motivated by a seminal result of Keldysh[8] for the ionization of atoms in a time dependent electric field $\mathcal{E}(t) = \mathcal{E}\cos(\omega t)$. This introduces a new physical scale, and Keldysh computed the ionization probability as a function of the dimensionless *adiabaticity parameter* $\gamma_K = e\mathcal{E}/\sqrt{2mE_b}$, that characterizes the fast $[\gamma_K \gg 1]$ and slow $[\gamma_K \ll 1]$ regimes. Remarkably, Keldysh's WKB result interpolates smoothly between the non-perturbative tunnel-ionization regime where $\gamma_K \ll 1$, and the perturbative multi-photon regime where $\gamma_K \gg 1$. This formalism was generalized to the Schwinger effect in QED,[9,10] with an analogous "adiabaticity parameter"

$$\gamma \equiv \frac{mc\omega}{e\mathcal{E}}, \tag{10}$$

$$P_{\text{pair prod.}} \sim \begin{cases} \exp\left[-\frac{\pi m^2 c^3}{e\mathcal{E}\hbar}\right] & , \gamma \ll 1 \text{ (nonperturbative tunneling)} \\ \left(\frac{e\mathcal{E}}{m\omega}\right)^{2mc^2/\hbar\omega} & , \gamma \gg 1 \text{ (perturbative multiphoton)} \end{cases} \tag{11}$$

In the perturbative multi-photon regime, this QED pair production effect has been observed in a beautiful experiment (E-144) at SLAC,[11] in which a laser pulse collided with the (highly relativistic) SLAC electron beam, leading to nonlinear Compton scattering involving 5 photons, producing a high energy gamma photon that decays into an electron-positron pair. By contrast, it is hoped that in future laser facilities it will be possible to probe deep into the *nonperturbative regime* where $\gamma \ll 1$, to see the truly nonperturbative Schwinger effect.

The Keldysh approach captures an enormous amount of important physical information. The bottom line is that various methods have been developed which can be used to compute the pair production probability when the background electric field depends on just one coordinate.[4,12] The problem can be understood as a one-dimensional scattering problem,[9,10] based on Feynman's interpretation of positrons as electrons propagating backwards in time.[13] Then the probability can be extracted from the reflection coefficient for a "Schrödinger" problem of scattering in the time domain. The reflection probability can be computed exactly [numerically], as in the quantum kinetic approach,[14–16] or estimated using semiclassical WKB arguments.[9,10,17] We are then led to ask: can we engineer the pulses in a way to enhance the pair production or to make it more distinctive? Such pulse engineering techniques are fundamental tools in the new field of attoscience.[18]

3.1. *Dynamically Assisted Schwinger Mechanism*

In Ref. 19 it has been shown that a superposition of two time-dependent pulses, one strong but slow, and the other weak but fast, can lead to a significant enhancement of the tunneling process associated with the Schwinger effect. This "dynamically assisted Schwinger mechanism" was introduced in Ref. 19 with the example of fields:

$$\mathcal{E}_{\rm slow}(t) = \mathcal{E}\,\mathrm{sech}^2(\Omega t) \quad ; \quad \mathcal{E}_{\rm fast}(t) = \epsilon\,\mathrm{sech}^2(\omega t) \tag{12}$$

with parametric hierarchies:

$$0 < \epsilon \ll \mathcal{E} \ll \mathcal{E}_c \quad ; \quad 0 < \Omega \ll \omega \ll m \tag{13}$$

Surprisingly, even though both field amplitudes, $\mathcal{E}$ and ϵ, are below the critical field $\mathcal{E}_c$ in (1), there is significant enhancement of the pair production rate when the frequencies are also related as in (13). The non-perturbative pair production process that we would associate with the slow strong field interacts with the perturbative multiphoton pair production process that we would associate with the fast weaker field to produce a stronger impact than each process separately. A specific realization of this idea was proposed recently,[20] involving a strong, slow optical laser pulse and a weak, fast X-ray pulse. Particles in the Dirac sea can perturbatively absorb a high-frequency photon from the weak, fast field, thereby lowering the effective tunnel barrier for the non-perturbative process. This leads to promising predictions[20] of observable pair-production yields based on current technology.

3.2. *Momentum spectra for sub-cycle pulse structure*

In Ref. 15 it was found that the momentum spectrum of the electron-positron pairs produced by the Schwinger effect has interesting oscillatory structures for laser pulses described by a temporal envelope with sub-cycle structure:

$$\mathcal{E}(t) = \mathcal{E}_0 \cos(\omega t + \phi) \exp\left(-\frac{t^2}{2\tau^2}\right) . \tag{14}$$

When $\omega\tau \gtrsim 4$ [i.e., when the number of oscillations under the envelope exceeds 4], the longitudinal momentum spectrum has distinct oscillations, set by the frequency scale ω. These are not seen in the standard WKB approximations. Physically, they are due to resonance effects in the associated one-dimensional scattering picture.[16] These become even more dramatic when the carrier phase ϕ is nonzero. Indeed, when $\phi = \pi/2$ there are momenta at which there is no particle production. Again, this can be explained by the scattering picture, and moreover it shows an unusual relation to the quantum statistics of the particles, as described in detail in F. Hebenstreit's paper in these Proceedings.[21] This is a new distinctive signature of the Schwinger effect, which may help to unambiguously identify this process.

3.3. *Catalysis of Schwinger vacuum pair production*

In Ref. 22 another interpretation of the dynamically assisted Schwinger mechanism was proposed, viewing the problem as the propagation of photons in the X-ray probe pulse in an intense and effectively constant electric field provided by the stronger and slower optical laser pulse. There is a non-zero absorption coefficient for photon propagation in such a strong field, and from this one can deduce the rate of pair production. Technically, this requires the computation of the imaginary part of the photon polarization tensor in an electric field. As the X-ray frequency approaches the threshold of $2m$ there is a dramatic exponential enhancement

$$\mathrm{Im}(\Pi) \approx \frac{\alpha}{\sqrt{\pi(\pi-2)}}\, e\mathcal{E}\, \exp\left\{-\frac{m^2}{e\mathcal{E}}(\pi-2)\right\} \tag{15}$$

There is a window of opportunity in the range of laser intensity up to $I \approx 9 \times 10^{25} W/cm^2$ in which this catalyzed Schwinger mechanism is dramatically enhanced relative to the pure Schwinger effect with just the strong optical laser pulse, without the catalyzing X-ray pulse. Interestingly, this is closely related to the induced metastable decay process of Voloshin and Selivanov,[23] as has been recently discussed in Ref. 24 where the catalysis process is viewed in a complementary way as photon-stimulated pair-production.

4. Worldline Instantons

From these analyses, we learn that temporal inhomogeneities tend to enhance the rate, because it is easier to tunnel through an oscillating barrier. On the other hand, spatial inhomogeneities tend to suppress the rate, because the field falls off as the particles accelerate away from the nucleation point. This raises an interesting question: how do these competing effects play out in an ultra-short laser pulse that is tightly spatially focussed? This is a difficult question to answer, because the conventional WKB and QKE approaches have not yet been generalized to higher-dimensional inhomogeneities. However, the recent "worldline instanton" approach can be formulated, and in some cases has been solved, for such electric fields.[25]

The worldline instanton approach is a semiclassical approximation to Feynman's worldline formalism of the effective action. Feynman[26] formulated a first-quantized form of QED, which amounts to representing[27] the effective action as a quantum mechanical path integral over closed loops $x_\mu(\tau)$ in four dimensional spacetime, with the closed loops being parametrized by the proper time τ. The propertime parametrization had been developed earlier by Fock and Nambu,[28] and was also used by Schwinger, in operator form rather than in path integral form, in his landmark QED computation of vacuum pair production.[6] Feynman's worldline path integral formalism has since been extended significantly, primarily for applications in *perturbative* quantum field theory,[27] building on analogies and motivation from the Polyakov formulation of string theory. This rebirth has led to many beautiful advances in our understanding of perturbative scattering amplitudes, but here I describe an application to *non-perturbative* processes. For simplicity, consider scalar QED. The effective action for a scalar charged particle in a Euclidean classical gauge background $A_\mu(x)$ is:

$$\Gamma[A] = \int_0^\infty \frac{dT}{T}\, e^{-m^2 T} \int d^4x^{(0)} \int_{x(T)=x(0)=x^{(0)}} \mathcal{D}x \, \exp\left[-\int_0^T d\tau \left(\frac{\dot{x}_\mu^2}{4} + e\, A_\mu \dot{x}_\mu\right)\right]$$

The main technical difficulty is to compute the quantum mechanical path integral, a sum over closed trajectories in four-dimensional Euclidean space. A natural approach is to make a semiclassical approximation to the path integral, by solving the classical equations of motion (setting $2e = 1$)

$$\ddot{x}_\mu = F_{\mu\nu}(x)\, \dot{x}_\nu \quad , \quad (\mu, \nu = 1 \dots 4) \tag{16}$$

In this semiclassical approximation, the path integral is dominated by a classical loop called a "worldline instanton" [a closed-loop solution to the classical Euclidean equations of motion], together with quantum fluctuations about this loop. The dominant exponential factor in the imaginary part of the effective action is just

$$\exp\left(-S[x_{\text{classical}}]\right) \tag{17}$$

with the action evaluated on the worldline instanton trajectory. This idea was first applied to the vacuum pair production problem for a constant electric field in Ref. 30, and later extended to inhomogeneous background field configurations.[25] The prefactor contributions, which can be physically significant, are most efficiently computed using an analogy[31] to the Gutzwiller trace formula of nonrelativistic quantum mechanics, viewing the closed loop as a closed trajectory in phase space.

5. The Nambu Gauge

An interesting perspective on the worldline instanton problem is provided by the proper-time formalism of Fock[28] and Nambu.[29] We can now understand their analysis as a semiclassical illustration of the relation between the second-quantized description of the pair production process and the first-quantized worldline form. Nambu[32] studied the nonlinear gauge $\tilde{A}_\mu^2$ = constant. This gauge dramatically simplifies the solution of the worldline instanton equations, reducing them to first-order equations:

$$\dot{x}_\mu = -\tilde{A}_\mu(x) \tag{18}$$

To see this, note that in this Nambu gauge ($\tilde{A}_\mu^2$ = constant)

$$\begin{aligned}\ddot{x}_\mu &= -\dot{x}_\nu \partial_\nu \tilde{A}_\mu = -\dot{x}_\nu(\partial_\nu \tilde{A}_\mu - \partial_\mu \tilde{A}_\nu) - \dot{x}_\nu \partial_\mu \tilde{A}_\nu \\ &= \dot{x}_\nu F_{\mu\nu} + \frac{1}{2}\partial_\mu\left(\tilde{A}_\mu^2\right) = \dot{x}_\nu F_{\mu\nu}\end{aligned} \tag{19}$$

On the other hand, in order to put the gauge field into the Nambu gauge, by the gauge transformation $A_\mu \to \tilde{A}_\mu = A_\mu - \partial_\mu \Lambda$, we need to solve the gauge equation:

$$(A_\mu - \partial_\mu \Lambda)^2 = \text{constant} \tag{20}$$

This is now a partial differential equation, as distinct from the worldline instanton equations which are ordinary differential equations . This equation is a property of the fields, and does not obviously have anything to do with specific classical trajectories. It is, however, the Hamilton-Jacobi equation for the motion of a charged particle in this electromagnetic field, with Λ having the interpretation of the classical action of the trajectory.[28,29] This purely classical statement is a reflection of the semiclassical approximation to the worldline path integral representation of the quantum field theory effective action.

6. Conclusions

The observation of the Schwinger effect presents a fundamental challenge both theoretically and experimentally. Theoretically we need new non-perturbative techniques to provide efficient and precise calculations of the expected pair production rate in realistic short-pulse focussed laser fields. The approach must be sufficiently flexible and powerful to be able to optimize the pulse shape to maximize the pair production rate. On the experimental side, the main challenges are to obtain higher laser intensity, as close as possible to the critical field limit (1), and to be able to focus and shape the laser pulse(s) in both the space and time domain. Recent progress at ELI, HiPER and with XFEL lasers suggests that we may be very close to entering this new physical regime of ultra-relativistic physics. Beyond QED, there are fundamental questions to be answered concerning back-reaction effects, the physics beyond the Schwinger critical field, and concerning the possible simulation of gravitational effects such as Unruh and Hawking radiation, using the very large electric field acceleration to mimic strong gravitational fields.

Acknowledgments

I thank the organizers of the QFEXT09 conference, especially Kim Milton, for an inspiring and fascinating meeting. I also thank R. Alkofer, C. Dumlu, H. Gies, F. Hebenstreit, C. Schubert, R. Schützhold and T. Tajima for collaborations and discussions. I acknowledge support from the DOE through the grant DE-FG02-92ER40716.

References

1. W. Dittrich and H. Gies, *Probing the quantum vacuum. Perturbative effective action approach in quantum electrodynamics and its application*, Springer Tracts Mod. Phys. **166**, 1 (2000).
2. W. Greiner, B. Müller and J. Rafelski, *Quantum Electrodynamics Of Strong Fields*, (Springer, Berlin, 1985).
3. W. Heisenberg and H. Euler, "Consequences of Dirac's Theory of Positrons", Z. Phys. **98**, 714 (1936).
4. G. V. Dunne, "Heisenberg-Euler effective Lagrangians: Basics and extensions," Ian Kogan Memorial Collection, *'From Fields to Strings: Circumnavigating Theoretical Physics'*, M. Shifman et al (ed.), vol. 1, pp 445-522; [arXiv:hep-th/0406216].
5. The Extreme Light Infrastructure (ELI) project: http://www.extreme-light-infrastructure.eu/eli-home.php
6. J. Schwinger, "On gauge invariance and vacuum polarization", Phys. Rev. **82** (1951) 664.
7. W. Dittrich and M. Reuter, *Effective Lagrangians In Quantum Electrodynamics*, Lect. Notes Phys. **220**, 1 (Springer, Berlin, 1985).
8. L. V. Keldysh, "Ionization in the field of a strong electromagnetic wave", Sov. Phys. JETP **20**, 1307 (1965).
9. E. Brézin and C. Itzykson, "Pair Production In Vacuum By An Alternating Field," Phys. Rev. D **2**, 1191 (1970).
10. V. S. Popov, "Pair Production in a Variable External Field (Quasiclassical approximation)", Sov. Phys. JETP **34**, 709 (1972).
11. D. L. Burke *et al.*, "Positron production in multiphoton light-by-light scattering," Phys. Rev. Lett. **79**, 1626 (1997).
12. H. Kleinert, R. Ruffini and S. S. Xue, "Electron-Positron Pair Production in Space- or Time-Dependent Electric Fields," arXiv:0807.0909 [hep-th].
13. R. P. Feynman, "The theory of positrons," Phys. Rev. **76**, 749 (1949).
14. Y. Kluger, J. M. Eisenberg, B. Svetitsky, F. Cooper and E. Mottola, "Pair production in a strong electric field," Phys. Rev. Lett. **67**, 2427 (1991); "Fermion Pair Production In A Strong Electric Field," Phys. Rev. D **45**, 4659 (1992).

15. F. Hebenstreit, R. Alkofer, G. V. Dunne and H. Gies, “Momentum signatures for Schwinger pair production in short laser pulses with sub-cycle structure,” Phys. Rev. Lett. **102**, 150404 (2009) [arXiv:0901.2631 [hep-ph]].
16. C. K. Dumlu, “On the Quantum Kinetic Approach and the Scattering Approach to Vacuum Pair Production,” Phys. Rev. D **79**, 065027 (2009) [arXiv:0901.2972 [hep-th]].
17. S. P. Kim and D. N. Page, “Schwinger pair production via instantons in a strong electric field,” Phys. Rev. D **65**, 105002 (2002) [arXiv:hep-th/0005078].
18. F. Krausz and M. Ivanov, “Attosecond physics”, Rev. Mod. Phys. **81**, 163 (2009).
19. R. Schützhold, H. Gies and G. Dunne, “Dynamically assisted Schwinger mechanism,” Phys. Rev. Lett. **101**, 130404 (2008) [arXiv:0807.0754 [hep-th]].
20. A. Di Piazza, E. Lotstedt, A. I. Milstein and C. H. Keitel, “Barrier control in tunneling e^+e^- photoproduction,” arXiv:0906.0726 [hep-ph].
21. F. Hebenstreit, R. Alkofer, G. Dunne and H. Gies, “Quantum statistics effect in Schwinger pair production in short laser pulses,” arXiv:0910.4457 [hep-ph].
22. G. V. Dunne, H. Gies and R. Schützhold, “Catalysis of Schwinger Vacuum Pair Production,” arXiv:0908.0948 [hep-ph].
23. M. B. Voloshin and K. G. Selivanov, “On Particle Induced Decay Of Metastable Vacuum,” Sov. J. Nucl. Phys. **44**, 868 (1986).
24. A. Monin and M. B. Voloshin, “The spontaneous breaking of a metastable string,” Phys. Rev. D **78**, 065048 (2008) [arXiv:0808.1693 [hep-th]], “Photon-stimulated production of electron-positron pairs in electric field,” arXiv:0910.4762 [hep-th].
25. G. V. Dunne and C. Schubert, “Worldline instantons and pair production in inhomogeneous fields,” Phys. Rev. D **72**, 105004 (2005) [arXiv:hep-th/0507174]; G. V. Dunne, Q.-h. Wang, H. Gies and C. Schubert, “Worldline instantons. II: The fluctuation prefactor,” Phys. Rev. D **73**, 065028 (2006) [arXiv:hep-th/0602176]; G. V. Dunne and Q. h. Wang, “Multidimensional worldline instantons,” Phys. Rev. D **74**, 065015 (2006) [arXiv:hep-th/0608020].
26. R. P. Feynman, “Mathematical formulation of the quantum theory of electromagnetic interaction”, Phys. Rev. **80** 440, (1950).
27. C. Schubert, “Perturbative quantum field theory in the string-inspired formalism,” Phys. Rept. **355**, 73 (2001) [arXiv:hep-th/0101036].
28. V. Fock, “Proper Time In Classical And Quantum Mechanics,” Phys. Z. Sow. **12**, 404 (1937).
29. Y. Nambu, “The Use Of The Proper Time In Quantum Electrodynamics,” Prog. Theor. Phys. **5**, 82 (1950).
30. I. K. Affleck, O. Alvarez and N. S. Manton, “Pair Production At Strong Coupling In Weak External Fields,” Nucl. Phys. B **197**, 509 (1982).
31. D. D. Dietrich and G. V. Dunne, “Gutzwiller’s Trace Formula and Vacuum Pair Production,” J. Phys. A: Math. Theor. **40**, F825-F830, (2007), [arXiv:0706.4006 [hep-th]].
32. Y. Nambu, “Quantum electrodynamics in nonlinear gauge”, Supp. Prog. Theor. Phys. Extra No., p190 (1968).

ON THE GRAVITATIONALLY INDUCED SCHWINGER MECHANISM

GUGLIELMO FUCCI

Department of Mathematics, Baylor University,
Waco, TX 76798, USA
E-mail: Guglielmo_Fucci@Baylor.edu

IVAN G. AVRAMIDI

Department of Mathematics,
New Mexico Institute of Mining and Technology,
Socorro, NM 87801, USA
E-mail: iavramid@nmt.edu

In this paper we will present very recent results obtained in the ambit of quantum electrodynamics in curved spacetime. We utilize a newly developed non-perturbative heat kernel asymptotic expansion on homogeneous Abelian bundles over Riemannian manifolds in order to compute the one-loop effective action for scalar and spinor fields in curved spacetime under the influence of a strong covariantly constant electromagnetic field. In this framework we derived, in particular, the gravitational corrections, up to linear terms in Riemannian curvature, to Schwinger's result for the creation of particles in a strong electric field.

Keywords: Heat kernel expansion; Effective action; Schwinger mechanism.

1. Introduction

It is generally recognized that the effective action is a tool of fundamental importance in quantum field theory and quantum gravity.[1–3] In fact, its knowledge allows one to compute the full propagator and the full vertex functions of the quantum theory under consideration and, in turn, the S-matrix.[2] Amongst other methods, the effective action can be computed by utilizing the heat kernel approach which was first developed by Schwinger,[1,4] and later generalized to include curved spacetime by DeWitt.[5–7] In particular Schwinger computed the effective action for constant electromagnetic fields in Minkowski spacetime.[1] He noticed that in presence of an electric field the effective action acquires an imaginary part interpreted as creation of pairs induced by the electric field. This effect has been since then known as Schwinger mechanism. From a formal point of view the presence of an imaginary part of the effective action can be understood as follows: the effective action is given in terms of a particular integral, properly regularized, over t of the heat kernel diagonal.[2,3] In the presence of an electric field the heat kernel diagonal becomes a meromorphic function with isolated single poles on the real axis. These poles are, then, avoided by deforming the contour of integration which leads to an imaginary part given by the sum of the residues of all the poles.[1]

In this paper we will utilize a newly developed non-perturbative heat kernel asymptotic expansion for homogeneous Abelian bundles in order to obtain the gravitational correction (up to the first order in the Riemannian curvature) to the Schwinger mechanism for a covariantly constant electric field.

2. Non-perturbative Heat Kernel Asymptotic Expansion

Let (M, g) be a n-dimensional Riemannian manifold without boundary and $\mathcal{S}$ be a complex vector bundle over M realizing a representation of the group $G \otimes U(1)$, where G is a compact semisimple Lie group. Let ∇ be the total connection on the vector bundle $\mathcal{S}$, $\mathcal{R}_{\mu\nu}$ be the curvature of the G-connection and $F_{\mu\nu}$ be the curvature of the $U(1)$-connection (which will be called the electromagnetic field).

Let $U(t; x, x')$ be the heat kernel of the Laplacian $L = -g^{\mu\nu}\nabla_\mu\nabla_\nu$ and $\Theta(t) = U(t; x, x)$ be the heat kernel diagonal. As $t \to 0$ the heat kernel diagonal has a well known asymptotic expansion

$$\Theta(t) \sim (4\pi t)^{-n/2} \sum_{k=0}^{\infty} t^k a_k \,, \tag{1}$$

where a_k are the well known local heat kernel coefficients, which are polynomials in the curvatures (both R and F) and their derivatives.

In Ref. 8 we have considered the case in which

$$R \ll F \,, \quad \nabla\nabla R \ll F^2 \,, \quad \nabla F = 0 \,. \tag{2}$$

In this situation the electromagnetic field cannot be treated as a perturbation and instead of (1) we obtained a *new, non-perturbative* expansion of the heat kernel diagonal[8]

$$\Theta(t) \sim (4\pi t)^{-n/2} \sum_{k=0}^{\infty} t^k b_k(t) \,. \tag{3}$$

The coefficients $b_k(t)$ are polynomials in the Riemann curvature and its derivatives with the coefficients that are some universal dimensionless tensor-valued analytic functions that depend on F only in the dimensionless combination tF. These new non-perturbative coefficients of the heat kernel asymptotic expansion have been computed in Ref. 8 by utilizing a promising algebraic approach developed in Refs. 9–12. The form of the coefficients of the asymptotic expansion for the heat kernel diagonal can be expressed as follows

$$b_2(t) = \frac{1}{6}R + \Psi^{\mu\nu}(t)R_{\mu\nu} , \qquad b_3(t) = 0 , \tag{4}$$

$$\begin{aligned} b_4(t) = &-\frac{1}{72}R^2 + \frac{1}{6}R\, b_2(t) + \Phi_1^{\mu\nu\alpha\beta\gamma\delta\tau\sigma}(t)\; R_{\mu\nu\alpha\beta}R_{\gamma\delta\tau\sigma} \\ &+ \Phi_2^{\mu\nu\alpha\beta\gamma\delta}(t)\nabla_\mu\nabla_\nu R_{\alpha\beta\gamma\delta} , \end{aligned} \tag{5}$$

where $\Psi(t)$, $\Phi_1(t)$ and $\Phi_2(t)$ are quite involved analytic tensorial functions of F which have been explicitly obtained in Ref. 8.

3. One-Loop Effective Action and its Imaginary Part

The heat kernel asymptotic expansion described in the previous section has been used in Ref. 13 in order to evaluate the one-loop effective action, up to linear orders in the Riemannian curvature, for scalar and spinor fields under the influence of a strong covariantly constant electromagnetic field in curved spacetime. In the framework of ζ-function regularization, one can write the one-loop effective Lagrangian in terms of the heat kernel diagonal as follows[14]

$$\mathcal{L} = -\sigma \int\limits_{\epsilon/\mu^2}^{\infty} \frac{dt}{t} e^{-tm^2}\Theta(t) , \tag{6}$$

where $\sigma = +1$ for bosons and $\sigma = -1$ for fermions, m is the mass of the field and ϵ and μ are, respectively, the regularization and renormalization parameters.

As pointed out before, in presence of an electric field the heat kernel diagonal $\Theta(t)$ yields poles on the positive real axis which contribute to the imaginary part of the effective Lagrangian through their residues as[13]

$$\text{Im}\,\mathcal{L} = -\sigma\pi \sum_{k=1}^{\infty} \text{Res}\left\{t^{-1}e^{-tm^2}\Theta(t); t_k\right\} . \tag{7}$$

By utilizing the spectral decomposition of the electromagnetic 2-form F, it was shown in Ref. 13 that the imaginary part of the effective Lagrangian for scalar and spinor fields under the influence of solely an electric field $\mathcal{E}$ has the form

$$\mathrm{Im}\,\mathcal{L} = \pi(4\pi)^{-\frac{n}{2}}E^{\frac{n}{2}}G_0(y) + \pi(4\pi)^{-\frac{n}{2}}E^{\frac{n}{2}-1}\big[G_1(y)R + G_2(y)\Pi_1^{\mu\nu}R_{\mu\nu} + G_4(y)\Pi_1^{\mu\nu}\Pi_1^{\alpha\beta}R_{\mu\alpha\nu\beta} + G_5(y)E_1^{\mu\alpha}E_1^{\nu\beta}R_{\mu\alpha\nu\beta}\big]\,, \tag{8}$$

where $y = m^2/\mathcal{E}$, E_k are antisymmetric matrices satisfying[13]

$$E_{k\,\mu\nu} = -E_{k\,\nu\mu}\,,\quad E^k_{\mu[\nu}E^k_{\alpha\beta]} = 0\,,\quad E_kE_m = 0\,,\quad (\text{for } k \neq m) \tag{9}$$

and the projectors Π_k are defined by[13]

$$\Pi_k = -E_k^2\,. \tag{10}$$

One can easily recognize that G_0 in the expression (8) represents the original term computed by Schwinger,[1] while the additional functions proportional to the Riemannian curvature have been explicitly evaluated in arbitrary dimensions in Ref. 13 and represent the *new* gravitational contribution to the Schwinger mechanism.

In the physically relevant case of a four-dimensional spacetime one obtains for scalar fields[13]

$$G_0^{\text{scalar}}(y) = -\frac{1}{\pi^2}\mathrm{Li}_2(-e^{-\pi y})\,,\quad G_1^{\text{scalar}}(y) = -\left(\frac{1}{6}-\xi\right)\frac{1}{\pi}\ln(1+e^{-\pi y})\,, \tag{11}$$

$$G_2^{\text{scalar}}(y) = \frac{1}{48\pi^3}\left\{\frac{2\pi^3 y e^{-\pi y}}{1+e^{-\pi y}} + 8\pi^2\ln(1+e^{-\pi y}) + 18\pi y\mathrm{Li}_2(-e^{-\pi y}) + 54\mathrm{Li}_3(-e^{-\pi y})\right\}, \tag{12}$$

$$G_4^{\text{scalar}}(y) = \frac{1}{384\pi^3}\left\{\frac{16\pi^3 y e^{-\pi y}}{1+e^{-\pi y}} + 4\pi^2(17-3y^2)\ln(1+e^{-\pi y}) + 192\pi y\mathrm{Li}_2(-e^{-\pi y}) + 504\mathrm{Li}_3(-e^{-\pi y})\right\}, \tag{13}$$

$$G_5^{\text{scalar}}(0,y) = -\frac{1}{256\pi^3}\{4\pi^2(1-3y^2)\ln(1+e^{-\pi y}) + 48\pi y\mathrm{Li}_2(-e^{-\pi y}) + 72\mathrm{Li}_3(-e^{-\pi y})\}\,, \tag{14}$$

where ξ represents the coupling constant, while for spinor fields one has[13]

$$G_0^{\text{spinor}}(y) = \frac{4}{\pi^2}\mathrm{Li}_2(e^{-\pi y})\,,\quad G_1^{\text{spinor}}(y) = \frac{1}{3\pi}\ln(1-e^{-\pi y})\,, \tag{15}$$

$$G_2^{\rm spinor}(y) = -\frac{1}{12\pi^3}\left\{\frac{2\pi^3 y e^{-\pi y}}{1-e^{-\pi y}} - 8\pi^2\ln(1-e^{-\pi y}) - 18\pi y {\rm Li}_2(e^{-\pi y}) - 54{\rm Li}_3(e^{-\pi y})\right\}, \tag{16}$$

$$G_4^{\rm spinor}(y) = -\frac{1}{96\pi^3}\left\{\frac{16\pi^3 y e^{-\pi y}}{1-e^{-\pi y}} - 4\pi^2(20-3y^2)\ln(1-e^{-\pi y}) - 192\pi y {\rm Li}_2(e^{-\pi y}) - 504{\rm Li}_3(e^{-\pi y})\right\}, \tag{17}$$

$$G_5^{\rm spinor}(y) = \frac{3}{16\pi^3}\{\pi^2(4+y^2)\ln(1-e^{-\pi y}) - 4\pi y {\rm Li}_2(e^{-\pi y}) - 6{\rm Li}_3(e^{-\pi y})\}, \tag{18}$$

where ${\rm Li}_j(z)$ in the above formulas represents the polylogarithmic function. It is interesting, at this point, to analyze the limit as y approaches zero, namely the situation in which $\mathcal{E} \gg m^2$. For a scalar field one obtains,[13] by taking the limit $y \to 0$ of the expressions (11)-(14),

$$G_0^{\rm scalar} = \frac{1}{12}, \quad G_1^{\rm scalar} = -\left(\frac{1}{6}-\xi\right)\frac{1}{\pi}\ln 2, \tag{19}$$

$$G_2^{\rm scalar} = \frac{1}{6\pi}\ln 2 - \frac{27}{32\pi^3}\zeta(3), \quad G_4^{\rm scalar} = \frac{17}{96\pi}\ln 2 - \frac{63}{64\pi^3}\zeta(3), \tag{20}$$

$$G_5^{\rm scalar} = -\frac{1}{64\pi}\ln 2 + \frac{27}{128\pi^3}\zeta(3), \tag{21}$$

where $\zeta(s)$ is the Riemann zeta function. $G_i^{\rm spinor}(y)$, for $i \neq 0$, in four dimensions represent a special case since there is an infrared divergence as $m \to 0$ (or $y \to 0$). This means that there is no well-defined value for the small mass limit. Instead, one finds a logarithmic divergence, $\log(\pi y)$, as follows[13]

$$G_0^{\rm spinor} = \frac{2}{3}, \quad G_1^{\rm spinor}(y) = \frac{1}{3\pi}\log(\pi y) + O(y), \tag{22}$$

$$G_2^{\rm spinor}(y) = \frac{2}{3\pi}\log(\pi y) - \frac{1}{6\pi} + \frac{9}{2\pi^3}\zeta(3) + O(y), \tag{23}$$

$$G_4^{\rm spinor}(y) = \frac{5}{6\pi}\log(\pi y) - \frac{1}{6\pi} + \frac{21}{4\pi^3}\zeta(3) + O(y), \tag{24}$$

$$G_5^{\rm spinor}(y) = \frac{3}{4\pi}\log(\pi y) + \frac{9}{8\pi^3}\zeta(3) + O(y). \tag{25}$$

From the last formulas, one can clearly see that there are *new infrared divergences* in the imaginary part of the effective action for spinor fields in four dimensions when $\mathcal{E} \gg m^2$. This means, physically, that the creation of massless spinor particles (or massive particles in supercritical electric field) is magnified substantially by the presence of the gravitational field.[13] We would like to mention, here, that similar infrared divergences appear in the real part of the effective Lagrangian for massless spinor fields under the influence of a pure magnetic field. This means that the vacuum energy of charged spinors with small mass (or equivalently massive charged spinors for which $m^2 \ll \mathcal{B}$) dramatically increases due to the presence of the gravitational field.[15]

4. Conclusions

In this paper we have briefly described interesting new developments in the ambit of quantum electrodynamics in arbitrarily curved spacetimes. In particular, by using a new non-perturbative heat kernel asymptotic expansion obtained in Ref. 8, we have found, in Ref. 13, the gravitational corrections to the creation of pairs in a strong electric field. The Schwinger mechanism has, in fact, lately gained importance in light of new experiments proposed in order to observe the creation of electron-positron pairs from vacuum.[16–18] It would certainly be of interest to specialize the gravitational corrections to the Schwinger mechanism to gravitational backgrounds which are relevant in astrophysical settings. This would be important in order to understand in which situations the gravitationally induced Schwinger mechanism might be observable [a]. This effect could also have important consequences for theories with spontaneous breakdown of symmetry when the mass of charged particles is generated by a Higgs field. These theories would exhibit an enhancement of created particles (in the massless limit an infinite amount) at the phase transition point when the symmetry is restored and the massive charged particles become massless. This seems to be an interesting *new physical effect* that, for the reasons mentioned before, deserves further investigation.

Acknowledgments

GF would like to thank the organizers of QFEXT09 for such an excellent conference and the Department of Mathematics at Baylor University for the financial support to attend the conference.

[a] We would like to thank Christian Schubert for suggesting this problem.

References

1. J. S. Schwinger, *Phys. Rev.* **82** 664 (1951)
2. B. S. DeWitt, *Dynamical Theory of Groups and Fields* (Gordon and Breach Science Publishers, 1965)
3. B. S. DeWitt, *The Global Approach to Quantum Field Theory* (Oxford University Press, Oxford, 2003)
4. J. S. Schwinger, *Phys. Rev.* **93** 615 (1954)
5. B. S. DeWitt, *Phys. Rev* **162** 1195 (1967)
6. B. S. DeWitt, *Phys. Rev* **162** 1239 (1967)
7. B. S. DeWitt, *Phys. Rep.* **19**, No. 6 295 (1975)
8. I. G. Avramidi and G. Fucci, *Comm. Math. Phys.* **291**, 543 (2009)
9. I. G. Avramidi, *Phys. Lett. B* **305** 27 (1993)
10. I. G. Avramidi, *J. Math. Phys.* **36** 1557 (1995)
11. I. G. Avramidi, *J. Math. Phys.* **36** 5055 (1995)
12. I. G. Avramidi, *J. Math. Phys.* **37** 374 (1996)
13. I. G. Avramidi and G. Fucci, *J. Math. Phys.* **50**, 102302 (2009)
14. S. W. Hawking, *Comm. Math. Phys.* **55** 133 (1977)
15. G. Fucci, *J. Math. Phys.* **50**, 102301 (2009)
16. G. V. Dunne, New Strong-Field QED Effects at ELI: Nonperturbative Vacuum Pair Production, Key Lecture at the *ELI Workshop and School on "Fundamental Physics with Ultra-High Fields"*, (Frauenworth Monastery, Germany)
17. G. V. Dunne, H. Gies and R. Schtzhold, *arXiv* 0908.0948 [hep-th]
18. F. Hebenstreit, R. Alkofer, G. V. Dunne and H. Gies, *Phys.Rev.Lett.* **102**, 150404 (2009)

QUANTUM STATISTICS EFFECTS FOR SCHWINGER PAIR PRODUCTION IN SHORT LASER PULSES

F. HEBENSTREIT and R. ALKOFER

Institut für Physik, Universität Graz, A-8010 Graz, Austria

G. V. DUNNE

Department of Physics, University of Connecticut, Storrs, CT 06269, USA

H. GIES

Theoretisch-Physikalisches Institut, Universität Jena, D-07743 Jena, Germany
Helmholtz Institut Jena, D-07743 Jena, Germany

We investigate non-perturbative pair production from vacuum (the Schwinger effect) in the focal region of two counter-propagating, ultra-short laser pulses with sub-cycle structure. We use the quantum kinetic formulation to calculate the momentum spectrum of created particles and show the extreme sensitivity to the laser frequency ω, the pulse length τ and the carrier-envelope absolute phase ϕ. We apply this formalism to both fermions and bosons to illustrate the influence of quantum statistics in this type of electric background field.

Keywords: Schwinger effect; quantum statistics; vacuum polarization.

1. Introduction

Non-perturbative electron-positron pair production due to the instability of the vacuum in the presence of strong external electric fields — the so-called Schwinger effect — has been a long-standing prediction of quantum electrodynamics (QED)[1–3] but has not been observed yet. This effect was first considered for spatially homogeneous and static electric fields. The rate is exponentially small, with the scale set by the critical field strength which is of the order of $E_{\rm cr} = m^2c^3/e\hbar \approx 10^{18}\,\mathrm{V/m}$. While the production of constant electric fields of this order is rather unrealistic, recent developments in laser technology have raised hopes to approach the Schwinger limit in the focal region of colliding laser pulses — either at optical high-intensity laser facilities such as ELI or in X-ray free electron laser (XFEL) systems.

In this investigation we model the electric field produced in the focal region of two counter-propagating laser pulses by assuming that the scale of spatial variation of the electric field is much larger than the Compton wavelength. Thus, we approximate the experimental situation by a spatially homogeneous electric field $\vec{E}(t) = (0, 0, E(t))$, represented by an oscillatory field with a temporal Gaussian envelope:

$$E(t) = E_0 \cos(\omega t + \phi) \exp\left(-\frac{t^2}{2\tau^2}\right) . \tag{1}$$

For this type of electric field there is a simple analytic expression of the time-dependent vector potential $\vec{A}(t) = (0, 0, A(t))$ in terms of complex error functions.[5] Due to the appearance of such a variety of physical parameters – the field strength E_0, the laser freqency ω, the pulse length parameter τ and the carrier-envelope absolute phase (carrier phase) ϕ – we are faced with a rather complicated interplay between various scales, which ultimately leads to distinctive signatures in the momentum distribution of produced pairs.[4,5]

2. Quantum Kinetic Equation

The Schwinger effect is a non-equilibrium, time-dependent quantum process and hence quantum kinetic theory provides an appropriate framework. The quantum kinetic formulation arises as a rigorous connection between kinetic theory and mean-field approximation to scalar QED (sQED) and QED.[6,7] The key quantity in this approach is the momentum distribution function $f_\pm(\vec{k}, t)$ which satisfies a non-Markovian quantum Vlasov equation including a source term for particle-antiparticle pair production:

$$\frac{\mathrm{d}}{\mathrm{d}t} f_\pm(\vec{k}, t) = \frac{W_\pm(t)}{2} \int\limits_{-\infty}^{t} \mathrm{d}t' W_\pm(t') \left[1 \pm 2 f_\pm(\vec{k}, t')\right] \cos\left[2 \int_{t'}^{t} \mathrm{d}t'' \, \omega(t'')\right] . \tag{2}$$

Denoting bosons with (+) and fermions with (−), $W_\pm(t)$ are given by

$$W_+(t) = \frac{eE(t) p_\parallel(t)}{\omega^2(t)} \qquad \text{and} \qquad W_-(t) = \frac{eE(t)\epsilon_\perp}{\omega^2(t)} , \tag{3}$$

with e being the electric charge. $\vec{k} = (\vec{k}_\perp, k_\parallel)$ is the canonical three-momentum vector and $p_\parallel(t) = k_\parallel - eA(t)$ is the kinetic momentum along the electric field direction. $\epsilon_\perp^2 = m^2 + \vec{k}_\perp^{\,2}$ is the transverse energy squared and $\omega^2(t) = \epsilon_\perp^2 + p_\parallel^2(t)$ characterizes the total energy squared. It is absolutely crucial to note that $f_\pm(\vec{k}, t)$ has physical meaning as the distribution function of real particles only at asymptotic times $t \to \pm\infty$.

3. Quantum Statistics Effect

We consider the subcritical field strength regime $E_0 = 0.1E_{\rm cr}$, with $\tau = 2\cdot10^{-4}\,{\rm eV}^{-1}$, first concentrating on the case of vanishing carrier phase $\phi = 0$. It has been shown in a previous publication[5] that the momentum distribution function $f_-(\vec{k},t)$ in QED exhibits a distinctive oscillatory structure for $\sigma \equiv \omega\tau \gtrsim 4$, with the oscillation scale set by the laser frequency ω. An analogous calculation in the framework of sQED gives a very similar result. However, due to the difference in quantum statistics, $f_-(\vec{k},\infty)$ shows a local maximum at momentum values at which $f_+(\vec{k},\infty)$ shows a local minimum, and vice versa, as shown in Fig. 1.

The Schwinger effect in the electric field (1) without carrier-phase ϕ has been investigated previously in the framework of a WKB approximation, together with a Gaussian approximation for the momentum distribution:[8]

$$\frac{{\rm d}^3\mathcal{P}}{dk^3} \sim \exp\left(-\frac{1}{eE_0}\left[\frac{1+\sigma^2}{\sigma^2}\gamma^2\, k_\parallel^2 + \vec{k}_\perp^2\right]\right) , \qquad (4)$$

with $\gamma \equiv m\omega/eE_0$ being the Keldysh parameter. In fact, this approximation is too crude in several aspects: First, it does not see the distinctive oscillatory structure found in the exact (numerical) treatment; second, the Gaussian shape is somewhat broader than the true distribution function. In order to explain this discrepancy, we apply the quantum mechanical WKB instanton method,[9] for which the momentum distribution is

$$\frac{{\rm d}^3\mathcal{P}}{dk^3} \sim \exp\left(-2S_{\vec{k}}\right) \quad \text{with} \quad 2S_{\vec{k}} = i\oint_\Gamma \sqrt{m^2 + \vec{k}_\perp^2 + [k_\parallel - eA(t)]^2}\,{\rm d}t \ , \quad (5)$$

with Γ being the contour around the branch cut. After a change of variable, from t to $T = -A(t)/E_0$ we expand the instanton action and obtain an approximate solution in terms of an infinite series in powers of the dimensionless variables $\epsilon = \epsilon_\perp/(eE_0\tau)$ and $\kappa = k_\parallel/(eE_0\tau)$:

$$2S_{\vec{k}} = \frac{\pi\epsilon_\perp^2}{eE_0}\sum_{i=0}^{\infty} S^{(2i)} , \qquad (6)$$

with the first term $S^{(0)} = 1$, and the next three terms being given by

$$S^{(2)} = \left[\kappa^2 - \frac{\epsilon^2}{4}\right]\frac{1+\sigma^2}{2} , \qquad (7)$$

$$S^{(4)} = \left[\kappa^4 - \frac{3\kappa^2\epsilon^2}{2} + \frac{\epsilon^4}{8}\right]\frac{7+14\sigma^2+9\sigma^4}{24} , \qquad (8)$$

$$S^{(6)} = \left[\kappa^6 - \frac{15\kappa^4\epsilon^2}{4} + \frac{15\kappa^2\epsilon^4}{8} - \frac{5\epsilon^6}{64}\right]\frac{127+381\sigma^2+463\sigma^4+225\sigma^6}{720} \qquad (9)$$

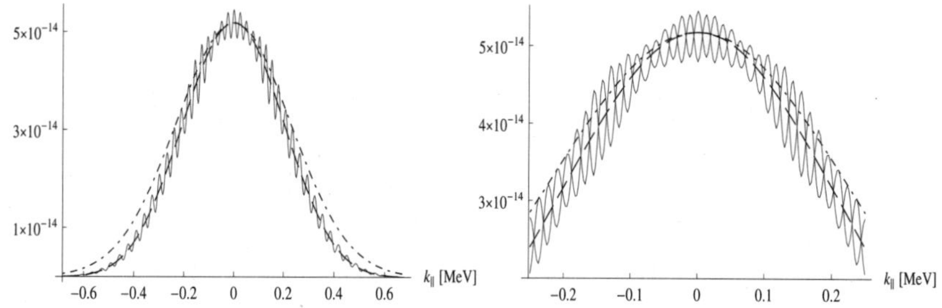

Fig. 1. Carrier phase $\phi = 0$; Left: Momentum distribution function $f_-(\vec{k}, \infty)$ for $\vec{k}_\perp = 0$ (solid line) in comparison with the WKB-Gaussian approximation Eq. (4) (dashed-dotted line) and the WKB instanton result Eq. (5) (dashed line). Right: Momentum distribution functions $f_\pm(\vec{k}, \infty)$ for $\vec{k}_\perp = 0$ in more detail (solid lines). At momentum values, where QED predicts a local maximum, sQED predicts a local minimum and vice versa. Note that the WKB instanton result Eq. (5) (dashed line) lies almost perfectly on the intersection points, while the WKB-Gaussian approximation (4) does not.

It is straightforward to calculate even higher order terms. But taking the first four terms into account, Eq. (5) already agrees very well with the averaged envelope of the exact momentum distribution. We point out that this averaged envelope lies almost perfectly on the intersection points between the momentum distribution of scalar and spinor particles, as shown in Fig. 1. Note that none of the semiclassical estimates predicts the oscillatory structure in the momentum distribution.

Finally, considering a carrier phase shift $\phi = -\pi/2$, there are momentum values at which no particles are expected to be produced,[5] which is due to a resonance phenomenon in the equivalent scattering picture.[10] We emphasize that this behavior can be expected for any time-antisymmetric electric field, corresponding to a time-symmetric vector potential $A(t)$. Again, comparing the results for scalar and spinor particles, we observe that $f_-(\vec{k}, \infty)$ shows a local maximum at momentum values at which $f_+(\vec{k}, \infty)$ shows a local minimum, and vice versa, as shown in Fig. 2.

4. Summary

The momentum distribution of produced particles is extremely sensitive to the physical parameters of a laser pulse. The same qualitative behavior is obtained for both scalar and spinor particles, but, due to quantum statistics, the oscillatory structure is interchanged. This new effect is complementary to the interchange of statistics found in the analogue-thermal distribution properties of the QED effective action in electric field backgrounds.[11,12]

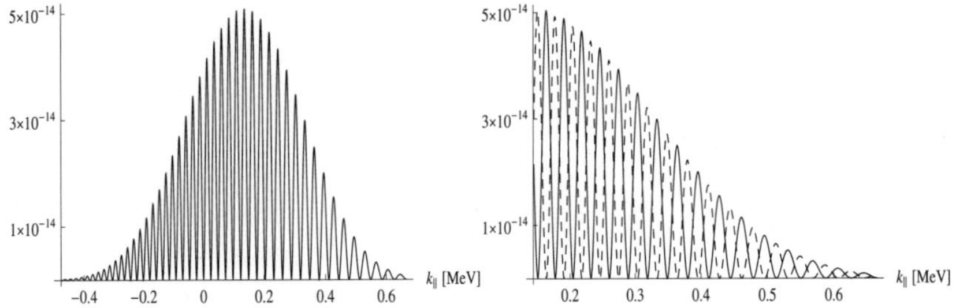

Fig. 2. Carrier phase $\phi = -\pi/2$; Left: Momentum distribution function $f_-(\vec{k}, \infty)$ for $\vec{k}_\perp = 0$. Right: Comparison of the momentum distribution functions $f_-(\vec{k}, \infty)$ (solid line) and $f_+(\vec{k}, \infty)$ (dashed line), in more detail.

Here we have considered two identical, colliding laser pulses. Recent proposals consider more complicated situations to overcome the strong suppression of the Schwinger effect.[13–15] Applying the quantum kinetic formulation, it would be possible to determine not only the total rate but also the momentum distribution.

Acknowledgments

We acknowledge support from the DOC program of the Austrian Academy of Sciences and from the FWF doctoral program DK-W1203 (FH), from the US DOE grant DE-FG02-92ER40716 (GD), and from the DFG grant Gi328/5-1 and SFB-TR18 (HG).

References

1. F. Sauter, *Z. Phys.* **69**, 742 (1931).
2. W. Heisenberg and H. Euler, *Z. Phys.* **98**, 714 (1935).
3. J. Schwinger, *Phys. Rev.* **82**, 664 (1951).
4. F. Hebenstreit, R. Alkofer and H. Gies, *Phys. Rev. D* **78**, 061701 (2008).
5. F. Hebenstreit et al., *Phys. Rev. Lett.* **102**, 150404 (2009).
6. Y. Kluger, E. Mottola and J. M. Eisenberg, *Phys. Rev. D* **58**, 125015 (1998).
7. S. Schmidt et al., *Int. J. Mod. Phys. E* **7**, 709 (1998).
8. V. S. Popov, *JETP Lett.* **74**, 133 (2001).
9. S. P. Kim and D. N. Page, *Phys. Rev. D* **75**, 045013 (2007).
10. C. Dumlu, *Phys. Rev. D* **79**, 065027 (2009).
11. B. Muller, W. Greiner and J. Rafelski, *Phys. Lett. A* **63**, 181 (1977).
12. W. Y. Pauchy Hwang and S. P. Kim, *Phys. Rev. D* **80**, 065004 (2009).
13. R. Schützhold, H. Gies and G. V. Dunne, *Phys. Rev. Lett.* **101**, 130404 (2008).
14. A. Di Piazza et al., *arXiv:0906.0726* (2009).
15. G. V. Dunne, H. Gies and R. Schützhold, *arXiv:0908.0948* (2009).

SCALAR-FERMION BOUND STATE WITH FINITE SIZE CORRECTIONS

D. A. OWEN

Physics Department, Ben Gurion University, Beer Sheva, Israel E-mail:owen@bgu.ac.il

ROGER C. BARRETT

Physics Department, University of Surrey, Guildford, Surrey, U.K.
E-mail:rc.barrett@physics.org

Using our knowledge quantum electrodynamics for a two-particle bound system it is possible to take into account the finite size of each of the constitutents. Thus we are able examine the influence of the charge distribution on the observed energy levels. We have a new formulation for a system comprised of a scalar particle and a fermion in the context of a relativistic quantum field theory. Possibilities are kaonic hydrogen, pionic hydrogen and similar systems.

Keywords: Bethe-Salpeter equation, exotic atoms, kaonic hydrogen.

1. Theoretical difficulties

The bound system of two fermions has been treated in great detail but that treatment is not applicable when one of the particles is replaced by a scalar. We shall illustrate the additional difficulty by briefly reviewing the two-fermion bound state and observing what transpires when one of the fermions is replaced by a scalar particle.

To begin with, consider a two-fermion system, with particles a and b. Their Green's function is given by

$$G(1,2;3,4) = -\left\langle 0|T\left(\psi'_a(x_1)\psi'_b(x_2)\bar{\psi'}_a(x_3)\bar{\psi'}_b(x_4)\right)|0\right\rangle \tag{1}$$

where $\psi'_a(x)$ is the field operator of the fully interacting fermion field a, etc. The bound system is described by the homogeneous Bethe-Salpeter equation by[1]

$$\phi_K(x) = G_K(x,x')I_K(x',x'')\phi_K(x'') \tag{2}$$

where x, x', x'' are relative coordinates and K the energy eigenvalue.

The interaction kernel I is obtained by expanding the Green's function G using Wick's theorem in the inhomogeneous Bethe-Salpeter equation. The inhomogeneous Bethe-Salpeter equation is given by[1]

$$G(1,2;3,4) = G_0(1,2;3,4) + G_0(1,2;5,6)I(5,6;7,8)G(7,8;3,4) \quad (3)$$

Here G_0 is the Green's function of two non-interacting fermions. Neglecting the self-interaction contributions (which are of higher-order) we can expand G_K (where G_K is G_0 in C.M. system*) in partial fractions and write

$$G_K(x,x') = \int \frac{d^4p}{(2\pi)^4} e^{-ip(x-x')} \left(\frac{1}{(\eta_a K+p)\cdot\gamma_a - \mathbf{p}\cdot\gamma_a - m_a + i\epsilon} \frac{1}{(\eta_b K-p)\cdot\gamma_b - \mathbf{p}\cdot\gamma_b - m_b + i\epsilon} \right)$$

$$= \int \frac{d^4p}{(2\pi)^4} e^{-ip(x-x')} \frac{1}{K - H_a - H_b} \left(\frac{1}{\eta_a K - p_0 - H_a} + \frac{1}{\eta_b K - p_0 - H_b} \right) \gamma_a^0 \gamma_b^0 \quad (4)$$

where we have chosen the C.M. coordinates with $X = \eta_a x_1 + \eta_b x_2$, $x = x_1 - x_2$, $\eta_a + \eta_b = 1$, $K = p_1 + p_2$; $p = \eta_b p_1 - \eta_a p_2$. Using the partial fraction expansion with $\mathbf{K} = 0$ in Eq. 4, the homogeneous Bethe-Salpeter Equation becomes

$$[K - H_a - H_b]\phi_K(x) = \Lambda_K(x,x')I_K(x',x'')\phi_K(x'') \quad (5)$$

The factor $[K - H_a - H_b]$ is obtained because the denominator of each of the propagators depends linearly on the energy K. This is clearly not true if one of the particles is spinless. We shall examine this explicitly and show how the same treatment for a two-fermion bound state can be obtained for a system having a scalar particle as one of its constituents. This is achieved by finding a new representation for the Klein-Gordon propagator.

2. Fermion-Scalar case

When one of the particles is a scalar, then Eq. 4 is replaced by $G_K(x,x') = \int \frac{d^4p}{(2\pi)^4} \; e^{-ip(x-x')} \frac{1}{(\eta_a K + p_0) - H_a + i\epsilon} \gamma_0 \frac{1}{(\eta_b K - p_0)^2 - \mathbf{p}^2 - m_b^2 + i\epsilon}$
If one now makes a partial-fraction decomposition the result is more complicated and not very attractive but can still be done.[2]

3. Klein-Gordon propagator

The simplicity obtained in the two-fermion case can be also achieved in this case by making a decomposition of the scalar propagator into its positive

---

$^*G_K(x,x') = \int dX e^{K\cdot(X-X')} G_0(1,2;3,4)$

and negative energy parts. Consider the Klein-Gordon propagator $\Delta_F(x)$, where we factorize the denominator to separate the positive and negative energy parts[2,3,4] i.e.,

$$\begin{aligned}\Delta_F(x) &= \int \frac{d^4p}{(2\pi)^4} \; e^{-ip\cdot x}\frac{1}{p^2-m^2+i\epsilon} \\ &= \int \frac{d^4p}{(2\pi)^4} \; \frac{e^{-ip\cdot x}}{2E(p)}\left(\frac{1}{p_0-E(p)+i\epsilon}-\frac{1}{p_0+E(p)-i\epsilon}\right) \end{aligned} \tag{6}$$

Let us consider the first term acting on a wave function of energy K:

$$\begin{aligned}&\int \frac{d^4p}{(2\pi)^4}\int d^4x' \; \frac{e^{-ip\cdot(x-x')}}{2E(p)}\frac{1}{p_0-E(p)+i\epsilon}\phi_K(\mathbf{x}')e^{-iKt'} \\ &= -i\theta(t-t')\int \frac{d^3p}{(2\pi)^3}\int d^3x' \; e^{i\mathbf{p}\cdot(\mathbf{x}-\mathbf{x}')}e^{-iE(p)t}2\pi\delta(K-E(p)) \end{aligned} \tag{7}$$

We see that this is non-zero only for $K > 0$ otherwise it vanishes. In a similar manner, it can be seen that the second term of Eq. 7 is non-zero only for $K < 0$. These observations can be summarized by writing the propagator as a matrix:

$$\Delta_F(x)\Phi = \int \frac{d^4p}{(2\pi)^4)}\frac{e^{-ip\cdot x}}{2E(p)}\begin{pmatrix}\frac{1}{p_0-E(p)+i\epsilon} \\ -\frac{1}{p_0+E(p)-i\epsilon}\end{pmatrix}\begin{bmatrix}\phi_+ \\ \phi_-\end{bmatrix} \tag{8}$$

where ϕ_+ and ϕ_- are the positive and negative energy wave functions respectively. Introducing the matrix, β_s

$$\beta_s = \begin{pmatrix}1 & 0 \\ 0 & -1\end{pmatrix} \tag{9}$$

we can write the Klein-Gordon propagator in a more succinct form.

$$\Delta_F(x) = \int \frac{d^4p}{(2\pi)^4}\frac{e^{-ip\cdot x}}{2E(p)}\beta_s\begin{pmatrix}\frac{1}{p_0-E(p)+i\epsilon} \\ \frac{1}{p_0+E(p)-i\epsilon}\end{pmatrix}$$

Hence using β_s we can write the Klein-Gordon propagator as

$$\Delta_F(x) = \int \frac{d^4p}{(2\pi)^4}\frac{e^{-ip\cdot x}}{2E(p)}\beta_s\frac{1}{p_0-\beta_s(E(p)-i\epsilon)} \tag{10}$$

4. Bethe-Salpeter equation for a scalar-fermion system

Returning to the Bethe-Salpeter equation we now use for the two-particle Green's function for a scalar-fermion system

$$G_k(x,x') = \int \frac{d^4p}{(2\pi)^4}\frac{e^{-ip(x-x'')}}{2E_b(p)}\frac{1}{\eta_a K+p_0-H_a(p)+i\epsilon}\gamma_0\frac{1}{\eta_b K-p_0-\beta_s(E_b(p)-i\epsilon)}\beta_s$$

$$= \int \frac{d^4p}{(2\pi)^4}\frac{e^{-ip(x-x'')}}{2E_b(p)}\frac{1}{K-H_a-\beta_s E_b(p)}\Lambda_K(x'',x')$$

where

$$\Lambda_K(x,x') = \int \frac{d^4p}{(2\pi)^4}e^{-ip\cdot(x-x')}\left(\frac{1}{\eta_a K+p_0-H_a(p)+i\epsilon}+\frac{1}{\eta_b K-p_0-\beta_s(E_b(p)-i\epsilon)}\right)\gamma_0\beta_s$$

and we have expanded in partial fractions using the linearity of the denominators of both propagators. Now the homogeneous Bethe-Salpeter equation can be written as

$$[K-H_a(p)-\beta_s E_b(p)]\phi_k(x) = \Lambda_K(x,x')I_K(x',x'')\phi_K(x'') \tag{11}$$

We notice that the relative momentum p does not appear in this equation and hence the energy eigenvalue K may be calculated to any accuracy. How does a scalar particle in this representation interact with the electromagnetic field?

5. Interaction of the electromagnetic field in two-particle formulation

The Klein-Gordon equation now becomes:

$$(p_0-\beta_s E_b(p))\phi(x) = 0 \tag{12}$$

This is non-linear and is not convenient for writing the interaction with the electromagnetic field. With a Foldy-Wouthuysen transformation it can be changed to the following form[3,4,5]

$$i\frac{\partial\Phi}{\partial t} = \left\{-\begin{pmatrix}1 & 1\\ -1 & -1\end{pmatrix}\frac{\nabla^2}{2m_b}+\begin{pmatrix}1 & 0\\ 0 & -1\end{pmatrix}\right\}\Phi$$

which can be obtained from the following Lagrangian:

$$\mathcal{L}_{00} = \frac{i}{2}(\bar\Phi\dot\Phi-\dot{\bar\Phi}\Phi)-\sum_{k=1}^{3}\frac{\partial\bar\Phi}{\partial x_k}\frac{M}{2m_b}\frac{\partial\Phi}{\partial x_k}-m_b\bar\Phi\beta_s\Phi \tag{13}$$

where $$M = \begin{pmatrix}1 & 1\\ -1 & -1\end{pmatrix},\quad \beta_s = \begin{pmatrix}1 & 0\\ 0 & -1\end{pmatrix},\quad \bar\Phi = \Phi^*\beta_s$$

To make the Lagrangian invariant under a local gauge transformation requires the introduction of the gauge field $A_\mu(x)$ which transforms under a local gauge transformation as

$$A_\mu(x) \rightarrow A_\mu(x) + (1/e)\partial_\mu \alpha(x)$$

With the replacement of $\partial_\mu \rightarrow \partial_\mu + ieA_\mu$ in $\mathcal{L}_{00}$ we can now write

$$\mathcal{L} = \mathcal{L}_{00} + \bar{\psi}(\not{p} - m_a)\psi + -\frac{1}{4}F^{\mu\nu}F_{\mu\nu} + \mathcal{L}_{int}$$

The Hamiltonian density corresponding to $\mathcal{L}_{int}$ is given by $\mathcal{H}^{int}_{0-1/2} = ie_b\mathbf{A}\cdot \left[(\nabla\bar{\Phi}')\frac{M}{2m_b}\Phi' - \bar{\Phi}'\frac{M}{2m_b}\nabla\Phi'\right] + e_b^2\mathbf{A}^2\Phi'\frac{M}{2m_b}\Phi' - e_a\psi^\dagger\gamma\cdot\mathbf{A}\psi + \mathcal{H}_c$ where $\mathcal{H}_c = \mathcal{H}_{cs} + \frac{e_ae_b}{4\pi}\int d^3xd^3y\frac{\rho_a(\mathbf{x},t)\rho_b(\mathbf{y},t)}{|\mathbf{x}-\mathbf{y}|}$ with $\rho_a = \psi^\dagger\psi$, $\rho_b = \bar{\Phi}\Phi$

6. Finite size and lowest-order energy corrections

The above expressions are for point-particles. For particles having a finite size these expressions become[6†]:

$$\mathcal{H}^{int}_{0-1/2}(x) = ie_b\mathbf{A}'(\mathbf{x}'t)\rho_\pi(\mathbf{x}'-\mathbf{x})\cdot\left[(\nabla\bar{\Phi}'(x))\frac{M}{2m_b}\Phi'(x) - \bar{\Phi}'(x)\frac{M}{2m_b}\nabla\Phi'(x)\right]$$
$$+e_b^2\mathbf{A}'(\mathbf{x},t)\rho_\pi(\mathbf{x}'-\mathbf{x})\bar{\Phi}'(x)\frac{M}{2m_b}\Phi'(x) - e_a\psi'^\dagger(x)\gamma\cdot\mathbf{A}'(\mathbf{x}',t)\rho_N(\mathbf{x}'-\mathbf{x})\psi(x) + \mathcal{H}_c \quad (14)$$

$$H_c = \int d^3x\mathcal{H}_c(x) = H_{cs}$$
$$+\frac{e_ae_b}{4\pi}\int d^3xd^3y\frac{\psi'^\dagger(\mathbf{x},t)\psi'(\mathbf{x},t)\bar{\Phi}'(\mathbf{y},t)\Phi(\mathbf{y},t)\rho_N(\mathbf{x}-\mathbf{x}')\rho_\pi(\mathbf{y}-\mathbf{y}')}{|\mathbf{x}'-\mathbf{y}'|} \quad (15)$$

Expanding G(1,2;3,4) using our expression for $\mathcal{H}^{int}_{0-1/2}$ we can obtain the irreducible kernel I(5,6;7,8).

$$\begin{aligned} G(1,2;3,4) &= -\langle 0|T(\psi'(x_1)\Phi'(x_2)\bar{\psi}'(x_3)\bar{\Phi}'(x_4)|0\rangle \\ &= -\langle 0|T(\psi(x_1)\Phi(x_2)\bar{\psi}(x_3)\bar{\Phi}(x_4)S|0\rangle \end{aligned}$$

with $$S = \sum\frac{(-i)^n}{n!}\int dy_1...dy_nT(\mathcal{H}^{int}_{0-1/2}(y_1)...\mathcal{H}^{int}_{0-1/2}(y_N))$$

† H_{cs} is the Hamiltonian density for Coulomb self-energy.

For the Coulomb kernel we find

$$I_c(5,6;7,8) = -i\frac{e_a e_b}{4\pi}\gamma_0 \delta(x_5 - x_7)\delta(x_6 - x_8)\delta(x_5^0 - x_6^0)$$

$$\times \int dx' dy' \frac{\rho_N(x_7 - x')\rho_\pi(x_8 - y')}{|\mathbf{x}' - \mathbf{y}'|}$$

Using this expression we can now write the Bethe-Salpeter equation as

$$[K - \alpha \cdot \mathbf{p} - m_a\beta - M\frac{\mathbf{p}^2}{2m_b} - m_b\beta_s]\phi_K(x) = [\delta(x - x') + \Omega_K(x,x')]\beta_s\gamma_0$$
$$\times \frac{ie_a e_b}{4\pi}\delta(x' - x'')\delta(x_0') \int d^3u d^3v \frac{\rho_N(\mathbf{u})\rho_\pi(\mathbf{v})}{|\mathbf{x} + \mathbf{u} - \mathbf{v}|}\phi_K(x'') + I_K'(x,x')\phi_K(x')$$

where $\Lambda_K(x,x') = \delta(x - x') + \Omega_K(x - x')$

(16)

and $I_K'(x,x') = I_K(x,x') - I_c(x,x')$

(17)

We treat $\Omega_K(x,x')$ and $I_K'(x,x')$ as perturbations hence our unperturbed wave equation (after using a Foldy-Wouthuysen transformation) is

$$\left[-\frac{\nabla^2}{2\mu} - \alpha \int d^3u d^3v \frac{\rho_N(\mathbf{u})\rho_\pi(\mathbf{v})}{|\mathbf{x} + \mathbf{u} - \mathbf{v}|}\right]\phi_{K_c}(\mathbf{x}) = E\phi_{K_c}(\mathbf{x}) \tag{18}$$

$$E = K_c - m_a - m_b; \qquad \mu = \frac{m_a m_b}{m_a + m_b}$$

Defining $\Delta E = K - K_c$, $I_K'(x,x')$ as a perturbation and retaining all terms to order α^4 we have

$$\Delta E = \left\langle -\frac{\mathbf{p}^4}{8m_a^3} - \frac{\mathbf{p}^4}{8m_b^3} + \frac{\alpha}{4m_a^2}\int d^3u d^3v \rho_N(\mathbf{u})\rho_\pi(\mathbf{v}) \right.$$
$$\left. \times \left\{ \frac{\sigma_a \cdot (\mathbf{r} + \mathbf{u} - \mathbf{v}) \times \mathbf{p}}{|\mathbf{r} + \mathbf{u} - \mathbf{v}|^3} + \frac{\alpha\pi}{2m_a^2}\delta(\mathbf{r} + \mathbf{u} - \mathbf{v}) \right\} \right\rangle \tag{19}$$

(where the expectation value is calculated from the wave function obtained by solving Eq. 18.) This is accurate to α^4 and includes all recoil corrections.

7. Gaussian distribution

Assuming the nucleon has a Gaussian distribution, we write

$$\rho_N(\mathbf{u}, \sigma_N) = \frac{1}{\sqrt{(2\pi)^3\sigma_N^6}} \exp\left(-\frac{\mathbf{u}^2}{2\sigma_N^2}\right) \tag{20}$$

$$\sqrt{\langle r^2\rangle_N} = \sqrt{3\sigma_N^2}$$

and for simplicity we take the radius of the scalar particle to vanish $\sigma_\pi \to 0$ Note that $\sigma_\pi \to 0$ implies $\rho_\pi(\mathbf{v}, \sigma_\pi) \to \delta(\mathbf{v})$ The integral in the unperturbed wave equation reduces to:

$$\int d^3u \frac{\rho_N(\mathbf{v}, \sigma_N)}{|\mathbf{r}+\mathbf{v}|} = \frac{1}{r}\mathrm{erf}\left[\frac{r}{\sqrt{2}\sigma_N}\right]$$

and the non-perturbative wave equation reduces to

$$\frac{1}{r^2}\frac{d}{dr}r^2\frac{dR_{nl}(r)}{dr} - \frac{l(l+1)}{r^2}R_{nl}(r) + 2\mu\left(E_{nl} + \alpha\frac{1}{r}\mathrm{erf}[\frac{r}{\sqrt{2}\sigma_N}]\right)R_{nl}(r) = 0 \tag{21}$$

$$\begin{aligned}\Delta E = \Bigg\langle &-\frac{\mathbf{p}^4}{8m_a^3} - \frac{\mathbf{p}^4}{8m_b^3} + \frac{\alpha}{4m_a^2 r^2}\left(\frac{1}{r}\mathrm{erf}[\frac{r}{\sqrt{2}\sigma_N}] - \frac{1}{\sqrt{2}\sigma_N}e^{-(\frac{r^2}{2\sigma_N^2})}\right)\sigma_a \cdot \mathbf{r}\times\mathbf{p} \\ &-\frac{\alpha\pi}{2m_a^2}\frac{1}{\sqrt{(2\pi)^3\sigma_N^6}}e^{-\frac{r^2}{2\sigma_N^2}} - \frac{\alpha}{2m_a m_b r}\left(\mathrm{erf}[\frac{r}{\sqrt{2}\sigma_N}]\mathbf{p}^2 + \frac{1}{r}\left[\sqrt{\frac{8}{\pi}}\frac{1}{r^3}e^{-\frac{r^2}{2\sigma_N^2}}\right.\right. \\ &\left.\frac{1}{r^4}(r^2 - 2\sigma_N^2)\mathrm{erf}[\frac{r}{\sqrt{2}\sigma_N}]\right]\mathbf{r}(\mathbf{r}\cdot\mathbf{p})\mathbf{p} \\ &-\frac{1}{r^2}\left[\frac{1}{r}\mathrm{erf}[\frac{r}{\sqrt{2}\sigma_N}] - \frac{2}{\sqrt{2}\sigma_N}e^{-\frac{r^2}{2\sigma_N^2}}\right]\sigma_a\cdot\mathbf{r}\times\mathbf{p}\Bigg)\Bigg\rangle\end{aligned} \tag{22}$$

This numerical calculations should not prove to be difficult.

8. Summary

In the foregoing we have 1) showed that the bound state equation for a fermion-scalar system can not be obtained in the same fashion as that of two interacting fermions. The difficulty is due to the quadratic nature of the inverse of the Klein-Gordon propagator. 2) To bypass this difficulty we find that the inverse of the Klein-Gordon propagator can be written in a linear form if we write it in a two-component representation. 3) We use this form of the Klein-Gordon propagator to obtain the Bethe-Salpeter equation for the bound-state of a scalar particle and fermion. 4) We have determined by the means of a Foldy-Wouthysen transformation the electromagnetic interaction in this representation and discuss the perturbation theory through order α^4. 5) Next we have shown how to include the finite size of each particle in the energy levels and 6) a model calculation is included in which the fermion has a gaussian distribution while the scalar is point-like.

9. Outlook

First, numerical results should be obtained to compare with the measured kaonic hydrogen energy levels. Secondly, a comparison with S-states requires an effective potential to account for strong, quark-gluon interactions.

Acknowledgments

Some of this work was carried out in the Physics Department of the University of Surrey and in the Theoretical Physics Department at the University of Liverpool. One of the authors (D.A.O) wishes to thank them for their hospitality.

References

1. E. Salpeter & H. Bethe *Phys. Rev.* **84**, 1232 (1951); J. Schwinger, *Proc. Natl. Acad. Sci. USA* **37**, 452 (1951); M. Gell-Mann & F. Low, *Phys. Rev.***84**, 350 (1951).
2. D. A. Owen *Phys. Rev.* **D 42**, 3534 (1990)
3. M. Halpert & D. A. Owen *J. Phys.* **G 20**, 51 (1994)
4. D. A. Owen *Found. of Phys.* **24**, 273 (1994).
5. L. Foldy & S. Wouthuysen, *Phys. Rev.* **78**, 29 (1950).
6. S. Drell & F. Zachariasen *Electromagnetic Structure of Nucleons* (Oxford University Press, 1961).

THREE-LOOP EULER-HEISENBERG LAGRANGIAN AND ASYMPTOTIC ANALYSIS IN 1+1 QED

I. HUET[a], D. G. C. McKEON[b] and C. SCHUBERT[a]

[a] *Instituto de Física y Matemáticas, Universidad Michoacana de San Nicolás de Hidalgo, Edificio C-3, Apdo. Postal 2-82, C.P. 58040, Morelia, Michoacán, México*

[b] *Department of Applied Mathematics, The University of Western Ontario, London, ON N6A 5B7, Canada*

In recent years, the Euler-Heisenberg Lagrangian has been shown to be a useful tool for the analysis of the asymptotic growth of the N-photon amplitudes at large N. Moreover, certain results and conjectures on its imaginary part allow one, using Borel analysis, to make predictions for those amplitudes at large loop orders. Extending work by G.V. Dunne and one of the authors to the three-loop level, but in the simpler context of 1+1 dimensional QED, we calculate the corresponding Euler-Heisenberg Lagrangian, analyse its weak field expansion, and study the congruence with predictions obtained from worldline instantons. We discuss the relevance of these issues for Cvitanovic's conjecture.

Keywords: Quantum Electrodynamics, Euler-Heisenberg, Asymptotic analysis.

1. Cvitanovic's conjecture for g-2 in QED

In their pioneering calculation of the $g-2$ factor of the electron to sixth order in 1974, Cvitanovic and Kinoshita[1] found a coefficient which was much smaller numerically than had been expected by a naive estimate based on the number of Feynman diagrams involved. A detailed analysis revealed extensive cancellations inside gauge invariant classes of diagrams. This led Cvitanovic[2] to conjecture that, at least in the quenched approximation (i.e. excluding diagrams involving virtual fermions) these cancellations would be important enough numerically to render this series convergent for the $g-2$ factor. Although nowadays there exist a multitude of good arguments against convergence of the QED perturbation series (see, e.g., Ref. 3), all of them are based on the presence of an unlimited number of virtual fermions, so that Cvitanovic's conjecture is still open today. Moreover, should it hold true for the case of the $g-2$ factor, it is natural to assume that it extends to arbitrary QED amplitudes in this quenched approximation, and possibly even to any finite number of fermions. This in turn would strongly suggest that the QED perturbation series should, instead of the usual expansion in the fine structure constant, more naturally be reordered as a series in the number of virtual fermion loops (i.e., in powers of n_F). And it would certainly have implications also for more general gauge theories.

In previous work[4,5] the QED effective Lagrangian in a constant field was used for analyzing the N – photon amplitudes in the low-energy limit. Based on existing high–order estimates for the imaginary part of this Lagrangian, Borel dispersion relations, and a number of two–loop consistency checks, this very different line of reasoning makes "quenched convergence" appear quite plausible for the case of the N – photon amplitudes. Its central point is an all-order conjecture for the imaginary part of the constant-field effective Lagrangian for Scalar QED in the weak field limit due to Affleck, Alvarez, and Manton[6] (AAM). Here we present ongoing work towards a first three-loop check of this conjecture.[7]

2. The AAM conjecture

Let us start with recalling the representation obtained by Euler and Heisenberg[8] for the one-loop QED effective Lagrangian in a constant field,

$$\mathcal{L}^{(1)}_{\rm spin}(F) = -\frac{1}{8\pi^2}\int_0^\infty \frac{dT}{T^3}\,\mathrm{e}^{-m^2T}\left[\frac{(eaT)(ebT)}{\tanh(eaT)\tan(ebT)} - \frac{1}{3}(a^2-b^2)T^2 - 1\right] \tag{1}$$

Here T is the proper-time of the loop particle and a,b are defined by $a^2-b^2 = B^2-E^2, \quad ab = \mathbf{E}\cdot\mathbf{B}$. The analogous formula for Scalar QED was obtained by Weisskopf[9] but will also be called "Euler-Heisenberg Lagrangian" (EHL) in the following. Except for the magnetic case, these effective Lagrangians have an imaginary part. Schwinger[10] found the following representation for the imaginary parts in the purely electric case,

$$\begin{aligned}
{\rm Im}\mathcal{L}^{(1)}_{\rm spin}(E) &= \frac{m^4}{8\pi^3}\beta^2\sum_{k=1}^\infty \frac{1}{k^2}\exp\left[-\frac{\pi k}{\beta}\right] \\
{\rm Im}\mathcal{L}^{(1)}_{\rm scal}(E) &= -\frac{m^4}{16\pi^3}\beta^2\sum_{k=1}^\infty \frac{(-1)^k}{k^2}\exp\left[-\frac{\pi k}{\beta}\right]
\end{aligned} \tag{2}$$

($\beta = eE/m^2$). These formulas imply that any constant electric field will lead to a certain probability for electron-positron pair creation from vacuum. The inverse exponential dependence on the field suggests to think of this as a tunneling process in which virtual pairs draw enough energy from the field to turn real. In the following we will be interested only in the weak field limit $\beta \ll 1$, which allows us to truncate the series in (2) to the then dominant first "Schwinger exponential".

For the Scalar QED case, Affleck et al.[6] proposed in 1982 the following all-loop generalization of (2),

$$\mathrm{Im}\mathcal{L}_{\rm scal}^{(\rm all-loop)}(E) \overset{\beta\to 0}{\sim} \frac{m^4\beta^2}{16\pi^3}\exp\Big[-\frac{\pi}{\beta}+\alpha\pi\Big] \tag{3}$$

This formula is highly remarkable for various reasons. Despite of its simplicity it is a true all-loop result; the rhs receives contributions from an infinite set of Feynman diagrams of arbitrary loop order, including also mass renormalization counterdiagrams. Moreover, the derivation given in Ref. 6 is very simple, if formal. Based on a stationary path approximation of Feynman's worldline path integral representation[11] of $\mathcal{L}_{\rm scal}(E)$, it actually uses only a one-loop semiclassical trajectory, and arguments that this trajectory remains valid in the presence of virtual photon insertions.

An independent derivation of (3), as well as extension to the spinor QED case, was given by Lebedev and Ritus[12] through the consideration of higher-order corrections to the pair creation energy in the vacuum tunneling picture. At the two-loop level, (3) has also been verified by a direct calculation of the EHL[13] (for the spinor QED case), as well as been extended to the case of a self-dual field.[4]

3. Connection between the AAM and Cvitanovic conjectures

Writing the AAM formula (3) as

$$\mathrm{Im}\mathcal{L}_{\rm scal}^{(\rm all-loop)}(E) = \sum_{l=1}^{\infty}\mathrm{Im}\mathcal{L}_{\rm scal}^{(l)}(E) \overset{\beta\to 0}{\sim} \mathrm{Im}\mathcal{L}_{\rm scal}^{(1)}(E)\,\mathrm{e}^{\alpha\pi} \tag{4}$$

it states that an all-loop summation has produced the convergent factor $\mathrm{e}^{\alpha\pi}$, clearly an observation similar in vein to Cvitanovic's. Moreover, at least at a formal level it is not difficult to transfer this loop summation factor from $\mathrm{Im}\Gamma(E)$ to the QED photon amplitudes.[4,5] Consider the weak field expansion of the l-loop contribution to the electric EHL:

$$\mathcal{L}^{(l)}(E) = \sum_{n=2}^{\infty} c^{(l)}(n) \Big(\frac{eE}{m^2}\Big)^{2n} \tag{5}$$

Using Borel dispersion relations, (3) can be shown[4,13] to imply that, at any fixed loop order l, the weak field expansion coefficients have the same asymptotic growth,

$$c^{(l)}(n) \quad \stackrel{n\to\infty}{\sim} \quad c_{\infty}^{(l)}\, \pi^{-2n}\Gamma[2n-2] \tag{6}$$

where the constant $c_{\infty}^{(l)}$ relates directly to the prefactor of the corresponding leading Schwinger exponential in the weak field limit:

$$\mathrm{Im}\mathcal{L}^{(l)}(E) \quad \stackrel{\beta\to 0}{\sim} \quad c_{\infty}^{(l)}\ \mathrm{e}^{-\frac{\pi}{\beta}} \tag{7}$$

As is well-known, the nth term in the weak field expansion of the l - loop EHL carries information on the corresponding $N = 2n$ - photon amplitudes in the low energy limit. Let us assume that the asymptotic behaviour should not depend on the choice of photon polarizations ε_i (this is plausible and supported by two-loop results[5]). Since the kinematical structure of the N - photon amplitudes in this limit reduces to a prefactor which is the same at any loop order,[14] one can eliminate it by dividing the l - loop amplitude by the one-loop one. Expanding (3) in α and combining it with (7) and (6) one then arrives at a formula for the ratio of amplitudes in the limit of large photon number,

$$\mathrm{lim}_{N\to\infty} \frac{\Gamma^{(l)}[k_1,\varepsilon_1;\ldots;k_N,\varepsilon_N]}{\Gamma^{(1)}[k_1,\varepsilon_1;\ldots;k_N,\varepsilon_N]} = \frac{(\alpha\pi)^{l-1}}{(l-1)!} \tag{8}$$

If we could now sum both sides over l and interchange the sum and limit, we could reconstruct the $\mathrm{e}^{\alpha\pi}$ factor, and conclude that the perturbation series for the N - photon amplitudes, at least in this low energy limit, is perfectly convergent! But this is too good to be true, since so far we have nowhere made a distinction between quenched and unquenched contributions to the photon amplitudes, and convergence of the whole perturbation series can certainly be excluded. However, as was noted in Ref. 5 this distinction comes in naturally if one takes into account that in the path integral derivation of (3) in Ref. 6 the rhs comes entirely from the quenched sector; all non-quenched contributions are suppressed in the weak field limit. And since (switching back to the usual Feynman diagram picture) the importance of non-quenched diagrams is growing with increasing loop order, it is natural to assume that their inclusion will slow down the convergence towards the asymptotic limit with increasing l, sufficiently to invalidate the above naive interchange of limits. On the other hand, there is no obvious reason to expect such a slowing down of convergence inside the quenched sector, which led to the prediction[5] that Cvitanovic's "quenched convergence" will indeed be found to hold true for the photon amplitudes.

As a further step in this line of reasoning, one should now check that the convergence of (6) does not show a slowing down when going from two to three loops if one keeps only quenched diagrams. However, a calculation of any three-loop EHL, be it in Scalar or Spinor QED, for an electric or self-dual field, poses an enormous computational challenge.

Now, in 2006 M. Krasnansky[15] calculated the two-loop EHL in 1+1 dimensional Scalar QED and found it, surprisingly, to have a structure almost identical to the one of the corresponding self-dual EHL in the four-dimensional case:

$$\begin{aligned}
\mathcal{L}^{(2)(4D)}_{\rm scal}(\kappa) &= \alpha \frac{m^4}{(4\pi)^3}\frac{1}{\kappa^2}\left[\frac{3}{2}\xi^2 - \xi'\right], \quad \xi(\kappa) := -\kappa\Big(\psi(\kappa) - \ln(\kappa) + \frac{1}{2\kappa}\Big)\\
\mathcal{L}^{(2)(2D)}_{\rm scal}(\kappa) &= -\frac{e^2}{32\pi^2}\left[\xi^2_{2D} - 4\kappa\xi'_{2D}\right], \quad \xi_{2D} := -\Big(\psi(\kappa+\frac{1}{2}) - \ln(\kappa)\Big)
\end{aligned} \tag{9}$$

($\psi(x) = \Gamma'(x)/\Gamma(x)$, $\kappa := m^2/(2ef)$, $f^2 = \frac{1}{4}F_{\mu\nu}F^{\mu\nu}$). This led us to consider 2D QED as a toy model for studying the above asymptotic predictions.

4. Extension of the AAM conjecture to 1+1 QED

Of course, this will make sense only if the AAM formula (3) can be extended to the 2D case. The worldline instanton approach of[6] can be extended to the 2D case straightforwardly,[7] yielding the following analogue of (3):

$$\mathrm{Im}\mathcal{L}(E) \sim \mathrm{e}^{-\frac{m^2\pi}{eE} + \tilde{\alpha}\pi^2\kappa^2} \tag{10}$$

($\tilde{\alpha} := 2e^2/\pi m^2$). Note that, contrary to the 4D case, the second term in the exponent also involves the external field. This leads also to a somewhat more complicated form of the corresponding asymptotic limit statement:

$$\lim_{n\to\infty}\frac{c^{(l)}(n)}{c^{(1)}(n+l-1)} = \frac{(\tilde{\alpha}\pi^2)^{l-1}}{(l-1)!} \tag{11}$$

5. Three loop Euler-Heisenberg Lagrangian in 1+1 QED

At the one and two-loop level, we have obtained the EHL in 2D Spinor QED explicitly in terms of the gamma and digamma functions:[7]

$$\mathcal{L}^{(1)}(\kappa) = -\frac{m^2}{4\pi}\frac{1}{\kappa}\Bigl[\ln\Gamma(\kappa) - \kappa(\ln\kappa - 1) + \frac{1}{2}\ln\bigl(\frac{\kappa}{2\pi}\bigr)\Bigr] \tag{12}$$

$$\mathcal{L}^{(2)}(\kappa) = \frac{m^2}{4\pi}\frac{\tilde{\alpha}}{4}\Bigl[\tilde{\psi}(\kappa) + \kappa\tilde{\psi}'(\kappa) + \ln(\lambda_0 m^2) + \gamma + 2\Bigr] \tag{13}$$

Here $\tilde{\psi}(\kappa) := -\xi(\kappa)/\kappa$, and λ_0 is an IR cutoff for the photon propagator which becomes necessary at two loops in 2D. Curiously, in the 2D case the two-loop Spinor QED result (13) is simpler (just linear in the digamma function) than the corresponding Scalar QED one (9). Using the well-known large - x expansion of $\ln\Gamma(x)$ in terms of the Bernoulli numbers B_n one can then easily verify that (11) does indeed hold true for $l = 2$.

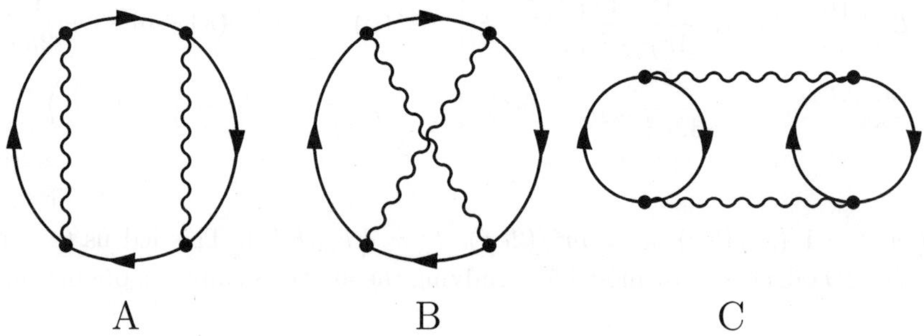

Fig. 1. Diagrams contributing to the three-loop EHL

At three loops our results are rather preliminary. There are three diagrams contributing to the EHL, depicted in fig. 1 (the solid line represents the electron propagator in a constant field). For all three we have obtained representation in terms of fourfold proper-time integrals. The first six coefficients $c^{(3)}(n)$ for the quenched part (diagrams A and B) were then obtained in part analytically, in part by numerical integration. As it turns out, at three loops all coefficients of the weak field expansion except the first one depend on the IR cutoff λ_0. Introducing the modified cutoff $\Lambda := \ln(\lambda_0 m^2) + \gamma$ the coefficients can be written in the form

$$c^{(3)}(n) = \tilde{\alpha}^2\Bigl(c_2^{(3)}(n)\Lambda^2 + c_1^{(3)}(n)\Lambda + c_0^{(3)}(n)\Bigr) \tag{14}$$

where the coefficients $c_{1,2}^{(3)}(n)$ are rational numbers, while $c_0^{(3)}(n)$ contains a $\zeta(3)$ already for $n=0$. Since the prediction (11) is cutoff-independent, it can involve only the $c_0^{(3)}(n)$'s, so that the $c_{1,2}^{(3)}(n)$'s must be subdominant. For the series $c_2^{(3)}(n)$ we have been able to compute a sufficient number of coefficients to verify that this is indeed the case. Therefore the IR cutoff-dependence should not prevent us from verifying the asymptotic relation (11). Unfortunately, at the time of this talk we have not been able yet to obtain a sufficient number of the coefficients $c_0^{(3)}(n)$ to present here a conclusive demonstration of (11), but we expect to be able to do so shortly.

6. Summary

Extending the worldline instanton method of[6] to 2D QED we have obtained a prediction for the asymptotic growth of the weak field expansion coefficients of the 2D EHL at any loop order. At two loops we have verified this prediction by an analytic calculation of the EHL. At three loops we have obtained an integral representation of the EHL suitable for a numerical calculation of the expansion coefficients, and we expect to be able shortly to verify (or refute) the three main facts relevant for the AAM conjecture, namely (i) that (11) holds at the $\tilde{\alpha}^2$ level, (ii) independence of spin, and (iii) asymptotic suppression of the non-quenched diagram C. On the slowing down issue, relevant for Cvitanovic's conjecture, it unfortunately seems not to be possible to get information from the 2D QED case, due to the dependence of the three-loop expansion coefficients on the IR cutoff Λ. This IR dependence seems to indicate a breakdown of perturbation theory for the photon S-matrix in 2D QED, a fact which is of independent interest, and appears not to have been noted before. Although the numerical value of the IR cutoff cannot affect the asymptotic limit and the AAM relation, it does have an influence on the rate of convergence towards it, which thus remains ambiguous. Therefore further progress in this line of attack on Cvitanovic's conjecture presumably has to await the calculation of the three-loop EHL in 4D.

References

1. P. Cvitanovic, T. Kinoshita, Phys. Rev. **D 10** (1974) 4007.
2. P. Cvitanovic, Nucl. Phys. **B 127** (1977) 176.
3. G.V. Dunne, Cont. Adv. in QCD, **478** (2002) [hep-th/0207046].
4. G.V. Dunne, C. Schubert, JHEP **0208** (2002) 053 [arXiv:hep-th/0205004]; JHEP **0206** (2002) 042 [arXiv:hep-th/0205005].
5. G.V. Dunne, C. Schubert, J. Phys.: Conf. Ser. **37** (2006) 59 [hep-th/0409021].

6. I.K. Affleck, O. Alvarez, N.S. Manton, Nucl. Phys. **B 197** (1982) 509.
7. I. Huet, D.G.C. McKeon, C. Schubert, in preparation.
8. W. Heisenberg, H. Euler, Z. Phys. **98** (1936) 714.
9. V. Weisskopf, K. Dan. Vidensk. Selsk. Mat. Fy. Medd. **14** (1936) 1.
10. J. Schwinger, Phys. Rev. **82** (1951) 664.
11. R.P. Feynman, Phys. Rev. **80** (1950) 440.
12. S.L. Lebedev, V.I. Ritus, Zh. Eksp. Teor. Fiz. **86** (1984) 408.
13. G.V. Dunne, C. Schubert, Nucl. Phys. **B 564** (2000) 591 [hep-th/9907190].
14. L.C. Martin, C. Schubert, V.M. Villanueva, Nucl. Phys. **B 668** (2003) 335 [arXiv:hepth/0301022].
15. M. Krasnansky, Int. J. Mod. Phys. **A 23** (2008) 5201 [hep-th/0607230].

CHARACTERISTICS OF GLUON PLASMA IN CHROMOMAGNETIC FIELD AT HIGH TEMPERATURE

M. BORDAG

Leipzig University,
Leipzig, 04109, Germany
E-mail: michael.bordag@itp.uni-leipzig.de

V. DEMCHIK* and V. SKALOZUB†

Dnipropetrovsk National University,
Dnipropetrovsk, 49000, Ukraine
**E-mail: vadimdi@yahoo.com*
†E-mail: skalozubv@daad-alumni.de

The results of analytic and numeric investigations of gluon field characteristics in chromomagnetic Abelian field at high temperature are present. It is noted that this field is spontaneously created at high temperature. The Debye mass is calculated from the one-loop polarization tensor of gluons. The magnetic mass is calculated from the polarization tensor and in Monte-Carlo simulations with Graphics Processing Units (GPU) on a lattice.

Keywords: Gluon magnetic mass, electric mass, Monte-Carlo simulations.

1. Introduction

Investigations of deconfinement phase of QCD is a hot topic nowadays. It was observed recently that in a non-Abelian gauge field theory at high temperature a spontaneous vacuum magnetization happens.[1,2] At zero temperature, it was discovered by Savvidy[3] that an Abelian chromomagnetic field is spontaneously created. The magnetized vacuum is unstable because of the tachyonic mode $p_0^2 = p_{||}^2 - gH$ in the gluon spectrum,

$$p_0^2 = p_{||}^2 + (2n+1)gH, \quad n = -1, 0, 1, ..., \tag{1}$$

where $p_{||}$ is a momentum component along the field. An evolution of this mode results in a condensate which completely screens magnetic field.

This situation is changed at finite temperature $T \neq 0$ when the spectrum stabilization happens due to either a gluon magnetic mass[2] or so-called

A_0-condensate which is proportional to the Polyakov loop.[1] These are the extensions of the Savvidy model to the finite temperature case. In this way the possibility of spontaneous generation of the strong temperature-dependent and stable color magnetic fields of order $gH \sim g^4T^2$, where g is a gauge coupling constant, is realized. Hence, investigation of gluon fields at this background is of interest. The most essential field characteristics are the electric (Debye) and magnetic masses responsible for screening of long range color electric and magnetic fields, respectively.

In field theory, the gluon magnetic mass can be determined in the imaginary time formalism through the polarization tensor (PT),

$$m^2_{magn.} =< n = -1, s|\Pi(k_4 = 0, k_3 \to 0, H, T)|n = -1, s > . \qquad (2)$$

Here, $k_4 = 2\pi lT$, $l = 0, \pm1, \pm2, ...$ is a Matsubara frequency, k_3 is a momentum component along the field $H = const.$ The mean value is calculated in the ground state of the spectrum (1), s marks the polarization of the state. The values $s = 1, 2$ correspond to the transverse polarizations (see[4,5]).

The Abelian chromomagnetic field directed in the third direction in coordinate and internal spaces marked by $a = 1, 2, 3$ in the Landau gauge is described by the potential

$$A^a_\mu = \delta^{a3}(0, 0, Hx^1, 0), \quad H = const. \qquad (3)$$

It is a solution to classical field equation without a source term. So, such field can be spontaneously created.

The Debye mass can be determined in the same way for Π_{44} components of the PT (see, for instance,[4,5]):

$$m^2_D = \Pi_{44}(k_4 = 0, k_3 \to 0, H, T). \qquad (4)$$

Thus, by calculating the PT one is able to derive these characteristics.

2. The Debye and magnetic masses

In Refs.[4,5] in $SU(2)$ gluodynamics, the polarization tensors of either neutral, $A_\mu = A^3_\mu$, or charged, $W^\pm_\mu = \frac{1}{\sqrt{2}}(A^1_\mu \pm A^2_\mu)$, gluons have been calculated in one-loop approximation and partially investigated. The Schwinger[6] operator formalism was applied and extended to the finite temperature case.

The Debye mass squared for both type of gluons has been calculated in the same form

$$m^2_D(H) = \frac{1}{4\pi^2} \int\limits_0^\infty \frac{dq}{q} \sum_{N=1}^{+\infty} \frac{N^2}{qT^2} \frac{gH\cosh(2gHq)}{\sinh(gHq)}\, e^{-\frac{N^2}{4qT^2}} . \qquad (5)$$

We investigate the case of the field and temperature ratio $r = \frac{gH}{4T^2} \ll 1$ and present the mass squared in the form: $m_D^2 = \frac{2}{3}T^2 f(r)$. The function $f(r)$ satisfies the initial condition $f(0) = 1$. Calculating the integral and sum, we obtain

$$
\begin{aligned}
m_D^2(H) &= \frac{2}{3}T^2 \left[1 - 0.8859\left(\frac{\sqrt{gH}}{2T}\right) + 0.4775\left(\frac{g^2H^2}{16T^4}\right)\right. \\
&\left. -i\,0.4775\left(\frac{\sqrt{gH}}{2T}\right) + O(\frac{g^3H^3}{T^6})\right], \qquad (6)
\end{aligned}
$$

where the numeric values of the coefficients are substituted.

To investigate the behavior of the transverse modes for neutral PT we have to calculate the mean value of the polarization tensor in the states with transverse polarizations $s = 1$ and $s = 2$ (for more details see Refs.[4,5]) in the limit of $k_4 = 0$, $\vec{k} \to 0$.

Here, we are interested in the static modes, $k_4 = 0$, perpendicular with respect to the field direction. So, we put $k_3 = 0$ and obtain

$$
\begin{aligned}
\langle s = 1|\Pi(k)|s = 1\rangle &= h^2\Pi_2, \\
\langle s = 2|\Pi(k)|s = 2\rangle &= h^2\left(\Pi_3 + \Pi_5\right), \qquad (7)
\end{aligned}
$$

where $h^2 = k_1^2 + k_2^2$.

We have to calculate the form factors Π_2, Π_3 and Π_5 also for this case. The explicit calculations yield

$$\Pi_5 = \frac{1}{(4\pi)^{3/2}}\frac{T}{\sqrt{gH}}\left[-4.21405 - 1.77245i\right], \qquad (8)$$

$$
\begin{aligned}
\Pi_2 &= \frac{1}{(4\pi)^{3/2}}\frac{T}{\sqrt{gH}}\left[-5.79894 - 7.08982i\right], \\
\Pi_3 &= \frac{1}{(4\pi)^{3/2}}\frac{T}{\sqrt{gH}}\left[1.04427 - 8.86227i\right]. \qquad (9)
\end{aligned}
$$

The sum of $\Pi_3 + \Pi_5$ equals, $\Pi_3 + \Pi_5 = \frac{T}{\sqrt{gH}}\left[-3.16978 - 10.6347i\right]$. The imaginary part is signaling the instability of the state because of the tachyonic mode presents in the spectrum, and the real one is responsible for the screening of transverse gluon fields.

Let us turn to the real part and substitute it in the Schwinger-Dyson equation

$$D^{-1}(k^2) = k^2 - \Pi(k) \qquad (10)$$

for the Green function. We obtain for the mean values

$$\langle\ s = 1\ |ReD^{-1}(h^2)|\ s = 1\ \rangle = h^2\left(1 + 5.79894\frac{T}{\sqrt{gH}}\right) \qquad (11)$$

and

$$\langle\, s=2\,|ReD^{-1}(h^2)|\; s=2\,\rangle = h^2\left(1+3.16978\frac{T}{\sqrt{gH}}\right). \qquad (12)$$

Two important conclusions follow from Eqs. (11), (12). First, there is no fictitious pole similar to that of in the zero field case.[7] Second, there is no the magnetic screening mass in one-loop order. This is in contrast to the case of charged gluons which acquire the magnetic mass in one-loop order.[2,8] The removing of the fictitious pole follows because the external field is accounted for exactly through the Green functions.

3. Magnetic mass of Abelian field on a lattice

In present section, we calculate the magnetic mass of the Abelian magnetic field on a lattice. In calculations we use the General Purpose computation on Graphics Processing Units (GPGPU) technology allowing to study the large lattices up to 32×64^3. Some details of Monte-Carlo simulations on the ATI GPUs can be found in Ref.[9]

In performed MC simulations, we use the hypercubic lattice $L_t \times L_s^3$ with hypertorus geometry. Lattice data are stored with the single precision, MC updating are performed with the single precision whereas all averaging measurements were performed with the double precision to avoid error accumulations. The constant homogeneous magnetic flux corresponding to the potential (3) is introduced on a lattice by applying the twisted boundary conditions. We have fitted by means of different functions the average

Table 1. Fit results for magnetic mass of Abelian magnetic field

Fit function	χ^2	C	m
$C\exp(-mr)$	901.8	0.063	$0.0244^{+0.0006}_{-0.0006}$
$C\exp(-m^2r^2)$	1924.4	0.035	$0.0157^{+0.0002}_{-0.0002}$
C/r	7.090	0.911	
$C/r\exp(-mr)$	7.086	0.912	$0.0000125^{+0.00052}_{-0.00054}$
$C/r\exp(-m^2r^2)$	7.090	0.911	$0.000155^{+0.00229}_{-0.00242i}$
C/r^2	31400	28.13	
$C/r^2\exp(-m^2r^2)$	7550	18.26	$0.0182i$
C/r^4	159500	248.9	
$C/r^4\exp(-m^4r^4)$	161000	10.0	0.0

magnetic flux through the lattice plaquette (perpendicular to the OZ axis):

$$\langle U_{untwisted}\rangle - \langle U_{twisted}\rangle = f(m, L_s). \tag{13}$$

The fitting functions, χ^2 and the magnetic mass value, are shown in Table 1. The function C/r^2 corresponds to the magnetic flux tube formation, the function C/r^4 describes the Coulomb like behavior and the function $C/r^2 \exp(-m^2r^2)$ is signaling the generation of the magnetic mass.[10] The functions $C/r, C/r\exp(-mr)$ have to be related to the increase of magnetic flux at finite temperature. This is because they describe the growth of total magnetic flux through the lattice $X-Y$ plane.

The best fit function is $C/r\exp(-mr)$ with small $m = 0.0000125$. The χ^2 is very close to the case of $m = 0$. Statistically these cases are indistinguishable because statistical errors are larger then the fitted value of m. Thus, we can conclude that the neutral component of the field is not screened. This result is in agreement with the one in the previous section.

4. Conclusion

We derived for neutral gluons that the magnetic mass is zero in both calculations. So, color Abelian magnetic fields are long range as usual $U(1)$ magnetic field. The Debye mass in the chromomagnetic field presence is smaller as compared to the zero field case. Therefore the range of electric fields at high temperature is larger as usually is assumed.

References

1. A. Starinets, A. Vshivtsev and V. Zhukovsky, *Phys. Lett.* **B322**, 403 (1994).
2. V. Skalozub and M. Bordag, *Nucl. Phys.* **B576**, 430 (2000).
3. G. K. Savvidy, *Phys. Lett.* **B71**, p. 133 (1977).
4. M. Bordag and V. Skalozub, *Phys. Rev.* **D75**, p. 125003 (2007).
5. M. Bordag and V. Skalozub, *Phys. Rev.* **D77**, p. 105013 (2008).
6. J. S. Schwinger, *Phys. Rev.* **D7**, 1696 (1973).
7. O. K. Kalashnikov, *Fortschr. Phys.* **32**, p. 525 (1984).
8. V. V. Skalozub and A. V. Strelchenko, *Eur. Phys. J.* **C33**, 105 (2004).
9. V. Demchik and A. Strelchenko, Monte Carlo simulations on Graphics Processing Units, *arXiv:0903.3053 [hep-lat]*, (2009).
10. T. A. DeGrand and D. Toussaint, *Phys. Rev.* **D25**, p. 526 (1982).

EXACT FORMFACTORS IN THE ONE-LOOP CURVED-SPACE QED AND THE NONLOCAL MULTIPLICATIVE ANOMALY

BRUNO GONÇALVES[1], GUILHERME DE BERREDO-PEIXOTO[2] and ILYA L. SHAPIRO[*,3]

Departamento de Física, ICE, Universidade Federal de Juiz de Fora, Juiz de Fora, CEP: 36036-330, MG, Brazil
E-mail: [1]*brunoxgoncalves@yahoo.com.br*
[2]*guilherme@fisica.ufjf.br*
[3]*shapiro@fisica.ufjf.br*

The well-known formula $\det(A \cdot B) = \det A \cdot \det B$ can be easily proved for finite dimensional matrices but it may be incorrect for the functional determinants of differential operators, including the ones which are relevant for Quantum Field Theory applications. Considerable work has been done to prove that this equality can be violated, but in all previously known cases the difference could be reduced to renormalization ambiguity. We present the first example, where the difference between the two functional determinants is a *nonlocal* expression and therefore can not be explained by the renormalization ambiguity. Moreover, through the use of other even dimensions we explain the origin of this difference at qualitative level.

Keywords: Multiplicative anomaly; QED; Proceedings; Formfactors.

1. Introduction

The one-loop calculations have a prominent role in Quantum Field Theory (QFT) and in many of its most relevant applications. In the background field method the one-loop contributions can be always reduced to the derivation of $\mathrm{Ln\,Det}\,\hat{H}$ of the operator $\hat{H}$, which is typically a bilinear form of the classical action with respect to the quantum fields. The operator $\hat{H}$ usually depends on the background fields (which may be just external fields). As a result the operation of taking the functional determinant of such an operator is mathematically nontrivial due to the infinite dimension of the

*Also at Tomsk State Pedagogical University, Russia.

corresponding matrix representation. In particular, relations such as

$$\begin{aligned} \mathrm{Det}\,(\hat{A}\cdot\hat{B}) &= \mathrm{Det}\,\hat{A}\cdot\mathrm{Det}\,\hat{B} \qquad \text{and} \\ \mathrm{Ln\ Det}\,\hat{A} &= \mathrm{Tr\ Ln}\,\hat{A}\,, \end{aligned} \tag{1}$$

which are certainly valid for the finite dimensional matrices should be, in principle, proved or taken by faith in QFT. There is indeed another possibility that these relations can be *disproved* and, according to mathematical logic, this can be done by means of at least one single nontrivial counter-example. For instance, that could mean a couple of operators, $\hat{A}$ and $\hat{B}$, for which the first relation in (1) would be violated. Such a situation was called *multiplicative anomaly* (MA).[1,2]

Considerable efforts have been applied to find an example where the first equality (1) would be violated, but until now in all cases the difference was likely caused by the renormalization ambiguity only.[3–5] This means that when one imposes the renormalization conditions to the three operators $\hat{A}$, $\hat{B}$ and $\hat{A}\cdot\hat{B}$, there may be a difference due to the independence of these renormalization conditions for the distinct operators. In particular, such a situation can take place when the functional determinants are defined by means of the generalized ζ-function,[6] because this approach "hides" the divergences and provides the regularized and renormalized result automatically. Then the μ-dependence should be implemented artificially and this opens the way for the MA. The example of such a situation has been analyzed in detail in Refs. 3 and 7. If we consider, for example, the Ln Det $\big(\Box+M_1^2\big)\cdot\big(\Box+M_2^2\big)$ on de Sitter background, the result will be a functional which depends on some constant parameters, namely on $M_{1,2}^2$ and on the scalar curvature Λ. Furthermore, this expression has dimension four. As a result it has exactly the same structure as the counterterms and, therefore, it is a subject of the renormalization ambiguity. Thus, it is very difficult to make positive conclusion on the existence of the MA based on such calculations. In order to establish the existence of the MA one needs to find it in such a finite sector of the effective action which can be clearly different from the counterterms.

The purpose of the present letter is to present an example of another sort, that means the *nonlocal* MA which is not reduced to the renormalization ambiguity. In order to construct such example we consider one of the most familiar theories, that is the usual spinor QED. We consider a curved space-time, but the effect can be observed even in flat space-time. This letter represents a short communication devoted to the MA and we leave technical details to the parallel publication,[8] devoted to the general investigation of quantum violation of conformal invariance for electromagnetic fields.

2. Photon Formfactors in the 1-Loop QED

Consider the problem of deriving the correction to the electromagnetic field propagation from the single loop of a Dirac fermion. The Euclidean action has the form

$$S = \int d^4x\sqrt{g}\Big\{\bar{\psi}\big(i\gamma^\mu\nabla_\mu + e\gamma^\mu A_\mu + M\big)\psi - \frac{1}{4}\,F^{\mu\nu}F_{\mu\nu}\Big\}.$$

The one-loop effective action (EA) in the metric and electromagnetic sectors can be defined via the path integral

$$e^{i\Gamma[g_{\mu\nu},\,A_\mu]} = \int D\psi D\bar{\psi}\, e^{iS}\,. \tag{2}$$

In the conventional form we find (see, e.g., Ref. 9)

$$\bar{\Gamma}^{(1)} = -\frac{1}{2}\,\text{Ln Det}\,\hat{H}\,, \quad \text{with} \quad \hat{H} = i\,\big(\gamma^\mu\nabla_\mu - ie\gamma^\mu A_\mu - iM\big)\,. \tag{3}$$

In order to use the heat kernel method, one has to multiply $\hat{H}$ by a conjugate operator $\hat{H}^*$, such that the product has the form $\hat{H}\hat{H}^* = \widehat{\Box} + 2\hat{h}^\mu\nabla_\mu + \hat{\Pi}$. The point is that the choice of the conjugate operator $\hat{H}^*$ is not unique. Here we consider the two following choices:

$$\hat{H}_1^* = -i\gamma^\mu\nabla_\mu + M - e\gamma^\nu A_\nu \qquad \text{and} \qquad \hat{H}_2^* = -i\gamma^\mu\nabla_\mu + M\,. \tag{4}$$

In what follows the difference between the $\hat{H}_1^*$ and $\hat{H}_2^*$ cases will be named "scheme dependence". The question is whether the $F^{\mu\nu}(...)F_{\mu\nu}$-type terms calculated via the expressions Ln Det $\hat{H}\hat{H}_1^*$ and Ln Det $\hat{H}\hat{H}_2^*$ are the same or not. In both cases we assume

$$\text{Ln Det}\,\hat{H} = \text{Ln Det}\,\big(\hat{H}\hat{H}^*\big) - \text{Ln Det}\,\hat{H}^*\,. \tag{5}$$

Now, in the first case the contributions of Ln Det $\hat{H}$ and Ln Det $\hat{H}_1^*$ are equal,[10] so in fact we can take

$$\text{Ln Det}\,\hat{H} = \frac{1}{2}\,\text{Ln Det}\,\big(\hat{H}\hat{H}_1^*\big)\,.$$

In the second case the expression Ln Det $\hat{H}_2^*$ does not depend on A_μ and therefore the $F^{\mu\nu}(...)F_{\mu\nu}$-type terms satisfy (using obvious notations) the relation

$$\text{Ln Det}\,\hat{H}\Big|_{FF} = \text{Ln Det}\,\big(\hat{H}\hat{H}_2^*\big)\Big|_{FF}\,.$$

So, if the first identity from (1) holds, we are going to meet the two equal expressions,

$$\frac{1}{2}\,\text{Ln Det}\,\big(\hat{H}\hat{H}_1^*\big)\Big|_{FF} = \text{Ln Det}\,\big(\hat{H}\hat{H}_2^*\big)\Big|_{FF}\,, \tag{6}$$

but if (6) does not hold, (1) is violated. We will show that in fact the two expressions have different finite parts despite the divergent parts being equal. Moreover in the case of $\hat{H}_2^*$ the gauge invariance is violated in the finite part of EA. Let us note that the last occurrence can be seen as one more confirmation of the MA. The reason is that the expression Ln Det $\hat{H}$ is gauge invariant by construction (we assume invariant regularization) and the expression Ln Det $\hat{H}_2^*$ does not depend on the gauge field A_μ and hence it is also gauge invariant. Hence, if Ln Det $\hat{H}\hat{H}_2^*$ is non-invariant, then

$$\text{Ln Det } \hat{H}\hat{H}_2^* \neq \text{Ln Det } \hat{H} + \text{Ln Det } \hat{H}_2^*$$

and we meet one more evidence of the MA.

Let us see whether the situation described above really takes place. In order to calculate Ln Det $\hat{H}\hat{H}_1^*$ and Ln Det $\hat{H}\hat{H}_2^*$ we use the heat kernel solution[11] which was earlier applied to the derivation of formfactors in the gravitational sector.[12,13] Let us note also that the same result can be achieved via the Feynman diagrams.[12]

The one-loop quantum correction for the $\hat{H}_1^*$ case has the form

$$\bar{\Gamma}^{(1)}\Big|_{FF} = -\frac{e^2}{2(4\pi)^2}\int d^4x\sqrt{g}F_{\mu\nu}\Big[\frac{2}{3\epsilon} + k_1^{FF}(a)\Big]F^{\mu\nu}, \tag{7}$$

$$\text{with} \quad k_1^{FF}(a) = Y\Big(2 - \frac{8}{3a^2}\Big) - \frac{2}{9},$$

where we used the following notations

$$Y = 1 - \frac{1}{a}\ln\Big(\frac{2+a}{2-a}\Big), \qquad a^2 = \frac{4\Box}{\Box - 4m^2}. \tag{8}$$

For the $\hat{H}_2^*$ case we meet a different result, namely (we do not use the notation $\big|_{FF}$ here because there are other $\mathcal{O}(A^2)$-terms)

$$\begin{aligned}
\bar{\Gamma}^{(1)}_{\sim A^2} &= -\frac{e^2}{2(4\pi)^2}\int d^4x\,\sqrt{g}\,\Big\{F_{\mu\nu}\Big[\frac{2}{3\,\epsilon} + k_2^{FF}(a)\Big]F^{\mu\nu} \\
&+ \nabla_\mu A^\mu\Big[Y\Big(\frac{8}{3a^2} - 2\Big) + \frac{2}{3}\Big]\nabla_\nu A^\nu + \nabla_\mu A^\nu\Big(\frac{16Y}{3a^2}\Big)\nabla_\nu A^\mu \\
&+ R_{\mu\nu}\Big(\frac{8Y}{3a^2}\Big)A^\nu A^\mu + A^\nu A^\mu\Big(\frac{8Y}{3a^2}\Big)R_{\mu\nu} \\
&+ A_\alpha A^\alpha\Big[Y\Big(\frac{1}{3} - \frac{4}{3a^2}\Big)\Big]R + R\Big[Y\Big(\frac{4}{3a^2} - \frac{1}{3}\Big)\Big]A_\alpha A^\alpha\Big\},
\end{aligned} \tag{9}$$

$$\text{where} \quad k_2^{FF}(a) = Y\Big(1 + \frac{4}{3a^2}\Big) + \frac{1}{9}.$$

In the expressions (7) and (9), ϵ is the parameter of dimensional regularization

$$\frac{1}{\epsilon} = \frac{2}{4-d} + \ln\left(\frac{4\pi\mu^2}{m^2}\right) - \gamma, \qquad \gamma = 0.5772\ldots .$$

It is easy to see that the divergences are exactly the same in the two expressions but, at the same time, the finite parts indicate the presence of MA. In fact, the situation is exactly as it was described above. In the divergent parts of the two formulas (7) and (9) there is no scheme dependence, while the finite nonlocal parts of these expressions do differ and, also, (9) is not gauge invariant. The scheme dependence can not be eliminated by adjusting the renormalization condition, because the last does not concern the nonlocal part of EA. So, we have confirmed the existence of MA for the Dirac operator. However, in this situation the interested reader has the right to ask natural questions like: "Is it all correct?" and "Why does the MA take place?" Of course, the first question can be addressed only through a clear answer to the second one, and we will present such an answer in the next section.

3. $\hat{a}_n$ Coefficients and the Origin of MA

In order to understand the origin of the MA, let us remember that the heat kernel solution of Ref. 11 is a sum of the series of the coincidence limits of the Schwinger-DeWitt coefficients $a_n(x,x')$. The equal divergences of the two effective actions (7) and (9) mean that the coefficients a_2 of the two operators do coincide in the four-dimensional space. The distinct finite parts mean that some other coefficients are in fact different. Therefore the natural way to check the correctness of the results (7) and (9) is to calculate the coincidence limit of some other coefficient, e.g., $a_1(x,x')$, or $a_3(x,x')$. Before we begin our calculations, let us imagine what should we expect as a possible output. For this end it is most interesting to consider an arbitrary dimension d of space-time. The $4d$ case considered above has shown that the divergent part of the effective action is scheme-independent and thus universal. Mathematically, there is nothing special about $4d$, so we can expect that this universality holds also in other even dimensions.

Let us note that the expression $\hat{a}_k = \text{Tr} \lim\limits_{x'\to x} a_k(x,x')$ with $k=1$ corresponds to the UV divergence of EA in $2d$, with $k=2$ in $4d$, with $k=3$ in $6d$ etc. Therefore the universality of the UV divergences implies that $\hat{a}_1$ is universal in $2d$, $\hat{a}_2$ in $4d$, $\hat{a}_3$ in $6d$ etc. The most interesting moment in this story is that the universality of the Schwinger-DeWitt coefficients in the "right" dimensions automatically implies the non-universality of the overall finite contributions in *any* particular dimension! The point is that the general expression for the coincidence limit $\lim_{x'\to x} a_k(x,x')$ does not depend on d, but the corresponding functional trace $\hat{a}_k$ does. As a result, if the two traces are equal in the "right" dimensions, they are unlikely to be equal in other dimensions. For instance, all terms except $\hat{a}_2$ are scheme-dependent in $4d$, and therefore the sum of the series made out of these terms is also not universal. Indeed, this is exactly what we observe in the formfactors (7) and (9) calculated within the two distinct schemes.

Let us verify that the considerations presented above are correct. We start from the evaluation of $\hat{a}_1$ in $2d$. We know that the $\hat{a}_1 = \int \sqrt{g}\hat{P}$, where $\hat{P}$'s in the two cases are given by the expressions

$$\begin{aligned} \hat{P}_1 &= -\frac{1}{12}\,R + M^2 - \frac{ie}{2}\gamma^\mu\gamma^\nu F_{\mu\nu}\,, \\ \hat{P}_2 &= -\frac{1}{12}\,R - \frac{ie}{4}\gamma^\mu\gamma^\nu F_{\mu\nu} + M^2 + eM\gamma^\mu A_\mu \\ &\quad + \frac{ie}{2}\,(\nabla^\mu A_\mu) - \frac{(d-2)}{4}e^2 A^\nu A_\nu\,. \end{aligned} \tag{10}$$

It is easy to see that the difference between the two traces is reduced to the total derivative in $2d$, while in other dimensions it is more significant. Furthermore, only in $2d$ the $\mathrm{Tr}\,\hat{P}_2$ is a gauge invariant expression. Let us note that the difference in total derivative may indicate some real thing for the finite part of EA, but not for renormalization. Therefore the general expectation described above is completely confirmed in the $\hat{a}_1$ case. We leave it as an exercise to the reader to check that the situation is the same for the $\hat{a}_2$ coefficients, where the two schemes give equal results in the $4d$ case and distinct results for $d \neq 4$ cases.

As a last test, let us now consider the $\hat{a}_3$ coefficient. Within the first calculational scheme with $\hat{H}_1^*$ of (4), we just confirm the known result of Ref. 14,

$$\begin{aligned} \hat{a}_3^{(1)}\Big|_{AA} &= \frac{d\,e^2}{360}\,\big(\,2\,R_{\mu\nu\alpha\beta}\,F^{\mu\nu}F^{\alpha\beta} \;-\; 26\,R^\alpha_\nu\,F^{\mu\nu}\,F_{\mu\alpha} \\ &\quad + 24\,\nabla_\nu F^{\mu\nu}\,\nabla_\alpha F_\mu{}^\alpha \;+\; 5\,R\,F^{\mu\nu}\,F_{\mu\nu}\,\big)\,. \end{aligned} \tag{11}$$

The expression $\hat{a}_3^{(2)}$ for the second scheme, with $\hat{H}_2^*$, is rather bulky $\big[$here we are using $(\nabla A) = (\nabla_\mu A^\mu)\big]$:

$$\hat{a}_3^{(2)}\Big|_{AA} = \frac{de^2}{2880}\Big\{120(\nabla A)\Box(\nabla A) - 60F_{\mu\nu}\Box F^{\mu\nu}$$

$$- 24\nabla_\nu F^{\mu\nu}\nabla^\alpha F_{\mu\alpha} + 24(\Box A^\alpha)\big[(d-3)(\Box A_\alpha) + 2\nabla_\alpha(\nabla A)\big]$$

$$- 24(\nabla_\alpha\nabla_\mu A_\beta)\,\big[(\nabla^\beta\nabla^\mu A^\alpha) - (\nabla^\alpha\nabla^\mu A^\beta)\big]$$

$$+ A^\mu A_\mu\,\big[(18-7d)R^2_{\mu\nu\alpha\beta} - 8(9-d)R^2_{\mu\nu} - 6(5-d)R^2\big]$$

$$+ 8R_{\mu\nu\alpha\beta}\big[4(\nabla^\alpha A^\nu)(\nabla^\mu A^\beta) - 8F^{\mu\nu}F^{\alpha\beta} - 3(d-4)(\nabla^\mu A^\alpha)(\nabla^\nu A^\beta)$$

$$- R^{\lambda\nu\alpha\beta}A^\mu A_\lambda + 10R^{\mu\beta}A^\alpha A^\nu\big]$$

$$+ 16R_{\mu\nu}\big[10(\nabla A)(\nabla^\mu A^\nu) + (\nabla^\alpha A^\mu)(5\nabla_\alpha A^\nu - 2\nabla^\nu A_\alpha)$$

$$- (d-5)(\nabla^\mu A^\alpha)(\nabla^\nu A_\alpha) - 2R^\mu{}_\alpha A^\alpha A^\nu\big]$$

$$+ 10R\big[2(d-5)(\nabla_\mu A_\nu)(\nabla^\mu A^\nu) - 2(\nabla A)^2 + 3F_{\mu\nu}F^{\mu\nu} + 2R_{\mu\nu}A^\mu A^\nu\big]$$

$$- 48(\nabla^\alpha R_{\mu\nu\alpha\beta})(\nabla^\nu A^\beta A^\mu) - 24(\nabla^\nu R)[(\nabla_\nu A^\alpha A_\alpha) - (\nabla_\alpha A^\alpha A_\nu)]$$

$$- 12(d-2)A^\alpha A_\alpha\Box R\Big\}\,. \tag{12}$$

It is easy to check that, in $4d$, the formulas (11) and (12) do coincide with the third orders of the expansions of the complete expressions (7) and (9), correspondingly. This correspondence serves as an independent verification for the correctness of our formfactors (7) and (9).

The comparison of the expressions (11) and (12) shows that, in the flat space limit, the $\hat{a}_3^{(2)}$ does coincide with $\hat{a}_3^{(1)}$ in $6d$ and only in $6d$. Furthermore, we could prove that the terms porportional to $R\,F^{\mu\nu}F_{\mu\nu}$ in two expressions $\hat{a}_3^{(2)}$ and $\hat{a}_3^{(1)}$ coincide (up to total derivatives) on dS/AdS background. In any other dimension the gauge invariance is broken even in the flat space background, as it was expected from general arguments given above. The difference between $\hat{a}_3^{(2)}$ and $\hat{a}_3^{(1)}$ is precisely the one which can be observed between the first terms of expansion of the general expressions (7) and (9). At that point we can say that our general arguments concerning the origin of the MA is very well supported by direct calculations of the first three Schwinger-DeWitt coefficients.

4. Appelquist and Carazzone Theorem

Let us look at the UV and IR limits of the physical β-functions for the charge e. Starting from the expressions (7) and (9), correspondingly, we arrive at the following expressions for the β-functions:

$$\beta_e^1 = \frac{e^3\left[48 - 20a^2 + 3(a^2-4)^2(1-Y)\right]}{6a^2(4\pi)^2}\,, \tag{13}$$

versus

$$\beta_e^2 = \frac{e^3\left[4(12+a^2) + 3(a^4-16)(1-Y)\right]}{12a^2(4\pi)^2}\,. \tag{14}$$

In the high energy limit, when

$$p^2 \gg m^2\,, \quad a \to 2$$

and the two expressions give identical results, which also coincides with the one from the minimal subtraction scheme (up to a small correction),

$$\beta_e^{UV} \;=\; \frac{4\,e^3}{3\,(4\pi)^2} + \mathcal{O}\Big(\frac{m^2}{p^2}\Big)\,, \tag{15}$$

However, at the low-energy end the results are different, namely

$$\beta_e^{1\;IR} \;=\; \frac{e^3}{(4\pi)^2}\cdot\frac{4\,M^2}{15\,m^2} \;+\; \mathcal{O}\Big(\frac{M^4}{m^4}\Big) \tag{16}$$

for the first scheme $\hat{H}_1^*$, and

$$\beta_e^{2\;IR} \;=\; \frac{e^3}{(4\pi)^2}\cdot\frac{1}{5}\,\frac{M^2}{m^2} \;+\; \mathcal{O}\Big(\frac{M^4}{m^4}\Big) \tag{17}$$

for the second one, with $\hat{H}_2^*$. Thus we met a scheme ambiguity, also, in the decoupling theorem.[15]

In order to better understand the sense of the MA and the above difference in the β-functions, we can look at the lowest order term in the EA, where the difference shows up,

$$\frac{1}{30}\cdot\int d^4x\sqrt{g}\,F^{\mu\nu}\Big(\frac{\Box}{m^2}\Big)F_{\mu\nu}\,. \tag{18}$$

In the flat space-time, one can easily use integrations by parts to show that this term is proportional to the Maxwell Eqs., $(\nabla_\mu F^{\mu\nu})^2$. Hence this term will not influence the Eqs. of motion in flat space in the $\mathcal{O}(e^2)$ approximation.[14] However, the situation gets changed when we deal with the curved space. In this case we meet a difference which is proportional to curvatures, due to the relation

$$\int d^4x\, F^{\mu\nu}\,\Box\, F^{\mu\nu} = -\,2\int d^4x\, \nabla_\nu F^{\mu\nu}\nabla_\lambda F_\mu{}^\lambda + 2\int d^4x\, R_{\lambda\nu}F^{\mu\nu}F_\mu{}^\lambda - \int d^4x\, R_{\alpha\beta\mu\nu}F^{\mu\nu}F^{\alpha\beta}\,. \tag{19}$$

It is important that this difference is also confirmed by the derivation of the $\hat{a}_3$ coefficient described in the previous section.

5. Conclusions

We have calculated the formfactor in the electromagnetic sector of QED in curved space-time and found that this quantum correction depends on the choice of the calculational scheme (4). Thus we have proven the existence of the nonlocal and renormalization independent MA in quantum field theory. One of the consequences of this anomaly is the ambiguity in the prediction of the decoupling theorem,[15] which provides two different coefficients of the quadratic decoupling law at low energies.

The MA in the electromagnetic formfactor means that the off-shell EA possesses some new important ambiguity. One can use the Maxwell Eq. and show that in the flat space the ambiguous terms disappear on shell. However, this does not happen in curved space where we meet a real ambiguity proportional to the RFF-terms. Another important aspect is that the UVlimit is free of ambiguities and is essentially controlled by the logarithmic diveregences and conformal anomaly.[8,18] The MA which we discussed here is the fenomenon which is only possible for the quantum contributions of massive fields, where it may be, in some cases, essential.[2]

How should we interpret the existence of MA? In fact, the EA is always ambiguous to some extent. For instance, there is a strong dependence on the choice of parametrization for the quantum field[16] which becomes relevant beyond the leading-log approximation. Let us note that this is just a qualitative analogy, because the MA is not reduced to the known ambiguities of the EA, such as gauge fixing dependence or the renormalization scheme dependence (see, e.g. Ref. 17 for discussion and further references). Perhaps, from the practical viewpoint the best option is to follow the most natural approach and, for instance, take the most natural parametrization of quantum fields and the most natural and symmetry preserving scheme of calculation. On the other hand, it is always good to be aware on the real features of the utilized formalism, and from this perspective it is indeed important to know that the MA is a real thing.

Acknowledgments

I.Sh. is grateful to the scientific organizers of the QFEXT-09 meeting for the invitation and possibility to present this work and to the local organizers for kind hospitality. Authors are grateful to CNPq, FAPEMIG, FAPES and ICTP (I.Sh.) for support of their work.

References

1. M. Kontsevich and S. Vishik, *Geometry of determinants of elliptic operators,* hep-th/9406140; In Functional Analysis on the Eve of the 21st Century, Progress in Math. 131, Birkhuser Verlag, 1995;
   Determinants of elliptic pseudodifferential operators, hep-th/9404046.
2. E. Elizalde, L. Vanzo and S. Zerbini, Commun. Math. Phys. **194** (1998) 613;
   G. Cognola, E. Elizalde and S. Zerbini, Commun. Math. Phys. **237** (2003) 507, hep-th/9910038;
   E. Elizalde, G. Cognola and S. Zerbini, Nucl. Phys. B532 (1998) 407;
   E. Elizalde and M. Tierz, J. Math. Phys. **45** (2004) 1168, hep-th/0402186.
3. T.S. Evans, Phys. Lett. **B457** (1999) 127;
4. J.S. Dowker, *On the relevance of the multiplicative anomaly*, hep-th/9803200;
5. J.J. McKenzie-Smith and D.J. Toms, Phys. Rev. D58 (1998) 105001.
6. E. Elizalde, *Zeta regularization techniques with applications*, (World Scientific, 1994).
7. I.L. Shapiro and G. Cognola, Class. Quant. Grav. **15** (1998) 787.
8. B. Gonçalves, G. de Berredo-Peixoto and I.L. Shapiro, *One-loop corrections to the photon propagator in the curved-space QED.* To be published in Phys. Rev. D.
9. I.L. Buchbinder, S.D. Odintsov and I.L. Shapiro, *Effective Action in Quantum Gravity*, IOP Publishing, Bristol, 1992.
10. G. De Berredo-Peixoto, Mod. Phys. Lett. A16 (2001) 2463.
11. A.O. Barvinsky and G.A. Vilkovisky, Nucl. Phys. **333B** (1990) 471;
    I. G. Avramidi, Yad. Fiz. (Sov. Journn. Nucl. Phys.) **49** (1989) 1185.
12. E.V. Gorbar and I.L. Shapiro, JHEP **02** (2003) 021;
13. E.V. Gorbar and I.L. Shapiro, JHEP **06** (2003) 004.
14. I.T. Drummond and S.J. Hathrell, Phys. Rev. D 22 (1980) 343.
15. T. Appelquist and J. Carazzone, Phys. Rev. 11D (1975) 2856.
16. I.V. Tyutin, Yad. Fiz. (Physics of Atomic Nuclei) 35 (1982) 222 (in Russian).
17. G.A. Vilkovisky, Class. Quant. Grav. 9 (1992) 895.
18. M. Giannotti, E. Mottola, Phys. Rev. D79 (2009) 045014, arXiv:0812.0351.

SPIN LIGHT IN NEUTRINO TRANSITION BETWEEN DIFFERENT MASS STATES

A. GRIGORIEV

Skobeltsyn Institute of Nuclear Physics, Moscow State University, 119991 Moscow, Russia
E-mail: ax.grigoriev@mail.ru

A. LOKHOV

Department of Quantum Statistics and Quantum Field Theory, Moscow State University, 119991 Moscow, Russia
E-mail: lokhov.alex@gmail.com

A. STUDENIKIN

Department of Theoretical Physics, Moscow State University, 119991 Moscow, Russia
E-mail: studenik@srd.sinp.msu.ru

A. TERNOV

Department of Theoretical Physics, Moscow Institute for Physics and Technology, 141700 Dolgoprudny, Russia
E-mail: a_ternov@mail.ru

The neutrino spin light is considered in the process of a neutrino radiative transition between two different mass states in presence of medium. Simultaneously, by this study we investigate influence of background matter on the initial and final neutrino states in the process of massive Dirac neutrino decay due to the non-zero transitional magnetic moment. We derive corresponding corrections to the total width of the process over the matter density in most important for applications cases.

1. Introduction

Electromagnetic properties of neutrino are among the key items of modern particle physics (see Ref. 1 for a recent review). It seems quite natural that a massive neutrino would have nonzero diagonal or transition magnetic moment. If a neutrino has non-trivial electromagnetic properties, then neutrino coupling to photons is possible and several important, for instance for astrophysics, processes can exist.[2] Recently we have proposed a new mechanism of neutrino radiation of photons that is realized when a relativistic neutrino with nonzero magnetic moment moves in dense matter. This mechanism was termed the "spin light of neutrino" $(SL\nu)$.[3] The quantum theory of this phenomenon was developed in Ref. 4 (see also Refs. 5 and 6).

In this paper we extend our studies of $SL\nu$ [4] and consider the $SL\nu$ in a more general case when the photon is emitted in the neutrino radiative decay. The $SL\nu$ considered in Ref. 4 was investigated under condition of equal masses for the initial and final neutrino states. Here below we examine the case when the neutrino transition between two different neutrino mass states is realized. Thus, we consider the $SL\nu$ mode in the neutrino radiative decay in matter originated due to the neutrino transition magnetic moment.

It should be noted that the neutrino radiative decay was considered before by several authors.[7] It was shown that the process characteristics are substantially changed if the presence of a medium is taken into account. In these calculations, the influence of the background matter was considered only in the electromagnetic vertex. Here we are going to discuss the impact of the medium also onto the state of neutrino itself. At the same time we will be interested in the another aspect of the problem and consider it from the point of view of light emission. Under the condition of equal initial and final particle masses the process becomes equivalent to the $SL\nu$ in matter . With different masses for the initial and final neutrino states, the spin light becomes only the constituting channel for the overall process corresponding to the change of the neutrino helicity. The mechanism of $SL\nu$ is based on helicity states energy difference of the particle arising due to the weak interaction with the background matter. Hence, our study makes sense, obviously, if the scale of neutrino mass difference is of the order of spin energy splitting owing to the interaction with matter.

Let us specify now the process under consideration. We are considering the decay of one neutrino mass state ν_1 into another mass state ν_2 assuming that $m_1 > m_2$, and restrict ourselves with only these two neutrino species and accordingly with two flavour neutrinos. Having in mind that conditions for the most appropriate application of the process under study can be found in the vicinity of neutron stars we will take for the background a neutron-rich matter. In this case a process with participation of antineutrinos is more appropriate and thus will study here. However for the convenience in what follows we will still refer to the particles as to neutrinos. Since the interactions of flavour neutrinos with neutron star matter are the same and governed by the neutron density we will take equal interactions for the initial and final massive neutrinos with the matter.

2. Modified Dirac equation

The system "neutrino ⇔ dense matter" depicted above can be circumscribed mathematically in different ways. Here we use the powerful method of exact solutions, discussed in a series of our papers.[6] This method is based on solutions[4] of the modified Dirac equation for neutrino in the background medium

$$\{i\gamma_\mu \partial^\mu - \frac{1}{2}\gamma_\mu(1+\gamma^5)f^\mu - m\}\Psi(x) = 0, \tag{1}$$

where in the case of unpolarized and nonmoving matter $f^\mu = G_f/\sqrt{2}\,(n, \mathbf{0})$ with n being matter number density. At this, the energy spectrum of neutrino is given by

$$E_\varepsilon = \varepsilon\sqrt{(p - s\alpha m_\nu)^2 + m_\nu^2} + \alpha m_\nu, \quad \alpha = \frac{1}{2\sqrt{2}} G_F \frac{n}{m_\nu} \tag{2}$$

where $\varepsilon = \pm 1$ defines the positive and negative-energy branches of the solutions, s is the helicity of neutrino, p is the neutrino momentum. The exact form of the solutions $\Psi_{\varepsilon,p,s}(\mathbf{r}, t)$ can be found in Refs. 4 and 6.

3. Spin light mode of massive neutrino decay

The S-matrix element for the decay has the standard form that of the magnetic moment radiation process:

$$S_{fi} = -(2\pi)^4 \mu \sqrt{\frac{\pi}{2wL^3}} \delta(E_2 - E_1 + w)\delta^3(\mathbf{p}_2 - \mathbf{p}_1 + \mathbf{k})\overline{u}_f(\mathbf{e}, \mathbf{\Gamma}_{fi})u_i. \tag{3}$$

Here $\mathbf{\Gamma} = i\omega\{[\mathbf{\Sigma} \times \varkappa] + i\gamma^5 \mathbf{\Sigma}\}$, $u_{i,f}$ are the spinors for the initial and final neutrino states, $\mathbf{e}$ is the photon polarization vector, μ is the transitional magnetic moment[1] and L is the normalization length.

In the process, we have the following conservation laws:

$$E_1 = E_2 + \omega; \quad \mathbf{p_1} = \mathbf{p_2} + \mathbf{k}. \tag{4}$$

It is useful to carry out our computations through non-dimensional terms. For that purpose we introduce the following notations: $\gamma = \frac{m_1}{p_1}$; $\kappa = \frac{\alpha m_1}{p_1} = \frac{\tilde{n}}{p_1}$; $\delta = \frac{\Delta m^2}{p_1^2} = \frac{m_1^2 - m_2^2}{p_1^2}$. To single out the the spin light part of the radiative decay process we should choose different helicities for the initial and final neutrinos. Keeping the analogy with the usual process of $SL\nu$ we take the helicity quantum numbers as $s_1 = -1$, $s_2 = 1$. Then the solution of the kinematic system (4) can be written in the form

$$w = \frac{-(KD + x\kappa^2) + \sqrt{(KD + x\kappa^2)^2 - (K^2 - \kappa^2)(D^2 - \kappa^2)}}{(K^2 - \kappa^2)} \qquad (5)$$

with the notations $D = s_1\kappa - \delta$; $\tilde{n} = \frac{1}{2\sqrt{2}} G_F n$, $K = \sqrt{(1 - s_1 \cdot \kappa)^2 + \gamma^2} - x$, here x stands for $\cos\theta$, θ is the angle between $\mathbf{p}_1$ and $\mathbf{k}$.

Performing all the calculations we obtain angle distribution of the probability for the investigated process:

$$\frac{d\Gamma}{dx} = \mu^2 p_1^3 \frac{(K - w + x)(wK - \kappa - \delta)w^3 S'}{\sqrt{(KD - w + x)^2 - (K^2 - \kappa^2)(D^2 - \kappa^2)}}, \qquad (6)$$

where $S' = (1 + \beta_1\beta_2)(1 - \frac{w - x - w\cdot x + w\cdot x^2}{\sqrt{1 + w^2 - 2w\cdot x}} x) - (\beta_1 + \beta_2)(x - \frac{w - x}{\sqrt{1 + w^2 - 2w\cdot x}})$ and $\beta_1 = \frac{1+\kappa}{\sqrt{(1+\kappa)^2 + \gamma^2}}$, $\beta_2 = \frac{\sqrt{1 + w^2 - 2w\cdot x} - \kappa}{K - w + x}$.

The total probability can be computed from the equation (6) by taking the integration over the angle θ range. However, manual calculations are not quite simple to carry through them. Even though the integral can be calculated exactly, the final expression is enormously complex and its explicit form is optional to be given here.

4. Results and discussion

It is worth to investigate the asymptotical behavior of the probability Γ in three most significant relativistic limiting cases keeping only the first infinitesimal order of small parameters. On this way we have,

$$\Gamma = 4\mu^2 \tilde{n}^3 (1 + \frac{3}{2}\frac{m_1^2 - m_2^2}{\tilde{n} p_1} + \frac{p_1}{\tilde{n}}), \text{ (ultrahigh density: } 1 \ll \frac{p_1}{m_1} \ll \frac{\tilde{n}}{p_1}); \qquad (7)$$

$$\Gamma = 4\mu^2 \tilde{n}^2 p_1 (1 + \frac{\tilde{n}}{p_1} + \frac{m_1^2 - m_2^2}{\tilde{n} p_1} + \frac{3}{2}\frac{m_1^2 - m_2^2}{p_1^2}), \text{(high density: } \frac{m_1^2}{p_1^2} \ll \frac{\tilde{n}}{p_1} \ll 1); \qquad (8)$$

$$\Gamma \approx \mu^2 \frac{m_1^6}{p_1^3}, \text{(quasi-vacuum case: } \frac{\tilde{n}}{p_1} \ll \frac{m_1}{p_1} \ll 1, m_1 \gg m_2). \qquad (9)$$

The obtained results (7) and (8) exhibit the power of the method of exact solutions since they establish clear connection between the case of massive weak-interacting particles when the masses of the initial and final particles differ with the previously investigated equal mass case. Indeed, it is easy to verify that these results transforms exactly into the results of $SL\nu$ calculation.[3] The asymptotic estimation (9) can not be reduced to the $SL\nu$ case and thus it is a new result, which is characteristic feature for the decay process under study. The above-mentioned asymptotical cases (7),

(8) and (9) where calculated with the assumption that the initial neutrino is relativistic ($\gamma = \frac{m_1}{p_1} \ll 1$). In particular the relativistic character of the initial neutrino propagation influences strongly on the emitted $SL\nu$ photon energy because of increase of the part of neutrino energy in it.

It is also interesting to investigate the spin light mode in the radiative decay of slowly moving massive neutrino (or even stationary initial neutrino). This process has been calculated several times.[7] We consider the vacuum case to find the interrelation of our results obtained using the method of exact solutions with the results of previous works. So, taking into account $\gamma = \frac{m_1}{p} \ll 1$, $\kappa = \frac{\tilde{n}}{p_1}$, $\delta \equiv \frac{\gamma^2}{2}$ for the probability of the process we finally get:

$$\Gamma \approx \frac{7}{24}\mu^2 m_1^3 \sim m_1^5. \tag{10}$$

We obtain here the same dependency of the probability from the mass of the decaying neutrino as in the classical papers on the radiative neutrino decay. By this means we justify usage of the modified Dirac equation exact solutions method.

Acknowledgments

One of the authors (A.S.) is thankful to Kim Milton for the invitation to attend the 9th Conference on Quantum Field Theory Under the Influence of External Conditions (Oklahoma, USA) and for the kind hospitality provided in Norman.

References

1. C. Giunti, A. Studenikin, *Phys. Atom. Nucl.* **72**, 2151, 2009.
2. G. Raffelt, *Stars as Laboratories for Fundamental Physics*, (Univ. of Chicago Press, 1996); G. Raffelt, Phys. Rep. **320**, 319 (1999).
3. A. Lobanov and A. Studenikin, *Phys. Lett. B* **564**, 27 (2003), A. Lobanov and A. Studenikin, *Phys. Lett. B* **601**, 171 (2004).
4. A. Studenikin, A. Ternov, *Phys. Lett. B* **608**, 107 (2005), A. Grigorev, A. Studenikin, A. Ternov, *Phys. Lett. B* **622**, 199 (2005).
5. A. Lobanov *Phys. Lett. B* **619**, 136 (2005).
6. A. Studenikin, *J. Phys. A: Math. Gen.* **39**, 6769 (2006); A. Studenikin, *J. Phys. A: Math. Theor.* **41**, 164047 (2008).
7. S. Petcov, *Sov. J. Nucl. Phys* **25**, 340 (1977), G. T. Zatsepin, A. Yu. Smirnov, Yad. Fiz. **28**, 1569 (1978) *Yad. Fiz.* **28**, 1569 (1978), P. B. Pal and L. Wolfenstein, *Phys. Rev. D* **25**, 766 (1982), J. C. D'Olivo, J. F. Nieves, and P. B. Pal, *Phys. Rev. Lett.* **64**, 1088 (1990), C. Giutni, C. W. Kim and W. P. Lam, *Phys. Rev. D* **43**, 164 (1990), A. Ternov, P. Eminov *J. Phys. G: Nucl. Part. Phys.* **29**, 357 (2005).

CHIRAL SYMMETRY BREAKING AND STABILITY OF THE MAGNETIZED VACUUM

SHANG-YUNG WANG

Department of Physics, Tamkang University, Tamsui, Taipei 25137, Taiwan
E-mail: sywang@mail.tku.edu.tw

The recent claim that there exists in QED a maximum magnetic field of 10^{42} G, above which the magnetized vacuum becomes unstable with respect to the so-called "positronium collapse" is critically examined and unequivocally refuted.

It is an observational fact that we are living in a magnetized universe, with planets, stars and galaxies all being magnetized.[1] The typical magnetic fields range from few μG in galaxies to few G on planets (such as the earth) and up to $10^{12} - 10^{16}$ G on the surface of neutron stars and magnetars. A fundamental question to ask is whether there exists in nature a maximum magnetic field. As we will see shortly, the answer to this question requires a nonperturbative understanding of quantum field theory under the influence of a strong magnetic field.

Conventional wisdom suggests that the magnetic field strength is not constrained by known physics. However, a maximum value for the magnetic field in QED, $B_{\max}$, has recently been conjectured:[2]

$$B_{\max} = \frac{m^2}{4e} \exp\left(\frac{\pi^{3/2}}{\sqrt{\alpha}} + 2C_{\mathrm{E}}\right) \simeq 10^{42} \text{ G}, \tag{1}$$

where m and e are respectively the electron mass and the absolute value of its charge in the absence of external fields, $\alpha = e^2/4\pi$ is the fine structure constant and $C_{\mathrm{E}} \simeq 0.577$ is Euler's constant. In obtaining (1), Shabad and Usov[2] considered a positronium (electron-positron bound state) placed in a strong magnetic field, and found that the magnetic field significantly enhances the Coulomb attraction between the constituent electron and positron. The Coulomb attraction becomes stronger and stronger until the electron and positron fall onto each other at the maximum magnetic field of 10^{42} G. This phenomenon is referred to by these authors as "positronium collapse" and could be a signal for possible vacuum instability.[2]

While the existence of a maximum magnetic field in QED appears novel and is potentially of fundamental importance, it is in fact in contradiction with many of the well-established results in QED in a strong magnetic field.[3–6] In particular, magnetic catalysis of chiral symmetry breaking has long been known as a universal phenomenon. A strong magnetic field acts as a catalyst for chiral symmetry breaking, leading to the generation of a dynamical fermion mass even at the weakest attractive interaction between fermions. The hallmark of this phenomenon is the dimensional reduction from $(3+1)$ to $(1+1)$ in the dynamics of fermion pairing in a strong magnetic field when the lowest Landau level (LLL) plays the dominant role. The massless fermion-antifermion bound state is the Nambu-Goldstone (NG) boson for spontaneously broken chiral symmetry.[3] The phenomenon of magnetic catalysis is universal in that chiral symmetry is broken in arbitrarily strong magnetic fields and for any number of fermion flavors.[3–5]

A powerful tool to study the electron-position bound states directly from QED is the Bethe-Salpeter (BS) equation. For the problem at hand, however, in order to consistently incorporate the effective electron mass and the screening effect in a strong magnetic field, the electron and photon propagators that enter the BS equation must be obtained by solving the corresponding Schwinger-Dyson (SD) equations in the same truncation. This is tantamount to solving the truncated SD and BS equations simultaneously, an important point that has gone unnoticed in Ref. 2.

It has recently been proved[5,6] that in QED (both massless and massive) in a strong magnetic field the bare vertex approximation (BVA), in which the vertex corrections are completely ignored, is a consistent truncation of the SD equations within the lowest Landau level approximation (LLLA). In particular, it can be shown that the truncated vacuum polarization is transverse and the dynamical fermion mass, obtained as the solution of the truncated fermion SD equation evaluated on the fermion mass shell, is manifestly gauge independent. Thus, for consistency with the results obtained in Refs. 5,6, the BS equation for the positronium has to be truncated in the BVA within the LLLA.

We choose the constant external magnetic field of strength $B > 0$ in the x_3-direction. The corresponding vector potential is given by $A_\mu^{\text{ext}} = (0, 0, Bx_1, 0)$. Note that the motion of the LLL electron and positron is restricted in directions perpendicular to the magnetic field, hence so is the motion of the bound state of the LLL electron and positron. We will

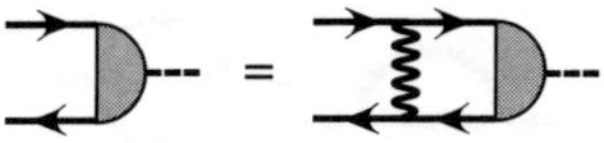

Fig. 1. BS equation for the positronium in the BVA. External lines are amputated.

henceforth refer to the latter as the LLL positronium. In the BVA within the LLLA, the BS equation for the LLL positronium in momentum space is given by[7] (see Fig. 1)

$$\chi(p_\parallel;P_\parallel)\,\Delta = -\,ie^2\int_q \exp\left(-\frac{q_\perp^2}{2eB}\right)\mathcal{D}_{\mu\nu}(q)\,\Delta\,\gamma_\parallel^\mu\,\Delta\,\frac{1}{\gamma_\parallel\cdot p'_\parallel+m_*}\,\chi(p'_\parallel;P_\parallel)$$
$$\times\,\frac{1}{\gamma_\parallel\cdot p'_\parallel+m_*}\,\Delta\,\gamma_\parallel^\nu\,\Delta, \qquad (2)$$

where $\chi(p_\parallel;P_\parallel)$ is the amputated BS amplitude, $p_\parallel^\mu=(p^0,p^3)$ is the momentum of the LLL electron, $P_\parallel^\mu$ is the momentum of the LLL positronium, m_* is the effective electron mass in a strong magnetic field, $p'_\parallel=p_\parallel-q_\parallel$, $q_\perp^2=q_1^2+q_2^2$ and $\int_q=\int d^4q/(2\pi)^4$. In (2), $\Delta=(1+i\gamma^1\gamma^2)/2$ is the projector onto the electron (positron) states with the spin polarized along (opposite to) the external magnetic field and $\mathcal{D}_{\mu\nu}(q)$ is the full photon propagator in the external magnetic field. The latter in covariant gauges is given by[4–6]

$$\mathcal{D}^{\mu\nu}(q)=\frac{1}{q^2+\Pi(q_\parallel^2,q_\perp^2)}\left(g_\parallel^{\mu\nu}-\frac{q_\parallel^\mu q_\parallel^\nu}{q_\parallel^2}\right)+\frac{g_\perp^{\mu\nu}}{q^2}+\frac{q_\parallel^\mu q_\parallel^\nu}{q^2q_\parallel^2}+(\xi-1)\frac{1}{q^2}\frac{q^\mu q^\nu}{q^2}, \qquad (3)$$

where $\Pi(q_\parallel^2,q_\perp^2)$ is the polarization function (see Refs. 4–6 for an explicit expression).

Because the LLL electron and positron always have their spins polarized in opposite directions along the external magnetic field, the LLL positronium in its ground state is in fact a parapositronium, i.e., a pseudoscalar state. This, together with symmetry arguments, implies that the amputated BS amplitude $\chi(p_\parallel;P_\parallel)$ takes the form $\chi(p_\parallel;P_\parallel)=A(p_\parallel,P_\parallel)\gamma^5$, where $A(p_\parallel,P_\parallel)$ is a scalar function of $p_\parallel^2$ and $P_\parallel^2$. Then, (2) simplifies to

$$A(p_\parallel,P_\parallel)=-\,ie^2\int_q \exp\left(-\frac{q_\perp^2}{2eB}\right)\mathcal{D}_{\mu\nu}(q)A(p'_\parallel,P_\parallel)\gamma_\parallel^\mu\,\frac{1}{\gamma_\parallel\cdot p'_\parallel+m_*}$$
$$\times\,\frac{1}{\gamma_\parallel\cdot p'_\parallel+m_*}\,\gamma_\parallel^\nu. \qquad (4)$$

When evaluated on the respective particle mass shells, $p_\parallel^2=-m_*^2$ and $P_\parallel^2=-M^2$, and supplemented with the effective electron mass m_* that is

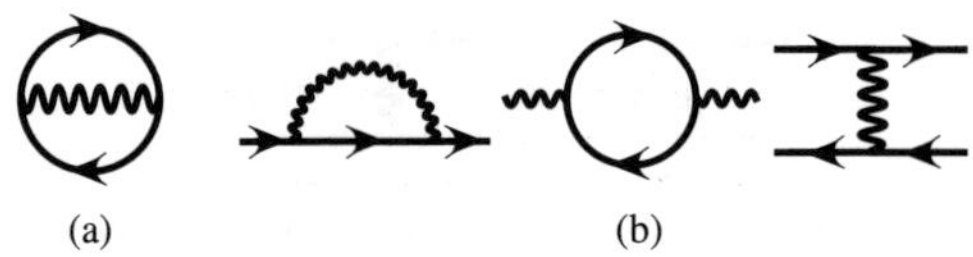

Fig. 2. (a) Contribution to the 2PI effective action at two-loop order. (b) The electron self-energy, vacuum polarization and electron-positron interaction kernel generated thereby. External lines are amputated.

obtained as the solution of the on-shell SD equations,[5,6] the BS equation (4) determines the mass of the LLL positronium, M.

A detailed analysis based on the Ward-Takahashi identity in the BVA within the LLLA reveals that contrary to the gauge independence of the on-shell SD equations,[5,6] the on-shell BS equation truncated in the BVA is inevitably gauge dependent. We now argue that in the BVA the on-shell BS equation has a controlled gauge dependence, thanks to a direct correspondence[5] between the SD equations truncated in the BVA and the 2PI effective action truncated at the lowest nontrivial (two-loop) order in the loop expansion. Let Γ_2 denotes the sum of all 2PI skeleton vacuum diagrams with *bare* vertex and *full* LLL electron and photon propagators. The contribution to Γ_2 at two-loop order is depicted diagrammatically in Fig. 2(a). The key point of the argument is to note that the direct correspondence can be generalized to include the BS equation truncated in the BVA. This is because, as shown in Fig. 2(b), the corresponding electron self-energy, vacuum polarization and electron-positron interaction kernel that enter the SD and BS equations in the BVA are the same as those generated by Γ_2. The argument is completed with the fact[8] that the truncated 2PI effective action evaluated at its stationary point has a controlled gauge dependence, i.e., the explicit gauge dependent terms always appear at higher order. Moreover, since the transverse components in $\mathcal{D}^{\mu\nu}(q)$ decouple and the gauge dependent contribution in (4) arises from the longitudinal components in $\mathcal{D}^{\mu\nu}(q)$ proportional to $q^\mu q^\nu/q^2$, we conclude that in the BVA and at the order of truncation only the first term in $\mathcal{D}^{\mu\nu}(q)$ proportional to $g_\parallel^{\mu\nu}$ contributes to the on-shell BS equation.

To prove the nonexistence of a maximum magnetic field in QED, we only need to consider the on-shell BS equation in an asymptotically strong magnetic field. For $B \gg B_0 \equiv m^2/e$, the explicit breaking of chiral symmetry associated with the perturbative electron mass m can be neglected. As per the fact[6] that $m_* \approx m_{\rm dyn}$ as $B \to \infty$ and the pseudo NG boson nature of the positronium that $M \to 0$ as $B \to \infty$, the BS equation (4) is evaluated on the mass shells $p_\parallel^2 = -m_{\rm dyn}^2$ and $P_\parallel^2 = 0$. Moreover, because of the strong screening effect in a strong magnetic field,[4–6] the integral in (4) is dominated by contributions from the region with $m_{\rm dyn}^2 \lesssim |q_\parallel^2| \ll eB$. Thus, $A(p'_\parallel, P_\parallel)$ in the integrand can be approximated by $A(p_\parallel, P_\parallel)$. We find that in the limit $B \to \infty$ the on-shell BS equation reduces to

$$1 = -2ie^2 \int_q \frac{\exp(-q_\perp^2/2eB)}{q^2 + \Pi(q_\parallel^2, q_\perp^2)} \frac{1}{(p-q)_\parallel^2 + m_{\rm dyn}^2}\Bigg|_{p_\parallel^2 = -m_{\rm dyn}^2} . \tag{5}$$

The proof is completed by noting that (5) is the same as the on-shell SD equation obtained in the BVA within the LLLA[5] that reliably determines the dynamically generated fermion mass, $m_{\rm dyn}$, in massless QED (see (4.20) in the first paper of Ref. 5). This also serves to justify *a posteriori* the controlled gauge dependence of the on-shell BS equation.

In conclusion, the positronium is unambiguously identified as the (pseudo) Nambu-Goldstone boson for spontaneous (explicit) chiral symmetry breaking in massless (massive) QED in a strong magnetic field. It is shown that the phenomenon of positronium collapse conjectured by Shabad and Usov never takes place. Consequently, there does not exist a maximum magnetic field in QED and the magnetized vacuum is stable for all values of the magnetic field.

Acknowledgments

The author would like to thank the organizers of the conference for their invitation and hospitality. This work was supported in part by the National Science Council of Taiwan under grant 96-2112-M-032-005-MY3.

References

1. M. Giovannini, Int. J. Mod. Phys. D **13**, 391 (2004).
2. A. E. Shabad and V. V. Usov, Phys. Rev. Lett. **96**, 180401 (2006); Phys. Rev. D **73**, 125021 (2006).
3. V. P. Gusynin, V. A. Miransky, and I. A. Shovkovy, Phys. Rev. D **52**, 4747 (1995); Nucl. Phys. **B462**, 249 (1996); D.-S. Lee, C. N. Leung, and Y. J. Ng, Phys. Rev. D **55**, 6504 (1997).
4. V. P. Gusynin, V. A. Miransky, and I. A. Shovkovy, Phys. Rev. Lett. **83**, 1291 (1999); Nucl. Phys. **B563**, 361 (1999); Phys. Rev. D **67**, 107703 (2003).
5. C. N. Leung and S.-Y. Wang, Nucl. Phys. **B747**, 266 (2006); Ann. Phys. (N.Y.) **322**, 701 (2007).
6. S.-Y. Wang, Phys. Rev. D **77**, 025031 (2008).
7. C. N. Leung and S.-Y. Wang, Phys. Lett. B **674**, 344 (2009).
8. A. Arrizabalaga and J. Smit, Phys. Rev. D **66**, 065014 (2002); M. E. Carrington, G. Kunstatter, and H. Zaraket, Eur. Phys. J. C **42**, 253 (2005).

LIST OF PARTICIPANTS

- Elom Abalo, OU ekabalo@ou.edu
- Brad Abbott, OU abbott@nhn.ou.edu
- Mauro Antezza, Kastler Borssel Lab. mauro.antezza@lkb.ens.fr
- James R. Babington, Imperial College j.babington@imperial.ac.uk
- Ryan Behunin, Maryland ryan_behunin@hotmail.com
- Giuseppe Bimonte, Napoli bimonte@na.infn.it
- Michael Bordag, Leipzig michael@bordag.com
- Caterina Braggio, INFN braggio@pd.infn.it
- Iver Brevik, NTNU iver.h.brevik@ntnu.no
- A Ya. Burinskii, NSI RAS bur@ibrae.ac.ru
- Antoine Canaguier-Durand, LKB Antoine.Canaguier@spectro.jussieu.fr
- Rodrigo Castillo-Garza, UCR jcast003@ucr.edu
- Claudio Ccapa, Balseiro, Argentina claudio.ccapa@gmail.com
- Arthur Champernowne, Seattle arthur@champernowne.net
- Ho Bun Chan, Florida hochan@phys.ufl.edu
- Chia-Cheng Chang, UCR cchan031@ucr.edu
- Hongbo Cheng, E China U. Science hbcheng@public4.sta.net.cn
- Joel Chevrier, Grenoble joel.chevrier@grenoble.cnrs.fr
- Hsiang-Chih Chiu, UCR hchiu001@ucr.edu
- Ana Contreras, Rio de Janerio anamaria@if.ufrj.br
- Olindo Corradini, Bologna Olindo.Corradini@bo.infn.it
- Diego Dalvit, LANL dalvit@tmail.lanl.gov
- Felipe Da Rosa, LANL darosa@lanl.gov
- Ricardo Decca, IUPUI rdecca@iupui.edu
- Maarten DeKieviert, Heidelberg maarten@physi.uni-heidelberg.de
- Babette Doebrich, Jena babette.doebrich@uni-jena.de
- Gerald Dunne, U Conn dunne@phys.uconn.edu
- Emilio Elizalde, Barcelona elizalde@ieec.fcr.es
- Simen Ellingsen, NTNU simen.a.ellingsen@ntnu.no
- Thorsten Emig, Köln te@thp.uni-koeln.de
- Raul Esquivel-Sirvent, UNAM raul@fisica.unam.mx

- Carlos Farina de Souza, Rio de Janeiro farina@if.ufrj.br
- Ignat Fialkovskiy, São Paulo ifialk@gmail.com
- Antonino Flachi, YITP flachi@yukawa.kyoto-u.ac.jp
- Larry Ford, Tufts ford@cosmos.phy.tufts.edu
- Guglielmo Fucci, Baylor Guglielmo_Fucci@baylor.edu
- Toshiyuki Fujii, Hiroshima fujix3000@hiroshima-u.ac.jp
- Steve Fulling, TAMU fulling@math.tamu.edu
- Luis Reyes Galindo, Cardiff ReyesGalindoLI@cardiff.ac.uk
- Dmitri Gal'tsov, Moscow State galtsov@phys.msu.ru
- Andrea Gambassi, Trieste gambassi@mf.mpg.de
- Holger Gies, Jena gies@tpi.uni-jena.de
- Marco Govoni, Modena marco.govoni@unimore.it
- Phil Gutierrez, OU gut@nhn.ou.edu
- Herald Haakh, Potsdam haakh@arete.quantum.physik.uni-potsdam.de
- Jon Harrison, Baylor Jon_Harrison@baylor.edu
- Noriyuki Hatakenaka, Hiroshima noriyuki@hiroshima-u.ac.jp
- Florian Hebenstreit, Graz florian.hebenstreit@uni-graz.at
- Carsten Henkel, Potsdam henkel@uni-potsdam.de
- Norman Horing, Stevens IT Norman.Horing@stevens.edu
- Davide Iannuzzi, Vrije Universteit iannuzzi@few.vu.nl
- Francesco Intravaia, Potsdam francesco.intravaia@qipc.org
- Steven Johnson, MIT stevenj.mit@gmail.com
- Ron Kantowski, OU ski@nhn.ou.edu
- Chung Kao, OU kao@nhn.ou.edu
- Goverdhan Khadekar, Nagpur, India gkhadekar@yahoo.com
- Klaus Kirsten, Baylor Klaus_Kirsten@baylor.edu
- Israel Klich, Virginia isklich@gmail.com
- Galina Klimchitskaya, Leipzig Galina.Klimchitskaya@itp.uni-leipzig.de
- Fernando Lombardo, Buenos Aires lombardo@df.uba.ar
- Valery Marachevsky, St. Petersburg maraval@mail.ru
- Jamir Marino, Palermo jamirmarino@gmail.com
- Juan Mateos Guilarte, Salamanca guilarte@usal.es
- Chris Markle, Washington U Cmarkle@physics.wustl.edu
- Francisco Diego Mazzitelli, Buenos Aires fmazzi@df.uba.ar
- Riccardo Messina, Paris/Palermo Riccardo.Messina@spectro.jussieu.fr
- Giuseppe Messineo, Trieste Giuseppe.Messineo@lnl.infn.it
- Kim Milton, OU milton@nhn.ou.edu
- Hyunsoo Min, Seoul hsmin@dirac.uos.ac.kr
- Aram Mkhitaryan, Yerevan aram.mkhitaryan@googlemail.com

- Umar Mohideen, Riverside umar.mohideen@ucr.edu
- Pedro Morales, Baylor pedro_morales@baylor.edu
- Gustavo Moreno, Buenos Aires/LANL gustavoarielmoreno@gmail.com
- Vladimir Mostepanenko, Leipzig Vladimir.Mostepanenko@itp.uni-leipzig.de
- Emil Mottola, LANL emil@lanl.gov
- Jeremy Munday, Caltech jnmunday@caltech.edu
- Paulo Maia Neto, Rio de Janeiro pamaianeto@gmail.com
- Roberto Onofrio, Dartmouth onofrio@northstar.dartmouth.edu
- David Owen, Ben Gurion daowen137@gmail.com
- Aalok Pandya, Rajasthan aalok_cosmos@yahoo.com
- Prachi Parashar, OU prachi@nhn.ou.edu
- Ines Cavero Pelaez, Zaragoza cavero@nhn.ou.edu
- Nikola Petrov, OU npetrov@aftermath.math.ou.edu
- Fabrizio Pinto, Interstellar Technologies fabrizio.pinto@verizon.net
- Irina Pirozhenko, JINR pirozhen@theor.jinr.ru
- Lev Pitaevskii, Trento lev@science.unitn.it
- Serge Reynaud, Paris Serge.Reynaud@spectro.jussieu.fr
- Giuseppe Ruoso, Legnaro ruoso@lnl.infn.it
- Aram Saharian, Yerevan saharian@ictp.it
- Martin Schaden, Rutgers mschaden@andromeda.rutgers.edu
- Christian Schubert, Michoacan schubert@ifm.umich.mx
- Bo Sernelius, Linkoping bos@ifm.liu.se
- Ilya Shapiro, UF Juiz de Fora ilyashapiro2003@yahoo.com.br
- Yury Sherkunov, Warwick y.b.sherkunov@warwick.ac.uk
- Vladimir Skalozub, Ukraine,DNU skalozubv@daad-alumni.de
- Mike Strauss, OU strauss@nhn.ou.edu
- Alexander Studenikin, Moscow State studenik@srd.sinp.msu.ru
- Lee-Peng Teo, Malaysia LeePeng.Teo@nottingham.edu.my
- Francesco Parisen Tolden, MPI Stuttgart parisen@mf.mpg.de
- Marin-Slobodan Tomas, RBI, Croatia tomas@thphys.irb.hr
- Carlos Villareal, UNAM carlos@fisica.unam.mx
- Jef Wagner, OU wagner@nhn.ou.edu
- Shang-Yung Wang, Taiwan sywang@mail.tku.edu.tw
- Alexej Weber, Heidelberg a.weber@thphys.uni-heidelberg.de
- Justin Wilson, Maryland jwilson.thequark@gmail.com
- Cem Yuce, Anadolu,Turkey cyuce@anadolu.edu.tr
- Saad Zaheer, MIT saadz@MIT.EDU
- Oleg Zaslavskii, Kharkov ozaslav@kharkov.ua